AF777402

Benchmark Papers in Geology

Series Editor: Rhodes W. Fairbridge
Columbia University

A selection from the published volumes in this series

Volume

32 MECHANICS OF THRUST FAULTS AND DÉCOLLEMENT / *Barry Voight*
34 CRYSTAL FORM AND STRUCTURE / *Cecil J. Schneer*
37 STATISTICAL ANALYSIS IN GEOLOGY / *John M. Cubitt and Stephen Henley*

A complete listing of volumes published in this series begins on p. 398.

P85-9849

QE 604 F32

APR 04 1985

Benchmark Papers in Geology / 75

A BENCHMARK® Books Series

FABRIC OF DUCTILE STRAIN

Edited by

M.R. STAUFFER

University of Saskatchewan

SHELL WESTERN E&P INC.
WOODCREEK LIBRARY
P. O. BOX 4423
HOUSTON, TX 77210

Hutchinson Ross Publishing Company

Stroudsburg, Pennsylvania

Copyright ©1983 by **Hutchinson Ross Publishing Company**
Benchmark Papers in Geology, Volume 75
Library of Congress Catalog Card Number: 82-23199
ISBN: 0-87933-442-8

All rights reserved. No part of this book covered by the copyrights hereon may be reproduced or transmitted in any form or by any means—graphic, electronic, or mechanical, including photocopying, recording, taping, or information storage and retrieval systems—without written permission of the publisher.

85 84 83 1 2 3 4 5
Manufactured in the United States of America.

LIBRARY OF CONGRESS CATALOGING IN PUBLICATION DATA
Main entry under title:
Fabric of ductile strain.
(Benchmark papers in geology ; 75)
Bibliography: p.
Includes index.
1. Rock deformation—Addresses, essays, lectures.
I. Stauffer, M. R. II. Series.
QE604.F3 1983 551.8 82-23199
ISBN 0-87933-442-8

Distributed worldwide by Van Nostrand Reinhold Company Inc.,
135 W. 50th Street, New York, NY 10020.

CONTENTS

SERIES EDITOR'S FOREWORD

The philosophy behind the Benchmark Papers in Geology is one of collection, sifting, and rediffusion. Scientific literature today is so vast, so dispersed, and, in the case of old papers, so inaccessible for readers not in the immediate neighborhood of major libraries that much valuable information has been ignored by default. It has become just so difficult, or so time consuming, to search out the key papers in any basic area of research that one can hardly blame a busy person for skimping on some of his or her "homework."

This series of volumes has been devised, therefore, as a practical solution to this critical problem. The geologist, perhaps even more than any other scientist, often suffers from twin difficulties—isolation from central library resources and immensely diffused sources of material. New colleges and industrial libraries simply cannot afford to purchase complete runs of all the world's earth science literature. Specialists simply cannot locate reprints or copies of all their principal reference materials. So it is that we are now making a concerted effort to gather into single volumes the critical materials needed to reconstruct the background of any and every major topic of our discipline.

We are interpreting "geology" in its broadest sense: the fundamental science of the planet Earth, its materials, its history, and its dynamics. Because of training in "earthy" materials, we also take in astrogeology, the corresponding aspect of the planetary sciences. Besides the classical core disciplines such as mineralogy, petrology, structure, geomorphology, paleontology, and stratigraphy, we embrace the newer fields of geophysics and geochemistry, applied also to oceanography, geochronology, and paleoecology. We recognize the work of the mining geologists, the petroleum geologists, the hydrologists, and the engineering and environmental geologists. Each specialist needs a working library. We are endeavoring to make the task of compiling such a library a little easier.

Each volume in the series contains an introduction prepared by a specialist (the volume editor)—a "state of the art" opening or a summary of the object and content of the volume. The articles, usually some twenty to fifty reproduced either in their entirety or in significant extracts, are selected in an attempt to cover the field, from the key papers of the last century to fairly recent work. Where the original works are in foreign languages, we have endeavored to locate or commission translations. Geologists, because of their global subject, are often acutely aware of the oneness of our world. The selections cannot therefore be restricted to any one country, and whenever possible an attempt is made to scan the world literature.

To each article, or group of kindred articles, some sort of "highlight commentary" is usually supplied by the volume editor. This commentary should serve to bring that article into historical perspective and to emphasize its particular role in the growth of the field. References, or citations, wherever possible, will be reproduced in their entirety—for by this means the observant reader can assess the background material available to that particular author, or, if desired, he or she too can double check the earlier sources.

A "benchmark," in surveyor's terminology, is an established point on the ground that is recorded on our maps. It is usually anything that is a vantage point, from a modest hill to a mountain peak. From the historical viewpoint, these benchmarks are the bricks of our scientific edifice.

RHODES W. FAIRBRIDGE

PREFACE

Ductile deformation is of interest to many different fields of pure and applied science: physics, metallurgy, civil engineering, mining, and geology, to name a few. Advances in our knowledge of the ways in which solids undergo strain rely on ideas from, and observations in, all these fields, merged to form part of modern materials science.

Like most fields of science, the study of deformed rocks has passed in stages from initial recognition, through description, to increasing experimentation and quantification. As with most aspects of earth science, the descriptive phase has been long and drawn out, and the study of deformed rocks has been quantitative only in the last two decades, despite a few earlier examples. Although experiments in rock deformation, especially those using rock-anologs such as clay, were run as early as the middle of the last century, such studies were not common and only in the last three decades has experimental work contributed to understanding ductile deformation in rocks. Also, early experiments did not always seem to yield results consistent with field observations, and quantitative work was often based on incomplete or inaccurate descriptive knowledge. In part, such inconsistencies were the result of a lack of communication between workers studying different aspects of deformed rocks. Reed, noting this in 1941 (p. 243), wrote

> One structural geologist may feel that his science is greatly advanced by well-conducted laboratory experiments, another that such experiments are generally misleading or of dubious value; one may rate highly any good structural map of a complicated area, while another may insist that only a dynamic interpretation of the complications constitutes science.

We are improving, however, and modern workers realize the importance of intermeshing all aspects of geology.

The purpose of this book is to serve as a guide to some of the main facts and concepts relating to the ductile deformation of rocks. There is considerable emphasis on those features seen in outcrop, hand specimen, and under the petrographic microscope. Historically, these features have provided our main clues to the mechanics of ductile strain, until recent use of the electron microscope. Regional analysis of strain, and the application of studies of ductile strain to the origin of folds and other structures are beyond the scope of this book. My first intention was to prepare a collection of papers that would trace historical development (with all its conflicts and disagreements), as well as provide a state-of-the-art exposé. However, such an ideal book would have been over 3,000 pages long, and in the present book, numerous works deserving of benchmark status have had to be reduced to the reference list. This final, trimmed version is at best a skeleton which the reader will need to fill out by copious reading of the literature.

A reader craving the excitement of conflict will not find it in this book, as most of the papers selected either form the basis of the main stream of thought or serve to summarize established knowledge. In fact, my choice of papers was made to a large degree on the basis of "benchmark ideas," whether the paper introduced or merely summarized them. Nonetheless, the choice of papers is subjective, partly because many ideas evolve through the work of numerous people.

I wish to dedicate this book to the young scientist, just beginning his study of deformed rocks, whose future work undoubtedly will serve to make sense of what has gone before.

Special thanks are extended to W. M. Schwerdtner and P. F. Williams for taking the time to criticize constructively both my choice of papers and my discussion of them, and especially to W. G. E. Caldwell for helping me straighten out my somewhat deformed English. The choice of papers remains mine.

REFERENCE

Reed, R. R., 1941, Structural Geology, in *Geology, 1888–1938,* Fiftieth Anniversary Volume, Geological Society of America, Boulder, Colo., 1941, 578p.

CONTENTS BY AUTHOR

FABRIC OF DUCTILE STRAIN

INTRODUCTION

Ever since James Hutton presented his "System of the Earth" to the Royal Society of Edinburgh in 1785 it has been realized that mountains are built through a variety of processes, including uplift, folding, faulting, intrusion, and of course, erosion (Hutton, 1788). Concepts of ductile deformation were undreamed of in Hutton's day, but the seeds had been sown. That some type of compression is necessary for the rise of mountain belts was first determined by De Saussure (1796) from his study of folds in the Swiss Alps, although he earlier believed them to be caused by subterranean explosions. It was not for nearly a century after De Saussure, however, that some glimmering of the extent of large-scale ductile deformation was realized, mainly through development of the Nappe Theory as an explanation of the structure of the Alps (Heim, 1878; Bertrand, 1884). But how are layers of solid rock able to be bent and contorted into folds? What kinds of smaller-scale adjustments occur to allow larger structures to form? Answers to these questions were beginning to come forth from studies of slaty cleavage (discussed later in this book) and metamorphic rocks (Rosenbusch, 1898, the father of dynamic metamorphism).

A comprehensive paper by Becker (1893) explained rock deformation by reference to the concepts of finite homogeneous strain, deformation of mineral grains, fracture, and rheological behavior of the rock. Though some of his conclusions have not stood the test of time, his work and that by other turn-of-the-century geologists (notably Bonney, 1886; Heim, 1900; Peach et al., 1907) advanced the understanding of ductile deformation considerably. Prior to this there had been considerable argument concerning the significance of deformation, otherwise known then as dynamic processes, dynamo-metamorphism, or dislocation-metamorphism, and numerous authors were still trying to explain metamorphic foliation as a feature of sedimentation, or of the cooling of igneous melts, or even of contraction of the Earth. It took over 100 years of post-Hutton scientific inquiry to establish firmly the role of ductile deformation in the evolution of the Earth's crust.

The facts that rocks deform like putty and that mountains come and go have been difficult to comprehend fully, and even by the 1930s many geologists still were somewhat loath to acknowledge the extent to which some seemingly solid rocks have flowed in a ductile manner, despite a century of mapping folds and observing deformed rocks under the microscope (e.g., Harker, 1932). Nevin (1936, p. 1) lamented that

> Physical states of matter and physical properties of rocks have always been a very fertile source of confusion. Through a wrong conception of the structure of matter arose ideas of unyielding character of solid bodies. To a lack of distinction between solids and viscous liquids may be attributed some of the misunderstanding over fluidity of movement and strength of earth's crust. From neglect of, or insufficient allowance for, the effect of geologic time on the properties of rocks under stress, originated many of the fallacies regarding structural geology.

The accumulation and sifting of ideas throughout the 1800s and early 1900s can be considered as normal scientific evolution. The work of Sander, mainly in the Austrian Alps (1930, 1948), however, represents a revolution in the study of rock deformation nearly as significant as that begun in petrology by Sorby in the mid-1800s (e.g., 1858) when he first applied Nicol's petrographic microscope to the study of rocks. Few features of deformed rocks and rock-deforming processes escaped Sander's attention: he studied many aspects, ranging from the geometry of polyphase folding to the optical alignment of grains in deformed and recrystallized rocks. The Alps once again provided the inspiration for a giant step forward in human knowledge. No less a minor revolution in the study of rock deformation was begun about the same time by Griggs (1936, 1938, 1940) through experimentally deforming rocks and minerals under conditions similar to those prevailing in nature.

Due in no small part to the pioneering work of Bruno Sander and David Griggs, we now have a firm basis for understanding how rocks behave as mountains are built. Following a war-induced ebb in the 1940s and 1950s, our knowledge again blossomed, especially in the early 1980s with ever-increasing use of knowledge from materials science, such as metallurgy (Lister et al., 1981), and ever-increasing quantification in both field and laboratory studies. Such work is forming an evolutionary trend whose results, in the words of Schwerdtner et al. (1980, p. vii) "may be as profound as those of the 'revolution' of plate tectonics."

Definitions of the words "ductile" and "ductility" vary according to the frame of reference. In ordinary usage they refer to a material capable of being fashioned permanently into a new form, drawn out, or hammered thin, without breaking. In the world of rock mechanics, "ductility" is often defined as the *amount of ductile strain* a material undergoes without breaking (or prior to breaking), regardless of the stress levels attained. Many geologists, however, use the term "ductility" to mean the *ease of ductile deformation*, or rock strength (competence), quite a different meaning from the definition of rock mechanics.

For field geologists, "ductile deformation" may be defined as any permanent deformation not involving discontinuities formed through brittle failure. Emphasis is placed on the fabric of the rock already deformed; the process of deformation is interpreted from that fabric rather than observed. One important deduction is that ductile deformation does not cause a loss of cohesion, as occurs, for example, during fault movement. The scale of reference is important, however. For example, a rock mass greater than several cubic meters may have undergone change in shape through a series of consecutive, temporary, and local cohesion-loss events, such as microcracking or grain-boundary sliding, with no general loss of cohesion of the rock mass. Likewise, a single crystal may undergo innumerable, consecutive, minute dislocations at the sites of migrating crystal defects during deformation by crystal glide, yet the crystal remains intact. One could draw a limited analogy between the displacement and stacking of gigantic thrust sheets, which has occurred in some mountain belts, and the formation of crystallographic glide planes in deformed mineral grains. The latter is one of the main processes of plastic (ductile) deformation; the former is a major example of large-scale brittle deformation. It is worthwhile to note that during thrusting such mountain belts seem to have behaved as single units, undergoing changes in shape through repeated brief movement on internal discontinuities (the thrust faults). Loss of cohesion is restricted to segments of the fault surfaces during the times of fault slip (stick-slip process), a mode of deformation very similar to translation glide in single crystals. Indeed, any permanent deformation could be considered ductile if one chooses a large enough frame of reference.

This book, however, follows the common convention of restricting the discussion of ductile strain to fabrics that appear to pervade the entire body of a rock, as observed in outcrop, hand specimen, or under low-power microscopy.

It has long been recognized that deformation occurs over a vast range of scales from global to ionic. A hierarchical list of processes involving ductile deformation is given below (allowing for some overlap in scale).

1. Mantle convection: rise of plumes, plate motion, subduction.
2. Deformation in mountain belts: folding, development of minor structures, faults (especially the so-called ductile faults).

3. Intrusion: deformation of plastic diapirs (evaporites or magmatic), deformation of country rock. Includes all the types of structures that can form during mountain building; folds, faults, and minor structures.
4. Flow of glaciers (ice or salt): includes the formation of all types of structures listed below.
5. Folding: may form with faults and occurs through, and concomitant with, numerous internal adjustments of the rock that result in the formation of many minor structures.
6. Ductile faulting: may occur with folds and involves the formation of minor structures.
7. Development of minor structures: including foliation, boudins, and change of shape of rock components such as grains, clasts, and fossils. Such deformation can have many causes.
8. Change of shape of grains (crystals or crystal fragments): processes discussed in Part II of this book but includes crystallographic glide, milling, pressure solution, dynamic recrystallization, and other processes.
9. Foliation development: discussed in Part I of this book but includes a change of shape of rock components, rotation, neomineralization, pressure solution, and other processes.

It is evident from this list that understanding the mechanics of plate tectonics, mountain building, folding, some types of faulting, intrusion, and glacial flow is dependent on knowing how the various minor structures develop. It is these minor structures that form the "fabric of ductile strain" in rocks, and an understanding of them is fundamental to understanding some of the more important aspects of the evolution of the earth.

REFERENCES

Becker, G. F., 1893, Finite Homogeneous Strain, Flow and Rupture of Rocks, *Geol. Soc. America Bull.* **4:**13-90.

Bertrand, M., 1884, Rapports de structure des Alpes de Glaris et du Bassin houiller du Nord, *Soc. Géol. France Bull.,* 3rd ser., **12:**318-330.

Bonney, T. G., 1886, Pressure Metamorphism, *Geol. Soc. London Quart. Jour.* **42:**38-115.

De Saussure, H. B., 1796, *Voyages dans les Alpes,* vol. 4, Neuchâtel.

Griggs, D. T., 1936, Deformation of Rocks under High Confining Pressures, *Jour. Geol.* **44:**541-577.

Griggs, D. T., 1938, Deformation of Single Calcite Crystals under High Confining Pressures, *Am. Mineralogist* **23:**28-33.

Griggs, D. T., 1940, Experimental Flow of Rocks under Conditions Favouring Recrystallization, Geol. Soc. America Bull. **51:**1001-1022.

Harker, A., 1932, *Metamorphism, a Study of the Transformations of Rock Masses,* Chapman and Hall, London, 362p.
Heim, A., 1878, *Untersuchungen Über den Mechanismus der Gebirgsbildung,* 2 vols., Basel.
Heim A., 1900, Gneissfaltenlung in alpinem Central Massive, *Naturf. Gesell. Zurich. Vierteljahrscrift* **45:**205–226.
Hutton, J., 1788, Theory of the Earth, *Royal Soc. Edinburgh Proc.* **1:**209–304. (This article is an expanded version of a dissertation concerning the system of the Earth, its duration, and stability, an abstract of a paper read to the Royal Society of Edinburgh in 1785. The 1788 paper was further developed into a two-volume work of the same title published in 1795 by William Creech, Edinburgh.)
Lister, G. S., H. J. Behr, K. Weber, and H. J. Zwart, ed., 1981, The Effect of Deformation on Rocks, *Tectonophysics* **78** (special issue) 698p.
Nevin, C. M., 1936, *Principles of Structural Geology,* 2nd ed., John Wiley, New York, 348p.
Peach, B. N., J. Horne, W. Gunn, C. T. Clough, and L. W. Hinxman, 1907, The Geological Structure of the Northwest Highlands of Scotland, *Geol. Survey Great Britain Memoir*, 668p.
Rosenbusch, H., 1898, *Elemente der Gesteinslehre,* 1st ed., Schweizerbart, Stuttgart.
Sander, B., 1930, *Gefügekunde der Gesteine.* Springer, Berlin, 352p.
Sander, B., 1948, Einfuhrung in die Gefügekunde der Geologischen Körper, Springer-Verlage, Vienna, translated by Phillips, F. C. and G. Windsor, 1970, to *An Introduction to the Study of Fabrics of Geological Bodies,* Pergamon, Oxford, 641p.
Schwerdtner, W. M., P. J. Hudleston, and J. M. Dixon, ed., 1980, Analytical Studies in Structural Geology, Tectonophysics **66** (special issue) 229p.
Sorby, H. C., 1858, On the microscopical structure of crystals, Geol. Soc. London Quart. Jour., **34:**453–500.

BIBLIOGRAPHY

Adda, Y., and J. Philibert, 1966, La diffusion dans les solides, Presses Universitaires de France, Paris.
Amelinckx, S., 1964, *The Direct Observation of Dislocations,* Academic, New York.
Ashby, M. F., and R. A. Verral, 1973, Diffusion Accommodated Flow and Superplasticity, *Acta Metallurgica* **21:**149–163.
Boullier, A. M., and Y. Gueguen, 1974, SP-mylonites. Origin of some Mylonites by Superplastic Flow, *Contr. Mineralogy and Petrology* **50:**93–104.
Brace, W. F., 1961, Mohr Construction in the Analysis of Large Geologic Strain, *Geol. Soc. America Bull.* **72:**1059–1080.
Breddin, H., 1964, Die tectonische deformation de fossilien und gesteinen der molasse von St. Gallen (Schweiz), *Geol. Mitt.* **4:**1–68.
Bridgman, P. W., 1964, *Studies in Large Plastic Flow and Fracture,* Harvard University Press, Cambridge, Ma.
Cahn, R. W., ed., 1970, *Physical Metallurgy,* North-Holland, Amsterdam, 1333p.
Choukroune, P., 1971, Contribution à l'étude des mechanismes de la deformation avec schistosité grâce aux cristallisations syncinématique dans les "zones abritées," "pressure shadows," *Soc. Geol. France Bull.* **3-4:**257–271.

Fairbairn, H. W., 1949, *Structural Geology of Deformed Rocks,* Addison-Wesley Press Inc., Cambridge, Ma., 344p.

Friedman, M., 1964, Petrofabric Techniques for the Determination of Principal Stress Directions in Rocks, in *State of Stress in the Earth's Crust,* W. R. Judd, ed., American Elsevier, New York, pp. 451-552.

Griggs, D., and J. Handin, eds., 1960, Rock Deformation, *Geol. Soc. of America Memoir 79,* 382p.

Heard, H. C., I. Y. Borg, N. L. Carter, and C. B. Raleigh, eds., 1972, Flow and Fracture of Rocks, *American Geophysical Union Monograph 16,* 352p.

Johnson, A. M., 1970, *Physical Processes in Geology,* Freeman, Cooper and Co., San Francisco, 577p.

Knopf, E. B. and E. Ingerson, 1938, Structural Petrology, *Geol. Soc. America Memoir 6,* 270p.

March, A., 1932, Mathematische theorie der regelung nach der Korngestalt bie affiner deformation, *Zeitschr. Kristallographie* **81:**285-297.

Nicolas, A. and J. P. Poirier, 1976, *Crystalline Plasticity and Solid State Flow in Metamorphic Rocks,* John Wiley & Sons, London, 444p. Especially recommended for its discussion of deformation processes and its list of French, German, and English language references.

Paterson, W. S. B., 1981, *The Physics of Glaciers,* 2nd ed. Pergamon, Oxford, 380p.

Paulitsch, P., ed., 1970, *Experimental and Natural Rock Deformation,* Springer-Verlag, Berlin, 525p.

Reid, C. N., 1973, *Deformation Geometry for Materials Scientists,* Pergamon, Oxford, 211p.

Spry, A., 1969, *Metamorphic Textures,* Pergamon Press, Oxford, 350p. The most complete discussion of textures in metamorphic rocks yet written.

Turner, F. J., 1948, Mineralogical and Structural Evolution of the Metamorphic Rocks, *Geol. Soc. America Memoir 30,* 342p.

Turner, F. J., and L. E. Weiss, *Structural Analysis of Metamorphic Tectonites,* McGraw-Hill, New York, 545p.

Voll, G., 1960, New Work on Petrofabrics, *Geol. Jour.* **2:**503-597.

Part I

SLATY CLEAVAGE AND ITS RELATIONSHIP TO STRAIN

Editor's Comments on Papers 1 Through 4

1 **WOOD**
Current Views of the Development of Slaty Cleavage

2 **SHARPE**
Excerpts from *On Slaty Cleavage (second communication)*

3 **SORBY**
Excerpts from *On Slaty Cleavage, as Exhibited in the Devonian Limestones of Devonshire*

4 **MEANS**
Natural and Experimental Microstructures in Deformed Micaceous Sandstones

The word "cleavage" means many different things in the English language. In the medieval mining industry it was used to refer to the splitability of a rock and is still used in this way by many quarrymen.

In structural geology, cleavage quickly came to refer to a secondary planar fabric in slaty and schistose rocks resulting from deformation, regardless of what other planes of easy splitability (such as joints) exist in the rock. Many terms have been used to refer to a bewildering variety of planar fabrics (or foliations), although a paper by Powell (1979), goes a long way toward devising a rational classification based primarily on morphology.

One of the early successes of thin-section petrology in the mid-1800s was to show that cleavage in slates and schists is caused by the alignments of mineral grains, particularly phyllosilicates. However, the origin of this grain alignment has been debated vigorously by nineteenth- and twentieth-century geologists. Does slaty cleavage result from crystalline or mechanical forces (or even electromagnetic, as suggested by some mid-1800s geologists, e.g., Sedgwick, 1835)? If mechanical, does it form perpendicular to applied compression, is it perpendicular to the maximum total strain-shortening direction, or does it develop parallel to planes of high shear stress?

The inaugural paper in this volume, by Wood (Paper 1), reviews the study of slaty cleavage and summarizes the field evidence in favor of an origin essentially by mechanical flow, with cleavage forming perpendicular to the direction of maximum total shortening whatever the effects of recrystallization. This paper also sets the tone for the entire volume in that it emphasizes the application of strain geometry to understanding the formation of structures in rocks (both kinematics and mechanics).

Careful and exciting work in the mid-1800s clearly established that deformation of rock masses is the major cause of slaty cleavage, and almost from the beginning, studies of slaty cleavage involved consideration of clast and fossil distortion (e.g., Phillips, 1844; Haughton, 1856; Hitchcock, 1861). Two papers are reprinted here that attack the problem in somewhat different ways. Paper 2 in this volume, by Sharpe, is one of the earliest discussions of the origin of slaty cleavage. In it, Sharpe presents an analysis of the relationship between folded bedding, grain shape, and slaty cleavage and gives the first recorded example of cleavage refraction. In Paper 3, Sorby discusses crystallizing versus mechanical action and also presents one of the earliest model experiments in structural geology, a simple but elegant display of the effects of pressure on grain rotation and reorientation. Sorby and Sharpe both agree with Phillips's (1844) original contention that mechanical forces ("creeping movements" according to Phillips) were responsible for the formation of slaty cleavage and agree with each other that pressure applied perpendicular to the cleavage caused both the cleavage and the alignment of the long dimensions of grains. They disagree, however, on the significance of rotation (Sorby) and changed shape of the elongate and aligned grains (Sharpe). Becker (1893) agreed with the importance of both rotation and grain deformation but believed that the forces were applied oblique to the cleavage.

Means (Paper 4) describes his attempts to duplicate experimentally some of the processes by which cleavage is formed. Means used salt-mica combinations as analogs of quartz-mica rocks in order to circumvent experimental problems related to the high strength of quartz-rich rocks. Although Mean's experiments are not merely scale-model experiments, they do not duplicate all the processes by which rocks are converted to slate or schist. An interesting set of experiments by Tullis (1976) also duplicated many of these processes.

Taken together, 150 years of field and microscopic work, model experiments, strain theory, and many rock-deformation experiments (e.g., Tullis, 1976) indicate that most types of slaty cleavage and schistosity develop perpendicular, or nearly perpendicular, to the direction of local total maximum shortening and do not form parallel to planes of high shear stress, nor necessarily perpendicular to the greatest principal compressive stress. However, this statement does not apply to all foliations, especially not to those defined by distinctive planar discontinuities, such as crenulation cleavage, initial orientations of which may be controlled by shear stress (Williams, 1976). Nor does it apply to foliations defined by the alignment of grains that were already aligned prior to deformation; such a foliation will be intermediate between the initial alignment and that predicted by the strain ellipsoid. Williams (1976) also argued that even slaty cleavage will be perfectly perpendicular to the direction of maximum shortening only if successive increments of strain are coaxial without simple shear or rotational components (see also Hudleston and Hook, 1980). Nonetheless, even in fold limbs which have undergone rotation, the deviation is commonly slight, especially where strain is large (Williams, 1976), although the effects of non-coaxial strain in them is undoubtedly important in the origin of fanned cleavage. Also, the fact that cleavage commonly ends up perpendicular to the direction of maximum total shortening (as deduced from strain studies) does not necessarily mean that it initiates that way (Williams, 1977), and its relationship to both stress and strain may be complex and change with time as the rock deforms.

It is now believed that slaty to schistose cleavage develops mainly through processes involving rotation and/or strain of old grains as well as nucleation and crystallization of new grains (e.g. Mancktelow, 1979). However, the relative importance of these and other processes (such as pressure solution) undoubtedly varies with rock type and conditions of deformation (Means, Paper 4; Alterman, 1973; Tullis, 1976; Alvarez et al., 1976; Groshong, 1976; Robin, 1979; Helmstaedt and Greggs, 1980; and Borradaile, 1981).

The origin of foliation (including slaty cleavage) probably has been the topic of more papers and reviews than any other single aspect of structural geology. It will likely remain a topic of considerable research, producing numerous divergent and occasionally contentious views for some time to come. The papers presented here do little more than outline the mainstream of thought on the subject of slaty to schistose cleavage. For a more complete understanding of the development of ideas the reader is referred to Leith (1905), Becker (1907), Dieterich (1969), Siddans (1972), Williams (1977), and Means (1977).

REFERENCES

Alterman, I. B., 1973, Rotation and Dewatering During Slaty Cleavage Formation: Some New Evidence and Interpretations, *Geology* **1**:33-36.

Alvarez, W., T. Engelder, and W. Lowrie, 1976, Formation of Spaced Cleavage and Folds in Brittle Limestone by Dissolution, *Geology* **4**:698-701.

Becker, G. F., 1893, Finite Homogeneous Strain, Flow and Rupture of Rocks, *Geol. Soc. America Bull.* **4:**13–90.
Becker, G. F., 1907, Current Theories of Slaty Cleavage, *Am. Jour. Sci.*, ser. 4, **24:**1–17.
Borradaile, G. J., 1981, Particulate Flow of Rock and the Formation of Cleavage, *Tectonophysics* **72:**305–321.
Dieterich, J. H., 1969, Origin of Cleavage in Folded Rocks, *Am. Jour. Sci.* **267:**155–165.
Groshong, R. H., 1976, Strain and Pressure Solution in the Martinsburg Slate, Delaware Water Gap, New Jersey, *Am. Jour. Sci.* **276:**1131–1146.
Haughton, S., 1856, On Slaty Cleavage, and the Distortion of Fossils, *London, Edinburgh and Dublin Philos. Mag. and Jour. Sci.*, ser. 4, **12:**409–421.
Helmstaedt, H., and R. G. Greggs, 1980, Stylolitic Cleavage and Cleavage Refraction in Lower Paleozoic Carbonate Rocks of the Great Valley, Maryland, *Tectonophysics* **66:**99–114.
Hitchcock, E., 1861, On the Conversion of Certain Conglomerates into Talcose and Micaceous Shists and Gneiss, by the Elongation, Flattening and Metamorphism of the Pebbles and the Cement, *Am. Jour. Sci. and Arts*, ser. 2, **31:**372–392.
Hudleston, P. J., and R. LeB. Hooke, 1980, Cumulative Deformation in the Barnes Ice Cap and Implications for the Development of Foliation, *Tectonophysics* **66:**127–146.
Leith, C. K., 1905, Rock Cleavage, *U.S. Geol. Survey Bull. 239,* 216p.
Mancktelow, N. S., 1979, The Development of Slaty Cleavage, Fleurieu Peninsula, South Australia, *Tectonophysics* **58:**1–20.
Means, W. D., 1977, Experimental Contributions to the Study of Foliations in Rocks: A Review of Research Since 1960, *Tectonophysics* **39:**329–354.
Phillips, J., 1844, On Certain Movements in the Parts of Stratified Rocks, *British Assoc. Adv. Sci. Rept. 1843,* pp. 60–61.
Powell, C. McA., 1979, A Morphological Classification of Rock Cleavage, *Tectonophysics* **58:**21–34.
Robin, P. Y. F., 1979, Theory of Metamorphic Segregation and Related Processes, *Geochim. et Cosmochim. Acta* **43:**1587–1600.
Sedgwick, A., 1835, Remarks on the Structure of Large Mineral Masses, and Especially on the Chemical Changes Produced in the Aggregation of Stratified Rocks during Different Periods after Their Deposition, *Geol. Soc. London Trans.* **3:**461–486.
Siddans, A. W. B., 1972, Slaty Cleavage—A Review of Research since 1815, *Earth-Sci. Rev.* **8:**205–232.
Tullis, T. E., 1976, Experiments on the Origin of Slaty Cleavage and Schistosity, *Geol. Soc. America Bull.* **87:**745–753.
Williams, P. F., 1976, Relationships Between Axial-Plane Foliations and Strain, *Tectonophysics* **30:**181–196.
Williams, P. F., 1977, Foliation: A Review and Discussion, *Tectonophysics* **39:**305–328.

1

Copyright ©1974 by Annual Reviews Inc.
Reprinted by permission from *Annu. Rev. Earth and Planetary Sci.* **2**:369–401 (1974)

CURRENT VIEWS OF THE DEVELOPMENT OF SLATY CLEAVAGE

Dennis S. Wood
Department of Geology, University of Illinois, Urbana, Illinois 61801

INTRODUCTION

A knowledge of the nature and origin of slaty cleavage is of paramount importance to the understanding of deformed rocks in general. This results from a number of facts: cleavage is a phenomenon of very widespread occurrence; it clearly has a close geometrical and genetic relationship to processes of rock folding; and the cleaved rocks of the lower metamorphic grades frequently contain indicators which may be used to determine the total strain entailed in the formation of particular geological structures. Slates are among the best materials available for accurate evaluation of natural ductile deformation of rocks. In addition, cleavage has considerable practical significance in terms of its use in the field for elucidation of large-scale geological structures and recognition of the polyphase deformational history of many regions. It is therefore not surprising that investigations of cleavages were undertaken by many of the more prominent earth scientists of the nineteenth century. Notable among these were Greenough, Sedgwick, De la Beche, Phillips, Darwin, Sorby, Harker, and Van Hise.

The axial planar relationship of cleavage to folding was fully appreciated by Sedgwick (50), Darwin (9), and Rogers (47), all of whom believed that slates had undergone recrystallization during deformation. Furthermore, Darwin (9) thought that slaty cleavage was at one end of a spectrum of continuous metamorphic alteration, to which coarser schistose and gneissose textures also belong. The timing relationships of cleavage formation were considered by Sedgwick (50), Phillips (39), and Sharpe (52), each of whom believed that cleavage was posthumous to folding. Tyndall (66), on the other hand, believed both features to be essentially contemporaneous. The intermediate viewpoint of Fisher (14) and Harker (22) was that cleavage and folding are both responses to the same compressional forces, but that cleavage does not appear until well after the initiation of folding.

Initial knowledge concerning the finite amount of shortening undergone by cleaved rocks was provided by studies of deformed fossils from southwestern England, North Wales, and Ireland. Phillips (39) was the first person to notice that there was a close relationship between fossil distortion and cleavage. The significance of this was appreciated by Sharpe (52) who stated the "general rule that the shells are most distorted in those beds which are most slaty," and who also recognized that "rocks affected by slaty cleavage have suffered a compression of their mass in a direction everywhere perpendicular to the plane of cleavage and an expansion of their mass in the direction of cleavage dip." This was the earliest referral of cleavage to a process of flattening and to the existence of one preferred direction of resultant extension. Sharpe (53) also clearly demonstrated the inequidimensional grain shapes that are characteristic of cleaved rocks. Sorby (54, 55) subsequently utilized ellipsoidal reduction bodies from the Cambrian rocks of North Wales to show that the ellipsoid of distortion or deformation ellipsoid had one of its principal planes precisely parallel to the cleavage, which is the plane normal to the direction of maximum finite shortening. In order to verify these relationships, Sorby (55) and Tyndall (67) conducted experiments with clay and wax, respectively, and both succeeded in producing fissility perpendicular to the applied force. Sorby (54) believed that the Cambrian rocks of North Wales, in which he found a typical deformation ellipsoid to have the form 1.6/1/0.27, had undergone shortening of 75% perpendicular to the cleavage.

The processes whereby preferred shape and crystallographic orientations of minerals are achieved and give rise to cleavage could only be seriously considered after Sorby's development of microscopic petrography. The microscope assisted Sorby in developing a hypothesis of mechanical rotation of initially inequidimensional grains such that their long dimensions come to lie in the cleavage. Sorby also believed that preferred orientation achieved by mechanical rotation could be further enhanced by the growth of new micaceous minerals in the plane of cleavage. Furthermore, he hinted that intracrystalline gliding could be a contributing mechanism (55) and he was first to demonstrate situations in which pressure-solution-transfer had been important during cleavage formation (56–58). Nevertheless, Sorby believed that during the formation of cleavage in argillaceous rocks, all other processes were subsidiary to rotation induced by compression. This was questioned by Van Hise (68) who followed Sorby in believing cleavage to have formed perpendicular to the principal compression direction, but differed in thinking that the primary factor in cleavage development was the growth of new minerals in a plane of flattening. Throughout the present century there have been many proponents of both Sorby's rotation hypothesis (e.g. Maxwell, 35) and Van Hise's crystallization hypothesis (e.g. Kamb, 29; Swanson, 62). Some workers have favored a combination of both hypotheses. For example, Ramsay (45) has suggested that mechanical rotation may have the dominant effect at the lowest metamorphic grades, with recrystallization being of greatest importance under higher temperature conditions. Oertel (38) has theoretically demonstrated how a mechanically produced preferred orientation can be enhanced by late-tectonic or post-tectonic crystal growth.

There has been more controversial discussion between those who believe cleavage to be a shearing phenomenon and those who believe it to be a flattening phenomenon which formed precisely perpendicular to the direction of maximum finite shortening. A shearing mechanism was advocated by Phillips (39), Laugel (32), Fisher (14), Becker (4, 5), Sander (48), Turner (66), Hoeppener (28), Wunderlich

(77), and Voll (69); whereas cleavage has been referred to a flattening process by Sharpe (52, 53), Sorby (54, 55), Tyndall (67), Haughton (24), Harker (21, 22), Van Hise (68), Leith (33), Ramsay (45), Dieterich (10), and Wood (75, 76). The substance of this controversy, within the present century, resulted from the incorrect extension by Becker (4) of the concept of the strain ellipsoid (63) or ellipsoid of distortion which was introduced into geology as a purely descriptive tool for describing deformational shape changes by Harker (22). Becker wrongly treated slaty cleavage as a fracture phenomenon and contended that only exceptional irrotational strains could produce flattening at right angles to the line of force, so that if fissility were produced by flattening it would be a mistake to infer that the direction of force was normal to the fissility. Becker considered that the circular sections of the strain ellipsoid were surfaces of no finite distortion and coincident with the surfaces of maximum shearing strain. He was incorrect with respect to both of these relationships. The circular cross sections will only be equivalent in area to the original spherical section provided that the intermediate axis of the ellipsoid remains similar in length to the original sphere diameter. As was pointed out by Griggs (19), the surfaces of no distortion in the triaxial ellipsoid are generally elliptical cones rather than planes of circular cross section. Furthermore, neither the planes of circular cross section nor the surfaces of no-finite distortion coincide with the surfaces of maximum shearing strain. As a result of these errors, Becker concluded that the direction of principal compression or maximum shortening would bisect the obtuse angle between the surfaces of maximum shearing strain. This is in direct contradiction to all experimental evidence, which shows that shear fractures are inclined at angles of less than 45° to the direction of principal compression. Thus even if Becker had been correct in his view that cleavage was a fracture phenomenon, his interpretation of its origin by shear is supported neither by theory nor experiment. Despite this, and notwithstanding all the evidence from natural strain indicators which clearly shows that cleavage is perpendicular to the short dimension of the deformation ellipsoid, the shear hypothesis has found continuing support.

The use of natural strain indicators in the evaluation of strain magnitudes relies upon the comparison of measurable axes of the deformation ellipsoid with some original parameter. Two alternative assumptions have commonly been made: either that the intermediate axis of the ellipsoid has remained unchanged (e.g. Cloos, 8), or that the volume has remained constant (e.g. Wood, 76). In some cases either or both of these assumptions may be valid, but in the general case both are invalid. It is clearly important to determine the volume changes that have occurred during deformation, but these are extremely difficult to evaluate. Apart from exceptional situations it is most unlikely that increase of volume occurs during natural deformational processes. The first consideration of volume change was by Sorby (54), who proposed that volume loss of almost 60% had accompanied the formation of cleavage in the Cambrian rocks of North Wales. Sorby arrived at this figure by assuming that the intermediate axis of the deformation ellipsoid had undergone no dimensional change. In the case considered this would have meant that compression parallel to the short axis of the ellipsoid was incompletely compensated by resultant extension parallel to the long axis of the ellipsoid. Fisher (14), on the other hand, believed that volume changes of the magnitude suggested by Sorby were completely unrealistic and, instead, considered the volume change to be totally negligible. Becker (5), from a consideration of the compressibilities of various rock materials, also thought that it was virtually impossible for significant volume changes to have occurred. Because ductile deformation is not dependent upon any elastic property, the relevance of Becker's argument is lost. Sorby's later viewpoint was that volume reduction of approximately 10% had occurred in the Cambrian slates of Wales (59), but Cloos (8) was of the opinion that there had been no deformational volume change in the central Appalachian fold belt. The somewhat novel view that cleavage is the result of a volume increase has been expressed by Kirillova (31) and explained in terms of expansion induced by the adsorption of water by clay minerals. More recently Ramsay (45) and Ramsay & Wood (46) have drawn attention to the consequences of ignoring volume change and have demonstrated that the deformation of lithified materials may well involve volume losses of up to 20%. Unless appropriate allowance is made for such changes, strains that are determined from natural indicators will be in error.

Considerations of volume change during deformation raise fundamental questions concerning the conditions of the materials at the time of deformation. Most workers have either considered or assumed that the rocks which now possess cleavage were completely lithified materials prior to their deformation. The person to first suggest that this might not be the case was Sedgwick (50, p. 486), who believed that cleavage developed as a last stage in alteration before the rock in question became completely lithified. The opinion that deformation occurred prior to complete consolidation was also expressed by Cloos (8) in connection with the folding at South Mountain. The possibility of cleavage being a prelithification deformation feature was more strongly suggested by Maxwell (35) for the Martinsburg slates in the Delaware Water Gap area of New Jersey and Pennsylvania. Maxwell's view has been followed in other areas by Moench (37), Powell (41), Braddock (6), and Clark (7). The idea that tectonic overpressures may create abnormally high pore water pressures such as to cause dewatering which may assist the formation of cleavage by facilitating mechanical rotation of grains, is at variance with most opinions of cleavage genesis. If such a process were to have been of universal significance, the effects of post-tectonic metamorphism would be much more widespread than has previously been suspected.

Although it is well over a hundred years since Sharpe observed that the most severely distorted fossils occur in the most intensely cleaved rocks, there has been little attempt to relate measurable variations in strain intensity to variations of either the microscopic or megascopic features of slaty rocks. Much information should be forthcoming concerning the relationship between variation in large-scale regional structure and variation in strain. Similarly, the relationship between variation in finite total strain and such microscopic features as the degree of preferred shape and crystallographic orientation need to be fully evaluated. These relationships, together with some of the other interesting aspects of rocks which possess slaty cleavage, will be discussed in the light of recent work.

STRAIN AND SLATY CLEAVAGE

Despite the stimulus which should have been provided by earlier workers such as Sorby, the systematic investigation of natural strain in rocks was long neglected. A major exception is the work of Cloos (8) which stands as one of the great contributions to structural geology. Other important additions to knowledge concerning the intensity and variation of natural strain included the works of Heim (26, 27) and Wettstein (72) using deformed oolites and fossils, and of Strand (61) and Flinn (15) using deformed conglomerates.

A great impetus to strain studies was provided by Flinn (16) in the paper entitled "On Folding during Three Dimensional Progressive Deformation" which dealt with strain from a theoretical standpoint and predicted the type of fold and boudinage structures that should result from all possible types of strain geometries. This work has had a deep influence upon subsequent theoretical and practical investigations of natural strain. Flinn (16) emphasized that the development of structures such as folds was subsidiary to the basic distortion of rocks. He also urged that the conventional tectonic axes (a, b or B, and c) of Sander should be discarded in favor of the deformation ellipsoid. In so doing, Flinn focused attention upon the true internal deformation of rocks, rather than on partial expressions of the deformation, and placed the onus for determining the deformation ellipsoid upon the field geologist.

The deformation ellipsoid is best obtained from measurements of formerly spherical bodies which possess mechanical properties that are similar to their enclosing matrix. Because such materials are relatively uncommon, various methods have recently been developed which enable objects of nonspherical and different initial shapes (8) to be used for strain determination (11, 13, 18, 45).

Finite strain states are most satisfactorily represented by plotting the deformation ellipsoid according to its axial ratio on a rectangular coordinate graph, termed the deformation plot (15, 16). Some terminological confusion has resulted from differing definitions of the three mutually perpendicular axes. Flinn used the system $Z > Y > X$, whereas Ramsay (45) and Wood (76) have used the system $X > Y > Z$. The only advantage of the former system is that it conforms to the practice used in describing the optical indicatrix, and if for any reason it is useful to know the angle between the circular cross sections of the ellipsoid, this is a convenient method. In view of the fact that these cross sections have virtually no mechanical relevance to rock deformation, the system $X > Y > Z$ is more logically preferable. The three axes of the deformation ellipsoid, X, Y, and Z define the principal finite strains, $1+e_1$, $1+e_2$, and $1+e_3$, where e_1, e_2, and e_3 are the principal extensions. Thus the ratio $(1+e_1)/(1+e_2)$ is plotted as the ordinate (X/Y); and the ratio $(1+e_3)/(1+e_2)$ is plotted as the abcissa (Z/Y). The simplest and most convenient deformation plot is that in which the ratios X/Y and Z/Y are plotted on logarithmic scales. This has the advantage that all lines representing equal changes of length of the three dimensions X, Y, and Z are straight. Figure 1 shows a logarithmic deformation plot for which a large number of lines of equal

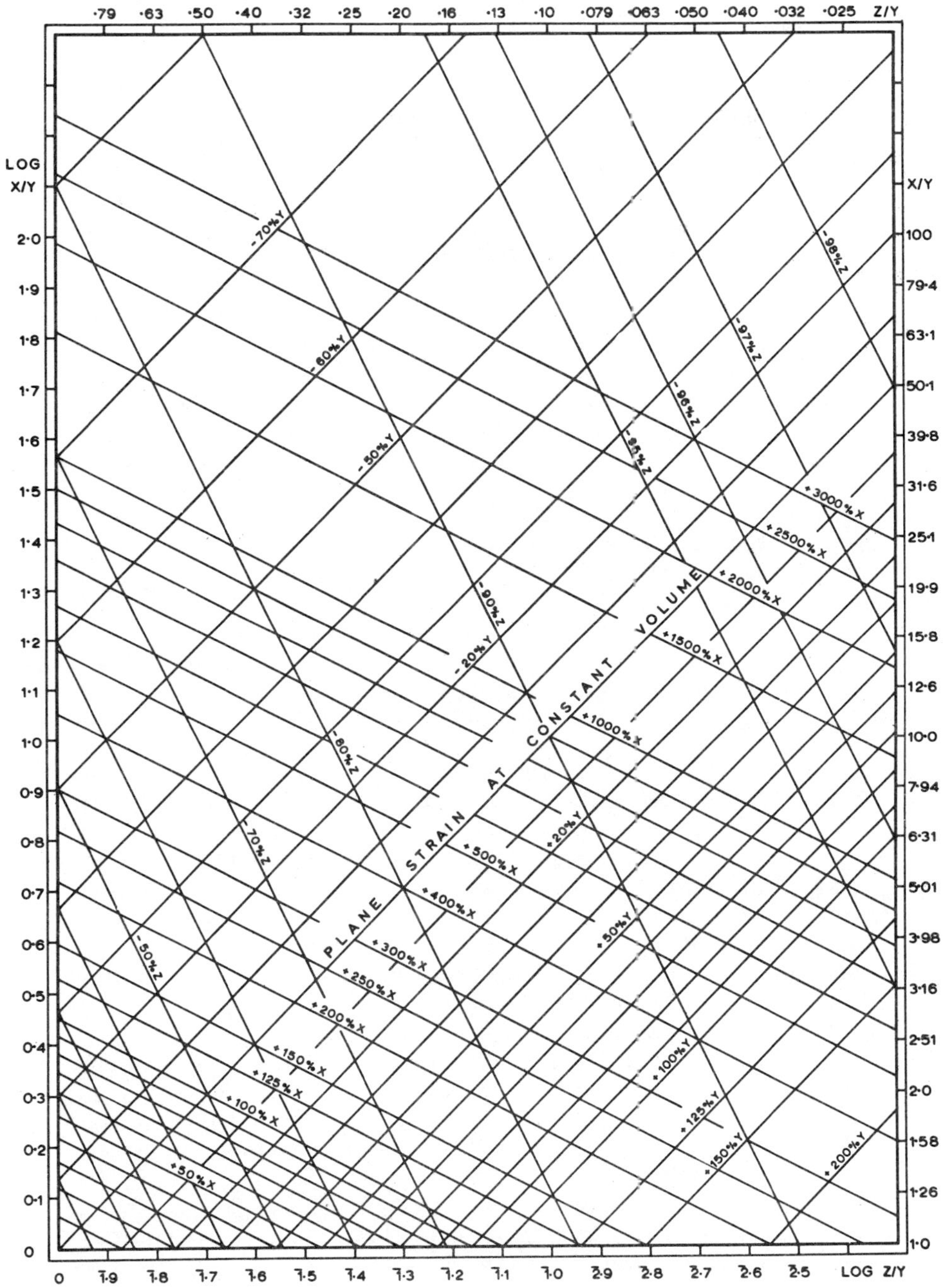

Figure 1 Logarithmic deformation plot ($\log_{10}$) for evaluation of finite strain parallel to the axes of the deformation ellipsoid.

dimensional change have been calculated (76). All deformation ellipsoids which are to be found in nature can be represented on this plot and, provided that volume changes have been insignificant, the strains in the three principal directions are immediately determinable. Oblate spheroids ($X = Y > Z$) of increasing oblateness plot progressively along the abcissa and prolate spheroids ($X > Y = Z$) of increasing prolateness plot along the ordinate. All general triaxial ellipsoids fall into two fields which are separated by the line of unit slope through the origin. Under conditions of deformation at constant volume these two fields are synonymous with deformation of flattening type and constriction type, respectively. Under similar conditions, a rock deforming by a process of plane strain would be represented by ellipsoids which progressively plot along the line of unit slope. Such ellipsoids have the special property that their intermediate axis (Y) would maintain constant length and be equivalent to the geometric mean of the two other axes. Rocks which have deformed by ideal plane strain appear to be relatively rare in nature, as are tectonically deformed rocks with either perfect linear fabrics of pure constrictional origin or perfect planar fabrics of pure flattening origin. Perfect linear fabrics are perhaps more frequently approached in nature than purely planar fabrics; possibly they are more noticeable.

All slates appear to fall in the flattening field of the deformation plot. Figure 2 shows some finite strain states for the Cambrian slate belt of North Wales, a region of essentially upright fold structures. For convenience, the lines of equal dimensional changes are indicated as percentages of the diameter of the equivalent volume sphere (d). In all cases, the principal plane XY of the deformation ellipsoid is parallel to the cleavage. There can be no question that cleavage is precisely perpendicular to the direction of maximum finite shortening.

The mean strain determinations represented in Figure 2 show considerable heterogeneity from one locality to another, although all ellipsoids at any single locality have remarkable shape homogeneity. The average length of the short dimension (Z) of the deformation ellipsoid is 35% of the equivalent sphere diameter, indicating a shortening across the cleavage of 65%. If volume change occurred during deformation, this would be a minimum shortening. The resultant principal extension is such that the long dimension (X) of the deformation ellipsoid has an average value of approximately 220% d. Consequently, the average principal elongation in the plane of cleavage has involved more than doubling of dimension. In this case, the measured extension would be in excess of the real extension in the event that significant volume change had occurred. The intermediate ellipsoid dimension (Y) has undergone apparent average extension by approximately 35%, which would also be reduced if allowance were needed because of volume change.

The variation in dimensional shortening parallel to Z has been between 55% ($Z = 45\%\ d$) and 75% ($Z = 25\%\ d$). Extension parallel to X has been between 60% ($X = 160\%\ d$) and 170% ($X = 270\%\ d$); and extension parallel to the intermediate axis (Y) has been between 20% ($Y = 120\%\ d$) and 45% ($Y = 145\%\ d$). These results from Wales are an interesting comparison with strain data for similar age rocks in the Taconic slate belt of the United States, where the associated fold structures are overturned and tight rather than upright and open as in North Wales. Therefore as might be expected, the deformation ellipsoids lie even farther into the flattening field of the deformation plot, with the average ellipsoid having the form 1.7/1/0.17 for the Taconic slate belt rather than 1.7/1/0.26 for the Welsh Cambrian slate belt. The finite strain data for the Taconic slate belt (Figure 3) shows that the short dimension of the average ellipsoid indicates a shortening of approximately 75% (Z = 25% *d*) across the cleavage, with concomitant extension of

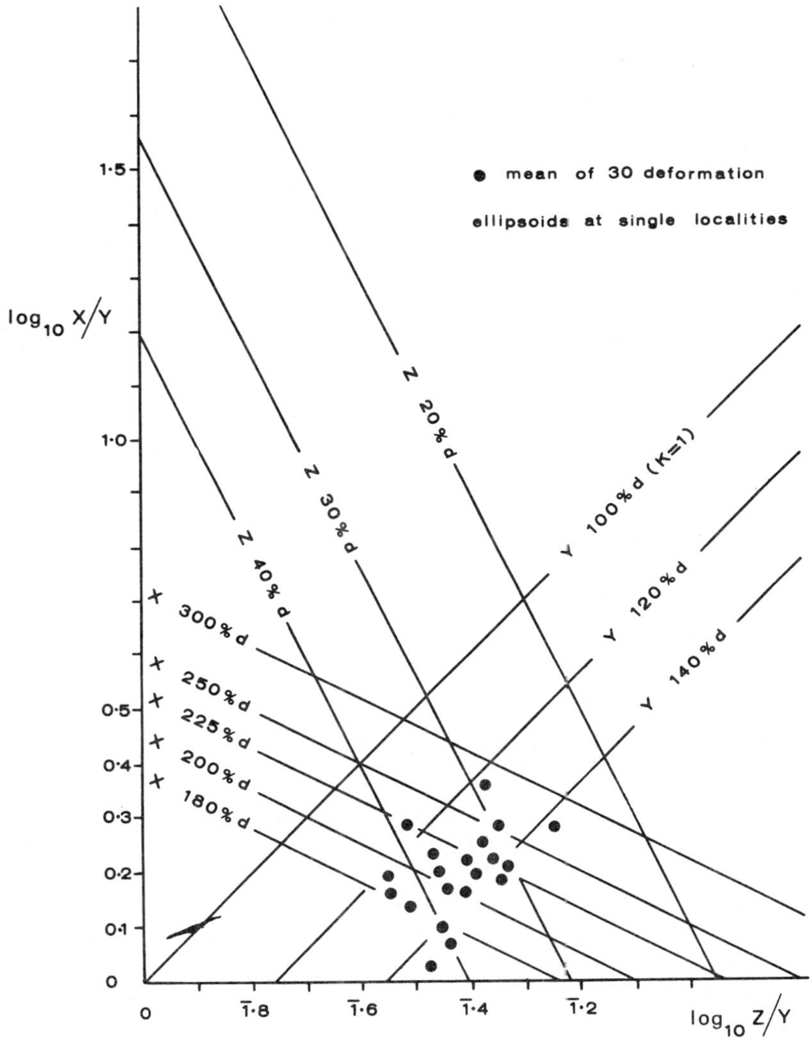

Figure 2 Finite strain data for the Cambrian slate belt of Wales.

the long dimension by 150% ($X = 250\%$ *d*) and of the intermediate axis by 50% ($Y = 150\%$ *d*). Although in this instance the intermediate dimension is reasonably constant from locality to locality, there is 10% variation in the amount of shortening across the cleavage, but almost 70% of strain variation in the direction of greatest finite extension within the cleavage.

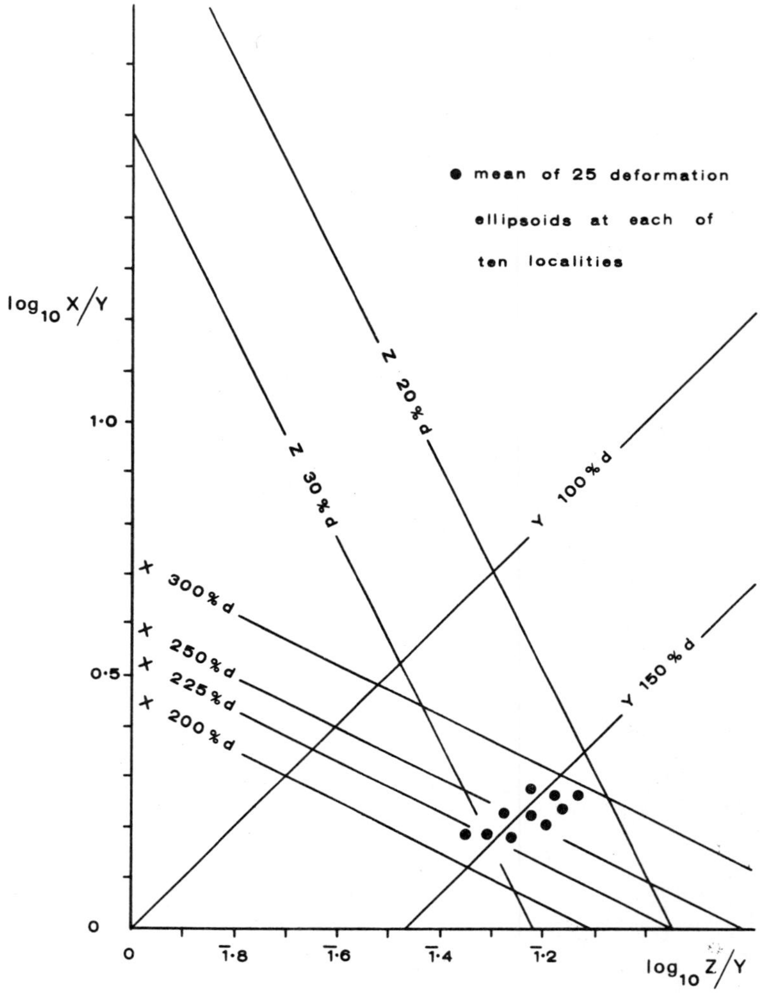

Figure 3 Finite strain data for the Taconic slate belt of the Appalachian region, Vermont, USA.

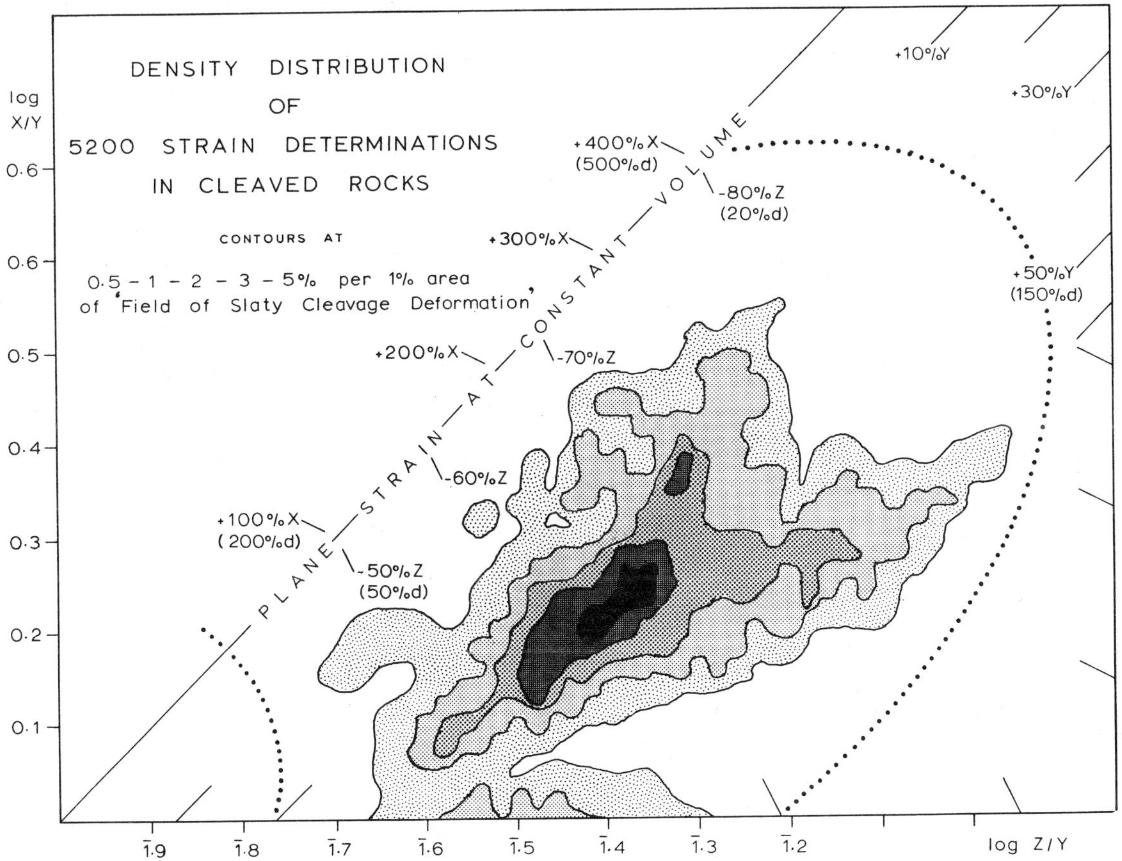

Figure 4 Contoured deformation plot of finite strain in slates of Cambro-Ordovician age from the European Caledonides and their North American equivalents.

When strain obtained from a large number of other slaty rocks is plotted together with that from the slate belts of Wales and the Taconic region, the deformation plot shows an extremely strong concentration of deformation ellipsoids, but represents a large variation of strain magnitudes (Figure 4). The data on this diagram consist of over 5000 strain measurements on slates of Cambrian age from New England, Wales, Scotland, and Norway; and of Ordovician age from New England, Wales, and Northern England. It is reasonable to place conservative limits to a "field of slaty cleavage deformation" (75), and this is outlined in Figure 4. The lower limit to this field depends upon the fact that a certain minimum amount of strain is required before cleavage appears. The upper limits to the field depend upon both geometrical and physical considerations. With continued deformation, cleavage eventually becomes deformed itself, frequently by folding upon axes which bear some close geometrical relationship to the previous strain pattern. Folding on axes parallel to the direction of principal extension in the previously formed cleavage is extremely common, particularly where the cleavage fabric contained an important linear element. In instances where the initial elongation component is less obvious, refolding may occur on axes parallel to the Y axis of the previous deformation ellipsoid; more commonly there is no such simple relationship. Another reason for assigning an upper limit to the field is that there will be a limit to the finite strain which any rock material can sustain and still retain a structure that could reasonably be termed cleavage. There will be no slaty rocks which have suffered shortening much in excess of 90% and resultant elongation in excess of 500%. For such strains to be achieved, ductilities would need to have been greater than those permitted by the temperature and pressure conditions of metamorphism under which slaty cleavage typically develops.

The average deformation ellipsoid for all cleaved rocks has the form 1.76/1/0.24, which implies a shortening perpendicular to the cleavage of approximately 70% ($Z = 30\%$ *d*), a principal elongation in the cleavage plane of rather more than 150% ($X = 250\%$ *d*) and a linear increase of 35% ($Y = 135\%$ *d*) parallel to the intermediate axis of the deformation ellipsoid. All of the best commercial roofing slates have strain values close to these and it is interesting to speculate whether this is a function of sampling or whether there is an optimum strain which gives rise to the premium slate.

VOLUME CHANGES DURING DEFORMATION

It is clearly important that the extent of volume change during deformation should be known. Since the time of the early suggestions by Sorby (54, 59) concerning this matter, it has received little serious attention. This was perhaps a consequence of the fact that Sorby (59) considered reduction in volume of only about 10% to have occurred in the slates which he examined. Such small volume changes would require little correction of the apparent strains measured. This results from the fact that such strains are obtained by comparing the three ellipsoid axes with the diameter of the equivalent volume sphere. If there was volume decrease, the measured ellipsoid must be compared with a greater than equivalent volume sphere. In the event of a 20% volume decrease, the correct comparative sphere diameter is only increased by approximately 6%. Therefore for volume changes of the order of 10%, only small errors are involved in neglecting volume change. If, however, the volume changes approach or exceed 20%, they must be taken into account. The effects of such volume change during deformation have recently been discussed by Ramsay & Wood (46).

Argillaceous slates have densities between 2.7 and 2.85 g/cc. These densities are considerably greater than those of the muds, and appreciably greater than those of the mudstones from which such slates were derived. Thus the magnitude of volume change during deformation depends to a considerable extent upon the condition of the materials at the commencement of deformation. This raises the fundamental question as to whether cleavage is a phenomenon involving the tectonic dewatering of unconsolidated sediments (35), or whether most slates were lithified prior to deformation. From a consideration of pre- and postdeformational densities, it appears possible that volume losses of 10–20% might be involved in the progression from lithified mudstone to slate (46). The relationship between slate densities and predeformational densities for various changes in volume are shown in Figure 5. The average densities of Cambrian slates from Wales and the Taconic region are 2.82 and 2.80, respectively. A decrease in volume of 20% during deformation would require that the initial material have had a density of 2.26. Materials that would fulfill this requirement would include average shales and mudstones with a porosity of about 25%, pure illite, or montmorillonite with a moisture content of about 16%. All of these are unlikely possibilities. No undeformed argillaceous sediments with porosities of 20% have been found from depths greater than 7000 feet in boreholes. Furthermore, Athy (1) showed that Pennsylvanian shales from Oklahoma possessed a density of 2.42 at 1000 feet, rising to 2.66 at 5000 feet. Similarly Hedberg (25) found that undisturbed shales of Tertiary age in Venezuela had a near surface density of about 2.0, rising to 2.35 at 3500 feet and 2.52 at 6100 feet.

It can be shown on structural and stratigraphic grounds that most argillaceous slates, at the time of deformation, were under high confining pressures with an overburden of several thousand feet. It is therefore very unlikely that predeformational densities could be less than 2.5, thereby placing an apparent limit of about 10% upon the amount of volume loss.

Consideration of density alone may be highly misleading. Shackleton (51) pointed out that removal of calcium carbonate and silica during deformation might be very considerable. He cited the fact that Cloos (8), in his study of oolite deformation, found that the more highly deformed oolites were of consistently smaller volume. This could mean that the size difference was a tectonically induced volume change rather than an original sedimentary feature. Volume loss as a result of processes such as continued alignment and closing up of enmeshed clay particles, and dehydration mineral transformations such as the transition from clay minerals to micas would all be taken into account by a simple consideration of densities. This would not be the case, however, for volume loss which had

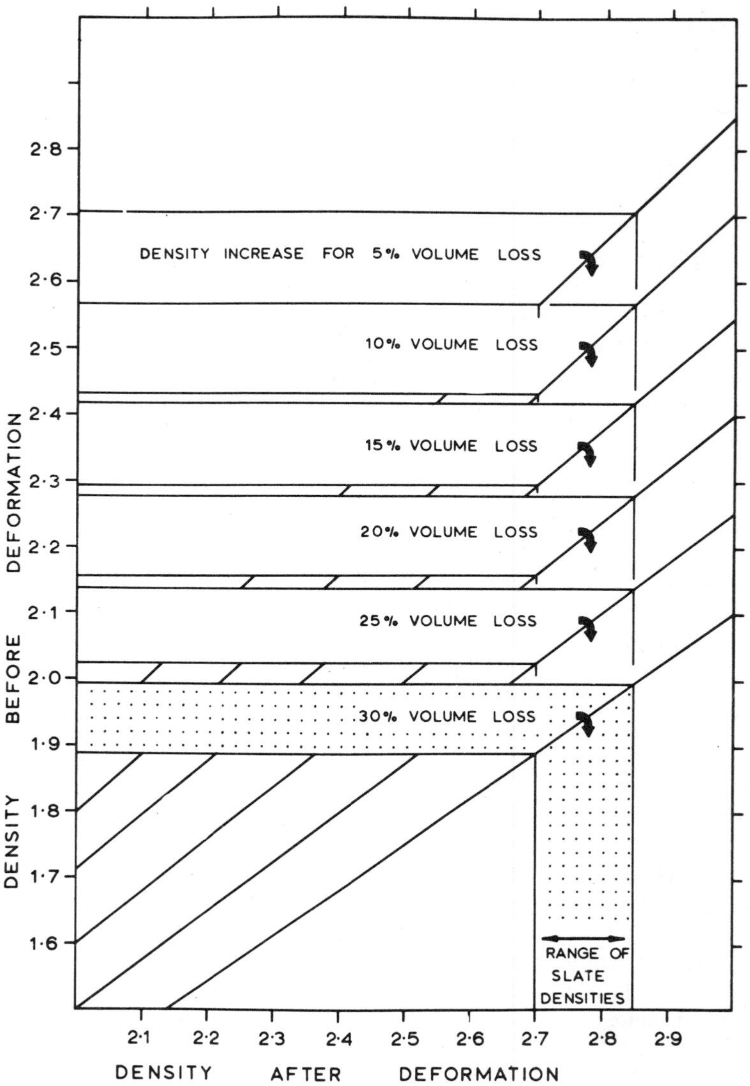

Figure 5 Relationship between slate densities and predeformational densities as a function of volume change.

occurred either as a result of the removal of soluble solids by permeating solutions, or as a result of the removal of solid material at stressed grain contacts by pressure solution transfer. The probability that pressure solution transfer has been an important deformation mechanism in all slates, and indeed in all appreciably deformed rocks, is very high. The fundamental problem is to decide if and where the dissolved material has been redeposited. At present there is no precise information concerning the scale upon which a deformed rock can be considered as a closed system with reference to pressure solution transfer. If some material is entirely removed from the rock body in question, the associated volume loss will not be revealed by density considerations. Observations on both calcareous and siliceous rocks suggest that most of the materials dissolved off surfaces parallel to the cleavage are locally redeposited as crystallographically continuous overgrowths, but there is no good quantitative information.

It would be most desirable to have information concerning the homogeneity of volume change in regions of deformed rocks. It appears that such changes may be significantly heterogeneous. This is suggested by the nature of strain variation in the Taconic slate belt (Figure 3). The mean deformation ellipsoids for ten localities show considerable variation of strain parallel to the direction of greatest finite extension for relatively small variations in finite shortening and virtually no strain variation parallel to the intermediate ellipsoid axis. This suggests that in response to reasonably constant shortening there has been variable compensating extension. This is readily explicable only in terms of differential volume change.

It is therefore likely that some volume change has occurred during deformation. Reduction of volume by 10% can be achieved without material loss. Additional reductions involving the complete removal of material are difficult to evaluate, but it is most unlikely that pressure solution transfer operates over long diffusion paths. It seems reasonable to propose that on all but the most microscopic scales, the total volume loss entailed in the acquisition of cleavage has been no greater than 20%. Ramsay & Wood (46) introduced the terms "field of apparent flattening" and "field of true flattening" to emphasize that, in the event of volume change, ellipsoids which have apparent positions below the line of unit slope through the origin of the deformation plot may in reality fall within the lower part of the constriction field. It is instructive to examine the effects of such lowering of the plane strain line on Figure 4. For a 10% volume loss none of the 5200 ellipsoids would come to lie outside the field of apparent flattening. For a 20% volume loss, some 3.5% of the ellipsoids would be in the field of true constriction rather than in the field of true flattening. If 20% is a realistic limit for volume change in most instances, then well over 90% of all slates fall within the field of true flattening. Also, the strains for any ellipsoid will, within acceptable limits of accuracy, remain as indicated on the deformation plot (Figures 2, 3, and 4).

STRAIN VARIATION AND GEOLOGICAL STRUCTURE

Few studies have attempted to establish the relationship between large-scale geological structures on the one hand and the patterns and variations in finite strain on the other hand. The nature of structural styles as related to both magnitudes of strain and patterns or modes of strain have recently received some preliminary treatment (76). The geometric effects of variation in finite strain

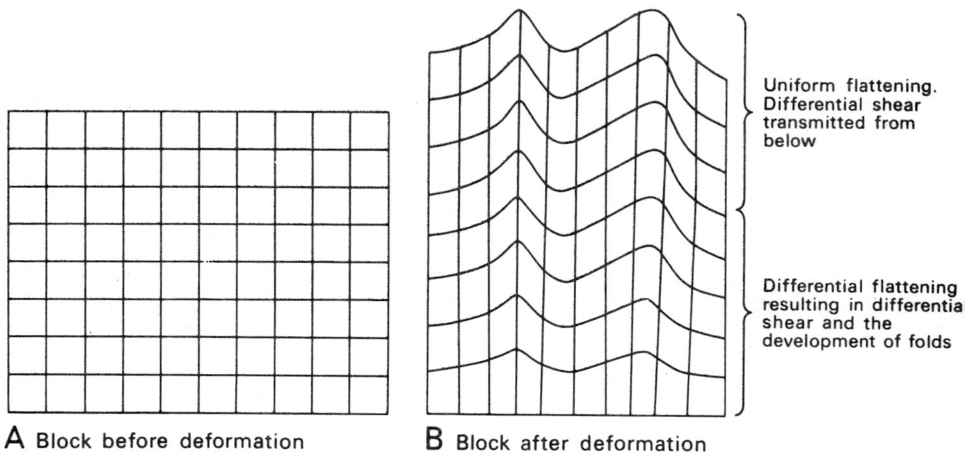

Figure 6 Folding as a result of differential flattening and differential flow (after Ramsay, 43).

related to folding have been more fully discussed by Ramsay (43; 45, p. 435) who demonstrated that differential flattening results in differential flowage which is megascopically expressed as increasingly amplified folding. He further showed that folding can develop in zones which themselves undergo uniform flattening, but where differential flow is transmitted from a neighboring zone (Figure 6). In addition Ramsay (43, 45) extended these theoretical considerations into three dimensions to possibly account for variations in fold plunge (Figure 7).

A practical demonstration of the manner in which inhomogeneous compressive strain along the length of a fold axial surface may lead to the formation of folds with variable plunge is afforded by the Cambrian slate belt of Wales. Along the length of this belt of rocks possessing vertical cleavage, there are four major culmination zones. Finite strain determinations reveal an extremely close relationship between the positions of fold culminations and depressions on the one hand and maximum and minimum values of the flattening component of strain on the other hand (Figure 8). This section, parallel to the cleavage of the belt in question, also shows the variation in vertical thickness of the Cambrian sequence and the variation in shape of the *XY* principal plane of the deformation ellipsoid. Variations both in fold plunge and in tectonic thickening of the stratigraphic section are clearly dependent upon variation in the amount of flattening across the cleavage. The lowest values of flattening (55%) coincide with the most obvious plunge depression, whereas flattening values in excess of 70% coincide with the major plunge culmination zones. Although all the deformation ellipsoids fall in the flattening field of the deformation plot (Figure 2), those which are more nearly oblate spheroids occur in the plunge depressions, whereas those which demonstrate

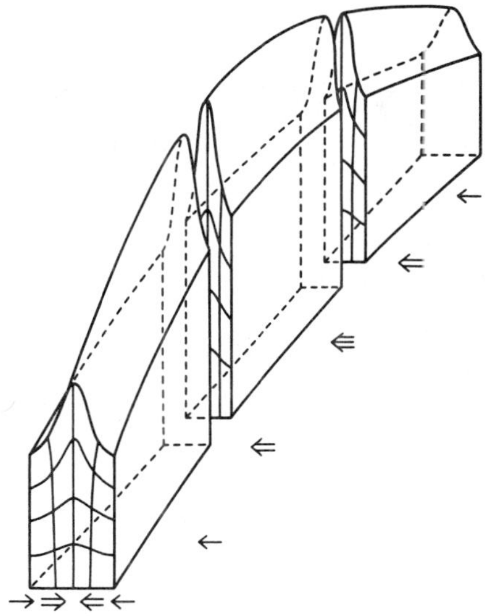

Figure 7 Variation in fold plunge as a result of heterogeneous compressive strain along the length of a fold axial surface (after Ramsay, 43).

the greatest principal extension occur in the culmination zones. Every direction within the plane of cleavage is one of extension, but this extension is more nearly equal in all directions in the case of the lower compressive strains that are characteristic of the plunge depression zones. The response to differential shortening has been such that the present vertical thickness in the depression zones is 50% greater than the initial thickness ($X = 150\%\ d$), whereas in the culmination zones it is 180% greater ($X = 280\%\ d$). The significance of such thickness variations for the stratigrapher is virtually self-evident and means that all isopach maps of strongly deformed regions require most careful evaluation.

STRAIN AND PREFERRED ORIENTATION

Anisotropic crystal growth in metamorphic rocks was clearly described by Becke (3) and Grubenmann (20). Harker (23) and subsequently Ramberg (42) considered that minerals with a growth anisotropy would orientate themselves during deformation so that their direction of most rapid growth would be parallel to the direction of least pressure. This preferred orientation would be thermodynamically controlled and would result in a minimization of internal stress (23). In a nonhydrostatic stress field such as pertains for almost all natural rock deformation, crystal lattices tend to be orientated so that their axes of maximum linear compressibility are parallel to the direction of maximum compressive stress (29, 30), thereby achieving the most thermodynamically stable position. Thus, as has been pointed out by

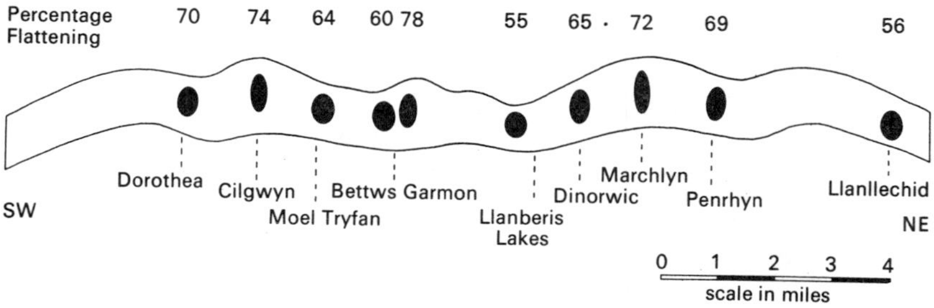

Figure 8 Relationship between variation in fold plunge and tectonic thickening as a function of differential strain along the length of the Cambrian slate belt of Wales.

Flinn (17), for a given mineral and stress condition, the preferred orientation resulting from growth anisotropy is precisely the same as that to be expected if orientation were dependent upon the elastic properties of minerals.

Because cleavage is defined by the preferred shape and crystallographic orientation of phyllosilicates together with the preferred shape orientation of other constituents, it is reasonable to assume that there should be some simple relationship between the degree of preferred orientation and the strain magnitude. In view of the fine grain size of slaty rocks and the fact that at least two crystallographic directions of a mineral must be measured in order to fully define its orientation, optical methods are insufficiently precise to permit the detection of small but significant fluctuations in mineral orientation. For slates, X-ray analysis provides the only satisfactory method for the investigation of preferred orientation as a function of strain. The value of such investigations depends upon whether the relationship between preferred orientation and independently measured total strain can be quantitatively expressed for any given environment.

The investigation of preferred mineral orientation in deformed rocks by X-ray photographic methods dates from the work of Sander & Sachs (49). Improved precision has been obtained in recent years by Wenk (70, 71) and by Starkey (60), the latter having obtained the first X-ray measurement of preferred orientation in fine-grained experimentally deformed quartz aggregates. The first quantitative evaluations of preferred orientation of platy elements in experimentally deformed materials were obtained by Means & Paterson (36). Using a modification of the commercial Norelco pole-figure goniometer, and applying spherical harmonic analytical methods, Baker, Wenk & Christie (2) have subsequently obtained fully quantitative data for quartz in fine-grained experimentally deformed aggregates. Texture or pole-figure goniometry has been applied to the investigation of preferred orientation of phyllosilicates in natural slates by Oertel (38) and by Tullis & Wood (65). In a preliminary single-specimen study of a slate containing natural strain indicators from the English Lake District, Oertel found that the basal planes of chlorite and muscovite had a 9 × random preferred orientation parallel to the cleavage and that the rock in question had undergone a tectonic shortening of 50%.

Oertel verified that greatest shortening had occurred precisely normal to the cleavage. He also found that the lineation parallel to the direction of principal extension within the plane of cleavage was reflected in the ultrafabric by a greater scatter of poles to phyllosilicate platelets in the plane normal to the lineation than in the plane containing both the lineation and the pole to cleavage.

The degree of preferred orientation to be expected for a given amount of shortening may be predicted according to the method of March (34). This assumes that the inequidimensional particles defining the cleavage anisotropy, were original constituents with initially random orientation which subsequently underwent passive rotation into a plane of flattening. The March model is basically a mathematical expression of Sorby's first prediction of fabric for a given magnitude of strain (55). Sorby considered the case in which the ratio of the greatest to smallest axes of the deformation ellipsoid was 6:1. He showed both geometrically and by experiment that grains which were initially inclined at any given angle to the plane perpendicular to the subsequent pressure would rotate such that the tangent of the angle after deformation would be reduced to one sixth of the tangent of the original angle. The validity of the assumption concerning the initially random orientation of components is highly questionable. Sorby (58, pp. 72–74) admitted that the major difficulty with the rotation hypothesis was that he had never been able to find an undeformed rock with random fabric, but he suggested that a parent rock possessing a tendency toward preferred orientation parallel to stratification could, with less than 10% of initial compressional shortening, be transformed into an essentially random starting orientation for the remainder of the strain history. It is doubtful whether this overcomes a problem which will not be satisfactorily solved until there is more precise information concerning both the present and previous mineralogy of slaty rocks.

It is usually assumed that the white mica of slates is muscovite. Many slates are both more sodic and more aluminous than would be expected if this were the case, and it seems likely that considerable amounts of paragonite are commonly present. This means that the starting phyllosilicate also contained sodium as the inter-layer cation. The precise nature of recrystallization of muscovite from illite and of paragonite from brammallite is unknown. Such knowledge would be most valuable to any explanation of the syntectonic achievement of preferred orientation. It is not known how or whether such preferred orientation involves the growth of muscovite, paragonite, and also chlorite by some process of replication upon the lattices of earlier clay minerals which have undergone tectonic rotation and collapse. The enhancement of rotationally achieved preferred orientation by late-tectonic or post-tectonic recrystallization is suggested by Oertel (38), who measured 9 × random preferred orientations for strains in which the March model prediction allowed for only an 8 × random orientation. Oertel has demonstrated that preferential post-tectonic growth of phyllosilicates parallel to their basal planes is more frequently stopped by grain impingement for those grains which deviate most strongly from the plane of preferred orientation. Valuable evidence bearing upon this problem has been provided by Tullis (64), who conducted experiments involving crystallization of micas during deformation and found that the preferred orientations obtained for a given strain value were precisely those predicted by the rotational model, with the greatest concentration of poles to mica basal planes always exactly parallel to the direction of maximum finite compressive strain.

The results of Oertel and Tullis clearly suggested that measurement of preferred orientation might allow not only the directions, but also the magnitudes of principal strains to be determined. Tullis & Wood (65) found that the degree of preferred orientation for three slates, for which the strains had been independently measured, was systematically less than that expected from the March model prediction for those strains. This suggests that other aspects of the total fabric, such as interactions between adjacent grains and the effect of the nonlayer silicates upon the theoretical orientation of phyllosilicates, cannot be neglected. An alternative explanation is that the preferred orientation was inhibited by an initial preferred orientation related to the primary layering.

Thus in the only two published comparisons between predicted strain and measured strain there is contradiction. Oertel's measured preferred orientation was greater than that predicted from the strain measured, whereas the preferred orientations measured by Tullis and Wood were lower than predicted. Accordingly, in the example considered by Oertel, the strains predicted from preferred orientation are higher than the measured strains, whereas the reverse is the case for the slates considered by Tullis and Wood. If significant volume loss had occurred during deformation, the measured compressive strains would need to be increased, and this would be more compatible with the result of Oertel.

A number of cleaved rocks of known strain have recently been examined by X-ray transmission pole-figure goniometry (G. Oertel and D. S. Wood, in preparation) and remarkable agreement is found between the measured deformation ellipsoid and the predicted deformation ellipsoid (Table 1). The largest difference

Table 1 Comparison of strains in slates obtained by measurement of natural strain indicators and strains predicted by the March theory from preferred orientation.[a]

Localities	Measured strain			Strain by March prediction			
	X	*Y*	*Z*	*X*	*Y*	*Z*	
Cambrian Slate Belt, Wales	180	131	43	181	138	40	(1)[b]
	200	138	37	174	138	42	(2)
	175	140	40	190	131	40	(3)
	156	147	45	150	150	44	(4)
	180	124	45	176	121	47	(5)
	218	133	35	200	124	40	(6)
	215	132	36	204	126	39	(7)
Ordovician slates, Kentmere	197	126	40	190	124	42	(8)
English Lake District	171	118	49	174	121	48	(9)

[a] Present axial lengths of deformation ellipsoids are given compared to 100 units of original length. The assumption is made that there has been no significant volume change during deformation.

[b] Numbering as on the deformation plot, Figure 9.

found between measurement and prediction of the principal finite compressive strain is 5%; for seven of the nine examples these strains agree to within 3%. The comparative strains in the two other principal directions also show good agreement. In considering strains in the direction of principal finite extension (X) the agreement is particularly significant because very small errors in measurement of the short axis of the strain indicators produce much larger errors in the strain computed for the X

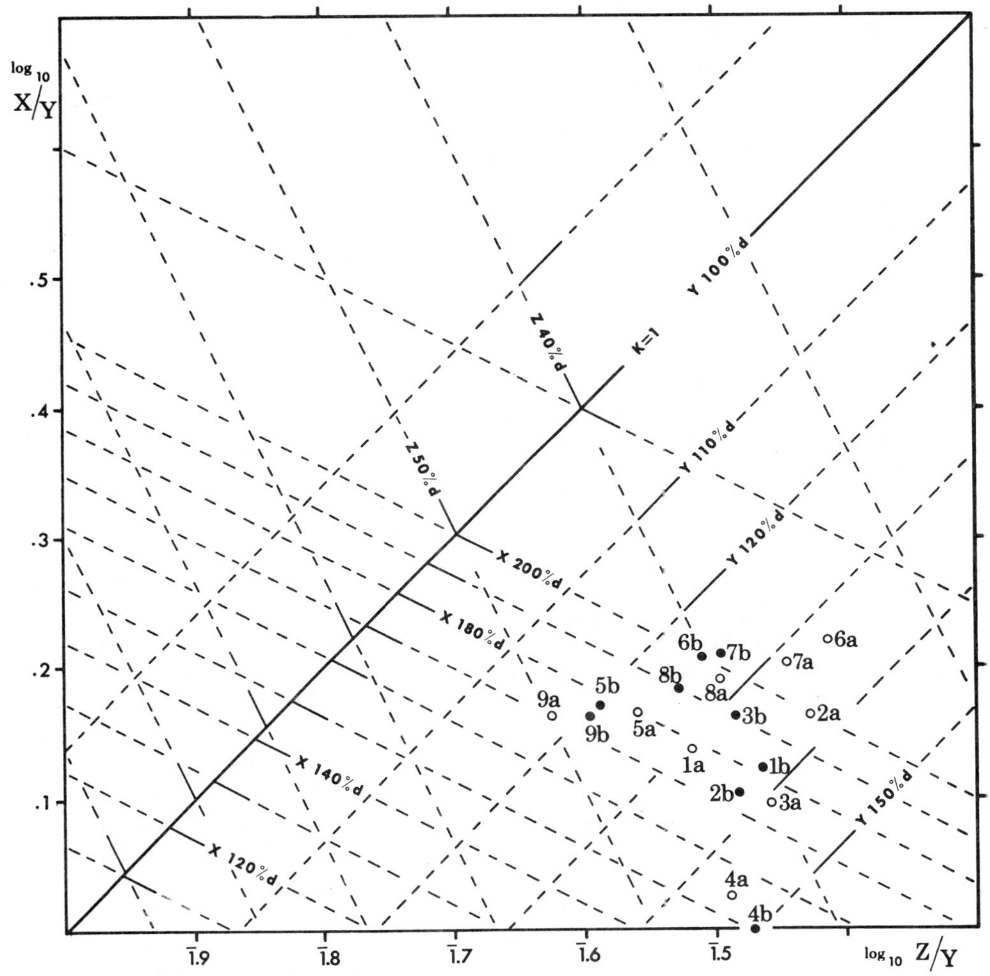

Figure 9 Deformation plot showing finite strains measured from deformed objects (a) and predicted by the March theory from preferred orientation of phyllosilicates (b) for slates from nine localities in North Wales and the English Lake District.

direction by reference to the diameter of the equivalent volume sphere. For the 27 strains compared by Oertel and Wood, 13 strains are virtually identical, 3 show somewhat higher predicted strains, and 11 show lower predicted strains than the measured apparent strains. These results are shown on the deformation plot in Figure 9.

If such agreement can be consistently obtained for a constant lithology in a given environment, it raises the possibility that strains predicted from preferred orientation, and checked against a minimal number of independently determined strains, can be readily and reliably obtained. Since good natural strain indicators are relatively rare in many deformed rocks, texture goniometry now allows detailed assessment of natural strain both on the regional scale and in relation to the evolution of individual geological structures. These methods will permit the production of strain contour maps on various scales.

PATHS OF PROGRESSIVE DEFORMATION IN CLEAVED ROCKS

Almost all measurements of strain in deformed rocks relate only to the finite state of strain. For a full understanding of deformational processes it is necessary to also consider the strain history. This requires knowledge of the incremental strains which have cumulatively resulted in the observed finite strain. Flinn (16) described the manner in which an original sphere deforms progressively through a series of intermediate ellipsoidal shapes until deformation is completed. The locus of the sequential intermediate ellipsoids is the deformation path, which may be considered as resulting from the cumulative effect of superposition of an infinitesimal number of incremental deformation ellipsoids. The simplest deformation paths result when the incremental deformation ellipsoids are of constant shape and the strain is irrotational. Such deformation paths define straight lines on the deformation plot in the three special circumstances where $X = Y > Z$; where $X > Y = Z$; and where $X/Y = Y/Z$ when Y has undergone no length change. For these three situations, Flinn (16, 17) introduced the now widely used terminology $K = 0$, $K = \infty$, and $K = 1$. The parameter K is defined from the deformation plot (Figure 1) as $a-1/b-1$, where $a = X/Y$ and $b = Y/Z$. In the general case all other deformation paths will not define straight lines on the deformation plot (44; 45, p. 329). The precise course of a deformation path is primarily dependent upon rheological behavior as a function of variability of stress values, strain rates, degree of anisotropy, and rotational components of deformation. The resulting complex nature of deformation paths has been theoretically considered by Ramsay (45, pp. 322–32) who pointed out that most natural strain histories have probably involved rotation and variable states of stress. In view of these considerations, and because there is no simple relationship between strain increments and finite strain, or between the finite strain and the deforming stresses, there is no practical justification for assigning intermediate values of K for general ellipsoids which fall within the fields of flattening or constriction. There is more value in assigning deformation fields (45, p. 141) for areas of the deformation plot which contain

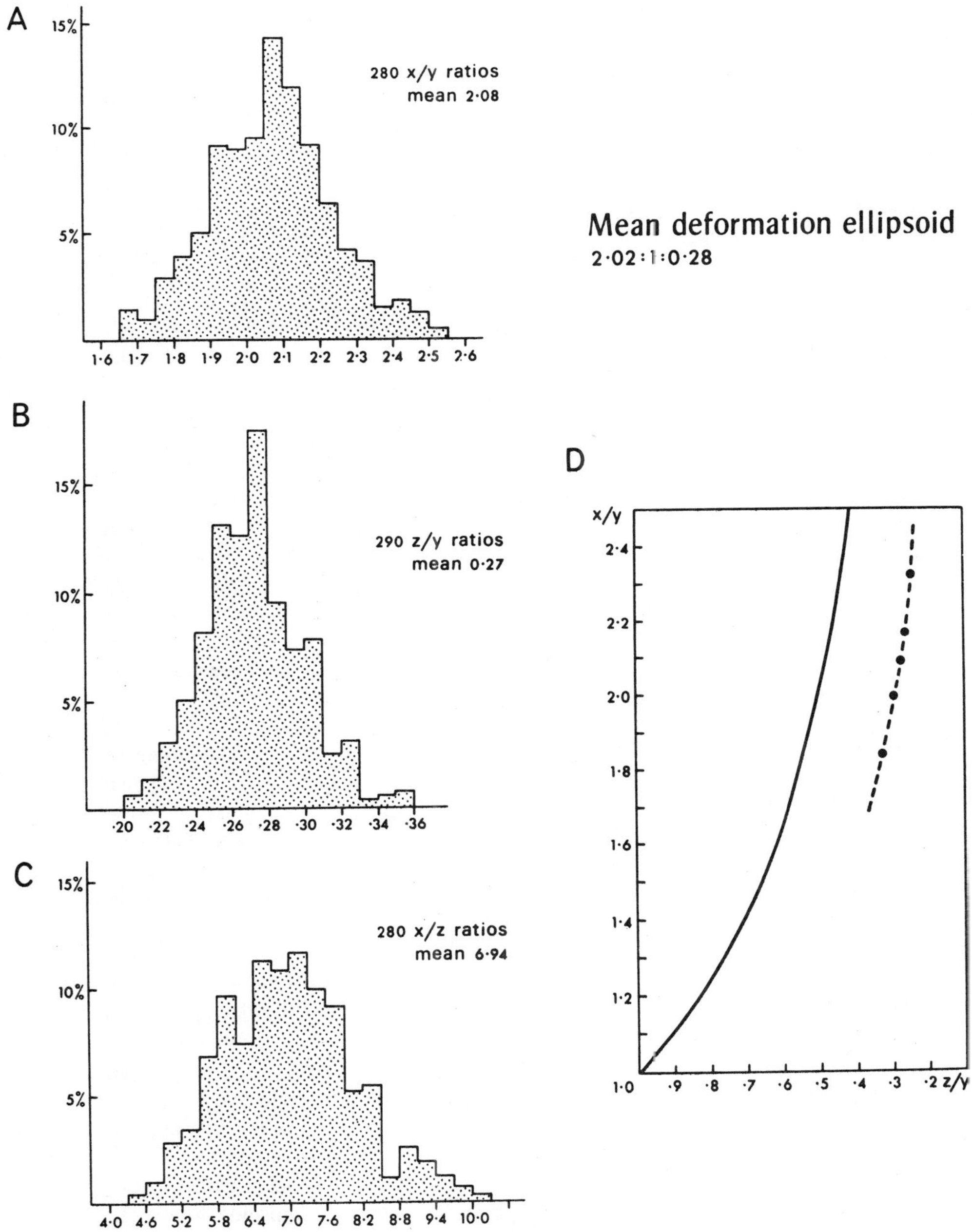

Figure 10 Variation of axial ratios in each plane of the deformation ellipsoid for a single rock specimen (a–c). Nonlogarithmic deformation plot with curve of plane strain at constant volume and deformation path predicted from strain heterogeneity.

various end points of many individual deformation paths relating to a particular structure or region (46).

There is no simple method for the derivation of natural deformation paths from deformation fields. The best approximation for obtaining natural deformation paths entails utilization of the normally heterogeneous nature of finite strain on any specified scale and requires one significant assumption. Figure 10 (a-c) shows three histograms illustrating considerable variation of axial ratios in each plane of the deformation ellipsoid for an Ordovician spotted slate from North Wales (75). The natural strain indicators show a mean deformation ellipsoid having an axial ratio of 2.02/1/0.28. Although the measurements were taken from a uniformly cleaved specimen of only ten cubic inches, it is evident that within this volume the rock has undergone highly nonuniform strain. The measurements indicated on the histograms together comprise the end points of a large number of individual deformation paths. Slight variations in rigidity among the indicators, size variation of the indicators, and variation in frequency of indicators relative to a matrix of slightly different ductility, would account for the observed strain heterogeneity. In view of the small specimen size, it seems reasonable to make the major assumption that despite variations in the state of strain of individual indicators, the general pattern of deformation for the volume in question would have been similar.

If all the components measured underwent strain according to a single overall pattern but achieved differing magnitudes of finite strain, they would lie on a single progressive deformation path. If this were the case, those objects with the highest strain value would have passed through stages represented by the objects with lower strain values. Conversely, if deformation had continued according to the same pattern, objects with the lower strain values would have attained the strain state of the objects with the highest measured strain values. Accepting the assumption, five points on a proposed deformation path may be obtained. By pairing the 20% least deformed ratios of Figure 10a with the 20% least deformed ratios of Figure 10b or 10c, the lowest point on the proposed path is obtained. This point and the remaining four mean 20% sample points are plotted on the nonlogarithmic deformation plot (Figure 10d). Some justification for this procedure is afforded by the fact that the proposed path tends toward the origin of the plot.

Application of the same procedure to the strain data of the Taconic slate belt (Figure 3) gives the result shown in Figure 11. In this example, complete measurements of individual ellipsoids removed the need to pair ratios from different symmetry planes of the ellipsoid. For each of the localities (1–4 on Figure 11), the mean values of the 20% samples of increasing departure from spherical are plotted directly. The resulting paths also tend strongly toward the origin and also tend to include the mean strain values for the remaining localities (5–7 on Figure 11).

This provides a reasonable general method for obtaining some idea of the form of actual deformation paths. It seems to work equally well both on the scale of the hand specimen and on the regional scale. The only justification for this or any other method of obtaining deformation paths will depend upon experimental

LOWER CAMBRIAN SLATES OF THE NEW YORK - VERMONT SLATE BELT

	DEFORMATION ELLIPSOID LOCALITY	SAMPLES	MEAN AXAL RATIO X Y Z	S.D. X/Y	S.D. Z/Y
CAMBRIAN	1 SCOTCH HILL Qy W CASTLETON Vt	30	1.70 - 1 - 0.15	.15	.02
	2 WELLS Qy POULTNEY Vt	44	1.84 - 1 - 0.15	.18	.04
	3 EUREKA Qy 3 Miles SE of FAIRHAVEN Vt	74	1.74 - 1 - 0.18	.15	.04
	4 CEDAR POINT Qy LAKE BOMOSEEN Vt	52	1.55 - 1 - 0.16	.08	.02
	5 MAHAR Qy 1.5 Miles SE of FAIRHAVEN Vt	30	1.59 - 1 - 0.22	.06	.03
	6 MIKES Qy POULTNEY Vt	28	1.62 - 1 - 0.31	.10	.03
	7 MATHEWS Qy (Ordovician red slate) WHITEHALL NY	25	1.55 - 1 - 0.35	.09	.03

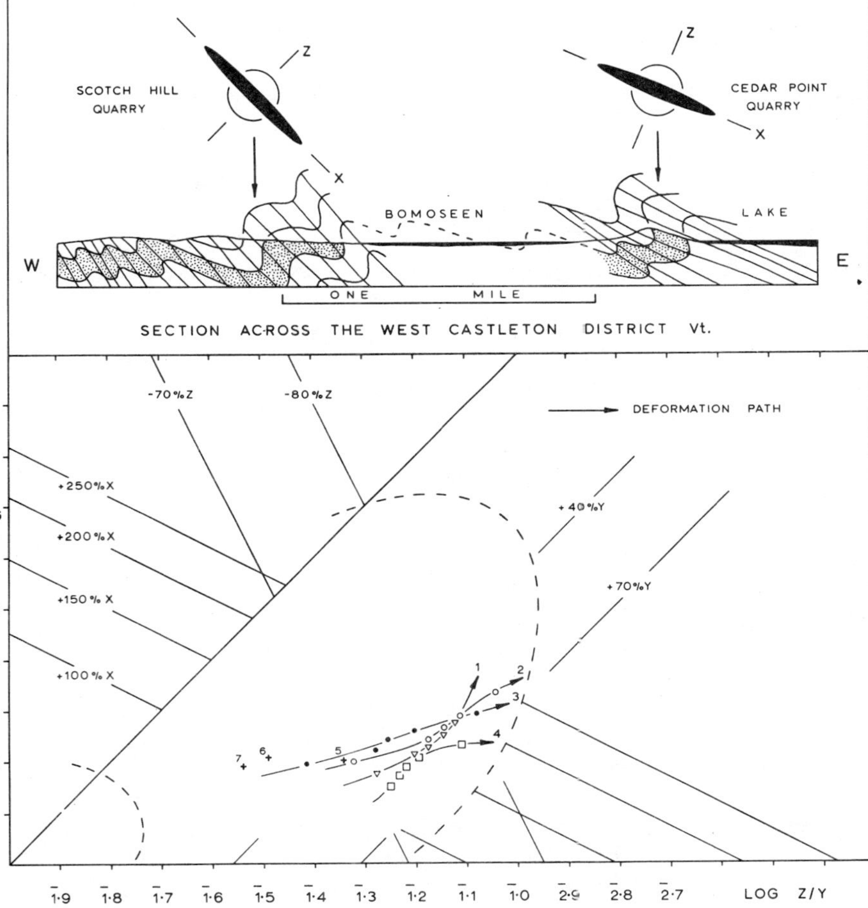

Figure 11 Deformation paths for the Taconic slate belt predicted from strain heterogeneity of single localities.

verification. Recently F. A. Donath and D. S. Wood (in preparation) have sought to do this by conducting high pressure experiments involving the deformation of oolitic limestone under identical conditions at 10% intervals of finite shortening. The results indicate a significant degree of inhomogeneity of strain among individual ooids for any given average finite strain. The considerable overlap in ooid shape from one deformed sample to another is such that the mean deformation ellipsoid of specimens shortened to 50% could have been precisely predicted from the heterogeneity of strain in samples shortened by 30% and vice versa. This leads to two tentative conclusions. The deformation paths determined by the method proposed above are reasonable approximations and, although natural deformation paths are not straight lines on the deformation plot, they are less complex than might be expected from theoretical considerations.

MECHANISMS OF CLEAVAGE FORMATION

Attempts to account for the origin of slaty cleavage have resulted in three principal controversies. The first of these concerns the respective viewpoints of Sorby who believed that mechanical processes were the dominant factor, and Van Hise who considered that chemical processes were of greater importance. Each admitted that processes of the alternative type had played some part in cleavage formation. The second controversy has been between those who considered cleavage to be essentially a fracture phenomenon produced by a shear mechanism, and others who believed it to be a flow phenomenon produced by a process of flattening. The ill-founded shearing hypothesis which has been previously discussed need not be considered further. A third and more recent controversy has appeared between those who favor tectonic dewatering of unconsolidated materials as an essential early stage in cleavage formation, and others who believe the materials to have been low porosity lithified rocks prior to deformation.

It must now be generally accepted that slaty cleavage is a flow phenomenon that has formed precisely parallel to the direction of maximum finite shortening. This has been fully substantiated by field studies (e.g. 75, 76), by computer-simulated model studies of folding (e.g. 10), and by experimental growth of phyllosilicates during deformation (64). Before considering whether the preferred orientations observed in slates were achieved by new crystal growth in the plane of flattening, or whether it is more likely that pre-existing grains underwent mechanical rotation into the plane of flattening, it is appropriate to consider the tectonic dewatering hypothesis. The unidirectional passage of large volumes of liquid through unconsolidated sediment would provide most efficient assistance for the rotation of inequidimensional particles.

On good evidence from the Orodovician Martinsburg slates, Maxwell (35) adopted the view that, in its early stages of development, cleavage may involve the dewatering of unconsolidated sediment under the influence of tectonically initiated pore water pressures of abnormal magnitude. By this hypothesis Maxwell sought to explain the high degree of preferred orientation of phyllosilicates as being produced by rotation contingent upon the through passage of water. He further suggested that such water would be likely to have effected some redistribution of both silica and carbonate. Selective removal of silica during deformation is strongly suggested by the work of Williams (73). It is difficult to refute Maxwell's explanation for the Martinsburg slates; it is well supported both by the mineralogy of the rocks in question and by the timing relationships between deposition and deformation. In view of the presence of clastic dikes subparallel and parallel to the cleavage of the Martinsburg slates, the occurrence of similar dikes in other cleaved rocks has been taken as proof of a similar origin for cleavage despite the absence of other supporting evidence (e.g. 37, 41). Clastic dikes, which are common in many sedimentary environments certainly demonstrate dewatering, but whether they indicate that this was in any way related to cleavage formation is highly questionable. If a sequence containing such dikes is subsequently affected by deformation leading to the formation of cleavage with high angle to the initial layering, most of the dikes will attain an orientation close to the plane of cleavage after moderate shortening has occurred. For the dikes to be genetically related to the cleavage, they would all need to be precisely parallel to that cleavage. In many instances where dewatering has been invoked to account for slaty cleavage this is clearly not the case. In the Siamo slate of Michigan (41) clastic dikes deviate up to 50° from the cleavage.

There are several reasons for rejecting the dewatering hypothesis as a general explanation for slaty cleavage. Three of these reasons refer to the evidence afforded by fossil distortion, the presence of cleavage in associated igneous rocks, and the development of cleavage in retrograde and retrogressive metamorphic rocks, respectively. In respect to the latter it is quite certain that water has been added rather than subtracted in many instances, for example where greenschist to amphibolite facies retrogressive metamorphism affects older granulite facies rocks.

The presence of distorted fossils in slaty rocks and the fact that they are most severely distorted in the most strongly cleaved rocks, raises one significant objection to the dewatering hypothesis. It is hardly conceivable that the organisms were deformed prior to fossilization and equally unlikely that they were lithified prior to the lithification of the enclosing sediment. Consequently the deformation related to the fossil distortion must have occurred subsequent to lithification of both fossil and matrix. Under such conditions, the expulsion of a certain amount of pore water during the early deformational stages would not be capable of significantly assisting the rotation of inequidimensional mineral components.

Most deformed regions contain pretectonic igneous rocks which possess cleavage or schistosity in common with their sedimentary host rocks. The Archaean greenstone belts consist of folded structures with associated slaty cleavage which passes uniformly through thick basic volcanic sequences as well as arenaceous, rudaceous, and ferruginous sedimentary materials (76). Similarly the extrusive and early intrusive members of the late Precambrian and earliest Paleozoic of the Scottish Caledonides possess the primary Caledonian cleavage structure of that region. The cleavage of the Cambrian slates of North Wales equally affects interbedded and underlying ignimbrites (74).

Because of the ductility contrasts normally involved, it is usually difficult to

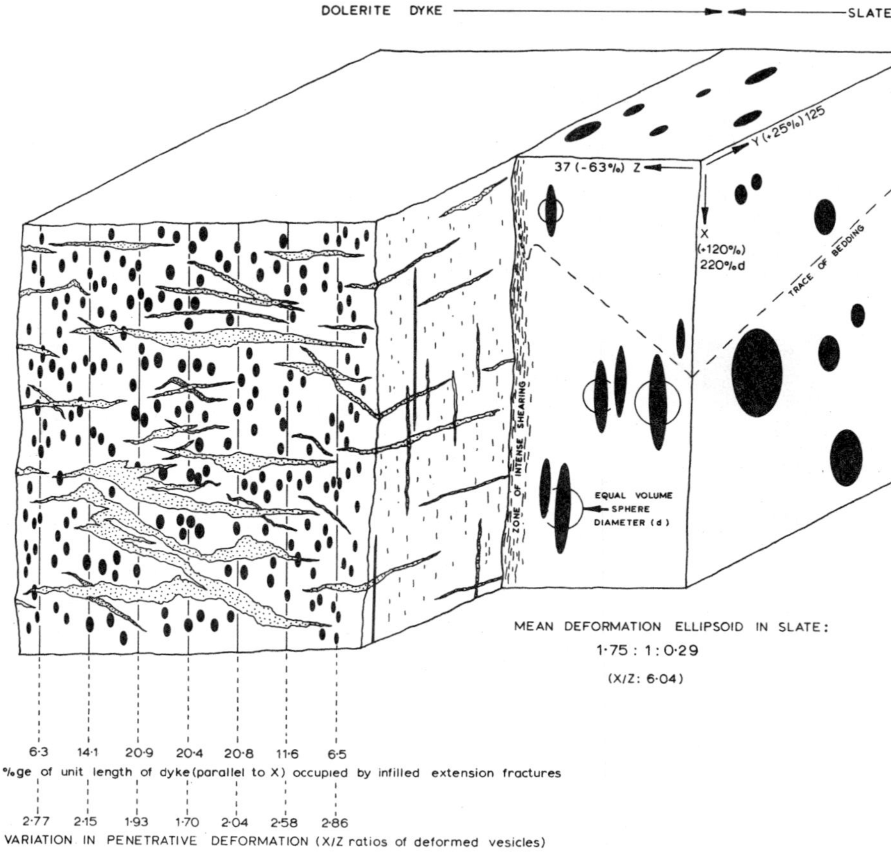

Figure 12 Variation of penetrative strain achieved by flow and discontinuous strain achieved by fracture across a dolerite dike included within slates of known strain (Penrhyn Quarry, Bethesda, North Wales).

obtain a precise relative assessment of the intensity of penetrative deformation in adjacent materials of igneous and sedimentary origin. An attempt to obtain such information is shown in Figures 12 and 13. In this example from Wales, Lower Cambrian spotted slates are cut by Ordovician diabase dikes. The deformation ellipsoid for the enclosing slates has the axial ratio 1.75/1/0.29, which indicates a shortening across the cleavage of 63% ($Z = 37\% \, d$) and a principal extension in the plane of cleavage of 120% ($X = 220\% \, d$). The enclosed dike which is about ten feet in width, possesses a cleavage that is everywhere less obvious than in the slates and which is better developed in the finer-grained marginal portions of the dike than in its coarser-grained interior. The dike, however, responded to the deforming stresses by

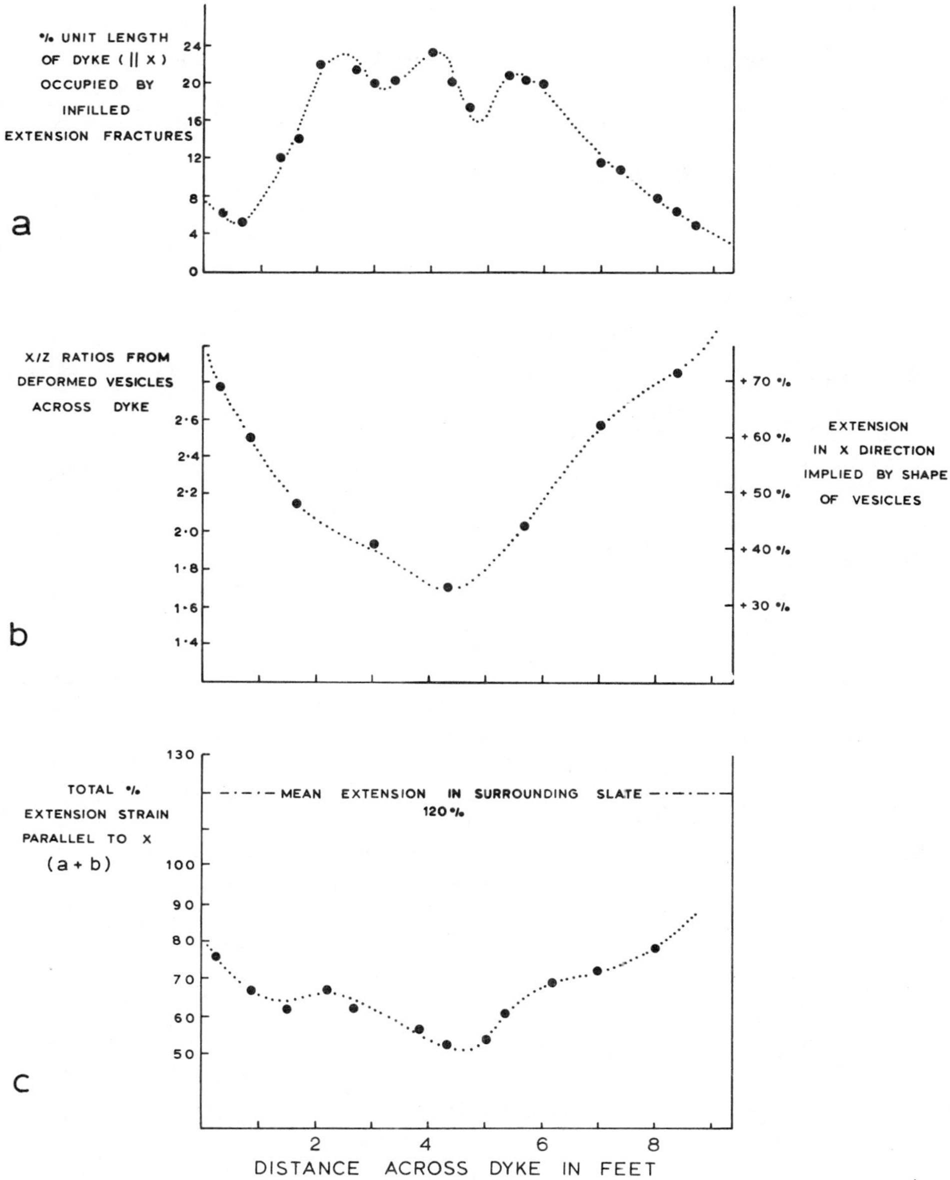

Figure 13 The extensional component of strain in a dolerite dike achieved by a combination of fracture (a) and flow (b), compared to the mean extensional component of strain in enclosing slates (c) (Penrhyn Quarry, Bethesda, North Wales).

adopting two different strain modes. It underwent penetrative nonuniform flow which is reflected by the formation of cleavage, and underwent additional non-penetrative discontinuous deformation which is reflected by the presence of extension fractures. The latter (Figure 12) are of more than one stage of development, with the earlier fractures having undergone folding prior to being cut by later fractures. These extension fractures probably represent periods of increased rate of strain, to which the dike was unable to respond by continuous flow. The variation in the elliptical cross sections of deformed vesicles across the dike is indicated in Figure 12 and the principal extension which this implies is shown in Figure 13b. In assessing the finite extensional strain, the width of the extension fractures must also be taken into account. The percentage of unit distance within the dike, measured parallel to the direction of principal extension, which is occupied by fracture infilling is shown in Figures 12 and 13a. The total principal extensional strain within the dike varies from a minimum of about 50% in the center to over 90% in the marginal zones (Figure 13c). This differential strain within the dike must be compensated elsewhere, either by major extension fractures or by boudinage which is locally common. The important comparison is that between the total strain in the dike margin and the total strain of the host slate (Figure 13c). There is a discrepancy of about 30% strain between these two principal extensional strains. This may be largely accounted for by the difficulty encountered in measuring vesicle distortion in the fine-grained zone of contact strain at the dike margin. It is clear that the strain within the dike increases considerably toward its margin and it is possible that it may increase to the strain level of the slates. It is more likely, however, that there is no continuous increase of strain up to the strain level of the enclosing slates. This is strongly suggested by the fact that the dike margins are always strongly sheared and are often faults. The example demonstrates that there is a considerable degree of compatibility between the strain states of both materials and offers an explanation for faulting as a function of differential ductility and heterogeneous strain. The host rocks to such commonly deformed intrusive rocks must have been considerably lithified to sustain the fracturing which permitted dike emplacement. In view of this, and because the mechanism of cleavage formation must have been generally the same for both materials, the dewatering hypothesis cannot be invoked to account for the cleavage of the Cambrian rocks of North Wales.

Having eliminated the hypothesis of tectonic dewatering as a universal explanation for slaty cleavage, the original mechanical and recrystallization hypotheses remain. Mechanical orientation of inequidimensional grains has undoubtedly contributed to the formation of cleavage in many instances. In a few situations it is difficult to prove that rotation has not been the sole mechanism involved. However, in the great majority of cases there is no doubt that chemical processes have been of incomparably greater importance. The textural relationships, mineralogical composition, and grain size of phyllosilicates in most slates demonstrate that they are the products of syntectonic recrystallization rather than being initial characteristics. The anisotropy possessed by cleaved rocks is only partially the result of preferred orientation of phyllosilicates. The majority of slates contain considerable proportions of highly inequidimensional grains of other minerals, the shape of which is certainly not a detrital feature. The ratio of longest to shortest dimensions of quartz grains in the Cambrian slates of Wales and the Taconic region is commonly 3:1; for the Valdres slates of Norway it is 4:1; and for the Ballachulish slates of Scotland it is about 6:1. Such grains possess syntaxial overgrowths in the direction of principal extension and have undergone dissolution along surfaces situated at high angles to the direction of principal shortening.

The effect of pressure solution as a mechanism for the formation of grain shape anisotropy in cleaved carbonate rocks, which was first described by Sorby (56, 57), has been more recently shown by Plessmann (40) to have been responsible for dimensional shortening of 50% perpendicular to the cleavage. The importance of pressure solution in effecting metamorphic differentiation was emphasized by Voll (69, p. 534), who noted its special significance at quartz-mica grain contacts.

All slates are characterized by the presence of phyllosilicates, carbonaceous material, ore minerals, and other heavy minerals which are concentrated in closely spaced dark seams parallel to the cleavage. These are interpreted as the insoluble residues remaining after pressure solution has occurred (40). These are strictly analogous to, but more uniform and regular than, the stylolite seams of diagenetic origin in sedimentary rocks. It is evident that this process facilitated concentration of primary fine-grained clay minerals, such that syntectonic recrystallization into coarser-grained muscovite and paragonite could occur more efficiently. Consequent upon the appearance of coarser-grained phyllosilicates, further pressure solution resulted in the profusion of plane quartz-mica boundaries that are so important a feature of the fabric of slates. The importance of pressure solution transfer as a deformational mechanism has been emphasized by Durney (12) and explained in terms of nonhydrostatic thermodynamic theory. The special significance of pressure solution for slaty rocks results from the fact that diffusion paths are shorter and rates of solution transfer consequently are higher than would be the case for coarser-grained rocks. The most important current problem is to determine the rock volume over which a particular system may be considered to be chemically closed. The solution to this problem will provide the only satisfactory evaluation of volume changes during deformation. On the scale of individual rock formations it is clear that the system has remained essentially closed, inasmuch as major concentrations of deposited silica and carbonate are not found marginal to such formations.

In the present state of knowledge it must be concluded that the recrystallization hypothesis of Van Hise is valid and that pressure solution transfer provides the mechanism. The fact that measured strain values agree with strain predictions which are based upon preferred orientations and which assume mechanical rotation raises interesting questions. Nevertheless, if the March model prediction of strain provides a sound empirical tool, its practical value is quite undiminished.

CONCLUDING REMARKS

Slaty cleavage is a flowage phenomenon involving flattening. It has formed precisely perpendicular to the direction of maximum finite shortening and has involved some loss in volume. Well-developed slaty cleavage has entailed shortening perpendicular to the cleavage in excess of 50% and commonly between 65 and 75%. In response to compression perpendicular to the cleavage, resultant extension has always occurred within the plane of cleavage. This extension has usually been distinctly greater in one preferred direction in the cleavage.

In some instances, where deformation closely followed deposition, mechanical rotation of inequidimensional particles may have been facilitated by tectonic dewatering which occurred as an initial stage of a continuous deformation from which slaty cleavage resulted. The majority of slates, however, were lithified rocks at the time of acquisition of cleavage. Some degree of rotation of existing inequidimensional components into the plane perpendicular to the direction of maximum finite shortening has undoubtedly occurred in all slates. Rotation has always been subsidiary to the metamorphic growth of new phyllosilicate minerals which are an essential feature of all rocks possessing slaty cleavage. Some redistribution of minerals such as quartz and carbonates by solution processes has occurred in all slates. Pressure solution is the mechanism most likely to have effected the grain shape changes which have occurred during deformation and to have caused the inequidimensional form of minerals other than phyllosilicates in slaty rocks.

Cleavage is therefore the result of a combination of various mechanically and chemically controlled processes which operated over extended periods of time. It is an early stage in a metamorphic continuum of which schistosity and gneissosity may be the end result under appropriate conditions.

Literature Cited

1. Athy, L. F. 1930. Density, porosity and compaction of sedimentary rocks. *Am. Assoc. Petrol. Geol. Bull.* 14: 1–24
2. Baker, D. W., Wenk, H. R., Christie, J. M. 1969. X-ray analysis of preferred orientation in fine-grained quartz aggregates. *J. Geol.* 77:144–72
3. Becke, F. 1903. Über Mineralbestand und Struktur der Kristallischen Schiefer. *Rep. Int. Geol. Congr. Vienna*, 4:553–70
4. Becker, G. F. 1896. Schistosity and slaty cleavage. *J. Geol.* 4:429–48
5. Becker, G. F. 1904. Experiments on schistosity and slaty cleavage. *US Geol. Surv. Bull.* 241:1–34
6. Braddock, W. A. 1970. The origin of slaty cleavage: evidence from Precambrian rocks in Colorado. *Geol. Soc. Am. Bull.* 81:589–600
7. Clark, B. R. 1970. Origin of slaty cleavage in the Coeur d'Alene district, Idaho. *Geol. Soc. Am. Bull.* 81:3061–72
8. Cloos, E. 1947. Oolite deformation in the South Mountain fold, Maryland. *Geol. Soc. Am. Bull.* 58:843–918
9. Darwin, C. 1846. *Geological Observations in South America.* London: Smith-Elder
10. Dieterich, J. H. 1969. Origin of cleavage in folded rocks. *Am. J. Sci.* 267:155–65
11. Dunnet, D. 1969. A technique of finite strain analysis using elliptical particles. *Tectonophysics* 7:117–36
12. Durney, D. W. 1972. Solution-transfer, an important geological deformation mechanism. *Nature* 235:315–17
13. Elliot, D. 1970. Determination of finite strain and initial shape from deformed elliptical objects. *Geol. Soc. Am. Bull.* 81:2221–36
14. Fisher, O. 1884. On cleavage and distortion. *Geol. Mag.* (Decade 3) 1:268–76, 396–405
15. Flinn, D. 1956. On the deformation of the Funzie conglomerate. Fetlar, Shetland. *J. Geol.* 64:480–505
16. Flinn, D. 1962. On folding during three-dimensional progressive deformation. *Quart. J. Geol. Soc. London* 118: 385–433
17. Flinn, D. 1965. Deformation in meta-

morphism. In *Controls of Metamorphism,* ed. W. S. Pitcher, G. W. Flinn, 46–72. Edinburgh: Oliver & Boyd
18. Gay, N. C. 1968. Pure shear and simple shear deformation of inhomogeneous viscous fluids, 2. The determination of the total finite strain in a rock from deformed objects such as deformed pebbles. *Tectonophysics* 5:315–39
19. Griggs, D. T. 1935. The strain ellipsoid as a theory of rupture. *Am. J. Sci.* 30:121–37
20. Grubenmann, U. 1904. *Die Kristallinen Schiefer.* Berlin: Gebrüder Borntraeger
21. Harker, A. 1885. The cause of slaty cleavage. *Geol. Mag.* 2:15–17
22. Harker, A. 1886. On slaty cleavage and allied rock structures. *Rep. Brit. Assoc. Advan. Sci.* 1885:813–52
23. Harker, A. 1939. *Metamorphism.* London: Methuen. 2nd ed.
24. Haughton, S. 1856. On slaty cleavage and the distortion of fossils. *Phil. Mag.* 12:409–21
25. Hedberg, H. D. 1936. Gravitational compaction of clays and shales. *Am. J. Sci.* 31:241–87
26. Heim, A. 1878. *Untersuchungen über den Mechanismus der Gebirgsbildung.* Basel: Schwarbe
27. Heim, A. 1921. *Geologie der Schweiz.* Leipzig: Tauchnitz
28. Hoeppener, R. 1956. Zum Problem der Bruchbildung, Schieferung and Faltung. *Geol. Rundsch.* 45:247–83
29. Kamb, W. B. 1959. Theory of preferred crystal orientation developed by crystallisation under stress. *J. Geol.* 67:153–70
30. Kamb, W. B. 1961. The thermodynamic theory of nonhydrostatically stressed solids. *J. Geophys. Res.* 66:259–71
31. Kirillova, I. V. 1962. Cleavage as an indicator of the character of mass movement during the folding process. In *Folded Deformations in the Earth's Crust, Their Types and Origins,* ed. V. V. Beloussov, A. A. Sorskii. Moscow: Izdatel'stvo Akad. Nauk S.S.S.R.
32. Laugel, A. 1855. Du clivage de roches. *C. R. Acad. Sci.* 40:182, 185, 978–80
33. Leith, C. K. 1905. Rock Cleavage. *US Geol. Surv. Bull.* 239:1–216
34. March, A. 1932. Mathematische Theorie der Regelung nach der Korngestalt bei affiner Deformation. *Z. Kristallogr.* 81:285–97
35. Maxwell, J. C. 1962. Origin of slaty and fracture cleavage in the Delaware Water Gap area, New Jersey and Pennsylvania. *Geol. Soc. Am. Mem.* Buddington Vol.: 281–311
36. Means, W. D., Paterson, M. S. 1966. Experiments on preferred orientations of platy minerals. *Beitr. Mineral. Petrogr.* 13:108–33
37. Moench, R. H. 1966. Relation of S_2 schistosity to metamorphosed clastic dikes, Rangeley-Phillips area, Maine. *Geol. Soc. Am. Bull.* 77:1449–62
38. Oertel, G. 1970. Deformation of a slaty, lapillar tuff in the Lake District, England. *Geol. Soc. Am. Bull.* 81:1173–88
39. Phillips, J. 1844. On certain movements in the parts of stratified rocks. *Rep. Brit. Assoc. Advan. Sci.* 1843:60–61
40. Plessman, W. von, 1964. Gesteinslösung, ein Hauptfactor beim Schieferungsprozess. *Geol. Mitt.* 4:69–82
41. Powell, C. McA. 1969. Intrusive sandstone dykes in the Siamo slate near Negaunee, Michigan. *Geol. Soc. Am. Bull.* 80:2585–94
42. Ramberg, H. 1952. *The Origin of Metamorphic and Metasomatic Rocks.* Chicago: Univ. Chicago
43. Ramsay, J G. 1962. The geometry and mechanics of formation of "similar" type folds. *J. Geol.* 70:309–27
44. Ramsay, J. G. 1964. Progressive deformation in tectonic processes. *Trans. Am. Geophys. Union* 45:106
45. Ramsay, J. G. 1967. *Folding and Fracturing of Rocks.* New York: McGraw-Hill
46. Ramsay, J. G., Wood, D. S. 1973. The geometric effects of volume change during deformation processes. *Tectonophysics* 16:263–77
47. Rogers, H. D. 1856. On the laws of structure of the more disturbed zones of the earth's crust. *Trans. Roy. Soc. Edinburgh* 21:431–72
48. Sander, B. 1930. *Gefügekunde der Gesteine.* Vienna: Springer
49. Sander, B., Sachs, B. 1930. Zur röntgenoptischen Gefügeanalyse von Gesteinen. *Z. Kristallogr.* 75:529–40
50. Sedgwick, A. 1835. Remarks on the structure of large mineral masses, and especially on the chemical changes produced in the aggregation of stratified rocks during different periods after their deposition. *Trans. Geol. Soc. London Ser. 2,* 3:461–86
51. Shackleton, R. M. 1962. In discussion of D. Flinn. *Quart. J. Geol. Soc. London,* 118, 430
52. Sharpe, D. 1847. On slaty cleavage. *Quart. J. Geol. Soc. London* 3:74–105

53. Sharpe, D. 1849. On slaty cleavage. *Quart. J. Geol. Soc. London* 5:111–29
54. Sorby, H. C. 1853. On the origin of slaty cleavage. *New Phil. J. Edinburgh* 55:137–48
55. Sorby, H. C. 1856. On slaty cleavage, as exhibited in the Devonian limestones of Devonshire. *Phil. Mag.* 11:20–37
56. Sorby, H. C. 1863. On the direct correlation of mechanical and chemical forces. *Proc. Roy. Soc.* 12:538–50
57. Sorby, H. C. 1879. Structure and origin of limestone. *Quart. J. Geol. Soc. London* 35:39–95
58. Sorby, H. C. 1880. On the structure and origin of non-calcareous stratified rocks. *Quart. J. Geol. Soc. London* 36:46–92
59. Sorby, H. C. 1908. On the application of quantitative methods to the study of the structure and history of rocks. *Quart. J. Geol. Soc. London* 64:171–232
60. Starkey, J. 1964. An X-ray method for determining the orientation of selected crystal planes in polycrystalline aggregates. *Am. J. Sci.* 262:735–52
61. Strand, T. 1944. Structural petrology of the Bygdin conglomerate. *Nor. Geol. Tidsskr.* 24:14–31
62. Swanson, C. O. 1941. Flow cleavage in folded beds. *Geol. Soc. Am. Bull.* 52:1245–63
63. Thomson, W., Tait, P. G. 1879. *Treatise on Natural Philosophy.* Cambridge: Cambridge Univ.
64. Tullis, T. E. 1971. *Experimental development of preferred orientation of mica during recrystallization.* PhD thesis. Univ. California, Los Angeles
65. Tullis, T. E., Wood, D. S. 1972. The relationship between preferred orientation and finite strain for three slates. *Geol. Soc. Am.* (Abstr. with programs) 4(7):694
66. Turner, F. J. 1948. Mineralogical and structural evolution of the metamorphic rocks. *Geol. Soc. Am. Mem.* 30:1–342
67. Tyndall, J. 1856. Observations on "The theory of the origin of slaty cleavage" by H. C. Sorby. *Phil. Mag.* 12:129–35
68. Van Hise, C. R. 1896. Principles of North American Pre-Cambrian geology. *Ann. Rep. US Geol. Surv.* 16:517–843
69. Voll, G. 1960. New work on petrofabrics. *Liverpool Manchester Geol. J.* 2:503–67
70. Wenk, H. R. 1963. Eine Gefüge-Röntgenkamera. *Schweizer. Min. Pet. Mitt.* 43, 707–19
71. Wenk, H. R., 1965. Eine photographische Röntgen-Gefugeanalyse. *Schweiz. Mineral. Petrogr. Mitt.* 45:517–50
72. Wettstein, A. 1886. Uber die Fischfauna des Tertiären Glarner Schiefers. *Schweiz. Palaontol. Ges. Abh.* 13:1–101
73. Williams, P. F. 1972. Development of metamorphic layering in low grade metamorphic rocks at Bermagui, Australia. *Am. J. Sci.* 272:1–47
74. Wood, D. S. 1969. The base and correlation of the Cambrian rocks of North Wales. In *The Precambrian and Lower Palaeozoic rocks of Wales,* ed. A. Wood. Cardiff: Univ. Wales
75. Wood, D. S. 1971. *Studies of strain and slaty cleavage in the Caledonides of northwest Europe and the eastern United States.* PhD thesis. Univ. Leeds, England
76. Wood, D. S. 1973. Patterns and magnitudes of natural strain in rocks. *Phil. Trans. Roy. Soc. Ser. A* 274:373–82
77. Wunderlich, H. G. 1959. Erzeugung engständiger Scherflächen in plastischem Materiel. *Neues Jahrb. Geol. Palaeontol. Abh.* 1:34–44

2

Reprinted from pages 111–115, 117–118, and 128–129 of *Geol. Soc. London Quart Jour.* **5**:111–129 (1849), by permission of the Geological Society of London

ON SLATY CLEAVAGE (SECOND COMMUNICATION)

Daniel Sharpe

In a paper read to the Geological Society on the 2nd of December 1846, on the cleavage of slate rocks†, I endeavoured to work out certain general laws relative to the compression which such rocks had undergone, to the position of the planes of cleavage, and to the connection between their direction and the elevation of the beds. The conclusions were founded upon observations chiefly made in North Wales, Devonshire and Cornwall: I have since visited parts of Westmoreland and Cumberland, with the view of enlarging the field of observation, and the result is contained in the following remarks, which are thrown into the order adopted in the former paper.

Compression of slate-rocks in a direction perpendicular to the planes of cleavage.—In the former paper, p. 87, I stated that in all the slaty fossiliferous rocks examined, the distortion of the fossils proved that the mass of the rocks had undergone considerable compression in a direction everywhere perpendicular to the planes of cleavage, and some expansion in the direction of the dip of the cleavage; but that there was no reason to suppose that the rocks had suffered any change of volume in the direction of the strike of the cleavage. And it was inferred that these changes must be general in all slaty rocks, although it might not be easy to find proofs that they had occurred, where organic remains were absent.

There is however more evidence of compression to be found among the beds of unfossiliferous slate than might have been expected; and the examination of their mechanical structure affords quite as strong proofs of pressure as those derived from the distortion of the organic remains.

In the neighbourhood of the roofing-slates there are frequently found beds of a brecciated structure which cleave readily, but from their irregular composition are liable to break. Such beds are

† Journal of the Geological Society, vol. ii. p. 74.

largely exposed in the upper part of the quarries of green slate at Patterdale, and in the Langdale quarries and many other places in Westmoreland and Cumberland. In all these slaty breccias, the included masses are flatter between the planes of cleavage than in any other direction. Their irregular forms prove them to have been imbedded pebbles or fragments, and cannot be explained by reference to crystallization: indeed they have rarely any crystalline character, but consist of slate, differing more or less in grain, colour and hardness from the matrix in which they are imbedded. In the localities mentioned, the bedding is most distinctly marked, and we can judge with certainty of the position of the fragments with reference to it: if they had been originally flat pieces of slate accidentally deposited, we should find them lying on their flat sides in the bed, but this is not the case; their flattest sides are always parallel to the cleavage-planes, and consequently where the cleavage cuts the bedding at a high angle, as is the case in all the Langdale quarries, and in some of those at Patterdale, these thin masses appear to be standing upright in the bed on their edges; a position which they never could have reached if their forms had been originally those we now find. We can only explain the circumstances by supposing that these masses were flattened when the rock was subjected to a pressure which acted in a direction perpendicular to the planes of cleavage.

When we examine the outline of these imbedded masses, as it is seen on the face of a sheet of slate, we usually find them to be rather longer down the dip of the cleavage than across it or along its strike: this confirms the opinion that the rocks have expanded in the direction of the dip of the cleavage.

As we do not know the original forms of the imbedded fragments, we can only guess at the amount of compression they have undergone; but we may be certain that it has been very considerable, for in many localities their thickness between the cleavage-planes is seldom equal to half their diameter as seen on the planes of cleavage.

To illustrate the forms of the fragments imbedded in the brecciated slates, I have added two sketches of a piece of slate from the Patterdale quarry.

Fig. 1 is taken from the surface of the plane of cleavage showing the largest and flattest side of all the fragments, which are longest in the direction of the dip of the cleavage *y y*. Fig. 2 shows the side of the same sheet of slate with all the fragments flattened between the planes of cleavage. The lines *a a* represent the planes of bedding, with a considerable dip.

Between the well-marked slates and the great bands of porphyritic rocks which traverse the "green slate" district of the Lakes, there are extensive masses of metamorphic rocks more or less crystalline, which appear to have been sedimentary deposits altered by the neighbouring igneous action. In these rocks the cleavage is usually well-marked, and the bedding can frequently be distinctly traced. But the nearer we approach the axis of igneous action, the more obscure do we find the bedding, till it is gradually lost; the cleavage then becomes more faint, and we soon reach a mass of crystalline rock without either bedding or cleavage. Many of these metamorphic slates appear to have been originally of a brecciated structure, and contain a confused mixture of crystals with what appear to be fragments of

Fig. 1.

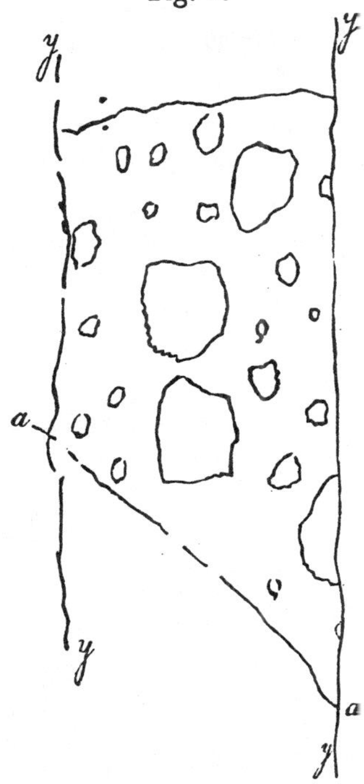

Fig. 2.

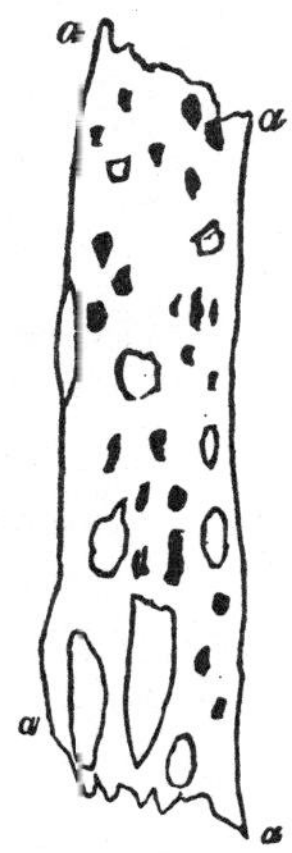

foreign origin. In slaty rocks of this description all the enclosed portions which can be supposed to have had a mechanical origin are flattened between the planes of cleavage, but nothing of the sort can be observed of the crystals, which are scattered through the mass without any reference to the cleavage-planes either in their form or position.

In good roofing-slates, which are distinguished by the fineness and uniformity of their grain, it is difficult to observe the forms of the parts of which they are composed: yet even in these, with the assistance of a lens, we can see that the constituent particles are flattest between the cleavage-planes, and longest along the dip of the cleavage. Thus, under whatever circumstances we are able to observe the forms of the component parts of slate, or of the foreign bodies contained in it, (whether these are pebbles, fragments of other rocks, or organic remains,) we find evidence that the mass has been compressed by a force which has acted in a direction perpendicular to the planes of cleavage.

The distinctive character of slate appears to be that it is composed of particles of a form more or less amygdaloidal, arranged in a similar direction. Its tendency to split with different degrees of resistance in different directions is the consequence of this arrangement and form of its constituent parts, and is therefore only a secondary character.

Two planes of cleavage in slate.—In preparing slates, the first process is to split them along the planes of cleavage to the thickness required, they are then to be cut to size; most slates of good quality can be split tolerably straight down the sides of the sheet by applying the tool at the top, from which we see that besides the true or principal planes of cleavage, there is also another direction in which the rock has a tendency to split in a slighter degree; this may be called the plane of secondary cleavage; the workmen call it the "side of the sheet." The ends of the sheet must be chipped or sawed to the size required, as the rock will not split at all in this direction, which is across both the other planes of cleavage: in fact it will give way in any other direction more easily than in this, which may be called the plane of greatest resistance.

A diagram will make this more intelligible, and will also explain the relation which these various planes have to the position of the constituent particles of the slate. Let us suppose fig. 3 to represent a block of slate, of which *a b c d* is one of the planes of cleavage, *a d* and *b c* showing the dip, *a b* and *d c* the strike of the cleavage; *b c g f* is the plane of secondary cleavage, *a b f e* is the plane of greatest resistance. The component parts of the rock are represented enlarged, of an almond shape, with their largest and flattest sides parallel to the true cleavage-plane *a b c d,* and with their longest diameters in the direction of the dip of the cleavage *a d.*

Fig. 3.

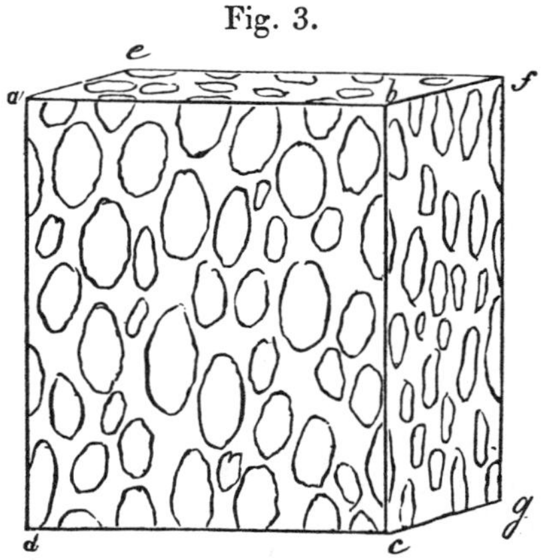

It is obvious that with such a distribution of its particles, the rock can offer less resistance to an attempt to split it down the plane of cleavage *a b c d,* than in any other direction; for that plane, and all planes parallel to it, pass along the flat surfaces of many of the particles, and intersect a smaller number of them than any other plane which can be drawn through the block.

Next to the true cleavage-plane, the direction in which the rock will offer least resistance is along the planes of secondary cleavage *b f g c,* for the particles being all lengthened in that direction, it will more readily separate down that plane than in any cross direction. Nevertheless the cleavage along these planes is very imperfect, and the sheets often fly off in a curve instead of splitting straight down.

In any other direction the particles must from their form and position offer greater resistance than along the two directions just mentioned, but the greatest resistance will be along the plane *a b f e* and all planes parallel to it, as they have to cut through a greater number of the constituent almond-shaped particles of the rock than any others: these therefore are the planes of greatest resistance. In the quarries in Langdale these planes are parallel to the bedding, nevertheless the rock will not split along them.

In splitting up slate, the workmen have observed that the sheets will bear a greater force without breaking, when the tool is applied "at the end" of the sheet, than if it is applied "at the side." This is an illustration of what has just been stated; when the tool is applied at the end of the sheet, it opens the slate down the dip of the cleavage-planes, and produces a strain on the sheet in the direction of its greatest power of resistance; but when the tool is applied "at the side" of the sheet, the strain falls along the planes of secondary cleavage, and the sheet easily breaks.

[*Editor's Note:* Material has been omitted at this point.]

Irregularities in the direction of the cleavage-planes.—There are occasional irregularities in the direction of the cleavage which should be studied, as they may assist our search after the general laws which have regulated its arrangement. Some of these are mentioned by Mr. Phillips in the Report of the British Association at Cork in 1843, p. 60, and other cases have been mentioned to me verbally by that gentleman.

In one of the slate-quarries at Patterdale, a sheet of quartz about an inch thick lies between two beds of slate: the cleavage does not pass into the quartz, and the planes of cleavage are slightly bent out of their course into a curve on each side of the quartz bed, as shown in figure 4.

Fig. 4.

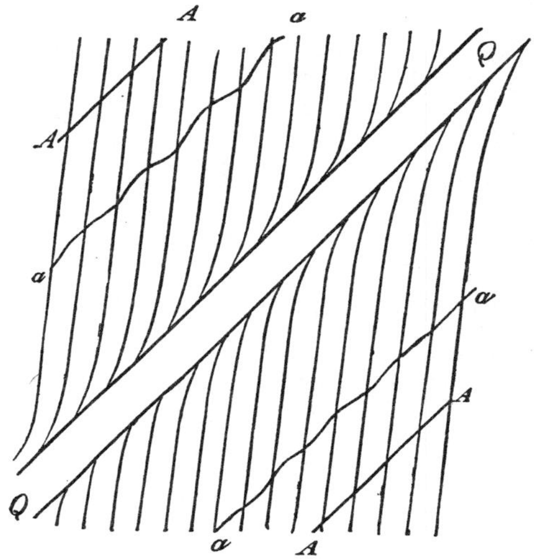

Here the lines A A and *a a* represent the surfaces of the different beds of slate dipping to the north at an angle of 45°; Q Q is a bed of quartz interstratified in the rock. The upright lines represent the cleavage-planes which dip N. 25° W. 85°: they do not enter the sheet of quartz, and appear slightly deflected by it on each side for the distance of about two inches. The surfaces of the beds *a a* are not flat, but are slightly waved in parallel lines coincident with the strike of the cleavage, a peculiarity to which we will return shortly.

In one of the Langdale quarries there is a similar bending of the cleavage against a sheet of quartz which cuts through the bedding along a joint in the rock: the circumstances are so nearly the same, that it is not necessary to add another drawing.

Such irregularities are of frequent occurrence, and their cause is obviously mechanical; the resistance offered by the oblique sheet of quartz to the pressure forward of the mass, has caused the bending of the sheets of slate at the ends which abut against the quartz.

Fig. 5.

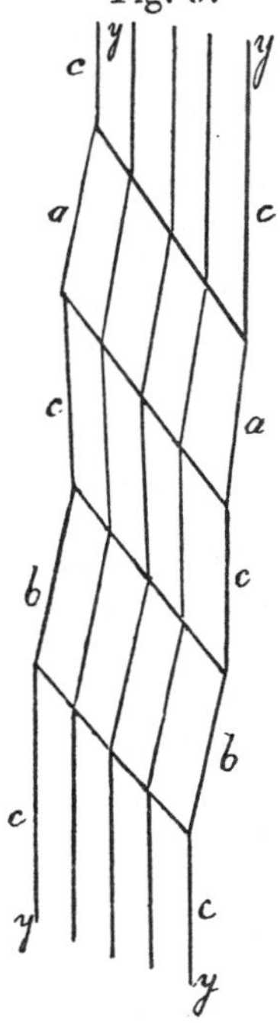

A slight change in the direction of the cleavage may frequently be observed in its passage from one bed of rock to another, arising probably from beds of different hardness offering different degrees of resistance to pressure.

Figure 5 represents the section of a fragment of slate from Langdale, in which the thin beds *a a* and *b b* are rather softer than the other beds *c c*; the lines *y y* show the direction of the planes of cleavage, which are perpendicular in passing through the beds *c c*, but dip 80° to the southward in the beds *a a* and *b b*, the beds dipping to the north at an angle of 60°.

Here the change in the direction of the cleavage in passing from bed to bed is fully 10°, which is an unusual amount of deviation; but a variation of two or three degrees under such circumstances is often met with. The cause of these irregularities must also be mechanical.

[*Editor's Note:* Material has been omitted at this point.]

In slaty rocks there has been a considerable compression of the mass of rock between the planes of cleavage; that is, in the direction corresponding to that of the elevating force; this compression being shown by the distortion of the included organic remains, and the flattening of the component portions of the rock, and bearing a proportion to the degree in which the cleavage is developed.

The compression of the mass in a direction perpendicular to the cleavage has been partially compensated by its expansion along the dip of the cleavage, in which direction only its expansion was permitted as the elevation of the beds enlarged the area occupied by them. The difference between the amount of compression in one direction and of expansion in another, is accounted for in the greater density of the rock after compression.

No connection has been detected between cleavage and crystallization, beyond a tendency of plates of talc and mica to arrange themselves along the planes of cleavage: but as on these planes there would be the least resistance to their intrusion or formation, this may have been a subsequent operation, and should not alter our opinion of the cause of the cleavage.

Thus all our observations and deductions ultimately converge to the conclusion that the cleavage must be attributed to pressure caused by the elevation of great masses of rock under conditions of which we are ignorant. And if to this conclusion it should be objected that no similar results can be produced by experiment, I reply that we have never tried the experiment with a power at all to be compared to that employed; and that this may be one of many cases where our attempts to imitate the operations of nature fail, owing to the feebleness of our means, and the shortness of the period during which we can employ them.

3

Reprinted from pages 20, 27-34, and 37 of *London, Edinburgh and Dublin Philos. Mag. and Jour. Sci.*, ser. 4, **11**:20-37 (1856), by permission of Taylor & Francis Ltd

ON SLATY CLEAVAGE, AS EXHIBITED IN THE DEVONIAN LIMESTONES OF DEVONSHIRE

Henry Clifton Sorby, F. G. S.

IS slaty cleavage the result of crystalline or mechanical action? This is a question I shall now attempt to answer by a description of the phænomena to be observed in the Devonian limestones of the neighbourhoods of Ilfracombe, Plymouth, and Torquay. I do not think any rocks could be found more suitable for deciding it, because they present us with cleavage in every state of development, from altogether absent to in great perfection; and also furnish us with examples of crystalline change of very varying character. A most careful examination of them, both in the field, microscopically, chemically, and by means of the polariscope, has convinced me that the structure on which their slaty cleavage depends may be completely explained on mechanical principles; and that, instead of chemical forces having given rise to it, they have had a contrary tendency; for whilst, other things being the same, the perfection of cleavage varies *directly* as the mechanical changes, it varies *inversely* as the chemical.

[*Editor's Note:* Material has been omitted at this point.]

If a thin section of an organic sand be examined with the microscope, it will be seen that the fragments of coral and shell are usually much longer than broad, as if derived from sections of more or less flat portions. If the rock be very thin-bedded, for instance like the Stonesfield slate, the greater part of these lie in the plane of stratification; but in such thick-bedded rocks as occur in the districts under consideration, when not cleaved, this is not the case, except with very large fragments. The smaller have their longer axes inclined in all positions, so that when a section cut perpendicular to the stratification is examined, there is seen to be no such arrangement as to give rise to any decided line of weakness, along which the rock would split in preference to any other. Let us now inquire what would be the effect on such a structure if the dimensions of the rock were changed by mechanical pressure.

If a rock has not been compressed, we may express this by saying that the ratio of the alteration in any two directions at right angles to each other is as 1 : 1; whilst if it had been compressed in such a manner that the proportion between lines of equal length before compression was changed so that in the line of pressure the length was one-sixth of that perpendicular to it, we may say that the ratio is as 1 : 6. If, for instance, before compression we had a circle, afterwards it would be an ellipse, whose axes were as 1 : 6. This is a very common amount of change in the line of dip in rocks that have a good cleavage, and so I take it by way of illustration. If, then, before compression a long-shaped fragment was inclined at any given angle to the plane perpendicular to the subsequent pressure, it may be seen from general mechanical principles, as well as proved by actual experiment, that the tangent of the angle, after the change of 1 : 6, would be one-sixth of the tangent of the original angle. Thus we have—

Angles of Inclination to the plane perpendicular to the pressure.

Originally.	After compression.
0°	0° 0′
10	1 41
20	3 28
30	5 30
40	7 58
50	11 14
60	16 6
70	24 36
80	43 23
90	90 0

According to these principles, if we suppose that in a mass of rock there were 600 particles having their longer axes lying in the space included within 5° on each side of positions inclined at 0°, 10°, 20°, &c. to the line of pressure, so that they were uniformly distributed, as is nearly the case in thick-bedded, uncleaved rocks, then, after compression so that the ratio was 1 : 6, their distribution would be changed, as shown in the following Table:—

Inclination to the direction of the pressure.	Original distribution.	Subsequent distribution.
0°	600	100
10	600	103
20	600	113
30	600	134
40	600	168
50	600	236
60	600	376
70	600	733
80	600	1825
90	600	3324

It will thus be seen that the effect is to produce a great diminution in the quantity that are inclined in the direction of the pressure, and a great increase in those nearly perpendicular to it, and thus to cause a very great preponderance nearly in the plane perpendicular to the pressure. In fact, as will be seen, in a space of 10°, there, we have thirty-three times as many as in an equal one in the line of pressure; and if very small spaces were taken, since the ratio between the arc and tangent of a very small angle is one of equality, in the exact direction of the pressure, the number of particles whose axes lay in that line would be spread over six times the space, whilst perpendicular to it they would be condensed into one-sixth, and hence the relative amounts in those two positions would be as $1^2 : 6^2 = 1 : 36$. If, then, the amount of the unsymmetrical fragments was very great compared with the rest of the rock, and if their strength was such that to break them was very much more difficult than to split along them without breaking them, the resistance to fracture in the two directions would be as 1 : 36; or in other words, the facility of cleavage in the plane perpendicular to the pressure would be far greater than in one inclined at any considerable angle to it. Of course for other changes of dimensions the same results would apply. For instance, if it was 1 : 3, the relative strength would be 1 : 9, and similarly for other values.

In order to confirm these results by experiment, as mentioned in my paper in the Edinb. Phil. Journ. cited above*, I mixed scales of oxide of iron with pipe-clay, so as to have as uniform a structure as could be produced, and then compressed it, when I found that the arrangement was exactly such as that indicated by calculation. This will be better understood from an inspection of the accompanying figures.

Fig. 1.

Fig. 2.

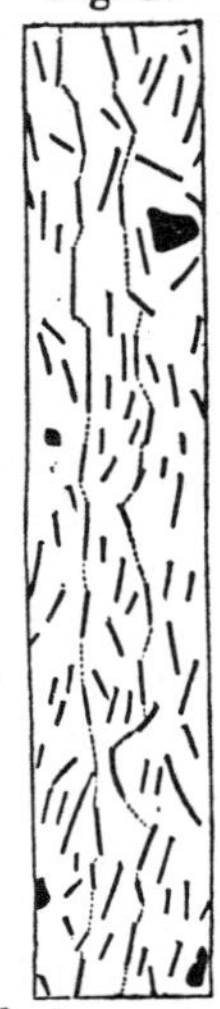

Fig. 1 is a representation of a portion mixed equally, then baked and rubbed to a smooth surface; and, like a thick-bedded, uncleaved rock, it has no decided line of weakness, due to the arrangement of the particles; whilst fig. 2 is a drawing of a portion originally of similar structure, which, having been compressed, clearly shows that it has been changed, precisely in the manner I have shown by calculation to be a necessary result. The dots indicate lines along which it could easily be split, without there being any fracture of the flakes; and I think no one could fail to perceive that it would be much easier to cleave it along that direction than in any other, and that a very decided line of weakness has been produced. If then an organic clay, containing fragments of coral or shell, had its dimensions changed in a similar manner, such should be their structural arrangement, and such it really is.

In studying the cleavage of rocks, it is best to make sections of such as have the cleavage inclined at a high angle to the stratification; for then there is no fear of confounding them together. I shall first describe the structure in the case of organic sandy clays. A very good example of this is *Plymouth No.* 3, of which the physical analysis is given above. In it the cleavage cuts the stratification at about 70°, and the fragments of coral, in place of lying in the plane of stratification, are chiefly inclined at low angles to the direction of the cleavage; in precisely the same manner as that shown to be the necessary result, if an uncleaved, thick-bedded rock of similar physical constitution had its dimensions altered by mechanical pressure. Now, in such a case as this, it is quite out of the question to refer the

* 1853, vol. lv. p. 137.

change of structure to any crystalline action. The particles are not crystals, but are clearly proved, by their structure and form, to be fragments of organic bodies. If the cleavage was due to the mechanical cause just described, their position is most easily accounted for; but I leave the explanation of the fact, on the supposition of the instrumentality of crystalline forces, to those who advocate that view of the subject. The same is the case with the grains of quartz sand, in such limestones as contain them. For instance, in *Ilfracombe No.* 5, of which the physical analysis is given above, and in which the cleavage is perpendicular to the stratification, as seen in a section cut perpendicular to the cleavage and stratification, the grains of sand are often two or three times as long as broad, and even in some cases five times; but in place of the longer axes lying in the plane of stratification, they are chiefly inclined at low angles to the cleavage. I may here remark that there is no difficulty in distinguishing grains of sand from small crystals of quartz, such as are met with in some cherty limestones, and that there is no doubt of them being sand in this, and that their arrangement is due to the cause I have just described.

When organic clays, with or without larger fragments, have undergone consolidation, it very frequently happens that the crystallization of the granules began in various parts, and forced away the bituminous and other impurities into detached, more or less spherical spaces, which afterwards crystallized in much finer grains. On this account they now appear in sections as darker patches, giving the rock a kind of oolitic structure, though quite distinct from the more genuine oolitic grains, visible to the naked eye. These are in many cases due to crystallization taking place from certain centres, and this being therefore the reverse kind of process, I call them *positive segregational* oolites, whilst the others may be called *negative*. When the limestone has no cleavage, these negative, segregational, oolitic grains, though often of irregular form, are usually more or less equiaxed, and not very much longer in one direction than in another, and the longer axes have no particular arrangement. This is, however, very far from being the case in such as have cleavage. As examples of this I may mention *Plymouth No.* 3, *Ilfracombe No.* 5, and *Paignton No.* 1, of which the composition has been given, and in all of which the cleavage is inclined to the stratification at a high angle. In them these granules, in place of being nearly equiaxed, are so compressed and elongated that they are several times longer in the line of cleavage than in the direction perpendicular to it; and as clearly show that there has been a change in the dimensions of the rock, affecting its ultimate constitution, as do other facts, seen on a large scale, prove it with respect to the great mass of the beds. They indeed, on a very small scale, present us with the same phænomena as are seen in the green spots in many of the Welsh slates, as briefly described by me in the paper already referred to. In *Ilfracombe No.* 8, whose constitution is given above, which is an organic clay, foliated with lenticular portions of dolomite, it is seen that the patches of crystals have been compressed and elongated in a similar manner, and hence they lie, not exactly in the plane either of stratification or cleavage, which are inclined to one another at from 10° to 20°, but in such a direction as is the resultant of their combined influence. In a section in the plane of cleavage, the elongation in the line of dip is most clearly exhibited.

In all the cases hitherto considered the organic fragments are not very numerous, and are so imbedded in an excess of organic clay, that, when the dimensions of the rock were changed, their position was altered, but they themselves did not suffer much compression. However, it is very far from being so when there was not much organic clay, or when the compression was very great; for then they are much altered in form and structure, and sometimes to so great an extent that it was some time before I could ascertain to what kind of organism they belonged. Mr. Sharpe has shown (Quart. Journ. of Geol. Soc., vol. ii. p. 74) that the larger organic bodies have had their form much altered in cleaved rocks, and I now proceed to prove that the alteration extends in some instances to the ultimate fragments, somewhat in accordance with what he supposed to occur in all slate rocks, as described in his second communication (Quart. Journ. of Geol. Soc., vol. v. p. 111), no account being taken of change of *position*, which, according to my own observations, is a more common cause of the cleavage. For this purpose I select the minute joints of encrinites, whose form and ultimate organic structure is so very determinate and distinct. In an uncleaved limestone, such as *Hope's Nose No.* 1, of which the composition is already given, the joints of encrinites, whose diameter is on an average about $\frac{1}{10}$th of an inch, have the proportion between their breadth and length as about 3 : 2, though in some cases they are equal, or even as 4 : 5. Being thus small short cylinders, they give rise to symmetrical sections, some rectangular, and others more or less entire circles or ellipses, whose axes vary on an average from being equal to the proportion of 6 : 7, but when very elliptic 2 : 3, or even in rare cases as 3 : 5.

Of all the limestones I have examined, one from Kingskerswell near Torquay, a coral-encrinitic sandy clay, possesses the most intensely developed cleavage, so much so that it required unusual precautions in preparing the thin sections; and it is a most instructive fact, that, of all I have seen, it also shows the greatest amount of compression. In uncleaved limestones the joints of encrinites have their longer axes either arranged promiscuously, or if anything they lie in the plane of stratification, but in this highly cleaved specimen they are very greatly compressed in the plane of cleavage. Their forms are quite distinct from those found in uncleaved limestones; they are often not symmetrical, but broken up irregularly; and, instead of being on an average nearly equiaxed, their greatest length, which always very closely coincides with the line of cleavage, is on an average about four times that in a direction perpendicular to it; and they show forms such as would result from the compression and distortion of the sections of the short cylinders, seen in the uncleaved specimens.

Fig. 3.

But besides their form being thus changed, the ultimate organic structure is altered in a corresponding manner. When not compressed, the structure of an encrinite joint is as shown in fig. 3, drawn from a fragment in a Devonian limestone, magnified 200 linear. It is cellular, the cells being somewhat angular, varying a little in size, but on the whole nearly equiaxed. This is the character when seen in a section cut perpendicular to the axes of the joints, and in many cases it is very nearly the same in every position; but in others the cells are arranged one over the other in the line of the axes of the joints, and the walls separating them in that direction are more or less absorbed, so as to give rise to a more or less irregular or perfect tubular structure. In some cases the cells are filled with dark material, whilst their walls are clear and transparent; in others they are filled with clear, crystalline calcareous spar and the walls are dark, so that the structure may be very readily seen; but when both are equally clear and transparent or uniformly dark, it cannot be recognized.

Fig. 4.

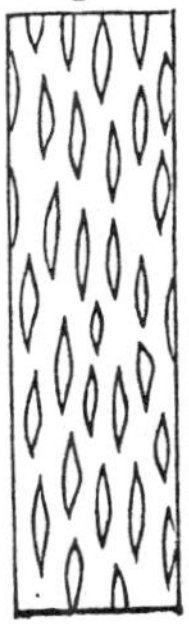

In the highly cleaved limestone from Kingskerswell the structure of the joints of the encrinites is very different from that just described, being usually as shown in fig. 4. This has quite a different character from what is seen when the tubular structure is cut obliquely; and besides, in uncleaved limestones the tubes in the different detached joints do not lie all in one direction, but promiscuously, whereas in this limestone the longer axes of the cells lie in the line of the compression of the joints, and nearly always in the plane of cleavage, those that do not quite coincide with it, differing from it only in such a manner as would result from irregular giving way, or from the compression of a tubular structure inclined to the line of cleavage. On an average the cells have their axes at least as 1 : 4, and are just of such a character as would occur from the compression of the joints to the extent which the alteration of their own form indicates. I could describe similar facts with respect to other organic bodies, but these appear to me so very clear and decisive that I think it unnecessary. In most cleaved limestones this compression of the solid organic fragments is only slight, and merely their position is altered; apparently on account of the materials by which they are surrounded having given way more readily than their own tissue; but yet it may be clearly seen that, other circumstances being the same, the amount of their compression is in proportion to the perfection of the cleavage.

Not only are the organic fragments thus compressed, but also crystals of calcareous spar and dolomite. In uncleaved limestones detached perfect rhombic crystals are often seen; and in some highly cleaved there are such forms as would result from the breaking up and compression of similar crystals. Moreover, in uncleaved limestones the calcareous crystals, filling the cavities in organic bodies or derived from the crystallization of organic clay, have their crystalline cleavage planes almost invariably straight; whilst in cleaved limestones they are often very considerably bent; and this is so particularly the case in the very highly cleaved specimen from Kingskerswell, that there is scarcely any calcareous spar that has straight cleavage planes, and much of it is so broken up and bent that it is only by comparing intermediate examples that the true nature of the structure can be ascertained. This then clearly proves that the compressing force acted so intensely and so gradually as to change the molecular arrangement even of calcareous spar and bend it, in the same manner as we may by the hand easily bend such flexible, unelastic crystals as those of talc or lead. It may perhaps be well to state that I do not think that this indicates that the rock was softened or melted by heat, but that it changed its form like a malleable substance, by the gradual movement of the ultimate atoms one over the other.

This compression and molecular rearrangment, combined with change of position, has been most effective in producing a line of weakness and cleavage in the finely crystallized organic clay of cleaved limestones. When a thick-bedded, uncleaved, fine-grained organic clay, such as the white lias of Radstock near Bath, is examined in a section cut perpendicular to the stratification, the structure is like fig. 5, which is magnified 200 linear. The stratification is in the line of the length of the figure, and, as will be seen, the longer axes of the crystals are arranged promiscuously, without relation

Fig. 5.

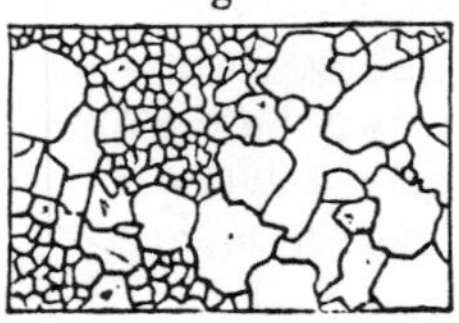

to it, and are nearly equiaxed. In some thin-bedded limestones they have their longer axes in the line of stratification. The organic clay occurring in the most highly cleaved limestone of Kingskerswell has, however, a structure as shown in fig. 6, magnified 200 linear, where the crystalline granules have a very unsymmetrical character, having their axes in the plane of cleavage very much longer than perpendicular to it, as though the compression indicated by the joints of the encrinites and larger crystals had affected the smallest, constituting the ultimate structure of the rock. I think no one could compare figs. 5 and 6 without perceiving that this compression would produce a line of weakness. In fig. 5 it will be seen that there is no particular direction along which the rock could break without fracturing the crystals; whereas in fig. 6 there is clearly one along which a fracture could extend by passing amongst them, whilst in that perpendicular to it such could not be the case without fracturing very many.

Fig. 6.

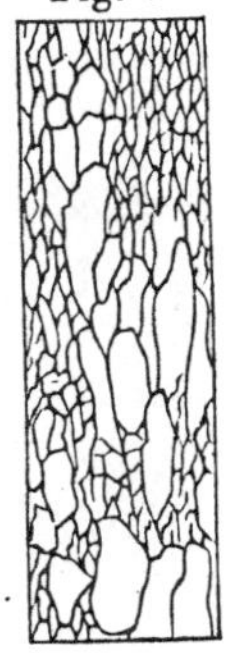

Taking then all the above facts into consideration, I do not see how we can arrive at any other conclusion than that the cleaved limestones have been very considerably compressed; and that this compression has so changed the position and form of the particles of which they are composed, that such a structural weakness has been produced in a plane perpendicular to the pressure, that they may be split along it in the manner characteristic of slaty cleavage.

[*Editor's Note:* Material has been omitted at this point.]

Summing up now the general facts contained in this paper, I show that, other circumstances being the same, the cleavage of the limestones varies directly as the amount of mechanical compression to which they have been subjected; and that the effect of this is such as would necessarily change the structure of uncleaved into that which occurs in those that are cleaved. Also, that cleaved limestones possess no crystalline polarity; and that, in place of crystallization producing slaty cleavage, it has a contrary tendency, and when perfect and complete obliterates it altogether. Though inorganic deposits do not present us with such decisive facts, yet I am persuaded that their slaty cleavage may be satisfactorily explained on similar principles, and that they agree, not only with what may be seen with high powers of the microscope, but with the structure of mountain masses. Can I therefore hesitate to conclude that *slaty cleavage is the result of mechanical and not of crystalline forces*?

4

Copyright ©1975 by the Geological Society of America
Reprinted from *Geol. Soc. America Bull.* **86:**1221–1229 (1975), courtesy of the Geological Society of America

Natural and Experimental Microstructures in Deformed Micaceous Sandstones

W. D. MEANS *Department of Geological Sciences, State University of New York at Albany, Albany, New York 12222*

ABSTRACT

Microscopic features associated with cleavage in micaceous sandstone, siltstone, and silty slate include mica films, mica beards, homogeneous distribution of oriented mica, and equant or inequant quartz. These features occur in rocks in a wide variety of combinations. In experiments, somewhat similar microstructures are obtained by slow folding of wet, salt-mica specimens. Points of resemblance between the artificial cleavage and natural examples are the presence of mica films, the axial-plane orientation of the cleavage, convergent and divergent cleavage fans, refraction of cleavage across lithological boundaries, evidence of possible solution effects, and total strains that are broadly similar in magnitude and orientation to strains estimated in rocks with slaty cleavage. *Key Words: cleavage, foliation, folding, structural geology.*

INTRODUCTION

An approach that has not received much attention in the study of slaty cleavage is to classify the microstructures associated with the cleavage and to seek strain mechanisms and displacement histories on a granular scale that can explain the common microstructures. That this is not an approach exploited very often can be seen from reviews of the slaty-cleavage problem by Siddans (1972) and Wood (1974). Emphasis has been placed instead on the relation of slaty cleavage to other structures seen in the field, such as folds and deformed objects, and the relation of the cleavage plane to the principal directions of total strain.[1] This yields the important result that slaty cleavage is commonly perpendicular, or nearly so, to the principal direction of greatest total shortening, but it leaves unanswered many questions about how the deformation proceeds on a granular scale and how the microstructure evolves during the deformation. Larger questions concerning the environments in which slaty cleavage forms and the lithic and tectonic requisites for cleavage development cannot be answered satisfactorily until the cleavage-forming processes are understood, or at least more certainly identified, on the microscopic scale.

Roofing slate is typically interbedded on various scales with more silty or sandy rock, and some commercial slate (such as the Martinsburg slate at East Bangor, Pennsylvania) is rich in quartz or carbonate of silt grain size. In thin sections, every gradation is seen between the microstructure of silty slate with quite perfect fissility and the microstructure of interbedded cleaved sandstone with inferior fissility. Thus some geologists extend the term "slaty cleavage" to include the cleavage in the sandstone (assuming, of course, that it is not better described by the term "fracture cleavage"). This can be a convenient practice in field work but is not the usage followed here. The term "slaty cleavage" is reserved in this paper for the foliation in micaceous rocks that have the marked fissility of roofing slate. This paper deals, then, not with slaty cleavage in the strict sense, but with closely allied structures in coarser grained rocks. The first part of the paper describes these structures as they appear in thin section. The second part presents the results of experiments designed to simulate some of the microstructures in rock analogues.

The experiments are part of a program of deformation studies (Williams and Means, 1971; Means and Williams, 1972, 1974) in which mineralogical similarity with natural rocks is sacrificed in order to obtain material in which some natural strain mechanisms can operate under conditions that are readily achieved in the laboratory. Quartz-mica rocks are represented by NaCl-mica analogues, the salt being chosen for its solubility in water and its ability to flow plasticly at room temperature and low confining pressure.

The results of the experiments are given in a form familiar to field geologists: the structures are shown in their final state only and not in various stages of development. Other studies have been carried out (Hobbs and others, 1973, 1975) and are continuing, in which the *progressive* development of folds and axial-plane foliations is observed.

Natural Microstructures

The microstructures of slate and interbedded cleaved sandstone are poorly known. This is particularly true of fine-grained roofing slate, in which it is difficult to resolve many details optically, even in extra-thin thin sections. The microstructures of cleaved sandstone and silty slate are easier to see in thin sections, but even in these rocks many details are obscure (for example, the grain-boundary configurations and intracrystalline structures of the fine-grained quartz and layer-silicate material between the silt or sand grains). For slate and cleaved sandstone alike, therefore, full description of the microstructure awaits extensive electron microscopy. Progress could be rapid in this field, with the availability of scanning electron microscopy (Gillott, 1969; O'Brien, 1970) and the development of a thinning technique by which selected areas of a rock sample can be examined directly by transmission electron microscopy (Oertel and Phakey, 1972; Phakey and others, 1972).

The simplified description of sandstone and siltstone microstructures that follows is based on two gross features that can be seen easily in thin sections: the spatial distributions of *oriented* layer-silicate grains and the shape of the coarse quartz or carbonate grains.

Figure 1 shows several microstructural components or end members associated with what the field geologist sees as cleavage in micaceous siltstone or sandstone. *Figure 1 is not a suitable basis for classifying all such cleavages;* it is drawn to emphasize that there is more than one microscopic aspect to the cleavage and to suggest that natural cleavages occur in a variety of intermediate forms between the following end-member types. Excellent descriptions of cleaved graywacke displaying these features are given by Williams (1972a, especially p. 21–30).

[1] "Total strain" is used here to denote the cumulative or net strain that relates the state of a rock at any given stage in its history (for example, as seen in outcrop) to its initial, undeformed state. The total strain at any time represents the combined, successive effects of an infinite number of infinitesimal strain increments that took place up to that time. The term "total strain" is used in preference to "finite strain" (Ramsay, 1967, p. 55) because, although the strain history of a rock is exactly expressed by a succession of infinitesimal strain increments, as a practical matter it is likely to be studied by considering a finite number of finite strain increments. It may, therefore, be confusing to continue using the term "finite strain" for the total strain because the total strain and convenient incremental strains may both be finite in magnitude.

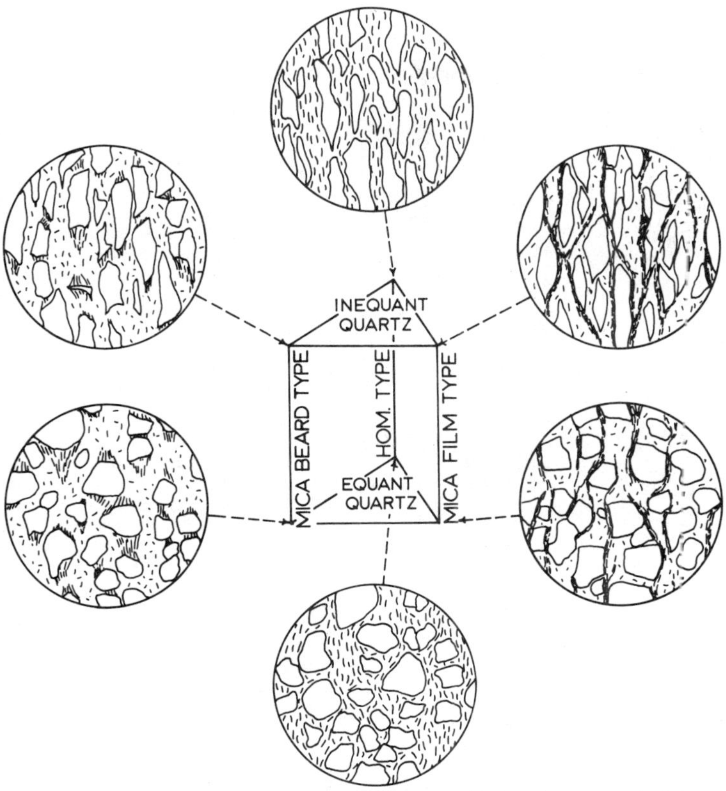

Figure 1. Microstructural end members that contribute to axial plane cleavages seen in deformed micaceous sandstone and siltstone. As in mineral groups, the pure end members do not necessarily exist in nature. End members shown are characterized by the distribution of oriented mica (or other layer silicates) and by the shape of the coarse quartz (or carbonate) grains. Again by analogy with mineral groups, some cleavages cannot be fully represented in this diagram because the number of end members is too small to represent all aspects. Aspects not represented include barren fractures (as in fracture cleavage), quartz beards, characteristics of the coarse detrital mica grains, and all aspects of the regions of fine quartz and of less oriented mica (represented schematically by random dashes in three of the diagrams). The trace of cleavage seen by the field geologist is vertical in all diagrams.

1. Mica Beard Type. The prominent oriented micas (or other layer-silicate grains) occur in aggregates between or adjacent to quartz grain boundaries where the latter make a high angle with the cleavage plane. The trace of {001} of mica is parallel or approximately parallel to cleavage. The term "mica beard" was used by Williams (1972a, 1972b), and examples are figured by him. Mica beards are similar geometrically and perhaps genetically to fibrous crystal growths of quartz and other minerals that grow in dilating cracks (Durney and Ramsay, 1973).

2. Mica Film Type. The prominent oriented mica occurs in irregular, anastomosing thin seams or films. The {001} planes of mica are parallel or approximately parallel to the films, and the films have a preferred orientation parallel to the cleavage. The length of the films in thin section is at least several times the diameter of the quartz grains. The term "mica film" was used by Voll (1960, p. 551). Good examples of films, together with beards, are shown in Figure 3 of Powell (1969). Mica films are also a very common microstructure in roofing slate, but they are not present in all slate.

3. Homogeneous Type. This type, commonly but incorrectly assumed to be *the* slaty-cleavage microstructure, is characterized by a uniform distribution of oriented mica flakes. The oriented layer-silicate grains are dispersed evenly through the spaces between the coarse quartz grains and are not concentrated in special regions or "domains" like the beards or films mentioned above. Microstructure types 1 and 2 can be referred to as "domainal" microstructures; type 3 is "nondomainal." This distinction presumably has implications for the microscopic displacement patterns and strain distributions during the deformation.

The microstructural end members described above differ in the spatial distribution of oriented layer-silicate grains. The quartz associated with each layer-silicate pattern may be equant or inequant with dimensional preferred orientation of its long dimensions parallel to cleavage. Quartz grain shape and layer-silicate distribution, taken together, give the six end members shown in Figure 1.

Figure 2A shows a siltstone with cleavage from the slate belt of eastern New York and western Vermont. There are pronounced mica films and some mica beards. The coarse quartz grains are fairly equant, but many of the finer grains are inequant and elongate parallel to the slaty cleavage. This example points up one of many inadequacies of Figure 1 as a comprehensive classification of cleavages in sandstone: quartz occurs in two habits, depending on the grain size considered, so that no single point inside the prism of Figure 1 could represent both the fine quartz and the coarse quartz. A similar problem would probably arise if one could determine the grain size of all the mica present.

Figure 2B shows another siltstone from the New York–Vermont slate belt. The cleavage is developed only in certain bands, even in a single thin section. Where it is developed, the microstructure is characterized by mica films and inequant quartz. Examples similar to this are described by Rowland (1939, Fig. 6) and Williams (1972a, Fig. 13d).

Mica films are emphasized here because these features are prominent microstructures not given general recognition in the literature and accordingly not accounted for in some theories of cleavage development. The discussions of Rowland (1939), Mosebach (1951), Voll (1960), Plessman (1964), Powell (1969), Braddock (1970), and Williams (1972a) are notable exceptions. In the experiments described below, mica films have been induced artificially by slow straining of wet, salt-mica specimens. Preliminary results of this work were discussed briefly by Means (1973).

Experiments

Experiments were carried out in the apparatus described by Means and Williams (1972). It was modified internally to accept oblong specimens about 8 by 3 by 2 cm in size that were encased in cylindrical lead jackets with external diameters of about 5 cm. The specimens were made up of layers of various compositions and were shortened parallel to the layering between oversize pistons (Fig. 3) by up to 50 percent of

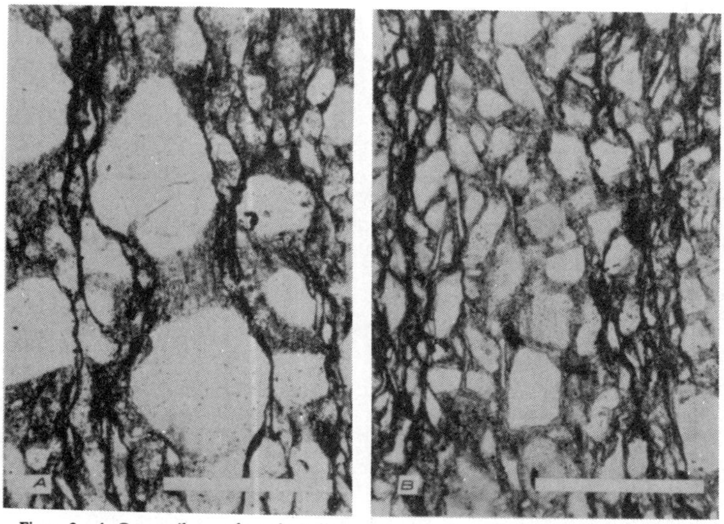

Figure 2. A. Coarse siltstone from the Bull Formation near Granville, New York. Dark mica films are prominent, and a conspicuous mica beard connects the two largest quartz grains. Scale bar is 100 μm. B. Siltstone from the same area as in part A showing two narrow bands (left and right) in which mica films are well developed. Scale bar is 100 μm.

their initial length in six or seven days (a shortening rate of 10^{-6} sec^{-1}) at room temperature. Many of the experimental procedures were developed with B. E. Hobbs and P. F. Williams (in prep.) and will be described more fully elsewhere.

Experiment 213. Specimen 213 (Figs. 4A, 4B) consists of six layers. Layers 1 and 6 are annealed copper. Experience has shown that the weak and isotropic salt-mica layers used in these experiments shorten primarily by homogeneous flattening unless stiff metal layers are included in the assembly. The metal layers buckle, and the salt-mica layers are constrained to take up a folded form between them. In specimen 213, the copper layers were shortened only about 10 percent (measured along their length), while the overall length of the specimen was reduced by about 45 percent. Layers 2 and 5 were made from a coarse salt-mica mix, show no microstructure of interest, and are not described further. Layers 3 and 4 were compacted with brine at a nominal compaction pressure (friction neglected) of 300 bars, from a mixture of 10 percent coarse salt, 67 percent medium salt, and 23 percent fine mica by weight. Further details of these components are given in Appendix 1. To provide some indication of the total strains in layers 3 and 4, each was provided with circular strain markers (see Appendix 1). The several layers were assembled into a pre-cast lead jacket and saturated with brine under vacuum. The brine was previously saturated with NaCl at room temperature. The whole assembly was then jacketed in rubber, placed in the deformation apparatus, surrounded by gas at a pressure of 70 bars, and shortened 45 percent in seven days.

After deformation the specimen was removed from its lead jacket, dried, potted in plastic, and sawn through its center perpendicular to the hinge lines of the folds. Figure 4A is a photograph of this saw cut, and Figure 4B is a map of the same surface that shows the trace of the cleavage in those parts of the specimen where it is discernible in thin section. Elsewhere the cleavage is not visible at all or is so weakly developed that its orientation is uncertain. Note in Figure 4B that the trace of the cleavage is parallel to the long dimension of the strain markers, except for one marker on the right-hand limb of the central antiform. Note also that the strain markers are not very good ellipses. The deformation is evidently inhomogeneous on the scale of the markers; furthermore, the markers may have become somewhat irregular due to anisotropic compaction before the deformation began. These strain markers are not useful as quantitative indicators of the total strain, but they do indicate the approximate relation between total strain and the cleavage; it is similar to what is observed in rocks with slaty cleavage (see Wood, 1974). In specimen 213 large extensions have also occurred *parallel* to the fold hinge lines. Some values of this extension are shown as percentage elongations in Figure 4B.

Figure 4C shows the microstructure of specimen 213 near one end of layer 3 (loc. A, Fig. 4B) where the strain has been small. The fabric is isotropic in all respects that can be judged optically at this scale. The fine mica grains make up an unresolvable paste or matrix that uniformly surrounds the salt grains. Both the medium and coarse salt grains are generally equant in shape. No trace of cleavage or of "bedding" are visible in such regions.

Figures 4D, 4E, and 4F show parts of specimen 213 where the cleavage is best developed. Figure 4D is a field (loc. B, Fig. 4B) where the total shortening normal to the cleavage has probably been 50 percent or more. Dark mica films can be seen that are deflected around the coarse salt grains (compare Fig. 2A). The coarse salt grains show cube faces that remain approximately at right angles to one another, suggesting that these grains have not been strongly deformed internally despite the large strain undergone by the material around them. The medium-sized salt grains present a different aspect; many are elongate in the foliation, with long dimensions exceeding the typical initial grain diameters shown in Figure 4C. These grains have become flattened

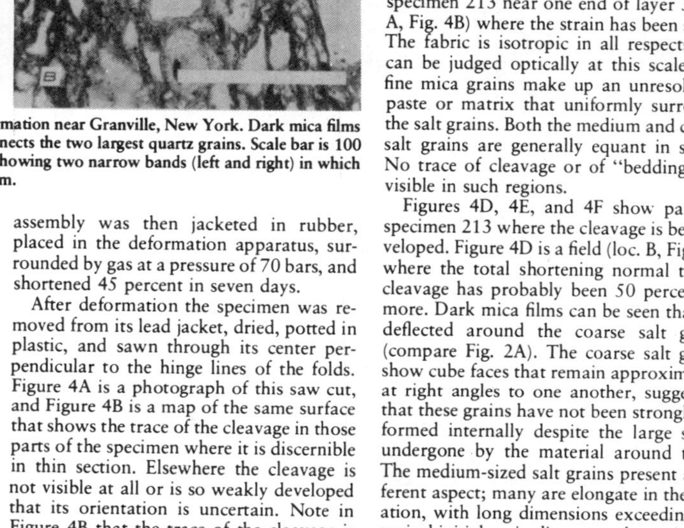

Figure 3. Interior of pressure vessel, showing fixed and moving internal pistons (left and right, respectively), the rubber jacket connecting them (black), the lead jacket (dotted), and the layered salt-mica specimen (gray and white). Small arrows represent gas confining pressure. Large arrow represents displacement of small-diameter external piston through which axial force is applied to the internal pistons and specimen. In operation, the apparatus is turned 90° and the external piston moves down vertically.

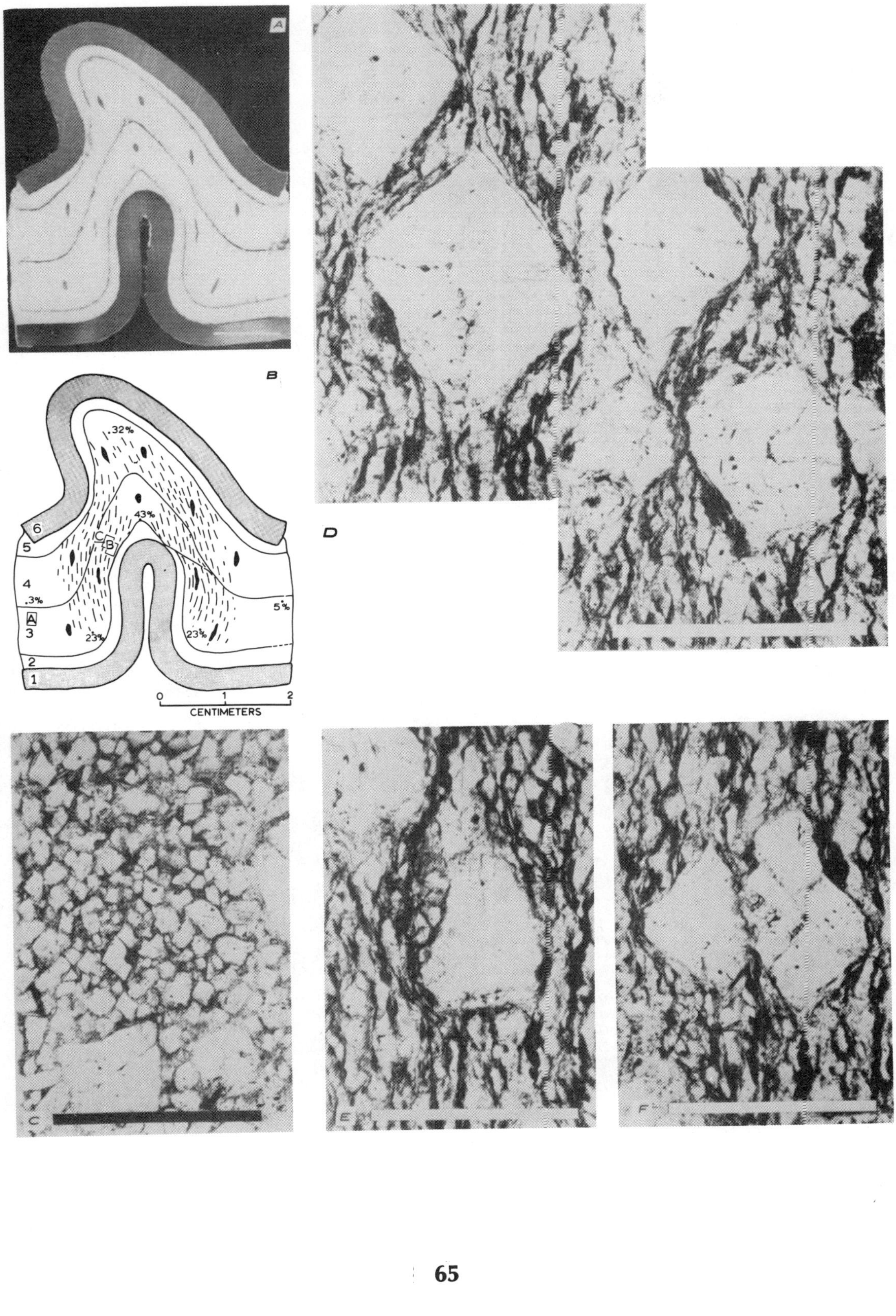
A
B
.32%
43%
6
5
4
.3%
A
3
2
1
C
B
23%
23%
5%
0
1
2
CENTIMETERS
D
C
E
F

Figure 4. A. Specimen 213. Profile section, cut perpendicular to the fold hinge line. Scale bar is 1 cm.

B. Specimen 213. A map of the surface shown in part A indicating the traces of cleavage (dashed) and local values of the percentage elongation parallel to the hinge line of the fold. Layers 1 and 6 are annealed copper. Layers 2 through 5 are salt-mica mixtures.

C. Specimen 213 detail at undeformed end of layer 3 (loc. A, part B). Scale bar is 1 mm.

D. Specimen 213 detail in region with good cleavage, cut perpendicular to fold hinge line (loc. B, part B). Scale bar is 1 mm.

E. Specimen 213 detail in region with good cleavage, cut parallel to fold hinge line (loc. C, part B). Note mica film extending full length of photograph. Scale bar is 1 mm.

F. Specimen 213 detail. Same region and orientation as part E. Possible solution boundary between two coarse salt grains. Scale bar is 1 mm.

in the foliation, either by slip or by some other mechanism. Other medium-sized salt grains are less flattened, and this appears to be true especially of grains for which the traces of the cube faces are inclined at about 45° to the slaty cleavage (as for the largest coarse salt grain in Fig. 4C). The most likely interpretation of this is that grains with cube faces at 45° to the cleavage plane are unsuitably oriented for dodecahedral slip ($T = \{110\}$, $t = \langle 1\bar{1}0 \rangle$, see Carter and Heard, 1970) so that these grains remain undeformed by slip, while grains with cube faces at a high angle to the cleavage have become flattened by dodecahedral slip. Figure 4D also shows a feature similar to those attributed to solution activity by some observers of slate (see Plessman, 1964; Durney, 1972; Groshong, 1974). The upper right-hand coarse salt grain has cube faces intact and still at approximately 90° on its right side but has a scallopy boundary against a mica film on its lower left boundary. This could be interpreted as evidence of removal of the lower left portion of the grain by solution along its contact with the mica film.

Figures 4E and 4F show details in a thin section cut perpendicular to the cleavage and parallel to the fold hinge lines (loc. C, Fig. 4B). Again, mica films are prominent, particularly in Figure 4E, where a more or less continuous example extends the full length of the photograph. The coarse salt grain in the center has {100} oriented at about 0° and 90°, respectively, to the cleavage. Figure 4F shows two coarse grains in the orientation that suppress dodecahedral slip, with what may be a solution boundary between them.

There is thus some evidence in specimen 213 for local solution effects, but no evidence has been seen of corresponding deposition effects, although some such deposition of salt must occur, at least during drying of the specimen following a run. Mica beards are understandably not present in any of these samples, as development of mica beards would clearly require conditions where mica can crystallize during the deformation. Salt beards might be expected here but none has been identified.

Experiments 215 and 216. These experiments were run with and without pore water, on five-layer assemblies, with aluminum layers on the outside, a single layer of pure salt in the center, and layers of salt-mica mix on either side of the center. The salt-mica layers were very rich in salt (10 percent coarse salt, 80 percent medium salt, 8 percent fine mica, 2 percent spectrographic carbon) and were compacted with brine at a pressure of 720 bars, then dried overnight. Specimen 215 was resaturated and run with interstitial brine; specimen 216 was run dry. Both were subjected to a confining pressure of 345 bars and were given an overall shortening of about 35 percent in about five days. Further details appear in Appendix 1.

Figure 5A shows specimen 215 after deformation, along with its deformed lead jacket, and a similar specimen and lead jacket before deformation. Note the exposed ends of strain-marker lines that are rubber-stamped on either side of the central salt layer in the undeformed specimen. These markers are initially about 2.5 mm apart. Similar rubber-stamped lines have been used by Hobbs and others (1973) to measure the total strain in hundreds of small regions throughout 12-layer folded specimens.

Figure 5B shows part of the surface of specimen 215 in detail. The central white layer is pure salt, and above and below it may be seen a ribbing caused by minute puckering of the interface between the lead jacket and the specimen. This puckering is parallel to the cleavage within the specimen, but it may or may not be closely related genetically to the cleavage. The black dots show the positions of strain markers initially spaced every 2.5 mm along each side of the salt layer. Near the hinge of the fold there has been about 65 percent shortening parallel to the "bedding."

Figure 6A shows the microstructure of

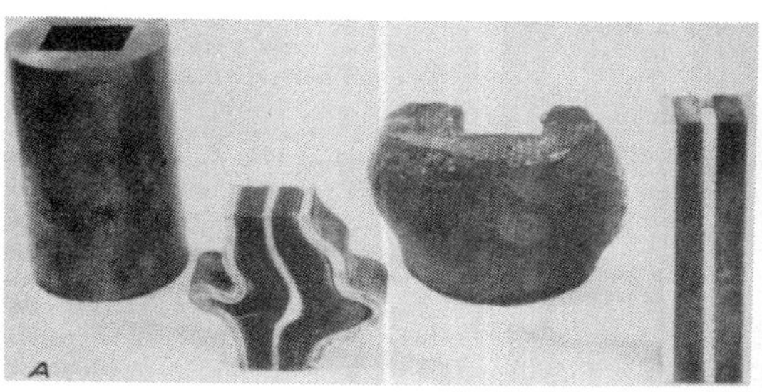

Figure 5. A. Specimen 215 after deformation (second from left) and specimen 216 before deformation (right). Behind the specimens are a lead jacket before deformation and part of the lead jacket in which specimen 215 was folded. Height of undeformed jacket is 7.9 cm. B. Specimen 215 detail. External view of the interface between the specimen and the jacket near the hinge of the main fold seen in part A. Black dots mark the positions of strain-marker lines initially equally spaced on the top and bottom of the white (pure salt) layer. Box indicates area shown in thin section in Figure 6B. White scale bar is 5 mm. Black scale bar shows initial spacing of strain markers.

the mix used in these three runs as it appears before deformation. As for specimen 213, the initial microstructure looks isotropic and the salt grains are generally equant.

Figure 6B shows the microstructure near the hinge of the fold in 215. Similar features to those described earlier for 213 are obvious, but the cleavage is less well defined in this very salt-rich material, just as it may be in very quartzose rocks. Note again the tendency for many salt grains, both coarse and medium, to be fairly equant in shape despite the fact that the material has been shortened 40 or 50 percent normal to the foliation.

Figure 6C, from specimen 216, shows the strikingly different aspect of the microstructure in a specimen run dry. Mica films are not prominent; instead, the conspicuous feature is general and quite homogeneous flattening of all the salt grains. The patches of interstitial mica and carbon present in the starting material are distorted and extended parallel to the foliation, but they do not make the characteristic wavy pattern of mica films in wet specimens, nor are they so continuous. A specimen deformed with kerosene has characteristics intermediate between those of specimens 215 and 216. The same intermediate character of microstructures developed in the presence of kerosene was seen in experiments on faulting and crenulation cleavage in salt-mica specimens by Means and Williams (1972). The general conclusion is that water exerts significant controls on microstructures in salt-mica specimens both as a pressure-transmitting medium *and* as a solvent. A similar dual role is expected of water in rocks and has commonly been cited.

Experiment 225. Specimen 225 is an eight-layer assembly of alternating coarse and fine materials. The coarse layers (pale gray, Fig. 7A) were made of the same mix used in specimen 213 with, in addition, 2 percent coarse mica (to represent coarse detritial mica in a graywacke) and 0.5 percent carbon. The fine layers (dark gray, Fig. 7A) were made of 74 percent fine salt, 24 percent fine mica, and 2 percent carbon (see Appendix 1). Both kinds of layers were compacted with brine at 300 bars, dried, rubber-stamped with strain markers, and resaturated with brine for the run. Specimen 225 was run directly in contact with the lead jacket, with *no stiff metal layers* to promote buckling. The intention here was to promote folding associated with large amounts of layer-parallel shortening. Portions of the sample near its center could be expected to shorten by a simple history of progressive, coaxial, pure strain increments. Total strains are indicated in this specimen both by the rubber-stamped markers on internal bedding interfaces and by paper grids fixed to the outer surfaces of the specimen (see Appendix 1).

Figure 7B is a map of specimen 225 showing the trace of the axial-plane cleavages seen in thin section. Note the convergent and divergent cleavage fans in the coarse-grained (light) and fine-grained (dark) layers, respectively. This is qualitatively similar to the familiar patterns of cleavage fanning in naturally deformed rocks. Note that although the familiar cleavage fans are present in the two lithologies, there is no substantial difference in the mica content of the two kinds of layers. The main differences are the grain sizes of the salt, the relative sizes of salt and mica, and the carbon content (see Appendix 1).

Figure 8 shows a detail of the cleavage developed in a region near the center of specimen 225, close to the interface between layers 4 and 5. Cleavage in the coarse layer (4) has characteristics previously mentioned for specimen 213. Examination at high magnification shows that the coarse mica grains, which tend to lie parallel to "bedding" in the undeformed material, now tend to lie parallel to the cleavage, as is seen in rocks (Williams, 1972a). Cleavage in the fine layer (5) is visible, but as with many fine-grained rocks, it is difficult to see exactly what defines the cleavage in this material on a granular scale. The coarsest mica grains tend to lie parallel to it, and there also appears to be a strong tendency for the fine mica to be so aligned, judging

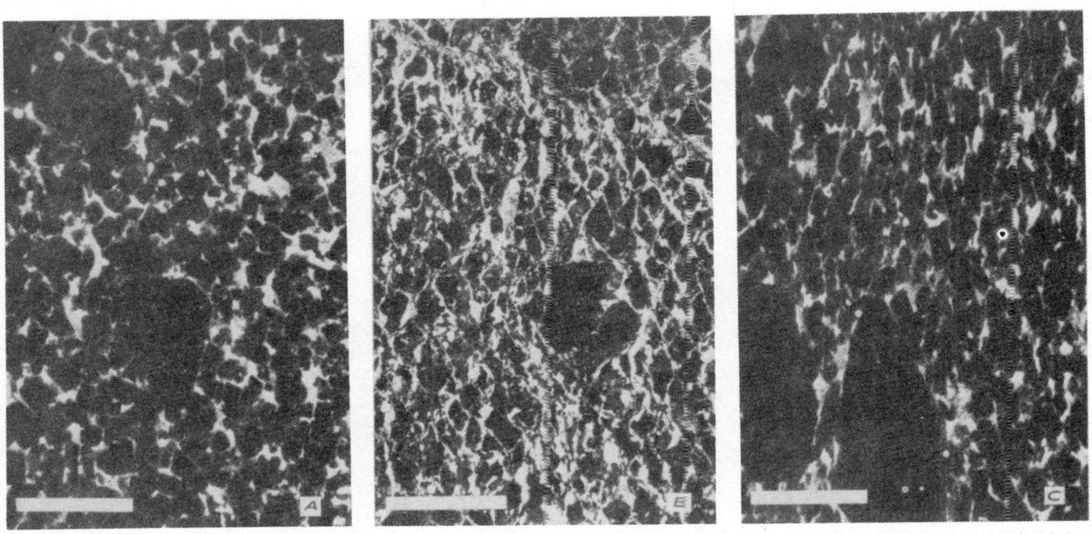

Figure 6. A. Appearance of an undeformed region in the mix used for specimens 215 and 216. Print of a Kodachrome transparency taken in plain light. Salt grains are black; micaceous areas, light. Scale bar is 500 μm. B. Specimen 215 detail (from locality outlined in Fig. 5B). Mica films (light) have a general vertical orientation, with much local variation where salt grains (dark) remain equant. Lower boundary of pure salt layer crosses upper right corner of photograph. Scale bar is 500 μm. C. Specimen 216 detail, showing the microstructure obtained in a specimen run without pore water (compare part B run with pore water). Scale bar is 500 μm.

from its bulk extinction behavior between crossed nicols.

The interface between layers 4 and 5 in Figure 8 has been shortened about 70 percent. The total principal strains in the central part of layer 4 are 110 percent extension parallel to the cleavage plane and perpendicular to the fold hinge line, 64 percent extension parallel to the cleavage plane and parallel to the fold hinge line, and 73 percent shortening perpendicular to the cleavage plane. The axial ratios of the total strain ellipsoid in the central part of layer 4 are 1.3/1.0/0.16. These ratios may be compared with axial ratios of 1.7/1.0/0.17 that are typical of reduction spots in roofing slate from the New York–Vermont slate belt (Wood, 1974, p. 376). The total strain in this part of specimen 225 approaches a case of pure flattening ($1.0/1.0/n < 1$), more nearly so than is common in slate, according to Wood. The volume change for the central part of layer 4 is -7 percent $\pm$ 5 percent, which may be compared with volume changes of -10 to -20 percent considered possible for slate by Ramsay and Wood (1973, p. 267). On the limbs of the folds in specimen 225, there is evidence of nonparallelism between the cleavage plane and the nearest principal plane of the total strain ellipsoid. This phenomenon is better demonstrated, however, in certain earlier experiments on dry salt-mica specimens by Hobbs and others (1973) and will be described elsewhere.

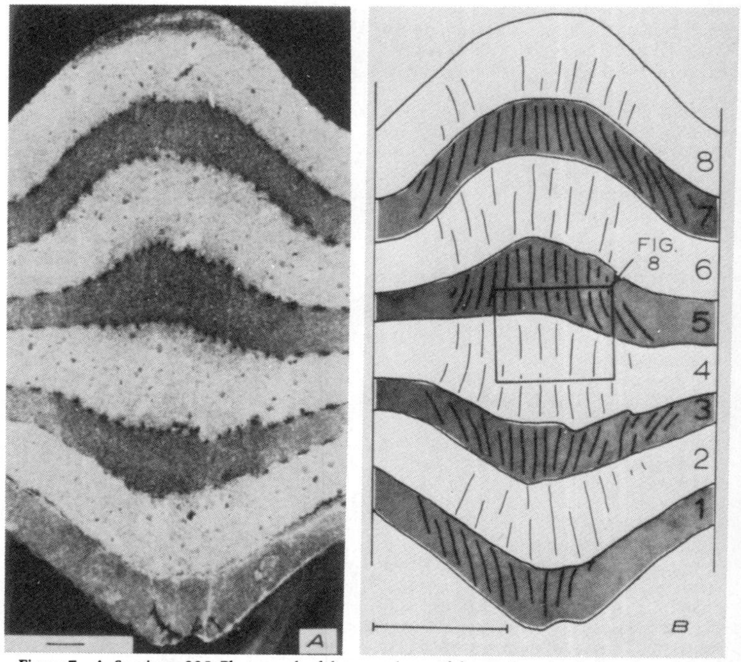

Figure 7. A. Specimen 225. Photograph of the central part of the specimen (excluding about 8.5 cm to the left and right of the area shown). White scale bar is 1 cm. Black scale bar shows initial spacing of strain markers (dots) on light-gray layers of coarse mix. B. Map of the area shown in part A, showing the trace of the cleavage. Note the convergent cleavage fans in the coarse layers (white) and the divergent cleavage fans in the fine layers (gray). Scale bar is 1 cm.

DISCUSSION

My purposes in this paper have been to point out the need for further study of cleavages and cleavage development on a granular scale and to show pictures demonstrating that cleavages resembling some natural ones in micaceous sandstone, siltstone, and silty slate can be induced in rock analogues. The next step in this program is to learn how the cleavages form in the analogues. This may also shed some light on how they form in nature, but there are distinct limitations here.

Limitations of the Salt-Mica Analogues

The present salt-mica analogues may permit operation of certain natural processes in cleavage development (bodily rotation of grains, deformation of grains by slip, solution or crystallization of quartz or calcite [salt], fracture of grains, sliding on grain boundaries, change in volume or shape of pore space), but these analogues do not permit other natural processes of *undoubted* importance for some kinds of cleavage development. In particular, the analogues do not permit solution, crystallization, or recrystallization of layer-silicate grains. For this reason alone there is no expectation that the salt-mica analogue deformed at room temperature can fully simulate all kinds of natural cleavage development. It is hoped, however, that these preliminary experiments with rock analogues will provide a useful basis for design of later experiments on real rocks.

Origin of the Mica Films

The origin of mica films is a fascinating problem both in rocks and in analogues such as these. Geologists have attributed mica films to several processes, and it may well be that there is more than one kind of mica film, just as there is more than one kind of cleavage in micaceous sandstone. It has been suggested, with good evidence, that some films form by local solution of quartz or calcite, so that the films are akin to stylolites (Plessman, 1964; Williams, 1972a). It has also been suggested that mica films develop by some kind of flowage differentiation or mechanical sorting, in which fine layer silicates are carried and deposited along channelways (the eventual films) by water that is moving through deforming sediment (Powell, 1969; Braddock, 1970). Beyond these questions of the role of water in the development of mica films, there are problems of where and why individual films nucleate and how they propagate into the relatively extensive sheets seen in thin sections. There is the related question of how the growth of a film is tied to the relative displacements of the grains around it. Do shear displacements occur parallel to the films during the deformation, despite the fact that they come to lie, ultimately, about parallel to planes of zero shear strain in the total strain ellipsoid?

In an effort to answer these questions for mica films in the salt-mica analogue, transmitted light experiments are now being carried out, in which a very thin salt-mica specimen is deformed between thick glass windows. The aim is to observe the incremental displacements of grains, the incremental strains of groups of grains, and the microstructural changes, step by step.

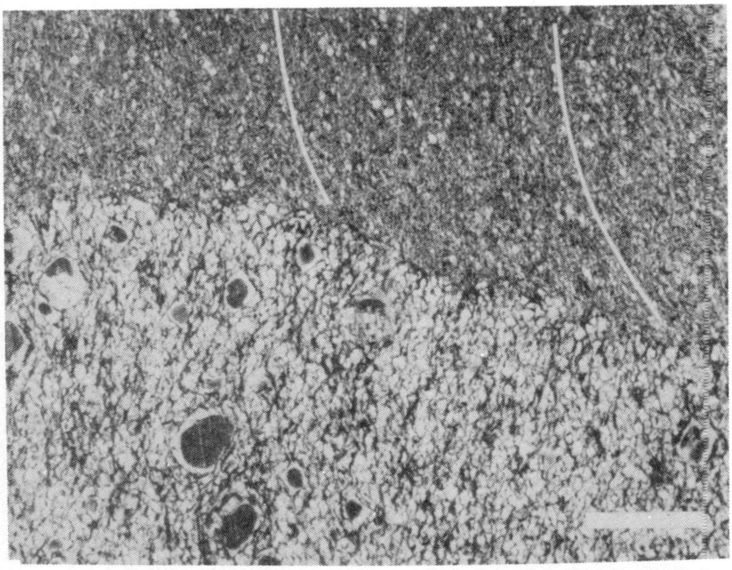

Figure 8. Specimen 225. Photomicrograph of the area indicated in Figure 7B. A "bedding" plane between coarse and fine layers passes across the center of the photograph. Note the refraction of cleavage as it passes from the coarse into the fine material. Local orientation of cleavage in the fine material indicated by the white lines. The dark areas on some of the coarse salt grains in layer 4 are parts of the crystals that were not completely wetted by the (kerosene) covering medium. Scale bar is 2 mm.

Outlook for Studies of Cleavage in Slate and Micaceous Sandstone

In my view, the slaty-cleavage problem is still a long way from solution. First, the classical picture of the microstructure as homogeneous preferred orientation of layer silicates is certainly incomplete. No conception of the origin of cleavage can be satisfactory until we know what the microstructures really are. It seems likely that there will prove to be "slates and slates."

Second, the microscopic behavior of sediment or rock developing cleavage is likely to be complex, in the sense that several different strain mechanisms may be operating simultaneously. The rocks or sediments involved contain at least two solid phases, each in a range of grain sizes, together with fluid-filled pore spaces. The experience of soil mechanics confirms that such systems exhibit complex microscopic behavior, even in short-term tests at surface temperatures. In cleavage-forming environments, where longer times and higher temperatures may permit widespread migration of grain boundaries and growth of new grains during deformation, the range of microscopic processes contributing to fabric development must be even broader than in soils. Geologists need to replace the vague descriptions of these phenomena now in use ("rotation," "recrystallization," "dewatering," and "pressure solution") with more refined descriptions that explain in detail the development of individual microstructures.

Third, even the alleged parallelism of cleavage with a principal plane of total strain needs to be reassessed. The essential parallelism of cleavage in certain roofing slates with a principal plane is not in doubt, but it would be surprising if all slaty cleavage is precisely parallel to a principal plane. This is because cleavage on fold limbs, for example, must often develop in circumstances where successive incremental strains have their principal directions parallel to different lines of grains in the rock. In such noncoaxial strain histories, the principal directions of total strain will rotate with respect to lines of grains. If cleavage is to track a principal plane of total strain exactly, each increment of strain must be associated with precisely the right small rotation of cleavage. Continuous perfect adjustment of cleavage to the new orientation may or may not be possible, depending on the nature of the microstructure existing prior to each increment of deformation and on the microstructural changes that occur during it. To discuss this question further, one has to specify, for many different individual cases, just what the macroscopic cleavage is defined by on the microscopic scale. Is it defined, for example, by the preferred orientation of all the mica films present, or by the preferred orientation of certain more continuous films, or perhaps by an orientation that bisects a small angle between two sets of films? Following this, one asks how the pattern of films changes during the next increment of strain. Do new films appear in certain orientations, or do suitably oriented segments of old films propagate at their edges, or take up additional displacements in their interiors? Questions such as these underscore the need for more detailed descriptions of the microstructures in rocks with cleavage and for experimental or theoretical approaches that will help to show how microstructural patterns evolve during deformation.

ACKNOWLEDGMENTS

I thank Bill Gregg, Bruce Hobbs, Bruce Nisbet, and Paul Williams for reading the manuscript, David Graham for photography, and Alton Hajeck for machine work. The work was supported by National Science Foundation Grant GA–15918.

APPENDIX 1. EXPERIMENTAL DETAILS

EXPERIMENT 213

Components of Layers 3 and 4

Coarse salt: NaCl, Fisher laboratory grade, granular. Typical grain size 700 μm.

Medium salt: NaCl, Diamond superfine popcorn salt. Typical grain size 125 μm.

Fine mica: Synthetic flourine phlogopite (Syntha-mica 202, Mycalex Corporation of America). Typical grain size 30 μm.

Circular Strain Markers

Layers 3 and 4 were provided with strain markers as follows: About half the powder used to make a layer was poured into the cavity of a rectangular pellet press and tamped lightly by hand. Five steel rods of diameter 1.6 mm were laid on the tamped surface perpendicular to the long dimension (7.9 cm) of the eventual layer. The other half of the powder was poured over the top of the rods, and the whole charge was saturated with brine and compacted at 300 bars. The layer was then removed from the pellet press, and the steel rods were extracted by pushing them out gently with a rod of smaller diameter. This produced perfectly circular holes which remained circular after drying. The holes were then filled with the same mix used to make the main part of the layer, with 2 percent carbon admixed. This was tamped into the holes by hand, at a pressure estimated to be about 20 bars, considerably lower than the pressure used to compact the layer as a whole.

EXPERIMENTS 215 AND 216

Components of Dark Layers

Coarse salt, medium salt as in specimen 213 above.

Fine mica: Illite from Morris, Illinois, A.P.I. clay mineral standard, Ward's Natural Science Establishment. Disaggregated and sieved; typical grain size 40 μm.

Spectrographic carbon: National Spectrographic Powder, Cat. No. L4160, Union Carbide Corporation. Typical grain size 40 μm.

Strain-Marker Lines

Rubber-stamped with water-insoluble Crown brand Super marking ink, R. A. Stewart and Company, New York, N.Y. Initial spacing of marker lines 2.5 ± 0.2 mm.

EXPERIMENT 225

Components of Coarse (Light) Layers

Coarse salt, medium salt, and carbon as in specimens 213 and 215 above.

Coarse mica: Synthetic fluorine phlogopite (Synthamica 202, Mycalex Corporation of America). Typical grain size 500 μm measured parallel to {001}.

Components of Fine (Dark) Layers

Carbon as in specimen 215 above.

Fine salt: NaCl, origin unknown. Typical grain size 30 μm but with several percent of 150 μm grains.

Fine mica: Synthetic fluorine phlogopite. A mixed powder containing mostly fine mica with typical grain size of 30 μm, together with a few percent of coarse grains as much as 500 μm in size, measured parallel to {001}.

Mineralogical Compositions of Coarse and Fine Layers (by Weight)

Coarse layers: 75.5 percent salt, 24 percent mica, 0.5 percent carbon.

Fine layers: 74 percent salt, 24 percent mica, 2 percent carbon.

Confining Pressure, Overall Shortening, Duration

70 bars, 45 percent, 7 days.

Paper Grid Strain Markers

It was discovered by accident that excellent strain-marker grids can be applied to external surfaces of wet specimens as follows: A sheet of Xeroxed graph paper is inserted between the lead and the specimen before a run, with the Xeroxed lines facing the specimen. Following the run, while the specimen remains wet, the paper is peeled off the specimen, but the Xeroxed lines remain on it in a deformed grid, indicating the total strains of the specimen surface. Evidently the black powder forming the lines adheres better to the specimen than to the wet paper, so that the graph-paper grid is effectively transferred to the specimen and distorted with it. Marker lines obtained in this way are finer and more evenly spaced than those obtained by rubber-stamping.

ALL EXPERIMENTS

Pressure Conditions

The confining pressures given above for these experiments are the pressure in the gas surrounding the specimen assembly (Fig. 3). The normal stresses across the lead-specimen interface will generally be higher than the gas pressure by an amount that is dependent on the configuration of the lead jacket. This extra pressure can be expected to increase during a run and to vary from lower values near the ends of the specimen (where the circumference of the lead jacket is stretched least) to higher values around the center of the specimen (where the circumference of the lead is stretched most). Where the mean radius of the specimen is about half the mean outer radius of the jacket, as in the present experiments, the extra pressure can be expected to be of the order of the tensile yield strength of the jacket, in this case about 100 bars.

The pore pressures in the wet experiments are thought to approximate the gas confining pressures because the specimens were saturated with brine at 1 atm before each run and excess brine was always found between the lead jacket and the rubber jacket following a run. Thus in experiment 225, for example, the pore pressure was probably around 70 bars while the total normal stress across the interface between the lead and the specimen may have risen to roughly twice this value near the center of the specimen. None of the pressure conditions, apart from the gas pressure, are well known in these experiments.

REFERENCES CITED

Braddock, W. A., 1970, The origin of slaty cleavage: Evidence from Precambrian rocks in Colorado: Geol. Soc. America Bull., v. 81, p. 589–600.

Carter, N. L., and Heard, H. C., 1970, Temperature and rate dependent deformation of halite: Am. Jour. Sci., v. 269, p. 193–249.

Durney, D. W., 1972, Solution-transfer, an important geological deformation mechanism: Nature, v. 235, p. 315–317.

Durney, D. W., and Ramsay, J. G., 1973, Incremental strains measured by syntectonic crystal growths, *in* DeJong, K. A., and Scholten, R., eds., Gravity and tectonics: New York, John Wiley & Sons, Inc., p. 67–96.

Gillott, J. E., 1969, Study of the fabric of fine-grained sediments with the scanning electron microscope: Jour. Sed. Petrology, v. 39, no. 1, p. 90–105.

Groshong, R. H., Jr., 1974, Martinsburg slaty cleavage caused by pressure solution [abs]: EOS (Am. Geophys. Union Trans.), v. 55, no. 4, p. 436.

Hobbs, B. E., Means, W. D., and Williams, P. F., 1973, Folding and microfabric development in experimentally deformed salt-mica specimens [abs.]: EOS (Am. Geophys. Union Trans.), v. 54, no. 4, p. 457.

——1975, An outline of structural geology: New York, John Wiley & Sons, Inc. (in press).

Means, W. D., 1973, Folding of wet salt-mica specimens at low strain rate [abs.]: EOS (Am. Geophys. Union Trans.), v. 54, no. 4, p. 457.

Means, W. D., and Williams, P. F., 1972, Crenulation cleavage and faulting in an artificial salt-mica schist: Jour. Geology, v. 80, p. 569–591.

——1974, Compositional differentiation in an experimentally deformed salt-mica specimen: Geology, v. 2, no. 4, p. 15–16.

Mosebach, R., 1951, Zur Petrographie der Dachschiefer des Hunsrück-Schiefers: Deutsch. Geol. Gesell. Zeitschr., v. 103, p. 368–376.

O'Brien, N. R., 1970, The fabric of shale — An electron-microscope study: Sedimentology, v. 15, p. 229–246.

Oertel, G., and Phakey, P. P., 1972, The texture of a slate from Nantlle, Caernarvon, north Wales: Texture, v. 1, p. 1–8.

Phakey, P. P., Curtis, C. D., and Oertel, G., 1972, Transmission electron microscopy of fine-grained phyllosilicates in ultra-thin rock sections: Clays and Clay Minerals, v. 20, p. 193–197.

Plessman, W. von, 1964, Gesteinslösung, ein Hauptfaktor beim Schieferungsprozess: Geol. Mitt., v. 4, p. 69–82.

Powell, C. McA., 1969, Intrusive sandstone dykes in the Siamo slate near Negaunee, Michigan: Geol. Soc. America Bull., v. 80, p. 2585–2594.

Ramsay, J. G., 1967, Folding and fracturing of rocks: New York, McGraw-Hill Book Co., 568 p.

Ramsay, J. G., and Wood, D. S., 1973, The geometric effects of volume change during deformation processes: Tectonophysics, v. 16, p. 263–277.

Rowland, R. A., 1939, A petrotectonic analysis of cleavage in otherwise unmetamorphosed sediments: Jour. Geology, v. 47, no. 5, p. 449–471.

Siddans, A.W.B., 1972, Slaty cleavage — A review of research since 1815: Earth-Science Rev., v. 8, p. 205–232.

Voll, G., 1960, New work on petrofabrics: Geol. Jour., v. 2, p. 503–567.

Williams, P. F., 1972a, Development of metamorphic layering and cleavage in low grade metamorphic rocks at Bermagui, Australia: Am. Jour. Sci., v. 272, p. 1–47.

——1972b, "Pressure shadow" structures in foliated rocks from Bermagui, New South Wales: Geol. Soc. Australia Jour., v. 18, p. 371–377.

Williams, P. F., and Means, W. D., 1971, Folding experiments on an artificial schist: Nature, v. 234, no. 48, p. 90–92.

Wood, D. S., 1974, Current views of the development of slaty cleavage: Earth and Planetary Sci. Ann. Rev., v. 2, p. 1–35.

Manuscript Received by the Society June 14, 1974

Revised Manuscript Received September 30, 1974

Manuscript Accepted October 25, 1974

Part II

DEFORMATION TEXTURES AND FLOW MECHANISMS

Editor's Comments on Papers 5 Through 11

5 **STAUFFER**
Deformation Textures in Tectonites

6 **KERRICH and ALLISON**
Flow Mechanisms in Rocks: Microscopic and Mesoscopic Structures, and Their Relation to Physical Conditions of Deformation in the Crust

7 **ELLIOTT**
Diffusion Flow Laws in Metamorphic Rocks

8 **CARTER and RALEIGH**
Principal Stress Directions from Plastic Flow in Crystals

9 **SPRY**
The Chronological Analysis of Crystallization and Deformation of Some Tasmanian Precambrian Rocks

10 **FERGUSON and HARTE**
Textural Patterns at Porphyroblast Margins and Their Use in Determining the Time Relations of Deformation and Crystallization

11 **MISCH**
Excerpt from *Paracrystalline Microboudinage of Zoned Grains and Other Criteria for Synkinematic Growth of Metamorphic Minerals*

Although the polarizing microscope was developed in the early 1800s, it was not until 1853 (Sorby) that the textures of deformed and metamorphosed rocks were first studied. Our understanding of the textures resulting from metamorphism and rock flowage have proceeded more or less hand-in-hand to the present day. Early work by Van Hise (1898*a*, 1898*b*), though mainly concerned with chemical metamorphism, is notable in that he clearly sets out some of the relationships between recrystallization and strain.

Of the many contributions to this field of geology only a few examples could be included here. Those chosen serve to introduce the various textures and their origins, to outline their geological interpretation, and to set the historical stage for the materials-science approach to the study of deformed rocks in the 1980s.

Stauffer (Paper 5) summarizes many of the textures found in deformed rocks and interprets these structures mainly in terms of crystal ductility, intergranular dislocations, and recrystallization. In this regard, it should be noted that mylonites, once considered to be largely the result of intense intergranular displacement and brittle grain-size reduction, are now realized to form mainly through recrystallization processes (dynamic recrystallization) not involving brittle breakup. Lapworth (1885) introduced the name "mylonite" and considered mylonites to be the result of crushing.

Paper 6 (Kerrich and Allison) briefly summarizes flow mechanisms with some discussion of textures and conditions under which the various mechanisms operate (for a good follow-up discussion of this, see Kerrich et al., 1977). The conditions leading to the dominance of one deformational process over another are not yet fully established, however, and some of the interpretations given in this paper are subject to change. For example, Kerrich and Allison claim that dislocation creep is the most common mechanism of plastic flow from upper greenschist conditions to higher grades of metamorphism. In my experience, dislocation creep has been a dominant process also in many lower greenschist rocks, whereas many high-grade rocks show evidence of having undergone mainly dynamic recrystallization. (Note also the switched legends between figures 5a and b in this paper.)

Elliott (Paper 7) elaborates on diffusion flow (pressure solution), long underestimated as a deformation process, and compares it with dislocation creep (intracrystalline flow). For further reading about deformation textures and processes, Kerrich (1978) gives a good review of pressure solution in rocks; Robin (1978) discusses the thermodynamics of pressure solution; Katz (1968) discusses the role of polygonization (or dynamic) recrystallization and the resulting reduction in grain size on the production of gneisses and mylonites; Higgins (1971) describes cataclastic rocks (with a good annotated bibliography), though he, like Stauffer (Paper 5), overestimates the role of brittle grain behavior and underestimates the role of pressure solution and dynamic recrystallization; and Etheridge and Wilkie (1979) discuss the role of dynamic recrystallization on diffusion flow and grain-boundary sliding.

A little-known fracturing process called "stress corrosion cracking" (Anderson and Grew, 1977) seems to have considerable significance in causing grain-size reduction at low states of deformation and is likely important in the production of mylonites and other cataclastic rocks (Kerrich et al., 1981).

Considerable work has been done on deformation by intracrystalline plastic flow (dislocation flow or creep), e.g. the production of deformation lamellae (twin lamellae, slip planes, microscopic kink bands, etc.) and some types of preferred crystallographic orientation. Most of our understanding of these features comes from experimental rock deformation, metallurgy, and electron microscopy of natural materials, all of which are beyond the scope of this book. However, one paper included here (Paper 8, Carter and Raleigh) outlines the experimental production of structures caused by intracrystalline flowage and discusses some aspects of their geological interpretation. For a fairly detailed overview of intracrystalline plastic flow, I recommend the following papers on experimental deformation: Turner et al. (1954), Carter et al. (1964), Raleigh and Paterson (1965), Heard and Carter (1968), Carter and Ave'Lallement (1970), Ave'Lallement and Carter (1971), Tullis et al. (1973), Willaime and Gandais (1977), Willaime et al. (1979), and Marshall and McLaren (1977), and the following field-based studies: Carter and Friedman (1965), Nicolas et al. (1971), White (1973), and Debat et al. (1978). Papers by Ave'Lallement and Carter (1970) and Kirby and Raleigh (1973) make attempts to apply what is known about plastic deformation to flow in the mantle and its control in plate tectonics. The rheological properties of the upper mantle as determined from the plastic characteristics of olivine is discussed by Ranalli (1980).

Three papers reprinted here discuss the chronological interpretation of deformation textures and metamorphic recrystallization. Spry (Paper 9) and Ferguson and Harte (Paper 10) deal mainly with the interpretation of porphyroblasts, whereas Misch (Paper 11) discusses microboudinage and microfolding of metamorphic mineral grains. The reader is especially invited to read also Becke (1924), Zwart (1963), Wilson (1971), Zwart and Calon (1977), Vernon and Flood (1979), and Ferguson et al. (1980).

Although it is well known that strain and metamorphic recrystallization are intimately related, little work has been done on how crystallization of new mineral phases affects the strength of rock. In one of the few papers on this topic, White and Knipe (1978) conclude that the ductility of a rock will be increased during metamorphic reactions and mineral phase changes, probably due to a loss of cohesion across grain boundaries, increased lattice defects, and more rapid recovery of the strained crystal lattice. Also, various deformation processes result in the differential migration and diffusion of elements, other than the pressure-solution effects mentioned earlier (Gratier, 1979; Kerrich et al., 1980). These result in a number of chemico-structural features in rocks including some types of veins, segregation bands in deformed crystals, metasomatic gneissic layering, and some types of migmatitic layering (Ramsay, 1980; Hanmer, 1981; Robin, 1979).

Unfortunately, space limitations also make it necessary to exclude papers discussing the important topic of crystallographic lattice reorientation through crystal deformation processes and recrystallization in deformed rocks. A general understanding of modern work can be obtained, however, from the following papers: Kamb (1958), Hara et al. (1973), Wilson (1975), Marjoribanks (1976), Lister et al. (1978), Lister and Paterson (1979), Lister and Williams (1979), and Lister and Hobbs (1980).

Rotated phenocrysts (e.g., garnets) have received considerable attention in the literature though no papers are included here. Especially recommended are papers by Rosenfeld (1970), Dixon (1976), Gosh and Ramberg (1978), Schoneveld (1977), and Schwerdtner (1979). Papers by Ramberg (1955), Stromgard (1973), Gosh (1975), and Gosh and Ramberg (1976) provide valuable discussions of textures related to rigid and brittle bodies enclosed in a ductile matrix (including boudinage).

REFERENCES

Anderson, O. L., and P. C. Grew, 1977, Stress Corrosion Theory of Crack Propagation with Application to Geophysics, *Rev. Geophys. Space Phys.* **15:**77–104.

Avé'Lallement, H. G., and N. L. Carter, 1970, Syntectonic Recrystallization of Olivine and Modes of Flow in the Upper Mantle, *Geol. Soc. America Bull.* **81:**2203–2220.

Avé'Lallement, H. G., and N. L. Carter, 1971, Pressure Dependence of Quartz Deformation Lamellae Orientations, *Am. Jour. Sci.* **270:**218–235.

Becke, F., 1924, Struktur und Klüftung, *Fortschr. Mineralogie, Kristallographie und Petrographie* **9:**185–220.

Carter, N. L., and H. G. Avé'Lallement, 1970, High Temperature Flow of Dunite and Peridotite, *Geol. Soc. America Bull.* **81:**2181–2202.

Carter, N. L., and M. Friedman, 1965, Dynamic Analysis of Deformed Quartz and Calcite from the Dry Creek Ridge Anticline, Montana, *Am. Jour. Sci.* **263:**747–785.

Carter, N. L., J. M. Christie, and D. T. Griggs, 1964, Experimental Deformation and Recrystallization of Quartz, *Jour. Geology* **72:**687–733.

Debat, P., J. C. Soula, L. Kubin, and J. L. Vidal, 1978, Optical Studies of Natural Deformation Microstructures in Feldspars (Gneiss and Pegmatites from Occitania, Southern France), *Lithos* **9:**132–145.

Dixon, J. M., 1976, Apparent "Double Rotation" of Porphyroblasts During a Single Progressive Deformation, *Tectonophysics* **34:**101–115.

Etheridge, M. A., and J. C. Wilkie, 1979, Grain Size Reduction, Grain Boundary Sliding and the Flow Strength of Mylonites, *Tectonophysics* **58:**159–178.

Ferguson, C. C., P. K. Harvey, and G. E. Lloyd, 1980, On the Mechanical Interaction between a Growing Porphyroblast and Its Surrounding Matrix, *Contr. Mineralogy and Petrology* **75:**339-352.

Gosh, S. K., 1975, Distortion of Planar Structures around Rigid Spherical Bodies, *Tectonophysics* **28:**185-208.

Gosh, S. K., and H. Ramberg, 1976, Reorientation of Inclusions by Combination of Pure Shear and Simple Shear, *Tectonophysics* **34:**1-70.

Gosh, S. K., and H. Ramberg, 1978, Reversal of the Spiral Direction of Inclusion-Trails in Paratectonic Porphyroblasts, *Tectonophysics* **51:**83-97.

Gratier, J. P., 1979, Mise en évidence de relations entre changement de composition chimique et intensité de deformation dans les roches à schistosité, *Geol. Soc. France Bull.* ser. 21, **7:**95-104.

Hanmer, S. K., 1981, Segregation Bands in Plagioclase: Non-Dilational En-echelon Quartz Veins formed by Strain Enhanced Diffusion, *Tectonophysics* **79:**T53-T61.

Hara, I., K. Takeda, and T. Kimura, 1973, Preferred Lattice Orientation of Quartz in Shear Deformation, *Hiroshima Univ. Jour. Sci.*, s.c. Geol., ser. C, **77:**1-10.

Heard, H. C., and N. L. Carter, 1968, Experimentally Induced "Natural" Intragranular Flow in Quartz and Quartzite, *Am. Jour. Sci.* **266:**1-42.

Higgins, M. W., 1971, Cataclastic Rocks, *U.S. Geol. Survey Prof. Paper 687,* 97p.

Kamb, W. B., 1958, Theory of Preferred Crystal Orientation Developed by Crystallization Under Stress, *Jour. Geology* **67:**153-170.

Katz, M., 1968, The Fabric of the Granulites of Mont Tremblant Park, Quebec, *Canadian Jour. Earth Sci.* **5:**801-812.

Kerrich, R., 1978, An Historical Review and Synthesis of Research on Pressure Solution, *Zentralbl. Geologie Palaeontologie,* Teil 1, **5/6:**512-550.

Kerrich, R., R. D. Beckinsale, and J. J. Durham, 1977, The Transition Between Deformation Regimes Dominated by Intercrystalline Diffusion and Intracrystalline Creep Evaluated by Oxygen Isotope Thermometry, *Tectonophysics* **38:**241-257.

Kerrich, R., I. Allison, R. L. Barnett, S. Moss, and J. Starkey, 1980, Microstructural and Chemical Transformation Accompanying Deformation of Granite in a Shear Zone at Mieville, Switzerland: With Implications for Stress Corrosion Cracking and Superplastic Flow, *Contr. Mineralogy Petrology* **73:**221-242.

Kerrich, R., T. E. La Tour, and R. L. Barnett, 1981, Mineral Reactions Participating in Intragranular Fracture Propagation: Implications for Stress Corrosion Cracking, *Jour. Structural Geol.* **3:**77-88.

Kirby, S. H., and C. B. Raleigh, 1973, Mechanisms of High-Temperature, Solid-State Flow in Minerals and Ceramics and Their Bearing on the Creep Behaviour of the Mantle, *Tectonophysics* **19:**165-194.

Lapworth, C., 1885, The Highland Controversy in British Geology; Its Causes, Course, and Consequences, *Nature* **32:**558-559.

Lister, G. S., and B. E. Hobbs, 1980, The Simulation of Fabric Development During Plastic Deformation and Its Application to Quartzite: The Role of Deformation History, *Jour. Structural Geol.* **2:**355-370.

Lister, G. S., and M. S. Paterson, 1979, The Simulation of Fabric Development During Plastic Deformation and Its Application to Quartzite: Fabric Transitions, *Jour. Structural Geol.* **1:**99-115.

Lister, G. S., and P. F. Williams, 1979, Fabric Development in Shear Zones: Theoretical Controls and Observed Phenomena, *Jour. Structural Geol.* **1:**283-298.

Lister, G. S., M. S. Paterson, and B. E. Hobbs, 1978, The Simulation of Fabric Development in Plastic Deformation and Its Application to Quartzite: The Model, *Tectonophysics* **45:**107-158.

Marjoribanks, R. W., 1976, The Relationship between Microfabric and Strain in a Progressively Deformed Quartzite Sequence from Central Australia, *Tectonophysics* **24:**31-67.

Marshall, D. B., and A. C. McLaren, 1977, The Direct Observation and Analysis of Dislocations in Experimentally Deformed Plagioclase Feldspars, *Jour. Mater. Sci.* **12:**843-903.

Nicolas, A., J. L. Bouchez, F. Boudier, and J. C. Mercier, 1971, Textures, Structures and Fabrics due to Solid State Flow in some European Lherzolites, *Tectonophysics* **12:**55-86.

Raleigh, C. B., and M. S. Paterson, 1965, Experimental Deformation of Serpentine and Its Tectonic Implications, *Jour. Geophys. Research* **70:**3965-3985.

Ramberg, H., 1955, Natural and Experimental Boudinage and Pinch-and-Swell Structures, *Jour. Geology* **63:**512-526.

Ramsay, J. G., 1980, The Crack-Seal Mechanism of Rock Deformation, *Nature* **284:**135-139.

Ranalli, G., 1980, Rheological Properties of the Upper Mantle in Canada from Olivine Microrheology, *Canadian Jour. Earth Sci.* **17:**1499-1505.

Robin, P. Y. F., 1978, Pressure Solution at Grain-to-Grain Contacts, *Geochim. et Cosmochim Acta* **42:**1383-1389.

Rosenfeld, J. L., 1970, Rotated Garnets in Metamorphic Rocks, *Geol. Soc. America Spec. Paper 129,* 105p.

Schoneveld, C., 1977, A Study of Some Typical Inclusion Patterns in Strongly Paracrystalline-Rotated Garnets, *Tectonophysics* **39:**453-472.

Schwerdtner, W. M., 1979, Natural Indicators of Solid-Body Rotation in Deformed Rocks, *Tectonophysics* **53:**T15-T20.

Sorby, H. C., 1853, On the Origin of Slaty Cleavage, *Edinburgh New Philos. Jour.* **55:**137-148.

Stromgard, K., 1973, Stress Distribution During Formation of Boudinage and Pressure Shadows, *Tectonophysics* **16:**215-248.

Tullis, J., J. M. Christie, and D. T. Griggs, 1973, Microstructures and Preferred Orientations of Experimentally Deformed Quartzites, *Geol. Soc. America Bull.* **87:**745-753.

Turner, F. J., D. T. Griggs, and H. Heard, 1954, Experimental Deformation of Calcite Crystals, *Geol. Soc. America Bull.* **65:**883-934.

Van Hise, C. R., 1898*a*, Metamorphism of Rocks and Rock Flowage, *Geol. Soc. America Bull.* **9:**269-328.

Van Hise, C. R., 1898*b*, Metamorphism of Rocks and Rock Flowage, *Am. Jour. Sci.,* ser. 4, **6:**75-91.

Vernon, R. H., and R. H. Flood, 1979, Microstructural Evidence of Time-Relationships between Metamorphism and Deformation in the Metasedimentary Sequence of the Northern Hill End Trough, New South Wales, Australia, *Tectonophysics* **58:**127-137.

White, S., 1973, The Dislocation Structures Responsible for the Optical Effects in Some Naturally Deformed Quartz, *Jour. Mater. Sci.* **8:**490-499.

White, S. H., and R. J. Knipe, 1978, Transformation and Reaction Enhanced Ductility in Rocks, *Geol. Soc. (London) Jour.* **135:**513–516.
Willaime, C., and M. Gandais, 1977, Electron Microscope Study of Plastic Defects in Experimentally Deformed Alkali Feldspars, *Soc. Francaise Minéralogie et Cristallographie Bull.* **100:**263–271.
Willaime, C., J. M. Christie, and M. P. Kovacs, 1979, Experimental Deformation of K-feldspar Single Crystals, *Bull. Mineral.* **102:**168–179.
Wilson, C. J. L., 1975, Preferred Orientation in Quartz Ribbon Mylonites, *Geol. Soc. America Bull.* **86:**968–974.
Wilson, M. R., 1971, On Syntectonic Porphyroblast Growth, *Tectonophysics* **11:**239–260.
Zwart, H. J., 1963, On the Determination of Polymetamorphic Mineral Associations, and Its Application to the Bosost Area (Central Pyrenees), *Geol. Rundschau* **52:**38–65.
Zwart, H. J., and T. J. Calon, 1977, Chloritoid Crystals from Curaglia; Growth During Flattening or Pushing Aside?, *Tectonophysics* **39:**477–486.

5

Copyright ©1970 by the National Research Council of Canada
Reprinted from *Canadian Jour. Earth Sci.* **7:**498–511 (1970)

Deformation textures in tectonites[1]

MEL R. STAUFFER
Department of Geological Sciences, University of Saskatchewan, Saskatoon, Saskatchewan

Received May 28, 1969
Accepted for publication August 14, 1969

A variety of deformation textures may be present in low temperature tectonites that have not been modified extensively by post-deformational recrystallization. These include gross grain shape, granular, breccia, pull-apart, strain-shadow, goatee, and shear-step textures. A study of these textures will reveal the behavior of the rock during deformation, specifically: (*a*) whether deformation took place under conditions of high or low intergranular cohesion, and (*b*) the relative importance of crystallographic gliding and intergranular movement between grains.

No study of deformed rocks is complete without a strain analysis. Strain analyses should include grain shapes, as well as description and interpretation of the various deformation textures.

Introduction

It has long been recognized that permanent strain can take place at the granular[2] scale by a number of different processes (macroscopic fracturing excluded). The types of processes considered most important are: brittle behavior of individual granules; intracrystalline dislocations[3] (principally crystallographic gliding); intergranular dislocations (differential movement between granules); and recrystallization (Fig. 1). A survey of theoretical and experimental work on rock and metal deformation (e.g. Barrett 1952; McLean 1957; and Turner and Weiss 1963) suggests to this author that these processes can be subdivided in the following manner (Table I).

The operation of these processes modifies and may completely obliterate the original texture of rocks. The resulting modified and superimposed textures are generally called *metamorphic textures*; these may be subdivided into *deformation textures*[4] (those caused by processes *a–h*, Table I) and *recrystallization textures* (processes *i–l*, Table I).

The recognition and interpretation of *deformation textures* is as important to understanding metamorphic rocks as the study of sedimentary and igneous textures is to understanding rock forming processes. In the author's opinion, deformatoin textures (particularly those resulting from intergranular dislocations) have not received the study they deserve. No existing textbook in structural geology or petrology contains an adequate description of deformation textures and only a few authors have discussed the relevance of such textures in understanding the processes of deformation (e.g. Cloos 1947; Zwart 1963; and Stauffer 1967). Most of the work done on naturally deformed rocks has been concerned with: (*a*) determining the relationship between the times of deformation and neo-mineralization (e.g. Spry 1963; Zwart 1963; Misch 1969); (*b*) the analysis of grain shape (e.g. Cloos 1947; Flinn 1956; Hossack 1968); and (*c*) the analysis of optical orientation, twinning and deformation lamellae (e.g. Nissen 1962; Friedman and Conger 1964; Carter and Friedman 1965). The purpose of this paper is to discuss the origin of various deformation textures and to demonstrate a method of interpreting deformed rocks.

[1]Presented at the Second Canadian Conference on Tectonics, held at McMaster University, Hamilton, Ontario, March 10–11, 1969.

[2]In this paper, the term *grain* will refer to a single crystal or fragment thereof. The term *composite-grain* will be used for distinct particles composed of two or more grains with stronger cohesion between themselves than with the surrounding grains or composite-grains. The term *granule* will be used as a general term to include both grains and composite grains, regardless of size.

[3]The term *dislocation* is used here in the usual geological way to mean *movement*, and does not refer to a crystal lattice defect as in solid state terminology.

[4]Throughout the remainder of this paper the term *deformation* will be restricted to include only those phenomena *not* listed under recrystallization in Table I.

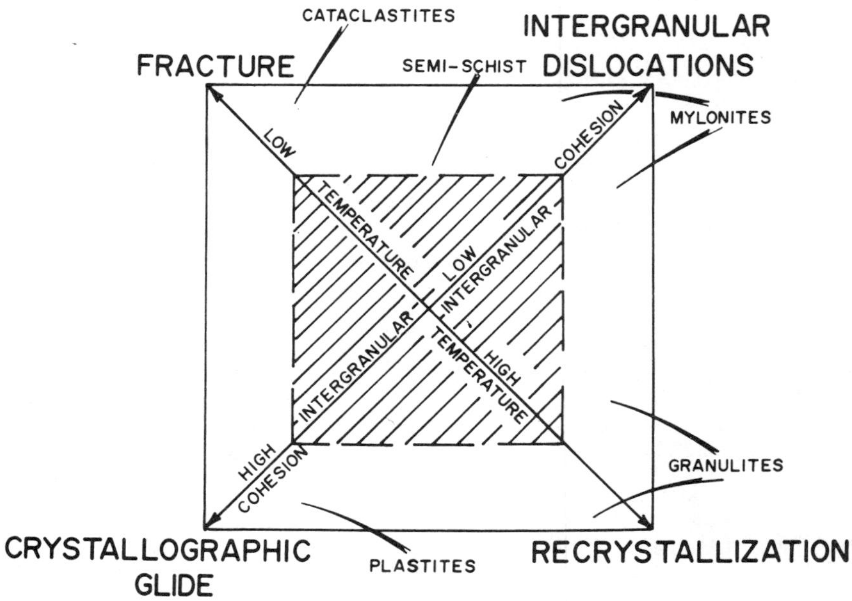

FIG. 1. Diagram showing the various types of deformational processes. A line connecting two different deformational processes (e.g. the line between fracture and intergranular movement or between crystallographic glide and intergranular movement) is scaled so that the deformation at each end has *no* contribution by the opposing process but is totally by the process labelling that end of the line. The midpoint of the line represents equal contribution by both processes.

The effects of temperature and intergranular cohesion are indicated by the diagonal lines. Fracture is primarily a low temperature feature and recrystallization is most important at high temperatures, regardless of the degree of intergranular cohesion. Intergranular movements occur best under conditions of low cohesion, whereas crystallographic gliding is more important if the cohesion is high (the effects of temperature are unknown, but probably crystallographic glide is enhanced over intergranular dislocations at higher temperatures). Other factors of the deformational environment will also affect the processes. High confining pressure will favor crystallographic glide and/or recrystallization. High water vapor pressure will favor fracture, intergranular dislocations and/or recrystallization, depending on temperature. High rates of strain will favor fracture and intergranular dislocations.

The area of diagonal shading indicates the deformational field for most tectonites. Slaty cleavages will be formed in the low temperature part of this area, grading to schistosity and gneissosity in the higher temperature parts. Various types of rocks (e.g. mylonite), which are the result of rather special deformational situations, are indicated.

Deformation Textures and their Origins

General

Deformation textures are best observed in rocks that have a considerable variation in granule size and which have not undergone significant post-deformational recrystallization. The larger granules in any rock tend to display deformation textures better than do the smaller granules. This is partly a function of ease of observation, but is probably also due to somewhat differential behavior of adjacent granules of different size. The deformation textures present in a rock will depend mainly on the relative importance of crystallographic gliding, granule rupture, and differential movement between granules. For any particular deformational environment (temperature, confining pressure, water vapor pressure, and strain rate conditions) this is determined by the ductility and brittility of the grains and the intergranular cohesion. Many of the textures seen in granules that have undergone deformation are due to recrystallization and not deformation *per se*; those recrystallization textures that are a direct

TABLE I

Types of deformation observable at the granular scale

Brittle behavior	(a)—granule boundary fracturing (b)—fracturing along crystallographically controlled planes (c)—fracturing along non-crystallographically controlled planes
Intracrystalline dislocations	(d)—translation gliding, (e)—twin gliding } crystallographic gliding (f)—kink banding on the crystallographic scale
Intergranular dislocations	(g)—movement between granules (including rotation) (granules may be rigid to brittle and slip over one another in a "brittle" manner or in the case of single crystal fragments, there may be a zone of slippage across which there is little loss of cohesion (grain boundary sliding in metallurgists terminology)) (h)—separation of newly formed subgrains (plucking, boudinage, differential rotation)
Recrystallization	(i)—polygonization (strain recovery involving a realignment of the previously distorted crystal lattice, this causes a distorted grain to be reduced to several smaller subgrains of slightly differing orientation without affecting the overall shape of the distorted host grain) (j)—neo-mineralization (k)—crystal growth of previously existing mineral species (syn-tectonic annealing) (l)—pressure-solution effects

result of mechanical deformation are included in the following discussion.

For convenience, textures are classified according to the importance of intergranular dislocations.

Textures Developed by Deformation with no significant Intergranular Dislocations

Plastite Texture

Rocks which have deformed totally, or almost totally, through crystallographic gliding within the component grains are here given the name *plastites* (see Fig. 1) to help separate them from rocks deformed by other processes (e.g. mylonites or granulites). The ideal rock to become a plastite would be a monomineralic crystalline rock (such as salt or marble), or a well-cemented monomineralic clastic rock (such as quartzite or some clastic limestones).

The details of crystallographic gliding are fairly well known for some minerals (e.g. Turner and Weiss 1963, pp. 335–360; Christie *et al.* 1964; Raleigh and Talbot 1967). Originally spherical grains deformed in this fashion will become ellipsoidal and euhedral grains take on new shapes that will approach ellipsoids depending on the original shape and the amount of deformation (Fig. 2*a–c*). An important feature of the texture resulting from this type of deformation is the lack of evidence of granulation along the grain margins.

In many rocks that have undergone this form of deformation there is evidence of brittle failure (either extensional or shear) of grains at a late stage in the deformation. If movement across such fractures is large, then considerable amounts of intergranular displacements within the matrix have occurred (e.g. Fig. 2*i*, *j*, *q*, and *r*). In some minerals, the fractures are crystallographically controlled (e.g. along cleavage planes in feldspar or calcite). Generally, crystallographically controlled fractures tend to be straighter than those not so controlled. Also, features such as undulatory extinction, deformation bands and deformation lamellae may be formed as a result of crystallographically controlled deformation (e.g. Christie *et al.* 1964, and Heard and Carter 1968).

Textures Developed by Deformation involving considerable Intergranular Dislocations

There are several different textures that form as a result of differential movement between granules (Fig. 2*d–k*) depending on the details of the individual granule behavior and on the intergranular cohesion.

Plucked-boundary Texture

This texture is clearly the result of matrix material moving past the granule in question (Figs. 2*d*, *e*, and 3*d*, *h*, *k*, *l*, *m*, and *n*). It may be preceded by a general fracturing of the granule rim (Figs. 2*m* and 3*s*) or may develop by fracture and granule removal as a continuous process (a sandpaper-like process). The plucked subgranules may remain identifiable, but in most rocks studied by the author they appear to have been ground down very small and thoroughly intermixed with the matrix. The margins of some plucked granules (presumably brittle and incompetent) are irregularly pitted (Fig. 3*h*, *j*, and *m*). Other granules (presumably brittle but strong) are

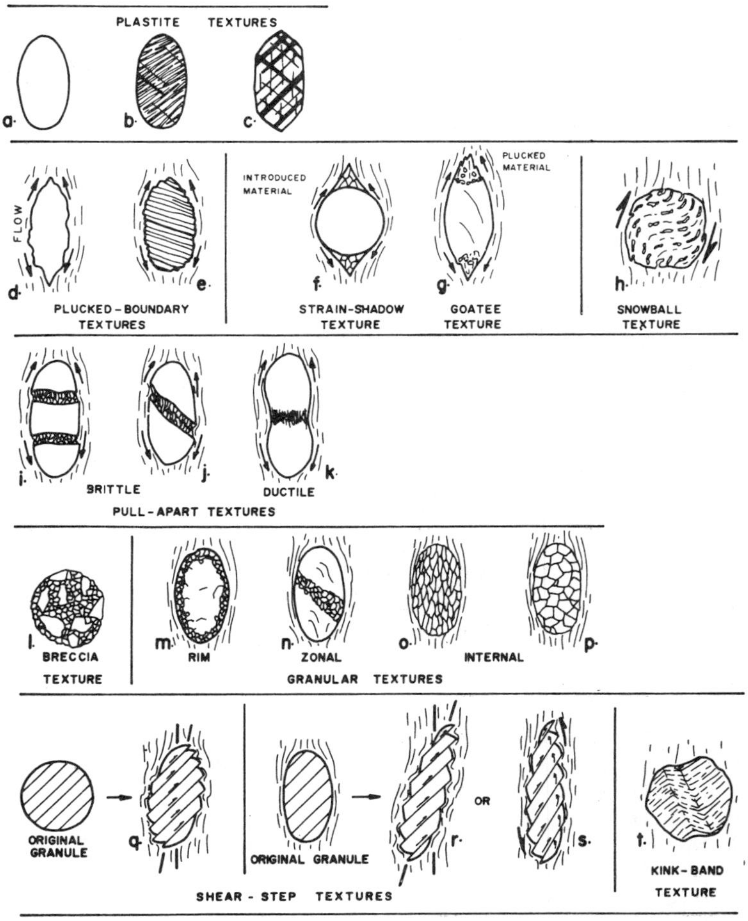

FIG. 2. Diagrammatic sketches and classification of deformation textures. The original grain is assumed to be spherical or equidimensional except where otherwise indicated.
a–c. Plastic textures, (*a*) deformation ellipse, (*b*) deformation lamellae, (*c*) twinned zones in deformed euhedral grain.
d–h. Textures resulting from intergranular movements under conditions of low to moderate intergranular cohesion. Crystallographic gliding and/or recrystallizations may occur concomitantly (e.g. *g* and *h*).
i–k. Pull-apart textures representing intergranular movements under conditions of moderate intergranular cohesion.
l. Breccia texture. Fracture and granulation resulting in a lowering of the intergranular cohesion.
m–p. Granular textures. Somewhat similar looking features may be formed by fracture or polygonization, and these may be followed by subgrain boundary adjustments (annealing). Whatever the details of their origin, granular textures cannot be used to determine the importance of intergranular movements.
q–s. Shear-step textures, showing the oblique relationships that will develop between granule shape and foliation (flattening plane), unless rotation occurs either after or during shear. Some intergranular movements must occur, but they may be minor.
t. Kink-band texture. Most commonly seen by this author in rocks that have undergone only slight deformation (<30% flattening). Some intergranular movement is required but it may be minor.
Deflected groundmass foliations are shown in most of the above examples and although this is a result of differential deformation behavior of the matrix and strain marker it probably does not necessarily indicate major intergranular movements.

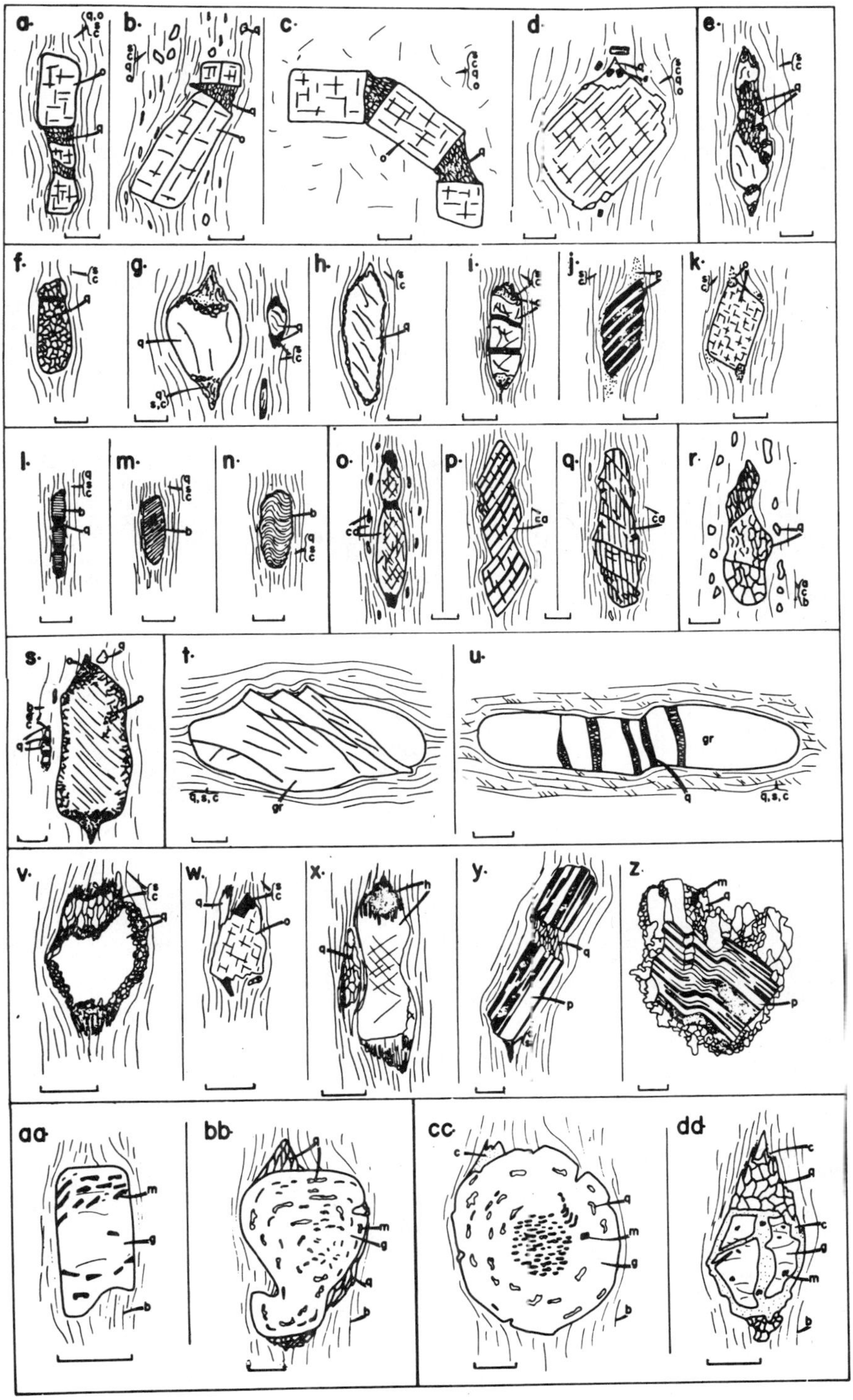
a.
b.
c.
d.
e.
f.
g.
h.
i.
j.
k.
l.
m.
n.
o.
p.
q.
r.
s.
t.
u.
v.
w.
x.
y.
z.
aa
bb
cc
dd
gr
q, s, c

FIG. 3. *a–c.* Pulled apart orthoclase phenocrysts from a Silurian porphyry dike in New South Wales, Australia. Sketches drawn from thin sections; *a* and *b* are perpendicular to foliation, *c* is parallel. The gaps (or necks) are filled with quartz (q) and the matrix consists mainly of small elongate quartz grains, equidimensional orthoclase grains (o), and minor sericite (s) and chlorite (*c*). In *b* and *c* one or more fragments have been rotated. Scale bar is 0.3 in. (.76 cm).
d. Orthoclase phenocryst from same rock type as *a–c*. Sketch drawn from thin section perpendicular to foliation. This phenocryst displays the effects of boundary plucking plus combined strain-shadow and goatee texture. The strain shadows contain quartz (q) and plucked orthoclase fragments (o). Scale bar is 0.3 in. (.76 cm) long.
e–i. Quartz phenocrysts from Silurian crystal tuff, New South Wales, Australia. *e* and *i* display somewhat different forms of pull-apart texture; in both, the gaps are filled with quartz (q), but in *i* they also contain chlorite. The goatees contain new-grown quartz as well as plucked quartz fragments. *f* displays internal-granular texture and is boudinaged. *g* shows well developed goatee texture with most of the material apparently plucked from the phenocrysts, although the goatees contain some chlorite and sericite. *h* displays a plucked boundary that has recrystallized after the deformation (the rim has a constant optic orientation that is different from that of the phenocryst itself and is strain free). The groundmass in this rock consists mainly of sericite, chlorite, and very small grains of quartz and feldspar. The quartz phenocrysts all display strain shadows. Sketches drawn from thin sections perpendicular to foliation. Scale bar is 0.3 in. (.76 cm) long.
j–k. Feldspar phenocrysts (plagioclase (p) and orthoclase (o)) from the same rock as *e–i*. These show plucked boundary and goatee textures with small fragments of plucked feldspar. From thin sections perpendicular to foliation. Scale bar is 0.3 in. (.76 cm) long.
l–n. Biotite (b) phenocrysts from same rock as *e–k*. *l* is pulled apart, *n* is folded and *l*, *m*, and *n* show ragged plucked boundaries. From thin sections perpendicular to foliation.
o–q. Calcite phenoclasts from limestone in the Tipperary region, Ireland. The groundmass consists of small, ellipsoidal calcite grains with some chert. *o* displays ductile pull-apart and goatee textures, the overall grain shape is probably due to a combination of crystallographic gliding and boundary milling. The neck of the pull-apart is highly strained, but does not appear to contain introduced material. *p* and *q* display shear-step texture developed on grains that were either initially elongate or became elongate during the shear faulting, but owe their elongation to some other process. The faulting is along twin planes in *p*, but there is no obvious crystallographic control of the faults in *q*. Drawn from thin sections perpendicular to foliation. Scale bar is $^1/_{10}$ in. (.25 cm) long.
r–s. K-feldspar and quartz phenocrysts in a stream pebble of Shuswap rock (gneiss) from south-central British Columbia. Quartz is either in angular, generally elongate, fragments or is pulled apart, with or without goatee texture. *r* shows granular texture, with a variety of subgrain shapes and sizes. The large K-feldspar phenocryst in *s* displays a fractured granular-rim texture. The groundmass consists of quartz, amphibole (a), biotite, feldspar, and some chlorite. Drawn from thin sections and a polished surface cut perpendicular to foliation. Scale bar is 0.2 in. (.50 cm) long.
t–u. Deformed granite pebbles (gr) from the Precambrian Missi Group, Saskatchewan. The shapes of quartz grains within the pebbles are approximately the same as the pebbles themselves. *t* displays shear-step texture and *u* is boudinaged; the gaps are filled with quartz. *u* also shows the weak development of a second foliation. The matrix is fine-grained, schistose graywacke. Drawn from photographs approximately perpendicular to foliation. Scale bar is 1 in. (2.5 cm) long.
v–x. Precambrian Missi Group pebbly schist, Saskatchewan. *v* shows a small quartz pebble with well developed granular-rim texture, a pull-apart feature (with a granular pulled-off fragment), and strain shadows containing new-grown chlorite and plucked fragments of quartz. The grain itself has angular zones of undulatory extinction the same size and shape as the subgrains in the pulled-off fragment. *w* shows a small orthoclase pebble (o) showing plucked-boundary and strain-shadow textures. The strain shadows contain chlorite. There is a smaller elongate quartz grain on the left-hand side, also with a chlorite-filled strain shadow. *x* shows a partly boudinaged hornblende grain (h). The pull-apart zone contains numerous shear fractures parallel to the crystallographic cleavage. The lower end of the grain is new-grown hornblende (strain shadow) with different pleochroism than the host grain and containing a faint textural pattern parallel to the external foliation. The upper end is a mosaic of chlorite, quartz, and fragments of plucked hornblende. Note the granular-textured quartz grain on the left-hand side, the left rim of which has undergone post-deformational recrystallization. Drawn from thin sections perpendicular to foliation. Scale bar is 0.1 in. (.25 cm) long.
y. Pulled-apart plagioclase grain oblique to foliation, from a Precambrian schistose granodiorite, Saskatchewan. The strain shadow on the bottom contains chlorite and sericite. The groundmass is mainly small grains of elongate quartz and feldspar with chlorite, sericite, and altered biotite and hornblende. Drawn from thin section perpendicular to foliation. Scale bar is 0.2 in. (.50 cm) long.
z. Kink-folded plagioclase in a Precambrian, breccia-textured quartz diorite, Saskatchewan. The sutured contacts probably indicate that a slight amount of post-brecciation recrystallization occurred. The kink bands are approximately parallel to the crudely developed foliation. The black angular grain is magnetite (m). Scale bar is 0.1 in. (.25 cm) long.
aa and *bb.* Garnet porphyroblasts (g) from the Precambrian Kisseynew Gneiss, Saskatchewan. *bb* shows strain shadows as well as inclusions of quartz and magnetite parallel to the crystal boundaries. In *aa* the garnet grew either parallel to the foliation or, more likely, has been milled down during or after growth, with the result that the inclusions no longer are parallel to the sides of the grain. All the garnets in this particular sample of Kisseynew Gneiss are rounded or elongate parallel to the biotite foliation, whereas in many places they are nearly perfect euhedral crystals. Scale bar is 0.1 in. (.25 cm) long.
cc. Garnet porphyroblast from Precambrian, garnetiferous biotite schist from the Mukteswar region in India. This shows garnet-in-garnet texture; the inside garnet contains inclusions of quartz that were probably originally parallel to an early foliation, but have been rotated either during or after the foliation development. The outer rim of the garnet probably grew under low stress conditions and as a result the inclusions parallel the crystal boundaries. Syn- to post-growth deformation has resulted in some deflection of the foliation. Scale bar is 0.1 in. (.25 cm) long.
dd. Same rock as *cc*, but from a different locality. This garnet probably correlates with the outer garnet of the garnet-in-garnet porphyroblasts and contains only scattered inclusions of magnetite. The groundmass has undergone considerable post garnet-growth deformation to form strain shadows and fracture the garnet (retrograde chlorite growth also). Scale bar is 0.1 in. (.25 cm) long.

smooth with rounded corners and original irregularities largely ground off (Fig. 3*k* and possibly *aa*). Many granules take on a lenticular shape (as opposed to an ellipsoidal shape) as a result of boundary plucking (Figs. 2*d* and 3*h*).

This texture probably represents a moderate cohesion between the strain marker and matrix granules (compared with plastite texture which represents a high intergranular cohesion).

Strain-shadow and Goatee Textures

Strain shadows (or pressure shadows) are fairly well-known textures (e.g. Pabst 1931); they form where extremely competent granules are surrounded by a matrix that has flowed largely by differential movement between granules. A gap is created around the margins of the competent granule in a plane perpendicular to the direction of flattening (Fig. 2*f*). This gap fills with precipitated material, commonly quartz, chlorite, or epidote.

Goatee texture (Fig. 2*g*) results when material is plucked from the ends of the competent granule and incorporated into the strain shadow. Every gradation may be found from *strain-shadows* composed entirely of precipitated material to *goatees* composed entirely of deformed original material mixed with plucked fragments (Fig. 3*e*, *g*, *i*, *j*, *k*, *o*, *s*, *v*, *w*, *x*, *bb*, and *dd*). In the latter case the intergranular cohesion was probably high and the granules concerned fairly brittle. Some goatees are concave towards the granule (Fig. 3*g*). This may indicate that rupture at the end of the granule was partly the result of stretching, due to a change of curvature through elastic flattening of the granule.

Microboudinage (pull-apart) Texture

This is a very obvious deformation texture and has been described numerous times (e.g. Johnson 1967; Stauffer 1967; and Misch 1968). The separation during microboudinage may vary from brittle to very ductile (Figs. 2*i*–*k*; and 3*a*–*c*, *e*, *i*, *l*, *o*, *u*, *v*, and *y*), in either case the most significant process is movement within the matrix material under conditions such that a moderate to high intergranular cohesion was maintained between the matrix and strain marker (though not necessarily between the matrix granules). Microboudinage may be accompanied or preceded by changes in grain shape as a result of crystallographic gliding or plucking from the granule boundaries (Figs. 2*i*–*k*; and 3*i*, *l*, *o*, and *u*), and may occur during syn-tectonic mineral growth (Misch 1968).

The brittle forms of microboudinage may occur by separation perpendicular to extension (tension?) cracks or oblique to shear fractures (Figs. 2*i* and *j*; and 3*a*, *e*, *f*, *i*, *l*, *o*, and *u*) and the gap fills with precipitated or recrystallized material derived from either the boudinaged granule (Johnson 1967) or other sources. In some grains, the planes of separation may be crystallographically controlled (Fig. 3*a*–*c*, and *l*). If the boudinaged granules were originally elongate and oblique to the principal strain directions then rotation of one or more of the separated segments may occur and the separation gaps (or necks) will be triangular in shape (Fig. 3*b* and *c*). Occasionally the separation gap or fracture extends slightly into the surrounding matrix (Fig. 3*b*).

Snowball Texture

The structural and metamorphic significance of this texture (Fig. 2*h*) has been adequately discussed in the geological literature (e.g. Ramsay 1962; Spry 1963; and Rosenfeld 1968). It forms as a result of concomitant rotation and grain growth. Grains displaying this texture are usually approximately equidimensional and must have had a considerably higher yield strength than the surrounding matrix. Deformation was largely by intergranular movements and cohesion between the matrix and the growing, rotating grain was low to moderate.

Textures not Necessarily a Direct Result of Intergranular Dislocations

Breccia Texture

This texture is typically displayed by cataclastites; it is well known and needs only little elaboration here. It is characterized by angular fragments of various sizes and shapes (Figs. 2*l* and 3*z*) and there is often no alignment of elongate granules. This texture presumably forms through ruptural breakdown of the rock, the end result of which may be a rock flour or "gouge rock". Although the new, fracture-formed subgranules must slide over one another if the rock undergoes any significant amount of deformation this has not markedly affected the texture. Whatever the initial intergranular cohesion, during deformation it was low enough that friction between the granules did not cause the plucking and stretching textures described previously (also see Borg *et al.* 1960). Many rocks displaying breccia texture have been "lithified" by syn- to post-deformation hydrothermal precipitation.

Granular-subgrain Textures

There is a variety of granular-subgrain textures, ranging from granular rims to granular zones to the entire grain consisting of a mosaic of subgrains (Figs. 2*m–p*; and 3*f* and *r*). The formation of subgrains has been variously described in the geological literature as due to cataclasis, polygonization, or growth of new, small, strain-free grains at the expense of the old, strained, host grain. In some rocks it may be possible to determine which of these processes occurred but in many cases it is not, especially if post-deformational grain growth took place. However, in polymineralic composite-grains, zones of granular subgrain texture are often mainly due to fracturing along grain boundaries, although other processes may have occurred as well. In many deformed rocks the subgrains have approximately the same shape as the host grain (Figs. 2*o* and 3*r*), whereas in other rocks the subgrains are more equant (Figs. 2*p* and 3*f*). The reasons for this are unknown.

Whatever the cause of the subgrains, their boundaries are often zones of weakness, and plucking and pulling-apart may occur if there is differential movement between the host grain (or composite-grain) and matrix.

Shear-step Textures

Faulting of individual granules may occur during, prior to, after, or in the absence of crystallographic gliding. If movement on the faults is large enough to result in a significant amount of strain, then the boundaries of the granules will be offset in a step-like manner (Figs. 2*q* and *r*; and 3*p*, *q*, and *t*). However, in some granules only one side is stepped (Fig. 3*t*), the strain probably being taken up by some other process on the other side.

Generally, if a granule deforms by shear on only one set of fault surfaces it will lie oblique to the plane of flattening (schistosity) unless it rotates after or concomitantly with the shear (Fig. 2*s*). However, there will be some special cases where an originally elongate granule, originally oblique to the flattening plane, will undergo just the right amount of shear displacement to become distorted into parallelism with the flattening plane without bodily rotation.

Shear-step textures probably form under conditions of fairly high intergranular cohesion with or without significant amounts of differential movement (there must be some such movement of course, or gaps would be left in the step notches).

Kink-band or Kink-fold Texture

Kink folds in individual crystal fragments are not as common as most of the other textures discussed, but do occur in some deformed rocks. For layered rocks, kink fold dynamics are fairly well known (e.g. Patterson and Weiss 1966). Kink folds in individual crystal fragments are undoubtedly analogous (e.g. Carter *et al.* 1964) and involve inhomogeneous slippage on crystallographic surfaces (such as twin planes) under conditions approaching or consistent with those necessary for crystallographic gliding.

Kink-folded grains have been noticed by the author in a wide variety of rock types (Figs. 2*t* and 3*z*). However, in most cases the rocks were weakly deformed and had undergone deformation mainly through fracturing and intergranular movements. It is possible, however, that the kinking occurred prior to most of the differential movement and that the fracturing was necessary before differential movement could occur.

Deflected Foliation

During flattening, the differential compaction of weak matrix material around a more competent granule causes the foliation to be deflected (e.g. Fig. 2*d–h*). If the granule is a porphyroblast the deformation occurred either during or after growth and was due to the applied stresses, not growth of the porphyroblast itself (see Zwart 1963, pp. 334–335). Although this texture indicates differential behavior between the matrix and the granule in question it cannot be considered indicative of significant intergranular dislocations in the absence of other features. This is because the amount of differential movement necessary is relatively small and the matrix could have deformed through any combination of crystallographic glide, recrystallization, and intergranular dislocations.

Combined Textures

Most rocks that have undergone deformation involving intergranular dislocations will contain granules showing several of the preceding textures. In some cases individual granules displaying two or more of these textures may be found (Fig. 3*d*, *e*, *f*, *i*, *j*, *k*, *l*, *m*, *n*, *o*, *s*, *v*, and *x*). Deformation involving intergranular movements does not exclude individual grains from deforming partially by crystallographic gliding processes, and in many rocks grains have apparently been flattened as well as milled down, sheared, and boudinaged (Fig. 3*e*, *i*, and *o*).

Also, deformation may occur by different processes in sequence, intermittently, or in an inhomogeneous and changing pattern throughout the rock. Such differential behavior with time would be controlled in a complex way by variations in temperature, confining pressure, water vapor pressure, mineral phases present due to the previous metamorphism, and strain rate.

In order to look at one rather simple example of a way in which such variations could occur, consider a rock with moderately high intergranular cohesion at the beginning of a deformation. The stresses will build up until the weakest surfaces undergo slip or rupture. In this case, deformation occurs by rupturing of granules and granule boundaries. The intergranular cohesion is thus weakened and intergranular movements may occur. As deformation progresses increased temperature and/or "strain hardening" through granule interlocking may facilitate crystallographic gliding. As deformation enters its waning stages the temperature will drop and the deformational process may revert to either brittle failure or intergranular movements. The resulting rock would contain a variety of deformation textures including: (*a*) pull-apart and goatee textures from the intergranular movement stage(s), (*b*) elongate grains as a result of crystallographic gliding and boundary plucking, and possibly (*c*) deformation lamellae, shear or extension fractures, possibly even shear-step texture, from the last stages of deformation.

Conversely, the same array of textures could be formed by continuous, slow, crystallographic gliding, with occasional, brief, periods of fracture and intergranular slippage when and wherever the local stress acceleration exceeded that which can be relieved by crystallographic gliding processes. Only in a case of clear-cut superposition of deformational phases could a reasonable interpretation of the sequence of deformational processes be made.

Interpretation of Deformed Granules

Three-dimensional Strain

The study of folded, faulted and internally deformed rocks should include not only a geometric, sequential analysis of folding and faulting but also analyses of the flowage processes, amounts, and directions. The analysis of flowage requires observations of granule shapes and deformation textures. As is well known, the shapes of individual granules (or the average shape of many granules) are representative of the amount of flow in different directions *only* if: (*a*) the granules were initially spherical (or originally had known shapes), (*b*) the granules had equal or less strength (but not more strength) than the rock as a whole, and (*c*) the deformation was entirely by crystallographic gliding processes (although a continued sequence of elastic strain – polygonization – strain – polygonization etc. may serve as well). If these conditions were not satisfied in a particular rock, then intergranular movement will have occurred and some combination of deformation textures indicative of such movement will be found. The relative proportions of different textures will depend on the degree of intergranular cohesion compared with the strengths of the granules and whether the granules were behaving in a brittle or ductile fashion (providing, of course, that recrystallization and neo-mineralization were relatively unimportant).

For most rock types, the individual granules will show less strain than the rock as a whole and no granule will be deformed more than the enclosing rock. Nevertheless, deformed granules that have undergone differential movement with the groundmass can be used in a general way as strain indicators. Their shapes will be indicative of the principal directions of flow and will usually give a general, though not accurate, measure of the relative amounts of flow in these principal directions (e.g. Cloos 1947; Hobbs and Talbot 1966; and Stauffer 1967).

Shear Director Indicators

Some deformation textures are caused (or modified during formation) by either different rates or opposite directions of movement of matrix material on opposite sides of the granule (Fig. 4). The shear stresses caused by such

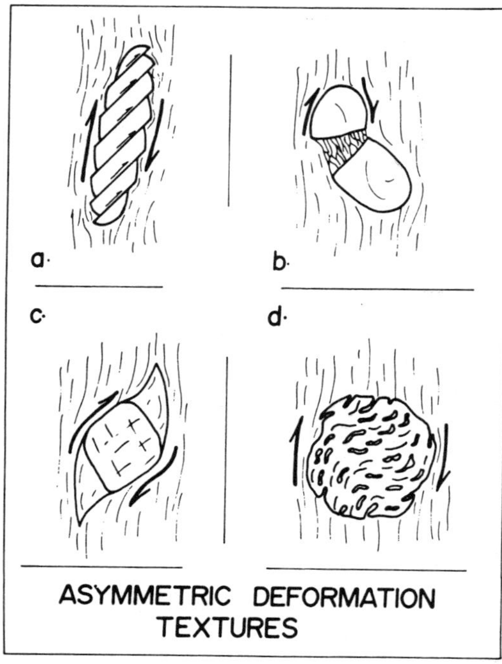

FIG. 4. Diagrammatic sketches of asymmetric deformation textures. (a) Shear-step texture, (b) rotated pull-apart texture, (c) sigmoidal strain-shadow texture, (d) snowball texture. Heavy arrow indicates sense of shear and/or rotation.

movement results in either a shearing of the granule (shear-step texture, Fig. 4*a*) or in rotation (rotated pull-apart, sigmoidal strain shadow, and snowball textures, Fig. 4*b*–*d*). Because these textures represent an asymmetry in the deformation they are here termed *asymmetric deformation textures*, all others being grouped as *symmetric deformation textures.*

Though relatively rare, asymmetric deformation textures are of special importance because the sense of rotation and/or shear can be determined and in some cases correlated with macroscopic structures (e.g. Fairbairn 1950; Ramsay 1962; and Rosenfeld 1968). In some rocks, however, individual granules have rotated or sheared in response to local stresses that were oblique to the stresses applied to the body of rock containing the granules and the asymmetry of the resulting strain cannot be correlated with macroscopic structures. This is particularly true where originally elongate granules are imbedded in a matrix that underwent flattening in a direction oblique to the granules. In this case the granules (or fragments of) will have rotated towards parallelism with the plane of flattening and different senses of rotation may have occurred for granules with different original orientations.

Interpretation of Deformational Process

Provided post-deformational recrystallization has not obliterated the deformation textures, some idea of the relative importance of crystallographic gliding versus intergranular movement can be obtained.

The deformation textures (or lack of them) shown by *all* mineral species and granule sizes must be studied and the probable relative strengths of adjacent granules considered in order to interpret the deformational processes. Intergranular movement may be indicated not only by pull-aparts, strain-shadows, goatees, deflected foliation, plucked boundaries, and snowball-texture (or other evidence of rotation), but also by such features as: round or little-deformed granules in a well-foliated matrix and differential shapes of different sized granules of the same mineral species.

It is generally considered that sharply pointed "pseudo ellipsoids" or lenticles are produced through fracturing and boundary milling of the granules, whereas true ellipsoids are produced by crystallographic gliding of originally spherical granules (e.g. Harker 1932, p. 167). This viewpoint is partly challenged here because both processes may be occurring simultaneously. Furthermore, many granules studied by the author that appear to be lenticles to the naked eye owe their pointed tapering ends to the accumulation of plucked material in the strain-shadow (goatee texture); the granules themselves may have deformed either through crystallographic gliding, boundary plucking, or some combination.

A few examples should serve to illustrate how I interpret deformational processes.

The first two examples are of Silurian igneous rocks in southeastern Australia. One rock type (quartz, feldspar, crystal tuff) probably had a low initial intergranular cohesion. The other (quartz, feldspar, porphyry dikes, which intrude various rocks including the above tuffs) had a high intergranular cohesion prior to deformation. Quartz phenocrysts in the dikes are ellipsoidal parallel to the regional slaty cleavage and oblique to the trend of the dikes; the amount of flattening varies locally from approximately 30% to 60% (Stauffer 1967). In the more strongly deformed rocks some quartz grains show granular, ductile pull-apart, and plucked boundary textures, otherwise the quartz deformation appears to have occurred primarily through crystallographic gliding. Most feldspar grains appear undeformed; some, however, show brittle pull-apart texture (Fig. 3*a–c*) although this is most common in the rocks with the most strongly flattened quartz phenocrysts. Some of the differential movement thus indicated may have occurred because the quartz (phenocrysts and matrix) was deforming whereas the feldspar was almost rigid. The fine-grained matrix displays the same features as the phenocrysts, although not so well. Deformation textures in the crystal tuff are quite different; many of the feldspar crystals appear as very elongate lenticles in hand specimen, and show well developed plucked boundary and goatee textures in thin section (Fig. 3*j–k*), the quartz varies from ellipsoids to rare lenticles, and many show internal-granular, plucked-boundary, brittle pull-apart and goatee textures (Fig. 3*e–i*). Quartz grains without these textures appear to have been flattened from 20% to 50% (Stauffer 1967). Biotite phenocrysts show a variety of deformation textures including pull-apart, plucked boundary, and kink-band textures (Fig. 3*l–n*). The matrix in the tuff has been recrystallized to fine-grained chlorite and sericite with minor, small, nearly round grains of quartz. My interpretation of these various features is that the dikes deformed primarily through crystallographic gliding within the quartz grains with some minor concomitant differential movement, whereas the crystal tuff deformed more than did the dikes (despite less flattening of the quartz phenocrysts) and this was achieved mainly through differential movement under conditions of low to moderate intergranular cohesion with some flattening of quartz grains through crystallographic gliding.

The next example is of Paleozoic limestones and marbles in the Tipperary region, Northern Ireland. There, undeformed limestones have a variety of grain sizes but, with the exception of some fossil fragments, most grains are roughly equidimensional. Those rocks, which have been strongly modified by deformation and recrystallization, have undergone a reduction in grain size with frequent relict large grains. All grains are elongate (most are approximately ellipsoidal but some are distinctly lenticular) and many of the larger ones show shear-step, strain-shadow, pull-apart, internal-granular, and plucked-boundary textures along with the usual one, two, or three sets of twin lamellae (Fig. 3*o–q*). If there is a preferred optic orientation it is not obvious under ordinary thin-section study. My interpretation of the deformational processes is that although crystallographic gliding and recrystallization occurred, differential movement was also important.

A stream boulder eroded from the Precambrian Shuswap terrain in south-central British Columbia shows several deformation textures. This boulder is of porphyroblastic gneiss in which deformation outlasted neo-mineralization. The porphyroblasts are approximately 50% pancake-shaped quartz (average about $\frac{1}{10}$ in. (0.25 cm) thick and $\frac{1}{2}$ in. (1.25 cm) in diameter) and 50% large K-feldspar grains (average about $\frac{1}{2}$ in. × 1 in. × 1 in. (1.25 cm × 2.54 cm × 2.54 cm)). The matrix consists mainly of approximately equal amounts of quartz, feldspar, biotite, and amphibole. Quartz porphyroblasts are parallel to the foliation in the matrix, wavy in outline, most contain internal-granular texture (Fig. 3*r*), some are boudinaged, and a few have goatee texture (Fig. 3*s*). The feldspar porphyroblasts are also parallel to the foliation, most display goatee texture and all have fractured rims. The crystal shape is still recognizable in most feldspar grains, although some are lenticular. Clearly, deformation involved considerable intergranular movements as well as recrystallization. Deformation of the quartz grains probably included crystallographic gliding and/or strain-polygonization sequences.

Other examples are from a group of deformed Precambrian sedimentary and igneous rocks near Flin Flon, Manitoba. Deformation textures are most clearly observed in conglomerate outcrops of the Missi Group. There are large local variations in the amount of deformation and even in areas of considerable deformation pebbles of different compositions have responded differently. However, cleavage formed of chlorite and muscovite grains is developed almost everywhere, and pebbles have been flattened into ellipsoids and are cigar-shaped in many places. In some places a superposed deformation has caused folding of the earlier, main foliation and pebbles have again been modified, either by folding or stretching to cause shear-step or brittle pull-apart textures (Fig. 3*t* and *u*). In places where the sporadic superposed deformation has not affected the rock, textures in the conglomerate matrix resulting from the main phase of deformation are seen to include primarily internal-granular, granular-rim, plucked-boundary, strain-shadow, and goatee textures with only minor pulling apart (Fig. 3*v–x*). Feldspar grains tend to be equidimensional, whereas most quartz grains are ellipsoidal (but form granular textured lenticles in the most strongly deformed rocks). The deformation associated with the main metamorphic foliation appears to have occurred through both intergranular movement and crystallographic gliding under conditions of moderate intergranular cohesion (though recrystallization and neo-mineralization also occurred). Local variations in processes probably occurred along with the variations in deformational intensity, with intergranular movement being more important where deformation was the most intense and presumably also the most rapid. Intrusive rocks in the area show even more variable deformation intensities than do the sedimentary rocks (possibly as a result of variable times of intrusion during the deformation). The most weakly deformed intrusive rocks studied to date display breccia texture, some of the feldspars are kink-folded and in some places the quartz grains are slightly elongate parallel to the regional, main foliation (Fig. 3*z*). Deformation of these rocks was primarily by rupture and granulation with only minor differential movement and crystallographic gliding. The more strongly deformed intrusive rocks (Fig. 3*y*) are similar to the conglomerate, though seldom as strongly deformed. Some of the more highly metamorphosed rocks in this area (Kisseynew Group) contain syn-tectonic garnets. In many places the boundaries of the garnets apparently have been smoothed and milled down by post-growth intergranular movements (Fig. 3*aa* and *bb*). My interpretation of these rocks is that all possible types of deformation were important; their relative importance varies from place to place and between rock types depending on the intensities of deformation and neo-mineralization. Textures indicating considerable amounts of intergranular movements are evident in all but the most weakly deformed rocks.

The last example is of Precambrian, garnetiferous, biotite schist from the Mukteswar region in the Lower Himalayas, India. These rocks have undergone several phases of deformation and recrystallization, and garnet-in-garnet texture is extensively formed (Fig. 3*cc*). The younger garnets contain quartz inclusions parallel to the crystal boundaries, whereas the older garnets contain parallel rows of quartz inclusions that extend across the entire grain (Fig. 3*cc*) and represent an early penetrative foliation that formed during or prior to the growth of the older garnets. The orientation of these early inclusions is not constant in any one thin section and many of the garnets containing them have rounded boundaries. Younger garnets generally have good crystal boundaries, although the matrix foliation is deflected around them. Also, in one thin section studied, some of the late garnets have developed strain-shadows apparently associated with retrograde metamorphism (Fig. 3*dd*). Deformation of these rocks involved at least three periods of neo-mineralization and intergranular movements. The importance of crystallographic gliding is unknown, although it could have occurred only in the micaceous matrix.

Discussion

In the past, the petrogenetic study of deformed and metamorphosed rocks has been concerned mainly with the metamorphic minerals present; much less consideration has been given to the processes of deformation than to those of recrystallization. This is a serious omission in the fields of structural geology and metamorphic petrology.

A major problem is the relative importance of intergranular movements, recrystallization, and crystallographic gliding in naturally deformed rocks. It should be obvious that this author considers all three to be important and that rocks deformed by only one process (e.g. plastites, mylonites, etc.) are special cases (see Fig. 1). This is based on observations of a wide variety of rock types from several continents (with the exception of granulites and rocks where recrystallization has obliterated any deformation textures that might have formed). In almost every rock where initial intergranular cohesion was high, strong deformations (greater than 50% flattening) involved considerable intergranular movement (possible exceptions included only some marbles where recrystallization and/or crystallographic glide were most important). In all rocks of initially low intergranular cohesion (or in rocks that underwent fracturing early in deformation), deformation was primarily through intergranular movement, though crystallographic gliding appears to have been important for some minerals. With only few exceptions, regardless of what other processes have occurred, intergranular movements are important, and their operation results in a number of diagnostic textures that can be used to help interpret the petrogenesis of deformed rocks. This viewpoint appears to be in opposition to statements made by several authors discussing experimental work (e.g. Turner and Weiss 1963; Carter *et al.* 1964), who seem to favor crystallographic gliding and "recrystallization" as the most important processes; strong intergranular movements apparently are considered by them to be of main importance only in rocks such as cataclastites and mylonites.

I end with a plea that future experimental work on rock deformation include studies of intergranular dislocations and their resulting textures.

Acknowledgments

I wish to thank Drs. I. LeMay, N. Wardlaw, and F. Langford, University of Saskatchewan, for reading and discussing the manuscript with me at various stages in its development. Also, I thankfully acknowledge the loan of thin sections and rock samples from Dr. N. Wardlaw, Mr. A. Mukherjee, and Mr. S. Ahuja, University of Saskatchewan. The bulk of this work was carried out under a National Research Council of Canada Grant.

BARRETT, C. S. 1952. Structure of metals. (2nd ed.) McGraw-Hill Book Co., New York.

BORG, I., FRIEDMAN, M., HARDIN, J., and HIGGS, D. V. 1960. Experimental deformation of St. Peter Sand: A study of cataclastic flow. Chapter 6 *in* Rock Deformation. Geol. Soc. Amer. Memoir 79, pp. 133–192.

CARTER, N. L., CHRISTIE, J. M., and GRIGGS, D. T. 1964. Experimental deformation and recrystallization of quartz. J. Geol. **72**, pp. 687–733.

CARTER, N. L. and FRIEDMAN, M. 1965. Dynamic analysis of deformed quartz and calcite from the Dry Creek Ridge Anticline, Montana. Amer. J. Sci. **263**, pp. 747–785.

CHRISTIE, J. M., GRIGGS, D. T., and CARTER, N. L. 1964. Experimental evidence of basal slip in quartz. J. Geol. **72**, pp. 734–756.

CLOOS, E. 1947. Oolite deformation in the South Mountain fold, Maryland. Bull. Geol. Soc. Amer. **58**, pp. 843–918.

FAIRBAIRN, H. W. 1950. Pressure shadows and relative movements in a shear zone. Trans. Amer. Geophysical Union, **31**, pp. 914–916.

FLINN, D. 1956. On the deformation of the Funzie Conglomerate, Fetlar, Shetland. J. Geol **64**, pp. 480–505.

FRIEDMAN, M. and CONGER, F. B. 1964. Dynamic interpretation of calcite twin lamellae in a naturally deformed fossil. J. Geol. **72**, pp. 361–368.

HARKER, A. 1932. Metamorphism, a study of the transformations of rock masses. Metheun and Co. Ltd., London.

HEARD, H. C. and CARTER, N. L. 1968. Experimentally induced "natural" intragranular flow in quartz. and quartzite. Amer. J. Sci. **266**, pp. 1–42.

HOBBS, B. E and TALBOT, J. L. 1966. The analysis of strain in deformed rocks. J. Geol. **74**, pp. 500–512.

HOSSACK, J. R. 1968. Pebble deformation and thrusting in the Bygdin area (Southern Norway). Tectonophysics, **5**, pp. 315–339.

JOHNSON, M. R. W. 1967. Belemnite deformation at Fernigen, Switzerland. Geol. Mag. **104**, pp. 268–273.

MCLEAN, D. 1957. Grain boundaries in metals. Oxford University Press, Toronto.

MISCH, P. 1969. Paracrystalline microboudinage of zoned grains and other criteria for synkinematic growth of metamorphic minerals. Amer. J. Sci. **267**, pp. 43–63.

NISSEN, H. U. 1962. Dynamic and kinematic analysis of deformed crinoid stems in a quartz graywacke. J. Geol. **70**, pp. 346–360.

PABST, A. 1931. "Pressure-shadows" and the measurement of the orientation of minerals in rocks. Amer. Mineralogist, **31**, pp. 55–70.

PATERSON, M. S. and WEISS, L. E. 1966. Experimental deformation and folding in phyllite. Bull. Geol. Soc. Amer. **77**, pp. 343–374.

RALEIGH, C. B. and TALBOT, J. L. 1967. Mechanical twinning in naturally and experimentally deformed diopside. Amer. J. Sci. **265**, pp. 151–165.

RAMSAY, J. G. 1962. The geometry and mechanics of formation of similar type folds. J. Geol. **70**, pp. 309–327.

ROSENFELD, J. L. 1968. Garnet rotations due to the major Paleozoic deformations in Southeast Vermont. Chapter 14 *in* Studies of Appalachian Geology: Northern and Maritime. (Zan, White, Hodley, and Thompson, *Eds.*), Interscience (Wiley), New York.

SPRY, A. 1963. The chronological analysis of crystallization and deformation of some Tasmanian Precambrian rocks. J. Geol. Soc. Australia, **10**, pp. 193–208.

STAUFFER, M. R. 1967. Tectonic strain in some volcanic, sedimentary and intrusive rocks near Canberra, Australia: a comparative study of deformation fabrics. New Zealand J. Geol. Geophys. **10**, pp. 1079–1108.

TURNER, F. J. and WEISS, L. E. 1963. Structural analysis of metamorphic tectonites. McGraw-Hill Book Co., New York.

ZWART, H. J. 1963. Metamorphic history of the Central Pyrenees, Part II, Valle de Aran, Sheet 4. Leidse Geol. Mededel. **28**, pp. 321–376.

6

Copyright ©1978 by the Geological Association of Canada
Reprinted from *Geosci. Canada* **5:**109–118 (1978)

Flow Mechanisms in Rocks:

Microscopic and mesoscopic structures, and their relation to physical conditions of deformation in the crust

R. Kerrich and I. Allison
Department of Geology,
University of Western Ontario,
London, Ontario N6A 5B7

Summary
Deformation of the crust is believed to occur dominantly by cataclasis at low temperatures and/or effective confining pressure, by pressure solution at intermediate temperatures, and by dislocation creep at high temperatures. Each flow mechanism gives rise to distinctive microscopic and small scale structures.

Brittle deformation with grain fracture leading to a reduction of particle diameter is characteristic of cataclastic flow.

Pressure solution produces grain shape fabrics by intercrystalline diffusion assisted by the presence of water. Grains may change shape at constant mass, or decrease in mass (and therefore in size) by long range diffusion: mass is then not locally conserved. Reduction of grain diameter leads to increased rates of deformation (strain softening).

Distinctive spaced cleavage zones form by pressure solution in which mineral species are redistributed due to different rates of deformation: the displacement field is discontinuous and deformation non-isochemical. Tectonic veins associated with pressure solution structures probably form by local mass transport; thus brittle and ductile mechanical behaviour coexist.

Dislocation creep produces grain shape fabrics by intracrystalline deformation, and may cause grain size reduction by subgrain formation and recrystallization. Preferred crystallographic orientations can arise from dislocation glide. Mass is conserved and deformation is believed to be essentially isochemical. Small scale structures formed by dislocation creep are ductile, with a continuous displacement field.

Introduction
Recent advances in materials science have led to a better understanding of the solid state flow mechanisms in crystalline materials. These concepts have been applied to deformation of rocks, with particular reference to flow in the mantle (Ashby and Verrall, 1977; Weertman, 1978).

The objective of this paper is to present a brief synthesis of the flow mechanisms believed to be important in crustal deformation. Microscopic and small scale structures typical of each mechanism are illustrated, and placed in the context of the physical conditions prevailing during deformation. A comprehensive treatment of flow mechanisms is not attempted: the appropriate works are referred to. We emphasise, and present observational evidence for, the relation between the physical environment of crustal deformation and the nature of the operative flow mechanism.

Flow Mechanisms
Crystalline aggregates may deform by a number of independent mechanisms. These are: 1) *Cataclasis,* involving repeated grain fracture and frictional sliding of particles; 2) *Dislocation creep,* in which strain is largely produced by the gliding motion of dislocations; 3) *Boundary diffusion creep,* involving mass transport at grain boundaries (generally known as Coble creep); 4) *Volume diffusion creep,* which takes place by means of the diffusion of point defects through grains (generally termed Nabarro-Herring creep). These flow mechanisms and the factors which influence them are reviewed by Ashby, (1972); Stocker and Ashby (1973); and Ashby and Verrall (1977). Other mechanisms of crystal plasticity such as kinking and twinning (see Gilman, 1969) do not in general contribute large strains.

Pressure solution (Sorby, 1863) is a term used by geologists to describe a process of diffusive mass transport by which rocks deform at low temperature, probably assisted by the presence of fluid at grain boundaries; and which therefore has similarities to mechanism (3) above. Nabarro-Herring creep has been identified in the deformation of fine grained metals at high values of homologous temperature ($T/T_{melting}$). This creep mechanism is not thought to be significant in deformation of the crust or mantle, and is not considered further here.

Each mechanism has a constitutive flow law which relates strain rate to rheological properties (such as defect diffusivities), and the state variables temperature and differential stress. The ensuing discussion is concerned principally with the deformation of quartz and calcite because the mechanical behaviour of these species is better understood than that of other minerals.

Cataclasis
Cataclastic flow involves repeated grain fracture and frictional sliding of rock particles: both of these processes are

dependent on the magnitude of normal stress (Stearns, 1968; Hamil and Sriruang, 1976).

At greater crustal depths progressively higher values of confining pressure inhibit cataclasis by increasing the normal stres across potential fracture and sliding surfaces, thereby favouring mechanisms of crystal plasticity.

The effect of pore fluids is to reduce the total confining pressure by an amount equal to the fluid pressure, yielding an 'effective' confining pressure ($P_{effective} = P_{total} - P_{fluid}$). Under conditions of normal fluid pressure ($P_{fluid} \sim \frac{1}{3} P_{lithostatic}$) this transition from brittle to ductile behaviour takes place at a depth of a few kilometres in the earth's crust. In the presence of abnormal fluid pressure where P fluid may approach or exceed the confining pressure brittle mechanical behaviour may extend to significantly greater crustal depths. (Price, 1970; Carter, 1976).

In proximity to the terrestial surface, where conditions of confining pressure and temperature are both low, deformation of soft sediments is accommodated by frictional sliding of rock particles. This is illustrated for folded sediments and buckled anisotropic tuff in Figures 1A and 1B respectively. The microscopic structure of these rocks exhibits a low degree of grain fracture. Soft sediment deformation is discussed by Hobbs *et al.* (1976) and Woodcock (1976).

At higher values of confining pressure, and in consolidated rocks, deformation occurs by means of repeated grain fracture, coupled with frictional sliding and rotation of rock particles. These processes are best developed along fault zones, in which repeated fracturing leads to grain size reduction and generation of a 'fault gouge' (Engelder, 1974; Fig. 1C).

A special case of brittle deformation occurs in 'dry' consolidated rocks at high confining pressure, subjected to displacement at fast strain rates. Under these conditions fracture and frictional sliding at high velocity lead to temperature rise and generation of a melt phase known as pseudotachylyte (Fig. 1D). This feature has been discussed by Sibson (1975), and is believed to represent a locus of palaeo-seismic faulting. The influence of strain rate on deformation is discussed by Price (1975) and White (1975).

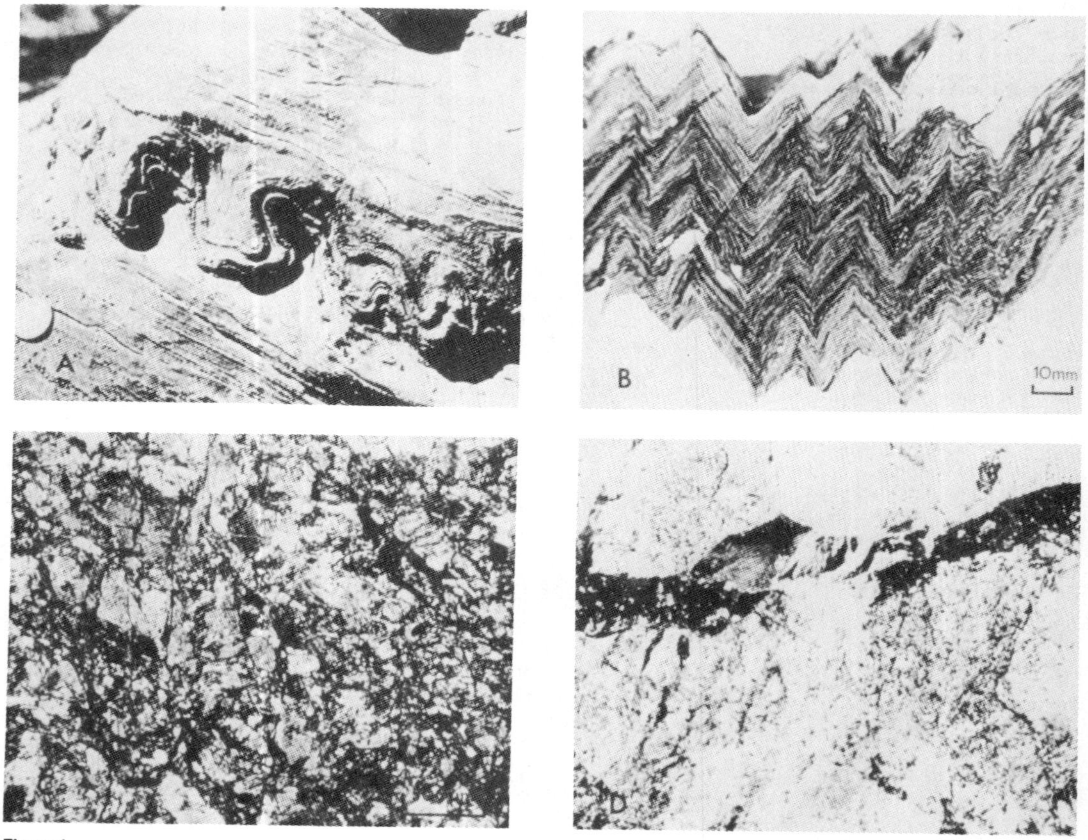

Figure 1
A: Folded layers in sea floor gravity slump. B: Crenulation folds formed by buckling of anisotropic sediment in a sea floor slump. C: Grain size reduction by fracture, along a fault. D: Pseudotachylyte vein in granite: rapid frictional sliding along a shear plane leads to melting.

(Courtesy of Dr. Bruce Bouley.)

Pressure Solution

Pressure solution is a deformation mechanism involving intercrystalline diffusive mass transport in response to applied stress. Mass transport is directed from grain boundaries at high normal stress to grain boundaries at low normal stress, permitting the shape change of grains. For the deformation of polycrystalline materials mass transport may occur by means of the surface diffusion of point defects (Coble creep, see Elliott, 1972), by diffusion through an adsorbed fluid layer (Weyl, 1959; Rutter, 1976), or by diffusion through a static pore solution. Rutter (1976) has emphasised that the presence of water at grain boundaries probably yields effective diffusivities in the aqueous phase at low temperatures (< 400°C) of the order of solid-state diffusivities at much higher temperatures, where boundary diffusion creep is observed to operate in the deformation of 'dry' metals and ceramics. This assumption is reasonable inasmuch as pressure solution is encountered principally in sediments which have undergone diagenesis, and in rocks deformed under conditions of low grade metamorphism. In these low temperature environments rocks generally have a relatively high porosity and high content of hydrated mineral phases. The subject of pressure solution has been reviewed by Kerrich (1978).

Grains change shape by removal of material along boundaries exposed to high normal stress, with diffusive transport to boundaries at lower normal stress and epitaxial overgrowth. In this state the mass of individual grains is approximately conserved (Fig. 2A). Alternatively, groups of grains under high normal stress may lose mass to groups of grains at lower normal stress, by long-range diffusion; or precipitate in structures such as pressure shadows (Fig. 2B). In this situation mass is not locally conserved.

It is possible that a preferred alignment of optic axes may develop by pressure solution due to differences in the crystallographically controlled rates of growth and solution. An attempt to detect this effect in a rock with textural evidence for dominant pressure solution met with a negative result (Allison and Kerrich, unpublished data).

Penetration of one grain into another by removal of mass along grain boundaries is illustrated by the deformed orthoquartzite in Figure 2C. The absence of intracrystalline deformation is indicated by the presence of rectilinear arrays of secondary fluid inclusions within grains. An example of mass removal at pebble contacts by pressure solution in a deformed conglomerate is given in Figure 2D.

The presence of layer silicates appears to enhance the rate of deformation of quartz and calcite relative to monomineralic rocks. Thus features of pressure solution are best developed in impure lithologies.

There is a large variation in the rate at which different minerals deform by pressure solution. For instance quartz and calcite deform faster under given conditions than phyllosilicates or meta oxides. Because of this differential response to pressure solution one mineral may be selectively removed from domains in a polymineralic rock. This results in an increase in the relative proportion of less mobile minerals, and permits immobile platy minerals to rotate into a parallel alignment. Deformation is then locally non-isochemical.

This effect typically develops along a series of subplanar zones which form a spaced pressure solution cleavage, or along tectonic stylolites in rocks with a high abundance of carbonate minerals. Structures formed by this process are illustrated in Figures 2E, 2F and 3A. Planar markers oriented obliquely with respect to the zones of pressure solution appear to be displaced because of volume loss (Fig. 2F), and this feature has been cited as evidence for the now contra-indicated shear hypothesis of cleavage formation.

Separation of minerals with an initially homogeneous distribution by pressure solution is a typical feature of buckled mineral fabrics. The separation process involves diffusion of a mobile phase from limbs into hinges in response to stress gradient, with concomitant enrichment in the limbs of layer silicates and other immobile phases (Williams, 1972; Stephansson, 1974). This is the typical crenulation cleavage structure, examples of which are given in Figures 3B and 3C. The displacement field in these structures is not continuous. Under certain conditions which are not well understood both quartz and micas are transported at approximately the same rate and may form quartz-mica "beards" at grain boundaries under low normal stress (see Fig. 3D).

Pressure solution zones are initiated at the boundaries of rigid particles which act as stress raisers in the adjacent matrix. Examples of these are pyrite crystals, fold hinges, pebbles and fossils in a fine grained matrix. The pressure solution zones are believed to propagate normal to the inferred direction of maximum principal stress.

Zones of pressure solution are frequently associated with tectonic vein arrays, and exhibit systematic geometrical relations to the veins. Tensile fractures initiate in a direction parallel to the maximum principal stress; pressure solution zones propagate normal to this direction. Such veins are generally composed of minerals that reflect the composition of the host rock (Figs. 3E, 3F). This observation, coupled with the structural relations, has led to the conclusion that the vein minerals are derived largely by diffusive transport from the adjacent rock (Durney and Ramsay, 1974). Control of vein composition by the host rock is shown in Figure 3F where quartz is precipitated in a fracture adjacent to the felsic part of a graded unit, and chlorite adjacent to pelite.

Tectonic veins are generally abundant in rocks deformed under conditions of low grade metamorphism where pressure solution is operative. The mechanism of mass transport into veins by pressure solution avoids difficulties inherent in the alternative hypothesis of precipitation from hydrothermal systems where the water/rock ratio is constrained to be high by mineral solubilities.

Large crustal volumes may undergo non-isochemical deformation by pressure solution if the mobile phase diffuses into sinks such as veins (Kerrich *et al.*, 1978).

The constitutive flow law for pressure solution given by Elliott (1972) and Rutter (1976) assumes diffusion to be rate controlling and is essentially similar to that for Coble Creep (Ashby, 1972), with a linear dependence of strain rate on differential stress and an inverse cube dependence on grain diameter. This accounts for the fact that pressure solution is favoured in fine grained lithologies; and also predicts that grain size reduction by mass loss leads to increased rates of deformation (strain softening).

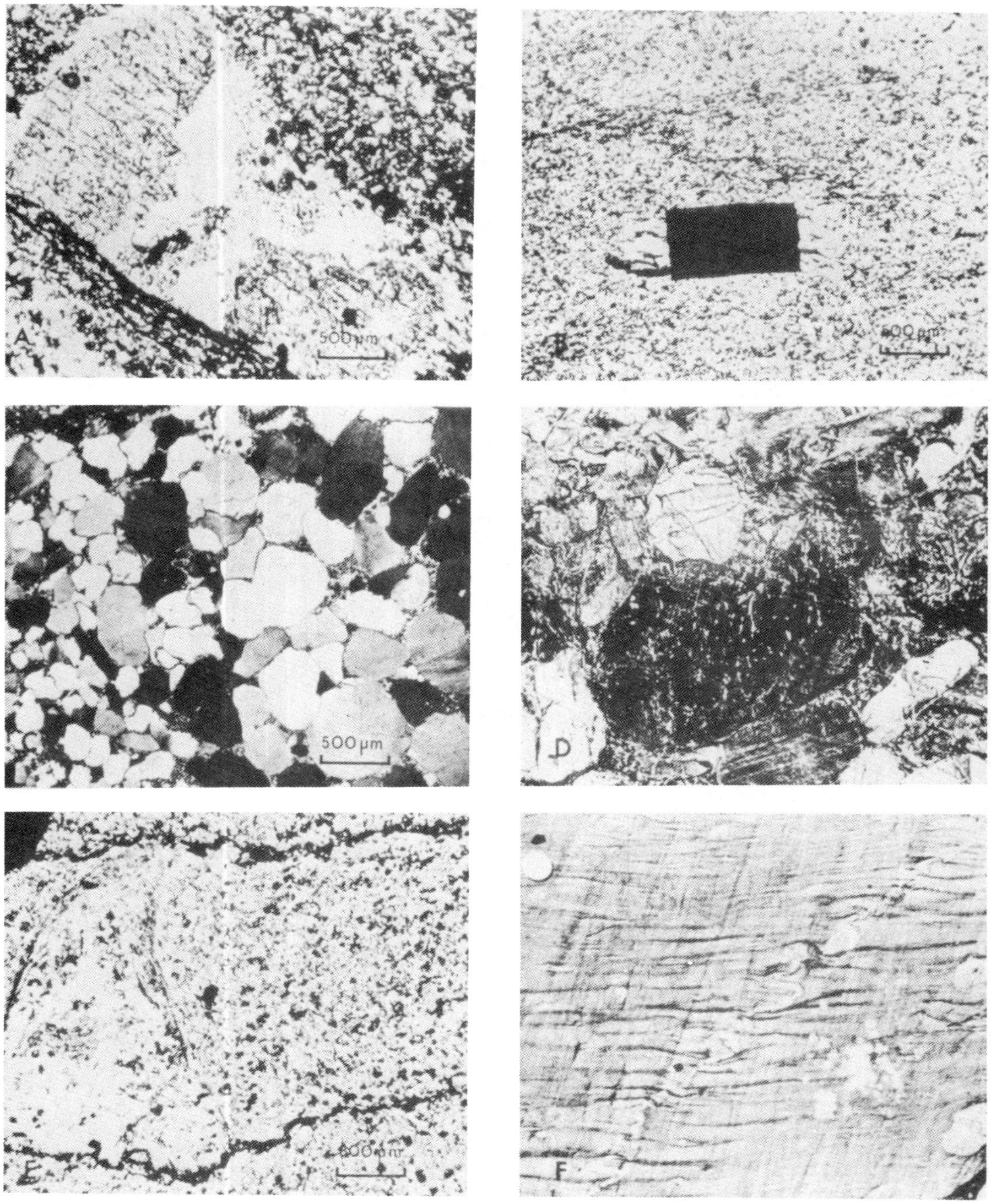

Figure 2
A: Mass loss of calcite in a tectonic stylolite with epitaxial overgrowth of pure calcite in pressure shadows. B: Redistribution of quartz around a rigid particle from boundaries exposed to high normal stress to boundaries at lower normal stress (pressure shadows). C: Mass loss at indented grain boundaries in a deformed orthoquartzite. D: Mass loss at indented boundaries in a deformed conglomerate. E: Tectonic stylolites transect a fossil and define spaced cleavage. F: Spaced pressure solution cleavage zones propagating from a buckled quartz vein. Volume change indicated by displacement of vein across cleavage zones.

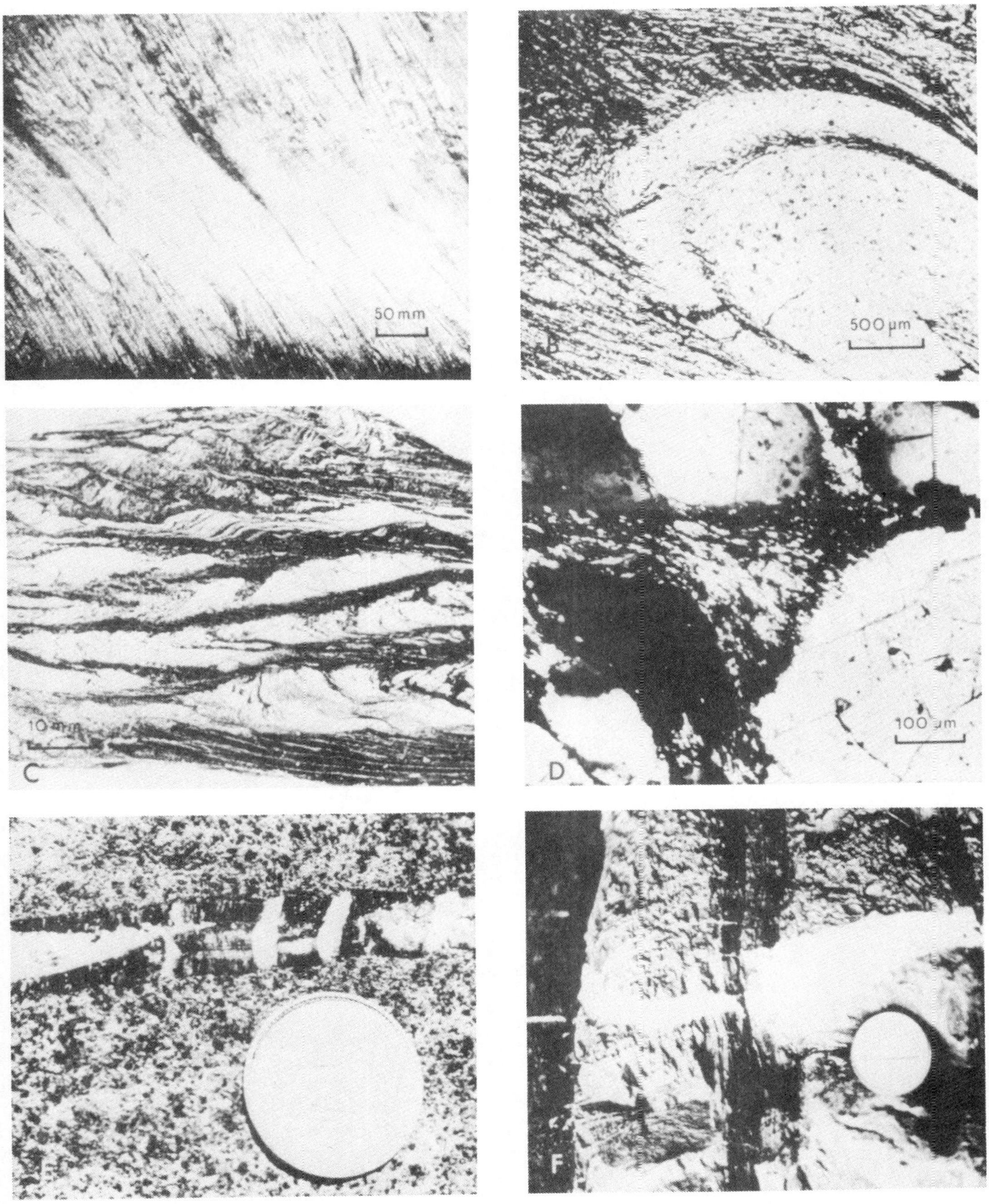

Figure 3
A: Folded bedding transposed into tectonic striping by pressure solution with removal of inverted limbs. B: Spaced pressure solution cleavage in the buckled matrix around a microfold. C: Crenulation cleavage formed by redistribution of quartz and micas by pressure solution in a buckled phyllite multilayer. D: Quartz and mica overgrowth in a domain of low stress. E: Boundinaged belemnite with migration of calcite from deformed matrix into pressure shadows. F: Deformed greywacke displaying local redistribution of quartz in grit layers and chlorite in shale.

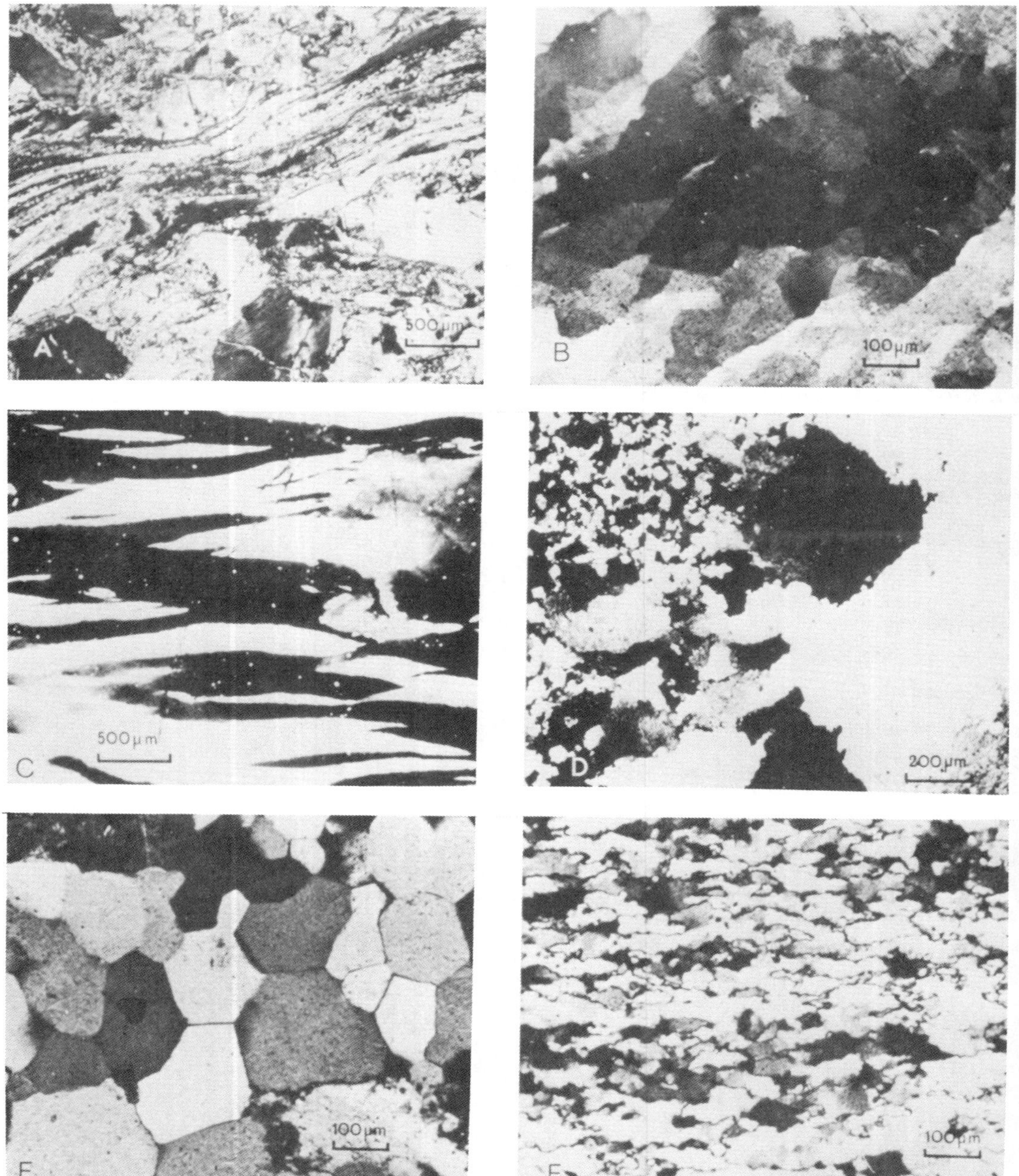

Figure 4.

A: Ribbon quartz produced by dislocation glide. Slight recrystallisation is evident on some quartz grain boundaries. B: Subgrains in quartz: the slight angular misorientation between subgrains is revealed by differences in interference colour. C: Deformation bands in quartz. D: Tectonic grain size reduction of mm size quartz grain into a mosaic of subgrains. Note the serrated grain boundaries. E: Equant polygonal quartz grains with equilibrium triple junctions (foam microstructure). Stable end point of annealing recrystallisation. F: Quartz mylonite. Irregular elongate grains with subgrains and undulatory extinction, pressured to have formed by dynamic recrystallisation.

Superplastic Flow. Superplastic flow is a term used to denote the phenomenon by which polycrystalline materials attain very high ductile strains. Several independent mechanisms have been implicated in superplastic flow, notably grain boundary sliding accommodated by diffusive transport of dislocation creep. Superplastic behaviour in metals and ceramics has been reviewed by Edington *et al.* (1976). Ashby and Verrall (1973) presented a model of superplastic flow in which essentially frictionless grain boundary sliding is accommodated by grain boundary diffusion.

The extent to which superplastic flow contributes to deformation of the crust is not known because of the difficulty of identifying switching of neighbouring grains in naturally deformed rocks. Boullier and Gueguen (1975) presented textural evidence for superplastic flow in some mylonites. This mechanism is favoured in materials having small equant grains, deformed at temperatures above 0.5 $T_{melting}$

Dislocation Creep

Dislocation creep is the most common mechanism of high temperature plastic flow in crystalline materials. Microstructures indicative of dislocation creep are abundant in rocks deformed under conditions of upper greenschist and higher grades of metamorphism. This mechanism involves a combination of glide and climb of dislocations. Under an applied stress dislocations are continuously produced (multiply) at Frank-Read sources and glide through the crystal producing a change of shape (see Nicholas and Poirier, 1976, for a review of dislocation creep).

At the lower temperature range of intracrystalline plasticity gliding dislocations may become pinned by obstacles such as impurities, inclusions and dislocation tangles. Grain boundaries also serve to inhibit the motion of dislocations (grain boundary hardening). Dislocation multiplication coupled with pinning causes the density of dislocations to rise. This has the effect of increasing the stored strain energy leading to work hardening, in which the applied stress must be continuously increased to produce a constant strain rate.

At higher temperatures ($T/T_{melting} >$ 0.5) dislocations may cross-slip or climb past obstacles from one slip plane to another, and may annihilate one another, thus reducing the dislocation density and strain energy. These processes by which dislocations are arranged into a lower energy configuration are termed *recovery*. Steady state dislocation creep is a balance between work hardening and recovery, permitting high ductile strains to be attained. Although the strain is due to the gliding motion of dislocations strain rate is dependent on the climb of dislocations by diffusion. During recovery dislocations are arranged into dislocation walls which separate slightly misoriented sub-grains containing fewer dislocations: this is termed polygonisation.

Experimental work has implicated structural 'water' as an important factor weakening quartz and other silicates relative to the strength of 'dry' crystals (Griggs, 1974; Blacic, 1975; Jones, 1975). The mechanism is not well understood, and the various hypotheses are reviewed by Nicholas and Poirier (1976).

Constitutive flow laws proposed for dislocation creep have the following general form (taken from White, 1976a):

$$\dot{e} = \frac{AD_v Gb}{kT} \left(\frac{\sigma}{T}\right)^n$$

where $\dot{e}$ is strain rate, A is a constant, D_v is the volume diffusivity, G is the shear modulus, b is the Burgers vector, k is Boltzmann's constant, T is temperature (°K), σ is differential stress, and n a constant.

The following features are characteristic of dislocation creep (as reviewed by Carter, 1976; Nicholas and Poirier, 1976): 1) Strain rate varies as a power (n) of the applied stress where n, determined empirically, varies between 3 and 6; 2) Dislocation creep processes result in a strain rate that is independent of grain size. However in polycrystalline aggregates the stress required to overcome grain boundary hardening is inversely proportional to the square root of grain size and thus fine grained aggregates require a higher flow stress than coarser ones. This is the Hall-Petch

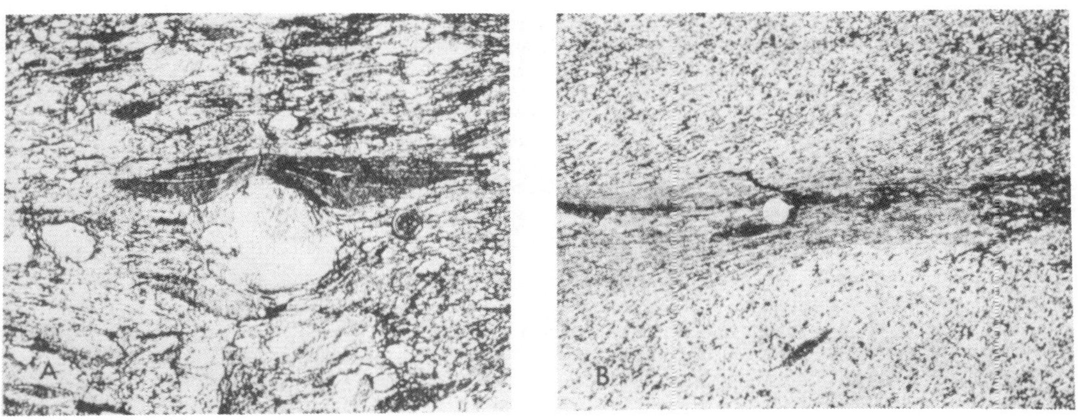

Figure 5.
Ductile shear zone: progressive development of schistosity at increasing states of strain from an initially undeformed granite. Schistosity defined by grain shape. B: Internal deformation of pebbles in conglomerate.

law (Nicholas and Poirier, 1976, p. 129); 3) There exists a substructure of polygonal cells whose size is inversely proportional to stress.

Microstructures. The high ductile strains which may be attained by dislocation motion are illustrated by the quartz ribbons in Figure 4A. Undulatory extinction is a feature generally present in deformed quartz (Figs. 4A, 4D, 4F), arising from binding of the crystal structure, accommodated by dislocation motion. Subgrains formed by polygonisation during recovery are shown in Figure 4B. During recovery dislocations may climb into planes of high dislocation density which separate domains of lower density. These kink bands or deformation bands are distinguished optically by their bonded extinction (Fig. 4C). Optical features arising from dislocation creep and their relation to dislocation structures are described and discussed by White (1973, 1976b).

Dislocation creep may be complicated by recrystallisation, a more severe process than recovery for reducing the internal strain energy of crystals. The presence of small, equant, strain-free grains is indicative of recrystallisation (Fig. 4D). Recrystallisation generally occurs first at grain boundaries of kink bands within crystals. Prior to the formation of discrete new grains the boundaries become serrated (Fig. 4D).

It is possible to distinguish between annealing and dynamic recrystallisation. The former occurs at elevated temperature in the absence of differential stress, following deformation, and leads to the formation of equant polygonal stain free grains (foam microstructure, Fig. 4E). Dynamic recrystallisation occurs during deformation at high temperatures. The newly recrystallised grains are themselves deformed and are generally elongate. The microstructures typical of quartz mylonites which exhibit grain size reduction (Fig. 4F), are the products of dynamic recrystallisation.

Although recrystallisation is not itself a deformation mechanism it does alter the microstructure and change grain size, and hence may influence the relative importance of competing flow mechanisms. White (1977) presents a comprehensive account of recovery and recrystallisation in quartz.

The presence of undulatory extinction alone implies that the state of strain is low and it is unlikely that steady state creep has been attained. Conversely, a well developed subgrain structure is indicative of steady state creep (White, 1976b). Preferred crystallographic orientations are believed to be the product of dislocation gliding.

Small Scale Structures. Grain shapes which arise from intracrystalline deformation may contribute to a grain shape fabric which defines mesoscopic L and/or S fabric elements (Fig. 5A). Zones of high deformation such as mylonites typically exhibit extreme grain size reduction as a result of dynamic recrystallisation; not as a product of grain fracture.

Ductile deformation produced when the strain is accommodated by dislocation creep leads to small scale structures that have a continuous displacement field, and in which the mass of original grains is conserved. Deformation by dislocation creep is believed to be essentially isochemical in the absence of fluid-rock interaction (Kerrich *et al.*, 1978). Rock structures reflect the nature of the dominant deformation mechanism which in turn has an overall correlation with the tectonic environment. Grain size reduction by fracture in a submetamorphic environment (Fig. 1C) may be distinguished from the fine grained microstructure of a mylonite produced by dynamic recrystallisation under amphibolite conditions (Fig. 4F). The crenulation folds illustrated in Figure 1B which formed on the seafloor have neither the penetrative schistosity that might be anticipated for high temperature deformation (c.f. Fig. 5A), nor pressure solution cleavage characteristic of deformation at intermediate temperatures (Fig. 3C).

Deformation of a quartzite by mass transport at grain boundaries, with no internal strain, under sub-greenschist facies conditions, is illustrated in Figure 2D. This contrasts with the high ductile stain of quartz grains deformed at amphibolite grade (Fig. 4A), where dislocation creep predominates. The difference between strain accommodated by mass removal at boundaries and by internal deformation is further illustrated by the mechanical behaviour of conglomerates deformed at low temperature (Fig. 2C) and high temperature (Fig. 5B) respectively.

Deformation Regimes

A large body of field evidence has led to the conclusion that deformation of the crust is dominated at low temperatures by cataclasis, at intermediate temperatures by pressure solution, and at high temperatures by dislocation creep. The basis of this generalization is the observation that pressure solution is abundant in rocks deformed under conditions of low grade metamorphism; with microstructures indicative of dislocation creep becoming increasingly important from conditions of upper greenschist to higher grades of metamorphism. Cataclastic flow appears to be dominant under sub-metamorphic conditions.

Because of the rheological complexity of rocks, and the large number of independent variables which influence each flow mechanism it is difficult to place quantitative limits on these three deformation regimes. A wide and variable degree of overlap is present.

The transition between the pressure solution and dislocation creep regimes for quartzites of 1000 μm, 100μm and 10 μm grain diameter has been estimated at 550°C, 450°C and 300°C respectively. Corresponding figures for calcite rocks of the grain sizes quoted are 360°C, 300°C and 200°C. This transition does not appear to be sensitive to total confining pressure (Kerrich *et al.*, 1977).

A temperature dependence of the dominant deformation mechanism is further supported by the observation that rocks retaining some record of high grade metamorphic mineral assemblages deformed at lower temperature may exhibit features of pressure solution. Examples of these features are not abundant, contrasting with the ubiquitous effects of pressure solution during low temperature prograde metamorphism. This may be attributed to two factors. Firstly, high grade metamorphic rocks are relatively coarse grained, thus inhibiting pressure solution; secondly, during retrograde metamorphism hydration reactions absorb water, in contrast to the excess water present during prograde metamorphism.

A transition in the creep behaviour of quartzites from dominant pressure solution to dislocation creep with increasing strain rate has been established by Mitra (1976).

In a number of highly deformed rocks grain size reduction by dislocation creep appears to cause a switch to pressure solution, which permits faster flow rates at smaller grain diameters (White, 1976; Kerrich *et al.*, 1978). These observations support the conclusion that transition between the flow regimes is dependent on temperature, grain size, and strain rate.

Deformation mechanism maps. The overall rate of deformation of a polycrystalline solid depends on which independent mechanism contributes the fastest strain rate: this in turn is influenced by the rheological properties of the material (i.e., defect diffusivities) and the state variables temperature and differential stress (Stocker and Ashby, 1973). The inter-relationship of the different creep mechanisms as a function of the state variables, for a material of specified grain diameter, may be depicted on Ashby deformation mechanism maps (Ashby, 1972). These maps are constructed by solving the constitutive flow laws simultaneously, and display the stress-temperature space in which mechanisms are dominant (i.e., contribute the fastest strain rate).

Deformation-mechanism maps have been constructed for several geological materials including olivine (Stocker and Ashby, 1973), quartz (Rutter, 1976; White, 1976a), calcite (Rutter, 1976), and some ore minerals (Atkinson, 1977). These maps provide a theoretical basis for the observations presented above that pressure solution is dominant over dislocation creep for fine grained materials, at low values of temperature and strain rate.

Conclusions

It is believed that natural deformation of rocks is dominated by dislocation creep at high values of differential stress and temperature, and that pressure solution is an important mechanism in fine grained lithologies under conditions of low temperature and low strain rate. At very low temperatures and/or low values of effective confining stress cataclasis probably constitutes the dominant mechanism of crustal deformation.

Microstructures of rocks may therefore be used to infer the broad environment of deformation. It is important for instance to distinguish between rocks deformed at low temperatures and subsequently subjected to static metamorphism at higher temperature, from rocks deformed exclusively under higher temperature conditions.

For example, a lack of awareness of this factor may have given rise to the epigenetic replacement hypothesis for certain deformed strataform massive sulfide ore deposits, whereas many such deposits contain evidence of seafloor slumping during or after emplacement. A later overprint of deformation during regional metamorphism may lead to features that could be interpreted in terms of replacement and interpretation of all the structures as the product of high temperature tectonics.

Acknowledgements

The authors have benefitted from extensive discussion of rock deformation with B. K. Atkinson, E. H. Rutter and S. White. We are grateful to B. E. Gorman and J. White for critical reading of the manuscript. I. A. acknowledges National Research Council of Canada grant A3555.

References

Ashby, M. F., 1972, A first report on deformation-mechanism maps: Acta Metallurgica, v. 20, p. 887-897.

Ashby, M. F. and R. A. Verrall, 1973, Diffusion accommodated flow and superplasticity: Acta Metallurgica, v. 21, p. 149-163.

Ashby, M. F., and R. A. Verrall, 1977, Micromechanisms of flow and fracture, and their relevance to the rheology of the upper mantle: Univ. Cambridge, Dept. Eng. CUED/C. Mat/TR30.

Atkinson, B. K., 1977, The kinetics of ore deformation: Its illustration and analysis by means of deformation mechanism maps: Geologiska Föreningens Stockholm Förh., v, 99, p. 186-197, Stockholm ISSN 0016-786.

Blacic, J. A., 1975, Plastic-deformation mechanisms in quartz: the effect of water: Tectonophysics, v. 27, p. 271-294.

Boulleir, A. M., and Y. Gueguen, 1975, SP-Mylonites: origin of some mylonites by superplastic flow: Contributions to Mineralogy and Petrology, v. 50, p. 93-102.

Carter, N. L., 1976, Steady-state flow of rocks: Geophys. Space Phys. Reviews, v 14, p. 301-359.

Durney, D., and J. G. Ramsay, 1974, Incremental strains measured by syntectonic crystal growths: *in* K. A. DeJong and R. Scholten, eds., Gravity and Tectonics. New York, John Wiley, p. 67-96.

Elliott, D., 1972, Diffusion flow laws in metamorphic rocks: Geol. Soc. Amer. Bull. v. 84, p. 2645-2664.

Edington, J. W., K. N. Melton and C. P. Cutler, 1976, Superplasticity: Progress in Materials Science, v. 21, p. 63-170.

Engelder, J. T., 1974, Cataclasis and the generation of fault gouge: Geol. Soc. Amer. Bull. v. 85, p. 1515-1522.

Gilman, J. J., 1969, Micromechanics of flow in solids. New York, McGraw-Hill. 294 p.

Griggs, D. T., 1974, A model of hydrolytic weakening in quartz: Journal of Geophysical Research, v. 79, p. 1653-1661.

Hamil, B. M., and S. Sriruang, 1976, A study of fracture induced by dynamic tensile stress and its application to fracture mechanics; *in* R. G. J. Strens, ed., The Physics and Chemistry of Minerals and Rocks London, Wiley, p. 151-196.

Hobbs, B. E., W. D. Means and P. F. Williams, 1976, An Outline of Structural Geology: New York, Wiley and Sons, 571 p.

Jones, M. E., 1975 Water weakening of quartz and its application to natural rock deformation: Geol. Soc. Jour. London v. 131, p. 429-431.

Kerrich, R. 1978, An historical review and synthesis of research on pressure solution: Zentralblatt für Geologie u. Päläontologie, in press.

Kerrich, R., W. S. Fyfe, B. E. Gorman, and I. Allison, 1978, Local modification of rock chemistry by deformation: Contributions to Mineralogy and Petrology, in press.

Kerrich, R., R. D. Beckinsale, and J. J. Durham, 1977, The transition between deformation regimes dominated by intercrystalline diffusion and intracrystalline creep evaluated by oxygen isotope thermometry: Tectonophysics, v. 38, p. 241-257.

Mitra, S., 1976, A quantitative study of deformation mechanisms and finite strain in quartzites: Contributions to Mineralogy and Petrology, v. 59, p. 203-226.

Nicholas, A., and J. P. Poirier, 1976, Crystalline Plasticity and Solid State Flow in Metamorphic Rocks: New York, Wiley and Sons.

Price, N. J., 1970, Laws of rock behaviour in the Earth's crust: *in* Somerton, ed., Rock Mechanics - Theory and Practice: Proc. 11th Symp. Rock Mechanics, Berkely A.I.M.M., p. 1-23.

Price, N. J., 1975, Rates of deformation: Geol. Soc. London Jour. v. 131, p. 553-575.

Rutter, E. H., 1976, The kinetics of rock deformation by pressure solution: Royal Soc. London Phil. Trans., v. 283A, P. 203-217.

Sibson, R. H., 1975, Generation of pseudotachylyte by ancient seismic faulting: Astronomical Society Geophys. Jour., v. 43, p. 775-794.

Sorby, H. C., 1863, On the direct correlation of mechanical and chemical forces: Royal Soc. London Proc., v. 12, p. 538-550.

Stearns, D. W., 1968, Fracture as a mechanism of flow in naturally deformed layered rocks: *in* A. J. Baer and A. K. Norris, eds., Proceedings Conference on Research in Tectonics: Geol. Survey Canada Paper 68-52, p. 79-90.

Stephansson, O., 1974, Stress-induced diffusion during folding: Tectonophysics, v. 22, p. 233-251.

Stocker, R. L., and M. F. Ashby, 1973, On the rheology of the upper mantle: Geophys. and Space Physics Reviews, v. 11, p. 391-426.

Weertman, J., 1978, Creep laws for the mantle of the earth: Royal Soc. London Phil. Trans., in press.

Weyl, P. K., 1959, Pressure solution and the force of crystallisation - a phenomeriological theory: Jour. Geophys. Research, v. 62, p. 2001-2025.

White, S., 1973, The dislocation structures responsible for the optical effects in some naturally deformed quartzes: Jour. Mater. Sci., v. 8, p. 490-499.

White, S., 1975, Estimation of strain rates from microstructures: Geol. Soc. London Jour., v. 131, p. 577-583.

White, S., 1976a, The effects of strain on the microstructures, fabrics and deformation mechanisms in quartzites: Royal Soc. London Phil. Trans., v. 283A, p. 69-86.

White, S., 1976b, The role of dislocation processes during tectonic deformations with particular reference to quartz: *in* R. G. J. Strens, ed., The Physics and Chemistry of Minerals and Rocks: London, Wiley, p. 75-91.

White, S., 1977, Geological significcance of recovery and recrystallisation processes in quartz: Tectonophysics, v. 39, p. 143-170.

Williams, P. F., 1972, Development of metamorphic layering and cleavage in low grade metamorphic rocks at Bermagui, Australia: Amer. Jour. Sci. v. 272, p. 1-47.

Woodcock, N. H., 1976, Structural style in slump sheets: Ludlow Series, Powys, Wales: Geol. Soc. London Jour., v. 132, p. 399-415.

MS received March 30, 1978.

ERRATUM

Page 117, line 3 of the Acknowledgements should read: "B. K. Atkinson, R. W. Hodder, E. H. Rutter and S. White."

7

Copyright ©1973 by the Geological Society of America
Reprinted from *Geol. Soc. America Bull.* **84:**2645–2664 (1973)

Diffusion Flow Laws in Metamorphic Rocks

DAVID ELLIOTT *Department of Earth and Planetary Sciences, The Johns Hopkins University, Baltimore, Maryland 21218*

ABSTRACT

Rocks which deform by pressure solution obey a diffusion flow law with a linear viscous or Newtonian stress to strain-rate relation. Undeformed relics of original grains preserved within newly grown crystals at grain boundaries under tension and presolved surfaces, together with accumulation of inert particles at grain boundaries under compression, are diagnostic evidence of a diffusion flow law. At a given stress, strain-rate is inversely proportional to the grain size to a power of two or three. A geologically useful plot has inverse temperature versus the logarithm of grain size as coordinates. Such a graph is separated into fields by three boundaries which meet at a triple point; within each field, either lattice diffusion, grain-boundary diffusion, or a dislocation flow law is predominant. It may be possible to calibrate this graph from naturally deformed rocks. Photomicrographs of isoclinally folded greenschist-grade quartzites and rhyolitic flows from the South Mountain–Blue Ridge area in Maryland demonstrate a diffusive mass transfer deformation mechanism, but estimates of effective diffusion coefficient compared to currently available laboratory diffusion data are insufficient to identify the diffusion path with certainty. However, the comparatively low ratio of metamorphic temperature to melting temperature and the physical nature of grain boundaries in metamorphic rocks, particularly concentrations of low-density impurities at grain boundaries, suggest the grain-boundary diffusion flow law.

INTRODUCTION

A major obstacle to the understanding of the development of geological structures is our lack of knowledge of the *flow laws*—or the relations between stress and strain-rate at the time of deformation.

Regardless of the dimensions considered in a problem, ductile rocks deform as aggregates of crystalline grains, and the flow laws are decided on the scale of grains. An illustration of the importance of this problem is the controversy over the mechanisms and material properties which allow plates of lithosphere to move over the surface of planets through geological time. The question is: does a diffusion flow law hold, giving a linear stress to strain-rate relation as maintained by Gordon (1965, 1967), McKenzie (1968), and others; or is a dislocation flow law more appropriate, so that the asthenosphere is a power law material, as advocated by Orowan (1967), Weertman (1970), and others? Our lack of knowledge of the flow laws is not a special feature of the inaccessible mantle. The flow laws of crustal rocks, readily available for direct observation, are just as much of a mystery.

Have most of the grains in a thin section of naturally deformed rock changed shape by diffusive mass transfer or by dislocation motion? One point of this paper is to demonstrate that familiar "pressure solution" is equivalent to diffusion creep, and to show that direct microscopic observation can distinguish diffusion from dislocation flow. Another purpose is to clarify the theoretical relations between the different mechanisms of granular readjustment to deformation and to examine them in a geological context. Many of the ideas on these subjects are not found in the usual geological literature; a partial review will be attempted.

As we shall see, purely diffusive mass transfer is referred to in geology as "pressure solution," expressing the general idea that crystals dissolve where intensely stressed and grow where the stress is less. This idea is at least 100 yrs old, for Sorby (1863, 1879, 1908) presented convincing thin-section evidence of pressure solution and discussed its importance for diagenesis and metamorphism, while Van Hise (1896) and Becke (1903) considered pressure solution of major importance in the formation of schistosity.

More recently, the outstanding textural interpretations of Voll (1960) suggest that pressure solution combined with growth in the stretching direction is a major mechanism of deformation in the greenschist-grade rocks of the Welsh and Scottish Caledonides, the Rheinishes Schiefergebirge, the Northern Molasse basins, and the phyllite zones of the Swiss Alps. Durney (1972) illustrates nice examples from the Morcles Nappe. Many thin sections of deformed low-grade metamorphic rocks which I have seen show clear evidence of extensive pressure solution.

The theoretical understanding of pressure solution is in a much more unhappy situation. Work by Riecke (1895) was predated by Sorby, and his results are generally misinterpreted (reviewed by Turner and Verhoogen, 1960, p. 476). Clearly pressure solution should *not* be referred to as "Riecke's Principle." A theoretical treatment by Weyl (1959) was entirely concerned with pressure solution in a diagenic context. Fundamental and often overlooked work was done, however, by Correns (1939, 1949).

In metallurgy, it was not until Nabarro (1948) that diffusive mass transfer as a possible flow law was suggested, and both Nabarro (1948) and Herring (1950) considered the bulk lattice as the diffusion path and derived a linear stress to strain-rate flow law. The possibility of the grain boundary as the diffusion path was first proposed independently by Coble (1963) and Lifshits (1963). Direct observation of metals deforming by diffusion creep was not reported until 1963 (Squires and others, 1963). At the outset, the reader is invited to compare the illustration of a metal deformed by diffusion creep (Fig. 3) with the photomicrographs of deformed metamorphic rocks (Figs. 7 and 8).

As shown later in this paper, each diffusion and dislocation flow law occupies a sector of an inverse temperature versus grain-size graph, and the sector boundaries meet at a triple point. The position of this triple point varies with certain fundamental properties of the rock, and it is possible to locate its position by observations of thin sections of naturally deformed rocks.

With a number of different processes operating to produce a final result, how is it possible to find the rate-determining step? This question crops up throughout this paper; in essence, it is similar to determining the "critical path" for a whole number of different processes, where all must be finished to accomplish a given job. The various processes may be classified into two types, dependent and independent.

Two events, A and B, are dependent if A cannot proceed until B has taken place, so that the processes occur in sequence. The over-all rate is the slower of the dependent processes.

Rates of independent processes are not related to one another, and each may proceed at its own speed. The events can occur concurrently and nonsequentially. The over-all rate is the simple sum of all the independent rates, and in a sense, each process is free to compete for the largest share. The independent process with the fastest rate effectively dominates the over-all rate.

There is the matter of suitable names for the various flow laws, because they all involve some diffusion in one way or another (Orowan, 1967). In this paper, a diffusion flow law means one dominated by purely diffusion mass transfer, as opposed to dislocation flow laws in which the grains change shape by dislocation movement. The diffusion flow laws are subdivided according to the type of path—lattice or grain boundary—through which the diffusion occurs.

To discover what observations are necessary to prove diffusion flow, a brief review must be made of the assumptions underlying the derivations of the diffusion flow laws and of the available experimental evidence.

SI units are used throughout this paper.

DIFFUSION FLOW LAWS

Discussions of pressure solution in the geological literature occasionally refer to an equation worked out by Correns (1939, 1949). What is the relation between pressure solution, Correns' work, and diffusion creep? We will now draw all these different ideas together, keeping the treatment as simple as possible to emphasize the physics of the processes.

The partial molar volume v_i of a component i at temperature T and number of moles n_i can be written

$$v_i = \left(\frac{\partial \mu_i}{\partial P}\right)_{T,\, n_i}$$

where μ_i is the chemical potential of component i at a normal stress P. Integrating with respect to P, keeping T and n_i constant, and neglecting elastic compressibility, the chemical potential at normal stress P is

$$\mu_i^P = \mu_i^0 + \nu_i P$$

here μ_i^0 is an integration constant, the chemical potential of i at $P = 0$.

$$\therefore \mu_i^{P_1} - \mu_i^{P_2} = \nu_i(P_1 - P_2) \quad (1)$$

The above equation holds within the crystal proper. For material within the diffusion path, assuming ideality at a given T, P,

$$\mu_i^P = \mu_i' + RT \ln c_i^P$$

where c_i is the concentration of component i, R is the gas constant, and μ_i' is the chemical potential at the standard state.

$$\therefore RT \ln c_i^{P_1} - RT \ln c_i^{P_2} = \mu_i^{P_1} - \mu_i^{P_2} \quad (2)$$

Now equilibrium holds between material in the crystal and in the adjacent diffusion path, so (1) and (2) may be equated

$$\therefore \mu_i^{P_1} - \mu_i^{P_2} = \nu_i(P_1 - P_2) = RT \ln\left(\frac{c_i^{P_1}}{c_i^{P_2}}\right)$$

This equation describes the equilibrium differences in concentration of component i under different normal stresses at opposite ends of a diffusion path. It is the modified form of van't Hoff's Law derived by Correns (1939, 1949), who correctly recognized that his equation described the driving force for "solution transfer" between crystal faces under different normal stresses.

Let the reference state be the stress free one, $P_2 = 0$, and the normal stress at a local spot on the grain be $P_1 = \sigma_n$. If we consider only a single component material with μ_o, c_o as the chemical potential and equilibrium concentrations of the component at the reference state, then Correns' equation becomes

$$\mu - \mu_o = \nu \sigma_n = RT \ln\left(\frac{c}{c_o}\right) \quad (3)$$

But is it possible to approximate complex rock-forming minerals by a one-component material? Certain conditions must be met to permit mass transfer from one part of a diffusion path to another. A Schottky defect is an anion vacancy balanced by sufficient cation vacancies to maintain charge balance. It is the only kind of point defect (1) created and destroyed entirely at the grain boundary, permitting mass transfer, (2) whose concentration is sensitive to the state of stress existing at the grain boundary, and (3) which have a self-regulating mechanism for the ionic flow so that newly deposited crystal retains stoichiometry. Consequently, Schottky defects are responsible for pressure solution and diffusion creep, and it can be shown that it is possible to consider the problem entirely in terms of the sublattice with the slowest moving vacancies, as if this component made up the entire crystal, by defining the effective atomic volume Ω as the volume of 1 Schottky defect divided by the number of slow moving vacancies in the defect (Fryer, 1972). In silicates, this is probably a vacancy in an oxygen sublattice. It is customary to discuss vacancies in terms of atomic quantities, and so Correns' equation (3) becomes, with k as Boltzman's constant,

$$\frac{\Omega\sigma_n}{kT} = \ln\left(\frac{c}{c_o}\right) = \ln\left(1 + \frac{c - c_o}{c_o}\right) \simeq \frac{c - c_o}{c_o}.$$

The error in this approximation is less than 10 percent when $\left(\frac{\Omega\sigma_n}{kT}\right)$ is less than 2×10^{-1}. For typical geological conditions with a normal stress of $3.5 \times 10^8 nm^{-2}$ (3.5 kb), temperature 625°K, and a vacancy the size of 0^{-2} we find that $\left(\frac{\Omega\sigma_n}{kT}\right)$ is 10^{-1}, so that the approximation appears to be good.

If 2 points on a grain boundary have a difference in concentration Δc as a result of a difference in normal stress $\Delta\sigma_n$,

$$\frac{\Delta c}{c_o} \simeq \frac{\Omega\Delta\sigma_n}{kT} \quad (4)$$

Now we can proceed to derive the diffusion flow law more or less along the lines set out by Nabarro (1948) and Herring (1950). The flow rate of vacancies $J = -D'\frac{\partial c}{\partial x}$ from Fick's first law at the quasi-steady state, where D' is the diffusion coefficient of vacancies. The vacancy influx is equal to the ion outflux, so $D' c_o = -D$ in terms of the ion's diffusion coefficient D.

$$\text{Now concentration } c = \frac{\text{number of vacancies}}{\text{number of sites}} \cdot \frac{\Omega}{\Omega} = \frac{\text{volume of vacancies}}{\text{volume of crystal}}.$$

As a result we are able to interpret J as the rate of volume flow of vacancies. The velocity v with which a crystal face of unit area moves out is equal to the volume rate of ions added on to the face, which is in turn equal to the volume rate of vacancies which leave this spot, so $v = J$. The strain-rate $\dot{e}$ is equal to the spatial rate of change of velocity,

$$\dot{e} = \frac{\partial v}{\partial x} = \frac{\partial J}{\partial x} = \frac{D}{c_0}\frac{\partial^2 c}{\partial x^2} \sim \frac{D\Delta c}{c_0 X_0^2}$$

where X is the diameter of an equant grain. Substituting in equation (4)

$$\dot{e} \simeq \frac{D\Omega}{X^2 kT} \Delta\sigma_n .$$

The mass flow of vacancies from one point to another within a given grain in response to the gradient in normal stress around the grain boundary changes the shape of the grain. Gradients in local normal stress around a given grain consist of components of both local mean stress and local deviatoric stress. It is necessary however to relate the strain rate to the stress on the scale of the polycrystalline aggregate rather than the local stresses around an individual grain of changing shape. Choose a volume of rock big enough to contain large numbers of grains yet small enough that the macroscopic stresses and strains on the boundaries of the volume are homogeneous. If the boundaries of the polycrystalline specimen were subjected to an entirely hydrostatic stress, the local normal stress gradients around grains would quickly relax as a result of diffusion, and the diffusion process would then cease. The local normal stress gradients are proportional to the larger scale macroscopic deviatoric stress σ acting on a polycrystalline sample. If K is a proportionality constant,

$$\sigma = K\Delta\sigma_n$$

$$\therefore \dot{e} \simeq \frac{KD\Omega}{X^2 kT}\sigma$$

The purpose of the above treatment was to show that there are only a few steps from the work of Correns on the geological problem to the work by Nabarro and Herring from the metallurgical point of view.

If the diffusion path is through the bulk lattice, as originally suggested by Nabarro (1948) and Herring (1950), then

$$\dot{e}_l = \frac{K D_l \Omega}{X^2 r^2 k^2 T}\sigma \qquad (5)$$

While if the diffusion path is along the grain boundary as Cobble (1963), Lifshits (1963), and Gifkins (1967b) suggested, then

$$\dot{e}_{gb} = \frac{K \pi D_{gb} h \Omega}{X^3 r^{3/2} k T}\sigma \qquad (6)$$

In these flow laws, lattice and grain-boundary strain-rates are $\dot{e}_l$, $\dot{e}_{gb}$ and diffusion coefficients are D_l, D_{gb}; r is ratio of the grain's longest to shortest dimensions, and π has its usual value. For grain ratio r up to about 3:1, K is 21, but for extremely elongate grains, K is $8r$.

What measure of strain rate and stress are we using in these equations? The intensity of the deviatoric stress acting on the polycrystalline sample may be measured by the octahedral shear stress σ. In terms of the principal values of the stress tensor σ_1, σ_2, σ_3, the octahedral shear stress is:

$$\sigma = \tfrac{1}{6}[(\sigma_1 - \sigma_2)^2 + (\sigma_2 - \sigma_3)^2 + (\sigma_3 - \sigma_1)^2]^{\frac{1}{2}}$$

Similarly, the deviatoric strain-rate can be represented in terms of the octahedral shear strain-rate

$$\dot{e} = \tfrac{1}{6}[(\dot{e}_1 - \dot{e}_2)^2 + (\dot{e}_2 - \dot{e}_3)^2 + (\dot{e}_3 - \dot{e}_1)^2]^{\frac{1}{2}}$$

The stress σ and strain-rate $\dot{e}$ are always positive.

The diffusion creep equations (5 and 6) have been rederived at least a dozen times in the past 25 yrs. We are using here results of the most complete and rigorous derivation to date, by Raj and Ashby (1971, equations 24 and 26).

It is instructive to trace the most general possible history of a vacancy from its creation to its annihilation. A vacancy can be created at a grain boundary or within the grain, with or without interference from impurities or inclusions. The vacancy then travels through the grain in a path along which the total number of vacancies may or may not be conserved at any instant. When the edge of the grain is reached, the disordered material of the grain boundaries collapses into the vacancy, eradicates it, and changes the shape of the grain. This sequence of vacancy source, diffusion path, and vacancy sink is a series of dependent processes so that the slowest must be the rate-determining step. Diffusion creep assumes that the operation of the vacancy sources and sinks is completely unhindered and entirely located at grain boundaries, or in other words, that grain boundaries function as *ideal* sources and sinks. The rate-determining step is assumed to be the diffusion between source and sink.

The conservation of vacancies means that the divergence of the flow of vacancies is zero within the crystal, or alternatively that the Laplacian of the vacancy chemical potential μ is zero:

$$\nabla^2 \mu = 0$$

The solution of this Laplace's equation must satisfy boundary conditions of mechanical equilibrium, continuity, and chemical equilibrium. The equation of mechanical equilibrium is the same as is used in elasticity and plasticity and requires no comment, but the equations of continuity and chemical equilibrium lead to some unusual aspects which we will now discuss.

There have been two main ways of dealing with the continuity equation. Lifshits (1963), Gibbs (1965), Gifkins (1967b), and Raj and Ashby (1971) calculate the diffusional strain-rate using a space-filling aggregate of grains which are deformed in such a way that the polycrystalline continuity is always maintained. The stress field within the individual grains can be calculated (Raj and Ashby, 1971, Append. 2) and is found to be variable with large stress concentrations—what one would expect in natural polycrystalline rock. Another consequence of satisfying the continuity equation is that there is a coupled and dependent relation between diffusive mass transfer and grain-boundary sliding (Fig. 1). Diffusional creep could be called grain-boundary sliding with diffusional accommodation.

The other procedure is to avoid the problem of maintaining grain-to-grain continuity by dealing with a single grain of spherical shape (Herring, 1950; Coble, 1963; Green, 1970). In this case, a homogeneous stress field is maintained throughout the grain. The equations which result from these unrealistic abstractions are nevertheless approximately the same as those deduced from the polycrystalline model.

The equation (3) we have used for the chemical potential is

$$\mu = \mu_o + \nu \sigma_n .$$

Is this equation exact? What is its relation to the "theory of nonhydrostatically stressed solids"? The full equation for nonhydrostatically stressed solids was derived by J. W. Gibbs in 1876 and has received attention from Yang and others (1959), Kamb (1959, 1961a, 1961b), and McLellan (1970). I find the treatment by Li and others (1966) particularly clear. Green (1970) applied it to diffusion creep. This exact equation is

$$\mu = \mu_o + \omega + \nu \sigma_n .$$

Here ω is the elastic strain energy per mole, a term assumed negligible in equation (3). For typical geological conditions, this term affects the value by less than 1 percent, and so the approximation appears sound.

Changes in Grain Size and Shape

At constant stress, both lattice and grain-boundary diffusion flow laws (equations 5 and 6) have the general form:

$$\dot{e} = \frac{K'}{X^n r^m}$$

Grain size X is to a power n of 2 or 3, grain shape r to a power of 2/3 or 2, and K' is constant for the constant stress.

Green (1970) and Weertman (1970) suggested that this steady-state situation can never be the case and illustrated their interpretation with a coaxial deformation path (Elliott, 1972). *If* the path is coaxial and grains do not rotate, then as the grains get more elongate, r gets larger, and the process is apparently self-exhausting. This combination of assumptions would result in the maximum possible departure from the steady state, but several processes will operate to produce a situation closer to a steady state.

1. Even in a coaxial deformation path, the grains will be continuously rotating relative to each other as a result of grain-boundary sliding. Grain-boundary sliding brings grains together and moves them apart and is constantly shuffling grains about, bringing different faces into contact. The newly precipitated part of a grain can subsequently be subjected to compressive stresses and dissolve, tending to maintain equiaxial grains.

2. A coaxial deformation path is a very special case which will only occur at local spots in the deforming rock (Elliott, 1972). With

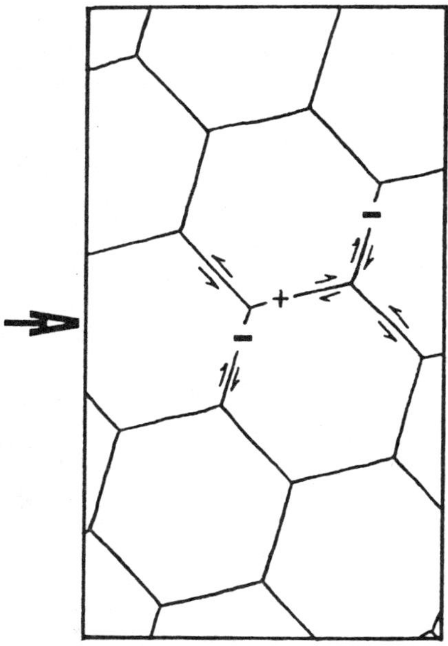

Figure 1. Space-filling array of hexagonal grains deformed by diffusion-accommodated grain-boundary sliding. For new crystal to precipitate (+) at a grain boundary under tension, the adjacent grains have to move apart. Grain-boundary sliding to give this grain separation must occur, and old crystal must be presolved (−) at those grain boundaries under compression. Precipitation, dissolution, and grain-boundary sliding all occur simultaneously and in a certain proportion, otherwise voids will open and continuity will be destroyed.

general noncoaxial paths, the tendency to maintain equiaxial grains will be enhanced.

3. Grain size X may decrease as the grain ratio r increases. This happens if the newly precipitated crystal forms a new grain, as is often observed. Conceivably this could even produce a constant ratio $K'/X^n r^m$ and lead to a steady state.

It appears that changes with time in grain shape and size have a complex effect difficult to evaluate precisely and are clearly in need of experimental work with both coaxial and noncoaxial deformation paths. I feel that these deviations from a steady-state creep are probably a minor correction, compared for example to changes in temperature or chemical environment during the course of a typical geological deformation episode.

Experimental Evidence of Diffusion Creep

Experimental determination of the stress–strain-rate relations for a number of materials provides unequivocal evidence of diffusion flow. Particularly convincing experiments are those of Burton and Greenwood (1970a, 1970b, 1971; see also Gifkins, 1971), who demonstrate both lattice and grain-boundary diffusion flow laws and show a linear viscous or Newtonian stress to strain-rate relation for pure copper.

Materials consisting of more than one phase can deform by diffusion creep under suitable stress levels, but the flow law may be more complex. Experimental work on magnesium alloys, polycrystalline corundum, and polycrystalline bromellite (BeO) has been reviewed by Karim (1970; see also Holt, 1970) and shown to follow a Bingham flow law for diffusion creep. Work by Burton (1971) on impure copper also suggests a Bingham flow law. Very fine grained, second-phase particles appear to inhibit the action of the grain boundaries from acting as ideal defect sources and sinks (Ashby, 1969; Greenwood, 1970; Burton, 1972). The rate of vacancy production now becomes the rate-determining step.

These second-phase particles must be extremely fine grained to have this blocking effect on grain-boundary vacancy production. Optically visible inclusions have a negligible effect on the diffusion flow law (Gibbs, 1965; Raj and Ashby, 1971). However, one important consequence of optically visible inclusions is to retard or even prevent grain-boundary migration.

For more details of experimentally determined stress to strain-rate relations, the reader is referred to the references cited above. The main concern of this paper is the possibility of demonstrating diffusion creep by direct microscopic observations in metamorphic rocks, as will be discussed next.

DIRECT OBSERVATION OF DIFFUSION CREEP

A clear idea of what precisely constitutes a grain boundary at any specific time is essential. Is the grain boundary always made up of the same material particles? What happens when part of an original grain boundary is presolved or a part is added on?

Imagine a thin marker surface defined by inert markers, and let this surface initially coincide with the grain boundary and completely enclose a grain (Fig. 2A). When the polycrystalline aggregate is deformed, the

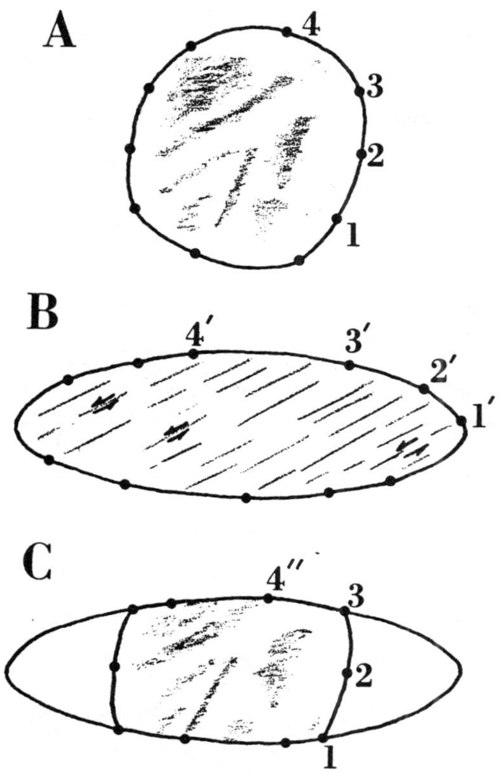

Figure 2. (A) Undeformed original grain with material points on grain boundary 1, 2, 3, 4. (B) Deformation of grain accomplished by dislocation glide. Transformed points 1′, 2′, 3′, 4′ still define the grain boundary. (C) Grain deformation by diffusion mass transfer. Relic of original grain boundary with material points 1, 2, and 3 now inside new crystal which is precipitating at each end. Solution surface cuts across top and bottom of old crystal. A material particle of some inert phase originally at 4 now accumulated at 4″. Grain boundary defined by new material particles.

grains must change shape, but there are only two geometric ways by which they can do this.

In the first method, the grain changes shape by a continuous transformation or strain. The material particles defining the grain boundary before the deformation lie on this boundary after the deformation (Fig. 2B). The mechanisms by which this grain deformation process is accomplished are dislocation processes such as slip or twinning.

In the second method (diffusion creep), material particles are removed (dissolved) from some parts of the original boundary and added elsewhere (precipitated) to the outside of the grain boundary. Inert particles *accumulate* at places where the crystal is being dissolved; and in regions where *new crystal* is building outward, the *relict grain boundary* may separate new crystal from *old crystal*—the remnants of the grain before deformation (Fig. 2C).

Chemical bonds of the old crystal must be broken and new ones formed in the new crystal. Sites of deposition of new crystal at any instant is the grain boundary. It is impossible for the old crystal to migrate out to the position of the new grain boundary.

Diffusion creep has been widely observed in a magnesium alloy optically and by transmission electron microscope (Squires and others, 1963; Rainey and Holt, 1971 and references cited there). The original old crystals are peppered through with tiny inert particles. During deformation, grain boundaries undergoing compression accumulate inert markers as the old crystal is dissolved away, while grain boundaries under tension develop zones of new crystal which are free of the inert markers (Fig. 3).

Several authors have suggested that the zones of new crystal in the magnesium alloy could be formed by grain-boundary migration rather than diffusion creep (Vickers and Greenfield, 1967; Lee, 1969; Dunlop and Taplin, 1972). What is the difference between these two diffusive processes, and what observations could be used to distinguish them?

Grain-boundary migration is an effort to minimize the elastic strain energy stored in the crystal or to minimize surface energy, whereas the driving force for diffusion creep is a normal stress gradient. Grain-boundary migration is a short-range diffusive readjustment of local atoms, but diffusion creep involves long-range transport of material from one side of a grain to another.

Although both diffusion creep and grain-boundary migration could produce optically continuous zones of new crystal, grain-boundary migration could not produce new grains of new crystal. *Dissolved parts of old crystal with inert particle accumulations* at grain boundaries under compression cannot be explained by grain-boundary migration—accumulations which were observed in the magnesium alloy by Rainey and Holt (1971).

The microscopic effects of diffusion creep can be seen in thin sections from the Eocambrian Weaverton Formation of the South Mountain–Blue Ridge region in Maryland. These quartzites were deformed into tight isoclinal recumbent folds under greenschist facies at about

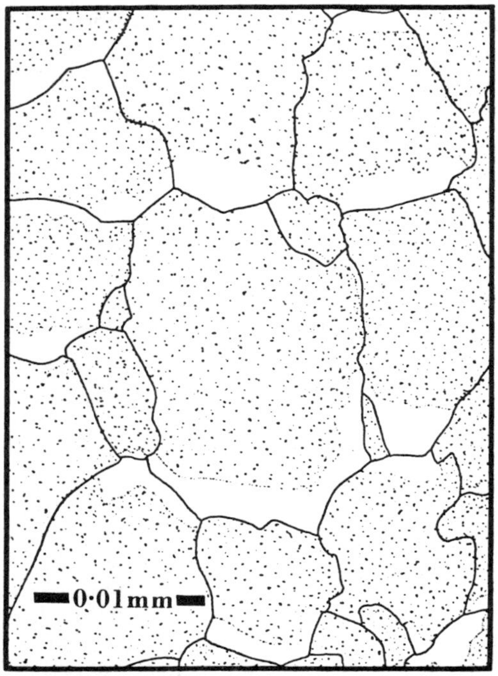

Figure 3. Original old crystal is full of finely disseminated inert particles (stippled). At grain boundaries under compression, old crystal is dissolved away leaving behind accumulations of inert particles. New crystal grows at grain boundaries under tension without inert particles (white). Vertical stretching direction, deviatoric natural strain *e* 28 per cent in a coaxial deformation path. Drawn from microphoto (of Karim, 1970, Fig. 2) of a magnesium alloy which underwent diffusion creep.

625°K, under an overburden equivalent to a hydrostatic pressure of $3.5 \times 10^8 nm^{-2}$ (3.5 kb).

In thin section, the dust rings which enclose the rounded detrital grains provide a recognizable marker surface enclosing the old crystals (Figs. 4, 5, 6, and 7). Where grain boundaries are under compression, the grains are presolved and the "dust" or "insoluble residue" forms zones of inert accumulations. At grain boundaries under tension, new crystal is formed outside the dust ring. This new crystal often has a diagnostic fibrous habit and frequently is full of inclusions and interleaved with minute white micas and chlorites. Occasionally, the new crystal is a new grain and not optically continuous with the old grain, so that X decreases as r increases with time.

The new quartz growing on the tensile side of opaque minerals such as magnetite and forming "pressure shadows" is particularly spectacular. Wherever magnetite/quartz grain boundaries are under compression, the magnetite deeply penetrates into the quartz grains (Fig. 7C; Elliott, 1972, Fig. 11).

Not all grains in a thin section show pressure solution or new quartz, nor was the depth of presolving or the width of overgrowths the same in all grains. This is not unexpected due to the inhomogeneity of stress on the scale of single grains.

Stratigraphically below the Weaverton quartzites is the Catoctin Formation, a pile of Precambrian volcanic rocks which includes rhyolitic flows. These flows frequently show *micropoikilitic texture*—single crystals of quartz which are choked with large volumes of fine feldspar inclusions. This texture is easiest seen with a gypsum plate, when the whole field of the microscope is broken up into different patches of vivid contrasting color, each representing one micropoikilitic quartz grain.

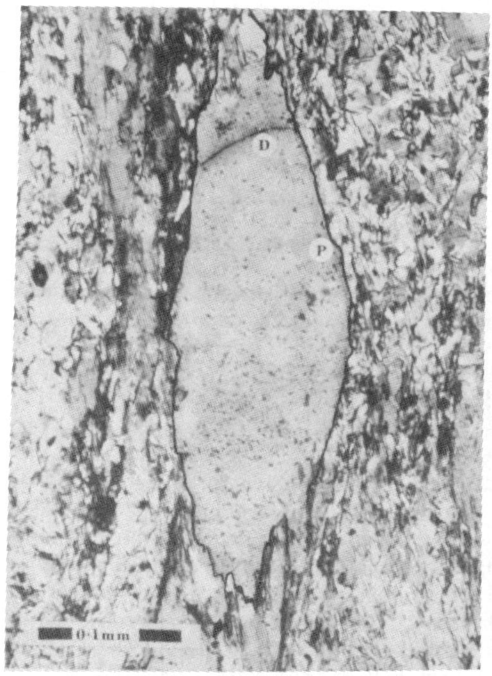

Figure 4. Remnant of a once much larger clastic grain still preserving a short gently curved arc of dust ring *D*. Grain shape a result of massive removal of material from presolved surfaces *P*. Weaverton Formation, crossed nicols, (all photomicrographs are of thin sections cut normal to cleavage and parallel to stretching direction).

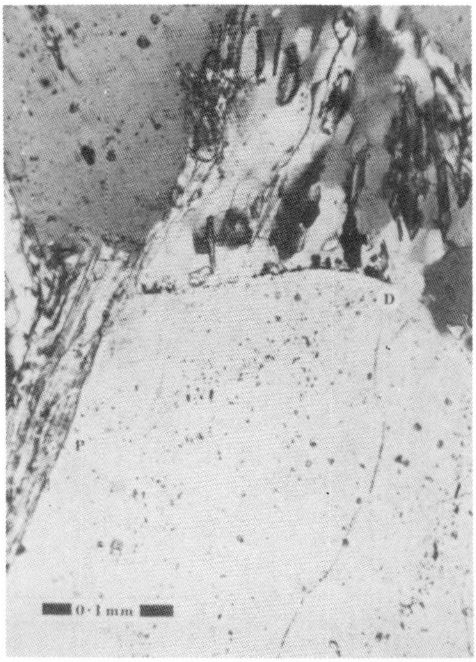

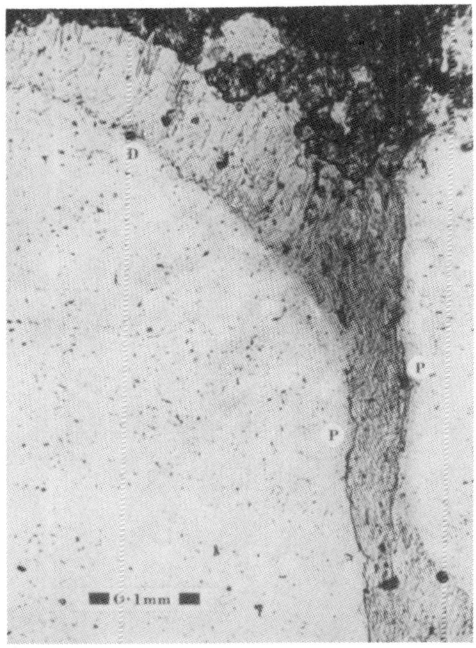

Figures 5 and 6. Sides of grains under compression show presolved surfaces outside of which are accumulations of inert grains. New quartz and some white mica is growing in extension direction outside of dust rings *D* marking relic grain boundaries enclosing old crystal. Weaverton Formation Figure 5 (left), crossed nicols; Figure 6 (right), plane light.

Micropoikilitic texture was produced by the initial crystallization and hydration of volcanic glass at subsolidus temperatures, probably during deuteric alteration of the volcanic pile toward the end of Catoctin volcanism. The term and its significance is a result of the classic work by Florence Bascom (1893, 1896) on the Catoctin volcanic rocks from the South Mountain area (the texture is reviewed by Lofgren, 1971).

Deformed micropoikilitic texture is particularly informative. The originally subspherical quartz grains are now ellipsoidal shapes parallel to the penetrative cleavage (Fig. 8). How did the deformation of these quartz grains come about? The homogeneous extinction colors with a gypsum plate indicate few subgrains and little dislocation glide or slip. The grains have drifted apart with new fibrous quartz and white mica growing at the ends. The sides have presolved and show occasional accumulations of inert hematite, which is scattered throughout the old crystal.

In both formations, it appears that diffusion creep is the mechanism which permits the deformation, but microscopic observations alone cannot establish whether grain boundary or lattice diffusion was dominant. It is necessary to specify the various paths along which diffusion may occur and discuss the physical nature of these paths.

DIFFUSION PATHS

What path do the vacancies travel on to accomplish pressure solution or diffusion creep? The type of path through which diffusion occurs permits a clear-cut distinction between the various diffusion processes. Four different diffusion paths can be recognized: through the bulk volume of the crystal lattice, grain boundaries ("high angle" $\gtrsim 10°$ unless otherwise designated), dislocations, and through pore fluids between the grains (Fig. 9).

Many petrologists would agree that paths must exist in deforming metamorphic rock with a diffusivity very much higher than is possible through the interior of the crystal lattice. These high diffusivity paths are implied by the term "dispersed phase" and the geological evidence suggesting the existence of

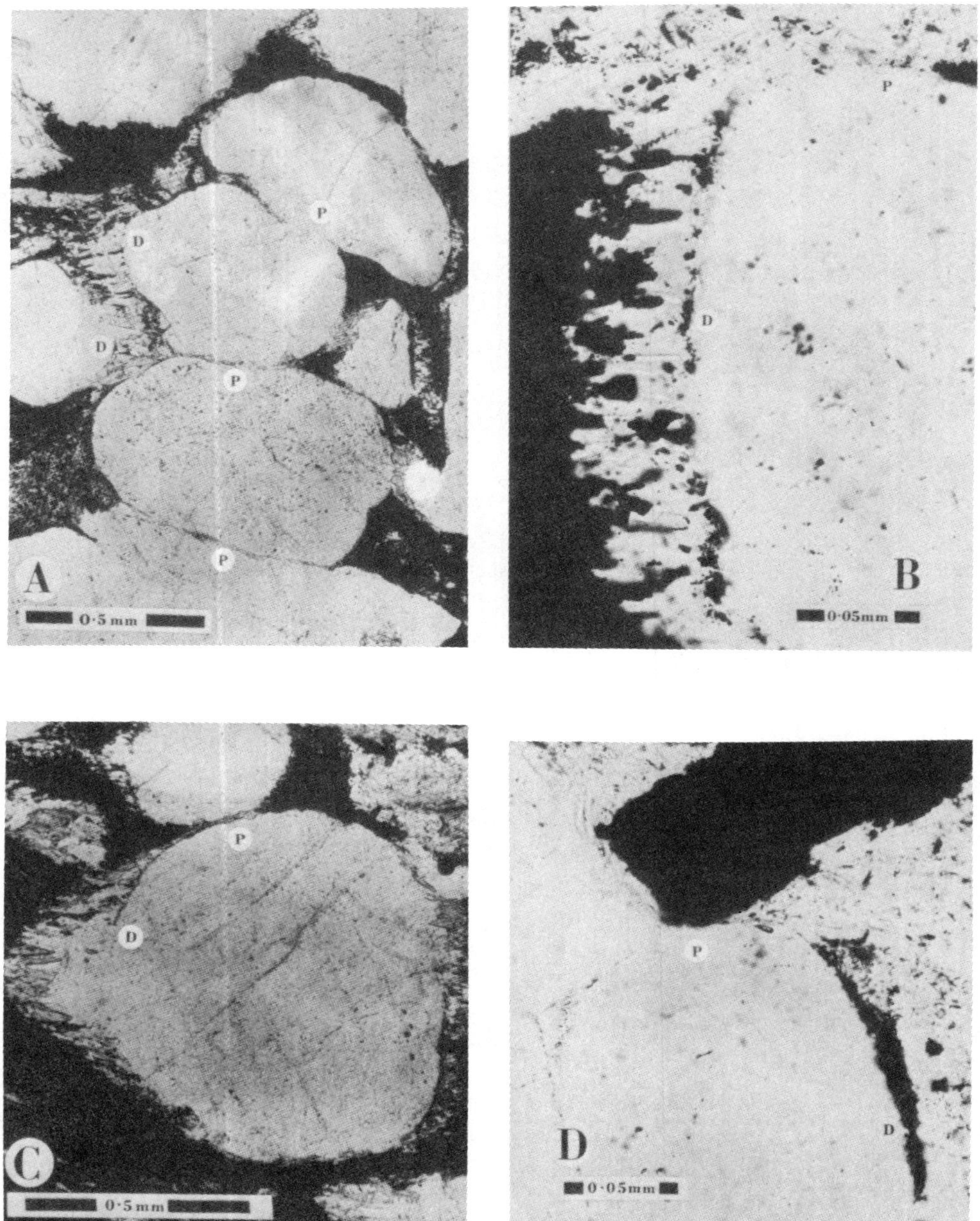

Figure 7. Hematite-cemented quartz sandstone at base of Weaverton, plane light. The dust rings indicate well-rounded, nearly circular clastic grains. Thin section is from upper limb of a recumbent isoclinal fold with local bedding/cleavage angle 14°. Straight fibers indicate an approximately coaxial deformation path. (A) and (C) Grains show dust rings *D* cut by presolving surfaces *P*. New quartz growing outside dust rings. (B) Detail of new quartz of fibrous habit growing outside dust ring *D*. Dust ring cut off by presolved surface *P* at top. (D) Ilmenite grain impressed into and piercing *P* dust ring *D* around old quartz grain.

rapid diffusion paths has been reviewed by Gresens (1966) and Mueller (1966).

The nature of grain boundaries has been deduced both from theoretical considerations and, in metals, direct observation with the field-ion microscope. This work essentially confirms the island theory due originally to Mott (1948) and summarized below (McLean, 1965; Gibbs and Harris, 1969; Gifkins 1967a, 1969).

(1) A grain boundary encloses each crystal. (2) The crystal lattice is very much more ordered within the grain than at the boundary. (3) The grain-boundary structure consists of "islands" and "channels." The "islands" are three to five atomic spaces thick with a distorted lattice, but comparatively good fit to crystal lattices of both adjacent grains. The "channels" separate the islands from one another and are regions of very poor fit, about one or two atomic spaces thick. (4) The channels form an interconnecting network lying in the grain boundary; the major part of the grain boundary energy and rapid diffusion is associated with the channels, which have a very high density of vacancy defects. (5) The greater the atomic misfit between adjacent grains, the wider the grain boundary. Interphase grain boundaries have the highest atomic disregistry and consequently the greatest grain boundary width and greater potential for diffusive mass transport.

It is found that the zone of enhanced diffusion is not all concentrated along the channels, and two simple observations indicate a much wider effective grain boundary than one or two atomic spaces.

1. The hardness of a large lead crystal is constant within the grain interior but suddenly declines by about 10 percent at the boundary between two grains (Westbrook, 1968). The width of the zone of decreasing hardness is tens of micrometers.

2. Etched and polished specimens do not show infinitesimally thin lines at grain boundaries, but optically visible bands of the order of microns wide (Headley and others, 1970).

A major difficulty with many crystalline materials is the extreme sensitivity of diffusion coefficients to even trace impurities in the host crystal lattice—a sensitivity so great as to account for orders of magnitude discrepancies between laboratories reporting diffusion data for the same substance. These "extrinsic" effects due to impurities become particularly important at temperatures below about 3/4 of the absolute melting point. Ionic and covalently bonded crystals are particularly sensitive to this effect. The observed enhanced diffusion along micrometer-wide grain boundaries in periclase is largely due to impurity concentrations (Wuensch and Vasilos, 1966). These impurities could be in the form of small crystals of another phase or increased impurity concentrations in solid solution near the grain boundary.

The mechanisms of diffusion processes are probably best understood in the halides. In halite, the presence of impurities such as H_2O on the grain boundary produces an enormous increase in grain boundary diffusion rates (Riggs and Wuttig, 1969). Ions of OH^- concentrate at grain boundaries and change either its structure or its charge distribution, or both.

In summary, it appears that grain boundaries have different diffusion properties than do grain interiors because of their different structures and their different chemistry. A *sandwich model* of a grain boundary appears to be most appropriate (Gibbs and Harris, 1969). The center of the sandwich has a thickness of atomic dimensions with extremely high diffusivity, and its structure is described by the island model. The two outer layers of the sandwich, of micrometer size, have a smaller enhancement of diffusion due to impurities.

Grain boundaries in metamorphic rocks are frequently peppered with concentrations of small crystals of second phases, and interphase grain boundaries are common. Micas may be especially important, for although essentially inert when in contact with quartz, they greatly increase the width of the diffusion paths (Weyl, 1959). Furthermore, it is widely held that grain boundaries in metamorphic rocks contain many substances not admitted into the crystal lattices, typically H_2O, CO_2, HCl, H_2S, and HF in various forms. These compounds all are well known to be capable of greatly disrupting the structures along crystalline surfaces. They would produce an effective grain boundary wider than normal. Metamorphic rocks appear to have the conditions favoring an enhanced diffusion along grain boundaries of the sandwich model.

Dislocations are believed to act as thin pipes of rapid diffusion through a crystal. The dislocation core is very poorly ordered and is associated with most of the diffusion. This core is surrounded by an outer layer in which diffusion is slower than in the core, but still at a higher rate than through the bulk lattice. The dislocation can therefore be imagined as a coaxial pipe.

Low angle ($\lesssim 5°$) grain boundaries and other semicoherent interfaces can be described by dislocation arrays and their diffusion behavior resembles coaxial pipes laid side by side in a slab. In ordinary grain boundaries, however, the dislocations lie so close together that individual dislocations lose their meaning. Dislocation pipe diffusion depends upon the dislocation density, a quantity which is very sensitive to the elastic strain energy present in the crystal lattice and the thermal and deformation history. Consequently, a dislocation pipe diffusion flow law is strongly history-dependent and difficult to treat rigorously.

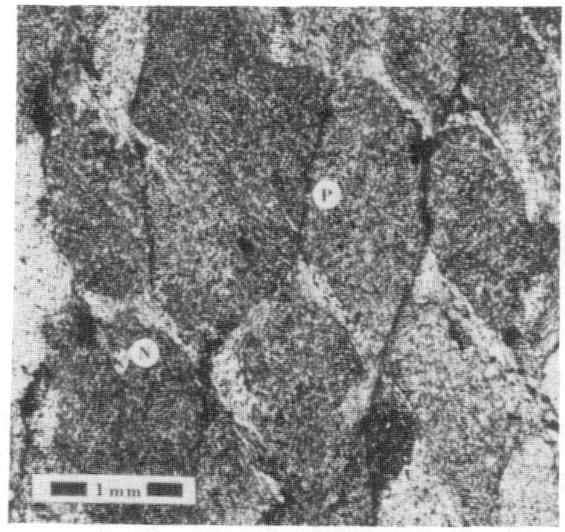

Figure 8. Dark-gray micropoikilitic quartz grains stuffed with inclusions and deformed by diffusion creep. Stretching direction is vertical. Presolved grain boundaries *P* with accumulations of inclusions at boundaries under compression; light-colored new fibrous quartz and white mica growing at grain boundaries under tension, for example above *N*. Catoctin rhyolitic flow, crossed nicols.

Pores

I am unaware of a definition of what constitutes a "pore." In this article, I will define a pore as having a radius large enough that the bonding forces of the atoms in the surrounding crystal have negligible effects on most of the pore volume. The pores are filled with liquid or gaseous phases, materials which behave as fluids of low viscosity and density. As a result, material within a pore will have negligible stress gradients and be essentially in a hydrostatic state.

Fissures are much larger still, but the behavior of fissure and pore fluids will be similar. Pores and fissures connect with the grain lattices by means of the high diffusivity paths provided by grain boundaries and dislocation pipes.

Sedimentary rocks may have had a high pore space immediately after deposition; but by the time cementation and diagenesis is completed, usually there is little pore space remaining. It is well established that the main pore reduction process during diagenesis is pressure solution (for example, Heald, 1955). This is accomplished by diagenetic overgrowths around the clastic grains and into the pores, filling them in. Presumably under metamorphic conditions, pressure solution is also an efficient pore reduction process.

The main distinction between diagenic and metamorphic pressure solution is the initial abundance of pore space in sedimentary rocks. However, even under conditions of diagenesis and deuteric alteration, pressure solution appears to proceed by diffusion other than through pore space; for example, stylolites found in dense glassy rhyolites (Bloss, 1954; Golding and Conolly, 1962) cannot have developed with the aid of an extensive pore space.

Pores and fissures (that is, veins) may be very effective in allowing mass transport into and out of a mass of rock. Progressive metamorphism is associated with intensive dehydration and decarbonation, and the large volumes which must be moved over considerable distances do so through a network of pores and fissures—but this is a larger scale than that of individual grains, with a typical dimension measured in meters.

Pores may very well be present (Brace and others, 1972), but I will now argue that for any reasonable amount of porosity in a deforming metamorphic rock, *mass transfer via pores cannot be a dominant process on the scale of individual grains.* Imagine a pore situated at a grain boundary (Fig. 9).

The mass transfer involves three distinct stages: (1) movement of material from that part of the grain which is under compression to the walls of the pore through the solid (via lattice or grain boundary); (2) dissolution of the material into the pore fluid; and (3) movement of the material through the pore fluid to its site of deposition. There are two mass-transport processes involved which are dependent on one another, and the very much smaller diffusion coefficients through solids than through pore fluids mean that diffusion through the solid will be the rate-limiting step.

The effect of pores is to reduce the length of the diffusion path. This is equivalent to increasing the effective grain size at which diffusion flow laws are important.

It is a common petrological assumption that large amounts of material are transported as associated or disassociated solute species of the sort found in aqueous solutions. Silica is often considered to move in this fashion as species such as H_4SiO_4. These types of particles are extremely large, and far too bulky to move along a grain boundary or dislocation pipe because they would jam into the crystal lattice (McLean, 1965). *If* transport of these large species occurs, it can do so only through pores. Possibly the species present in a pore and in a grain boundary are different because of this size control.

A monomineralic rock undergoing regional metamorphism should really be thought of as having three phases: the interior crystalline lattice phase, the phase of the grain boundaries and dislocation pipes, and the pore fluid phase. Very strong compositional gradients exist between these three phases.

Estimates of Paths

It is important to try to estimate an effective diffusion coefficient for the observed deformation of the Weaverton quartzite and Catoctin rhyolitic flows.

A general form of the diffusion flow laws (equations 5, 6) for a polycrystalline aggregate is

$$\dot{e} = \frac{K D_e \Omega}{X^2 r^2 k T} \sigma .$$

In general D_e, T, σ, X, and r all vary with time. If lattice diffusion operates, then $\dot{e} = \dot{e}_l$ and $D_e = D_l$; while if grain-boundary diffusion operates, $\dot{e} = \dot{e}_{gb}$, and the effective diffusion coefficient

$$D_e = \frac{D_{gb}\, h\, \pi\, r^{\frac{1}{2}}}{X} .$$

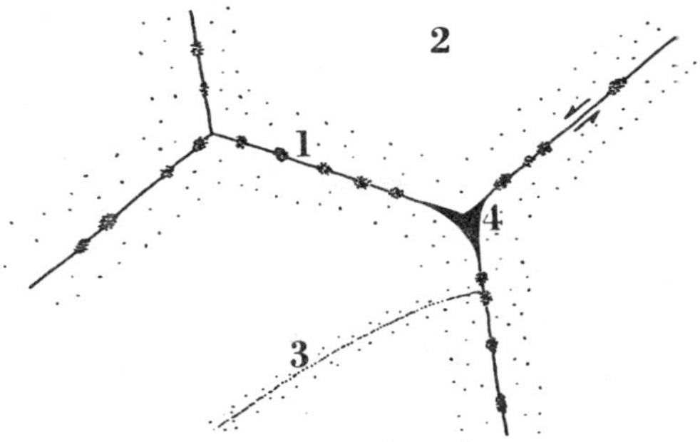

Figure 9. Possible diffusion paths in rocks. 1. Sandwich model of grain boundaries, diffusion mainly in channels (heavy stipple) but also through distorted lattice (open stipple). 2. Interior lattice of crystals. 3. Dislocation diffusion mainly through the core (dotted) but also in surrounding distorted lattice (open stipple). 4. Pore space (black). Here a pore is growing at a triple point due to grain-boundary sliding (arrows).

In a coaxial deformation path, the finite natural (or logarithmic) strain

$$e = \int_0^t \dot{e} \cdot dt$$

indicate maximum values by small m for T_m, σ_m, X_m, r_m. Maximum diffusion D_{em} will occur at T_m. Let t_m be the time interval from start to finish of the deformation event. Define $D_e = D_{em} \cdot \overline{D_e}$, $\sigma = \sigma_m \cdot \bar{\sigma}$, $X = X_m \cdot \overline{X}$, $r = r_m \cdot \bar{r}$, $t = t_m \cdot \bar{t}$.

Here $\overline{D_e}$, $\bar{\sigma}$, $\overline{X}$, $\bar{r}$, t, are normalized nondimensional quantities with values between 0 and +1.

Substituting into the general diffusion flow law and removing constants from behind the integral,

$$e = \frac{K D_{em} \Omega \sigma_m t_m}{k T_m X_m^2 r_m^2} \int_0^t \frac{\overline{D_e} \bar{\sigma}}{\overline{X}^2 \bar{r}^2 T} d\bar{t} .$$

Now the expression behind the integral is nondimensional and within an order of magnitude of unity, so that

$$D_{em} \approx \frac{e k T_m X_m^2 r_m^2}{K \Omega \sigma_m t_m} .$$

Finite natural strains e with coaxial deformation paths are in the order of 20 percent, the metamorphic temperature $T_m \sim 625°K$, the atomic volume of the rate-determining component Ω was probably an ion of 0^{2-}, the maximum tectonic stress σ_m was probably in the order of $10^7 nm^{-2}$ (100 bars), and diffusive flow is evident on grains with radii of 0.5 mm. The length of time for the deformation event responsible for these highly deformed rocks in the central Appalachians cannot be stated with any precision, but comparable deformation and syntectonic metamorphic events from the western Alps and New Zealand are no longer than 10 m.y. Using these figures, the *maximum* effective diffusion coefficient is $10^{-15} mm^2 s^{-1}$.

To date, remarkably few data are available for self-diffusion of oxygen through oxides and silicates. Some minerals which may be typical are summarized in Table 1. The large oxygen ion is usually the slowest moving and therefore the rate-determining component, and it may be reasonable to compare the self-diffusion of oxygen in different minerals at temperatures proportional to their melting points. A metamorphic temperature of 625°K is 0.32 of the melting point of pure silica at $10^5 nm^{-2}$ (1 bar); comparison of diffusion coefficients at such temperatures suggests that lattice diffusion is too small by at least three orders of magnitude (Table 1).

It is probable that water has a greater effect on grain-boundary than on lattice diffusion, because of concentration of water on grain boundaries as discussed earlier. This water probably increases the diffusion coefficient of oxygen through quartz. A rough way of estimating this might be to compare the minerals at 0.46 of their melting points, which is the ratio of metamorphic temperature to the melting point of silica at $3.5 \times 10^8 nm^{-2}$ (3.5 kb) in the presence of water (Table 1). If this comparison is correct, grain-boundary diffusion is a possible diffusion path, but the poor quality of such an estimate must be emphasized. Low temperatures also favor grain-boundary over lattice diffusion, as we shall see later.

TABLE 1. DIFFUSION THROUGH OXIDES AND SILICATES

O diffusion in:	At = 0.32 MP	At = 0.46 MP
Albite*	$10^{-23} \times 2.0$	10^{-18}
Vitreous silica†, SiO_2	$10^{-19} \times 2.1$	$10^{-15} \times 4.8$
Rutile, TiO_2†	$10^{-23} \times 4.2$	$10^{-17} \times 6.8$
Bromellite, BeO†	$10^{-22} \times 4.3$	$10^{-16} \times 1.5$
Cuprite, Cu_2O†	$10^{-20} \times 1.3$	10^{-15}
Bunsenite, NiO†	$10^{-22} \times 3.6$	10^{-15}

Note: All single-crystal data, except for silica.
** Merigoux, 1966.*
† Review of Harrop, 1968.

BROKEN-GRAIN DIFFUSION FLOW

The stress within a grain may reach the level at which fractures nucleate and grow, breaking up a larger grain into a number of smaller ones. Each smaller grain has its own grain boundary which is an ideal defect source or sink. Diffusion proceeds and the strain-rate will be much enhanced by the reduction in grain size.

Larger grains are particularly liable to transgranular fracture, either because they have a greater chance of containing a flaw or else because they have a greater chance of forming a microcrack from the stress concentrations around a pileup of dislocations (reviewed by Cottrell, 1964, p. 359). The result is that within a polycrystalline specimen, stress acts as an efficient grain-size reduction mill, systematically cracking all grains larger than a certain critical size.

The highly selective and systematic cracking of larger grains may be keeping all grains at a small enough size that the diffusion flow laws can dominate. Only a small percentage of grains would be cracking at any one time. Note, however, that diffusion mass transfer may cement the smaller fragments together, producing a larger elongate grain.

Grain fracturing can be seen easily if the grain and the fracture infilling are different minerals (Fig. 10). This is "microboudinage," important because of its suitability for measurement of deformation paths (Elliott, 1972). It is rare that grain fragments and fracture infilling can be distinguished if they have the same mineralogy, but a beautiful example is figured by Sander (1970, p. 363, Fig. 40). Sippel (1968) has presented very interesting evidence using cathode luminescence of widespread internal fracturing within quartz grains which is not visible with the petrographic microscope. This technique appears very promising, and broken-grain diffusion creep may be widespread.

Cataclastic rocks are generally considered to have developed in zones of particularly high stress which by itself would favor dislocation flow; but fine grain size is a distinctive feature of many of these rocks—is it possible that the grain size reduction permits such cataclastic rocks to deform by diffusion creep with a linear stress to strain-rate relation?

TRANSITION BETWEEN DISLOCATION AND DIFFUSION FLOW LAWS

In contrast to diffusion creep, dislocation-controlled creep has received considerable attention in geology (for example, Heard and Carter, 1968). Several different flow laws are possible depending upon which of the various possible dislocation mechanisms is operating. All are *independent* of grain size, but the stress dependence is a *power function* of the general form:

$$\dot{e}_d = \frac{B\, D_l\, \Omega}{kT}\,\sigma^n \qquad (7)$$

when $\dot{e}_d$ is the strain-rate due to dislocation creep, n and B are constant for a particular dislocation mechanism, and n takes a value between 3 and 6. The dislocation creep laws appropriate for rocks have been reviewed by Weertman (1970); all are special cases of the equation (7) above.

Subgrain boundaries are arrays of dislocations, and *if* these act as good sources and sinks of defects, then as Lifshits (1963) pointed out, the lattice diffusion creep equation must be interpreted with the *subgrain* radius rather than the grain radius. The subgrain size depends upon the stress to a power of 3, and so subgrain creep obeys a dislocation-type flow law (equation 7). Weertman (1970) and Green (1970) have suggested that subgrain creep may apply to rocks.

Figure 10. Tension fractures in magnetite grain, fractures infilled with new quartz. Weaverton; crossed nicols.

There is an important objection to the premises with which the subgrain creep law was derived. Edge dislocation components of subgrain boundaries can, in principle, emit or absorb defects, but to do this, they must increase local elastic strain energy and consequently are *not ideal* defect sources and sinks for defect concentrations which are near equilibrium (Gibbs and Harris, 1969). In an alloy known to undergo diffusion creep, no new crystal appears to be precipitating at subgrain boundaries; all occur at grain boundaries (Fig. 3).

Rock-forming minerals generally have fewer than the five independent slip systems which are needed to deform a polycrystalline aggregate compatibly and continuously by dislocation processes without grain-boundary sliding (the Von Mises criterion, see review of Paterson, 1969). We saw earlier that grain-boundary sliding was a necessary part of diffusion creep; consequently, grain-boundary sliding and grain deformation are *dependent* processes. Grain-boundary sliding is usually a faster and easier process than grain deformation (see Gibbs, 1967; Gibbs and Harris, 1969), so that grain deformation, by dislocation or diffusion creep, is the rate-determining step. This is so even though the total finite strain may be composed of about 60 percent strain due to grain-boundary sliding (Stevens, 1971)!

Grain-boundary sliding and grain deformation are treated as independent processes by Hahn and others (1967), and consequently their flow law is incorrectly derived.

How does the transition between diffusion creep and one of the dislocation creep laws occur? What is the total strain-rate affecting a specimen when the diffusion and dislocation flow laws are all operating on various grains? To answer these questions, it is necessary to briefly outline the rules to determine the appropriate flow law.

If a number of *independent* processes (A, B,

C . . .) are occurring during creep, then total whole-rock strain-rate component $\dot{e}_{ij}$ is

$$\dot{e}_{ij} = f_a \, \dot{e}_{a_{ij}} + f_b \, \dot{e}_{b_{ij}} + f_c \, \dot{e}_{c_{ij}} + \cdots$$

where $\dot{e}_{a_{ij}}$, $\dot{e}_{b_{ij}}$, $\dot{e}_{c_{ij}}$, are the strain-rate components corresponding to processes A, B, C, and so on. The total specimen strain-rate is the sum of the strain rates of the independent mechanisms, weighted according to the volumetric percent f_a, f_b, f_c of the grains dominated by each mechanism (Gifkins, 1970).

Process A requires both grain deformation and grain-boundary sliding; neither is sufficient on its own, so that $\dot{e}_{a_{ij}}$ are the strain-rate components due to both the grain deformation mechanism *and* the grain-boundary sliding necessary for coherence.

The total strain-rate in the cases considered in this article will generally be the sum of the dislocation, grain-boundary, and lattice-diffusion creep-rates:

$$\dot{e}_{ij} = f_d \, \dot{e}_{d_{ij}} + f_{gb} \, \dot{e}_{gb_{ij}} + f_l \, \dot{e}_{l_{ij}} \, .$$

In this relation, $\dot{e}_l$, $\dot{e}_{gb}$, and $\dot{e}_d$ are determined by equations (5), (6), and (7).

What is the general field within which one flow law *dominates* the others? How can these relations be plotted so as to be most readily usable for naturally deformed rocks? This can be answered by establishing the boundaries at which the strain-rates for any two different flow laws are equal. For cases when the contributions to the over-all specimen strain-rate $\dot{e}$ from both dislocation and grain-boundary diffusion creep are equal and lattice diffusion creep is negligible, then $\dot{e} \approx 2 f_d \, \dot{e}_d = 2 f_{gb} \, \dot{e}_{gb}$ and from equations (5) and (7)

$$\frac{D_l \, f_d \, B \, X^3 \, r^3 \, \sigma^{n-1}}{D_{gb} \, f_{gb} \, K \, \pi \, h r^{3/2}} = 1 \, . \qquad (8)$$

Similarly, when $\dot{e} \approx 2 f_d \, \dot{e}_d = 2 f_l \, \dot{e}_l$ and $f_{gb} \, \dot{e}_{gb}$ is negligible,

$$\frac{B \, X^2 \, r^2 \, f_d \, \sigma^{n-1}}{K \, f_l} = 1 \, ; \qquad (9)$$

also, when $\dot{e} \approx 2 f_l \, \dot{e}_l = 2 f_{gb} \, \dot{e}_{gb}$ and $f_d \, \dot{e}_d$ is negligibly small,

$$\frac{D_l \, f_l \, X \, r}{D_{gb} \, f_{gb} \, \pi \, h \, r^{\frac{1}{2}}} = 1 \, . \qquad (10)$$

Diffusion coefficients (such as D) are strongly temperature dependent and it may be shown that

$$D = D_o \, exp \, (-H/RT)$$

Here D_o and H are constants for a particular process and material but very different for different minerals and deformation processes. Because of the greater difficulty of forming and/or moving defects in the crystal than on the grain boundary, $H_l > H_{gb}$.

Now,

$$\frac{D_{gb}}{D_l} = \frac{D_{ogb}}{D_{ol}} exp \left(\frac{H_l - H_{gb}}{RT} \right)$$

$$= \frac{D_{ogb}}{D_{ol}} exp \left[\frac{H_l}{RT} (1 - H_{gb}/H_l) \right]$$

by substituting the above expression into equation (8) and taking natural logarithms,

$$\frac{H_l}{RT} = -\left(\frac{1}{1 - H_{gb}/H_l} \right) ln \left[\frac{D_{og} \, K \, \pi \, h \, r^{3/2} \, f_{gb}}{D_{ol} \, B \, \sigma^{n-1} \, f_d} \right] + \left(\frac{3}{1 - H_{gb}/H_l} \right) ln \, (Xr) \, . \qquad (11)$$

Similarly from equation (10)

$$\frac{H_l}{RT} = -\left(\frac{1}{1 - H_{gb}/H_l} \right) ln \left[\frac{D_{ogb} \, \pi h r^{\frac{1}{2}} f_{gb}}{D_{ol} f_l} \right] + \left(\frac{1}{1 - H_{gb}/H_l} \right) ln \, (Xr) \qquad (12)$$

and from equation (9)

$$ln \, (Xr) = \tfrac{1}{2} \left[\frac{K \, f_l}{B \, f_d \, \sigma^{n-1}} \right] . \qquad (13)$$

For a particular rock type and chemical environment, the variables are temperature T, grain size X, grain shape r, and stress σ. Temperature, grain size, and grain shape can be measured reasonably accurately, and the lattice-diffusion activation energy H_l of a wide variety of materials is available. It would seem that the most suitable representation is a semi-log plot with $\frac{H_l}{rT}$ for the vertical axis, versus grain size and shape $ln(Xr)$ for the horizontal axis (Fig. 11).

On such a graph, the relation $f_d \, \dot{e}_d = f_{gb} \, \dot{e}_{gb}$ (equation 11) is a straight line with slope $\frac{3}{1 - H_{gb}/H_l}$ and intercept with the *positive horizontal axis of*

$$ln \left[\frac{D_{ogb} \, K \, \pi \, f_{gb} \, h r^{3/2}}{D_{ol} \, B \, f_d \, \sigma^{n-1}} \right] .$$

The boundary along which $f_l \, \dot{e}_l = f_{gb} \, \dot{e}_{gb}$

(equation 12) is a line with slope $\frac{1}{1 - H_{gb}/H_l}$ and intercept on the horizontal axis of

$$ln\left[\frac{D_{ogb}\, \pi\, f_{gb}\, h r^{\frac{1}{2}}}{D_{ol}\, f_l}\right].$$

The boundary at which $f_d\dot{e}_d = f_l\dot{e}_l$ (equation 13) is a vertical line on the graph.

If $f_{gb}\dot{e}_{gb} = f_d\dot{e}_d$ and $f_d\dot{e}_d = f_l\dot{e}_l$, then $f_l\dot{e}_l = f_{gb}\dot{e}_{gb}$, and these three lines intersect at a common *triple point* (Fig. 4). At the triple point, the three equations (9, 10, 11) are satisfied for a particular rock type (D_{ol}, D_{ogb}, H_l, H_{gb}, f_{gb}, f_l, f_d, h, X, r) and metamorphic conditions T, but the *stress* is a difficult variable. Equation (12) is independent of stress, and so the triple point can only migrate along the straight line of equation (12). Increasing stress moves the triple point downward and to the left along this straight line.

The calibration of this semilog plot (Fig. 11) is of key importance and could be done by laboratory experiments. For example, using radioisotope self-diffusion experiments, it is possible to measure H_{gb}/H_l, D_{ol}, D_{ogb}, and h.

Another approach would be to directly calibrate positions of the triple point and slopes of the lines by varying temperature, grain size and shape, and stress in a series of creep tests. This may require faithful duplication of the chemical conditions of metamorphism, particularly trace impurities and grain-boundary concentrations of H_2O, CO_2, and so on. Parts of the graph may be experimentally unattainable because of difficulties in deciding the proportions of grain-deformation mechanisms at the steady state, or not achieving a true secondary creep stage.

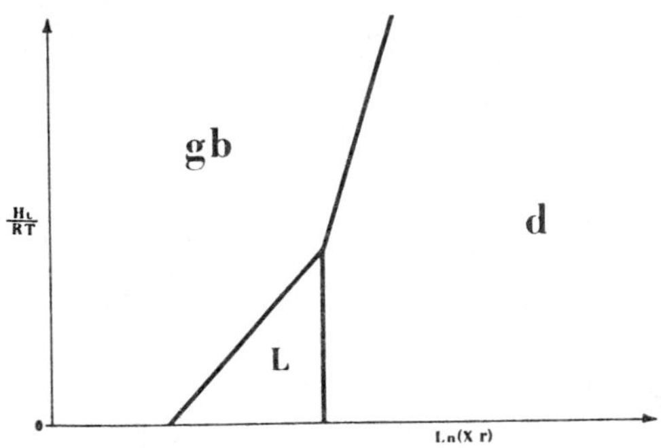

Figure 11. Semi-log plot of inverse temperature as Hl/RT versus grain size and shape as $ln(Xr)$. Lattice diffusion l, grain-boundary diffusion gb and dislocation d flow laws are each dominant in certain fields for a particular rock type.

It is apparent that direct extrapolation of laboratory creep tests to the low deviatoric stress levels typical of natural deformation under conditions of regional metamorphism requires an unjustified assumption of constant deformation mechanism, for as the stress is lowered, dislocation creep is *invariably* replaced by a diffusion flow law.

Another alternative is to try to calibrate the graph by careful observations in naturally deformed rocks, for the approximate temperature can frequently be estimated from the petrology of the mineral assemblages, and grain size and shape are readily measurable. Such work is in progress (see Elliott, 1970).

CONCLUSIONS

Familiar pressure solution is a diffusion mass transfer process, and if the grains in a metamorphic rock have changed their shapes mainly by pressure solution, a diffusion flow law holds with a *linear* stress to strain-rate relation.

It was demonstrated that only a few steps relate Correns' work on pressure solution to the family of diffusion flow laws of Nabarro-Herring type.

Microscopic observations exist which permit a clear distinction between deformation dominated by dislocation processes and deformation by diffusive mass transfer.

In photomicrographs of greenschist-grade quartzites and rhyolitic flows, pressure solution is the dominant process. But there are no microscopic criteria for identifying the path along which diffusion occurred. Estimates of the effective diffusion coefficient were attempted; available data is suggestive but does not permit conclusive calculations. On physical grounds, because of the low temperatures and the effects of water, one would expect grain-boundary diffusion to dominate.

The microstructure of grain boundaries in metamorphic rocks probably resembles a sandwich, a core of channel networks one or two atomic spaces thick having very poor atomic fit and extremely high diffusivity with micron-wide zones of enhanced diffusion on both sides. Diffusion through pores is not the rate-determining step on the grain scale, but may increase the grain size at which diffusion flow is effective.

Boundaries between the lattice, grain boundary, and dislocation flow laws can be drawn on a graph of inverse temperature versus grain size. Is it possible to calibrate this graph from observations of naturally deformed rocks of varying grain size and metamorphic grade?

Grain size occurs to a power of 2 or 3 in the diffusion flow laws and is an important variable. Diffusion flow is faster with small grains, and grain-size reduction by fracturing will tend to favor diffusion flow laws. Do cataclastic rocks undergo a large part of their strain history via broken-grain diffusion creep? More specifically, can the difference in grain-size dependence be used to distinguish grain boundary from lattice diffusion flow laws in naturally deformed rocks?

Within an active mountain-building belt, it is likely that temperature, grain size and shape, and deviatoric stresses vary with time and place. *No unique* flow law is proposed. Pressure solution or diffusion creep might dominate in an area, then grain coarsening combined with an increase in deviatoric stress could shift the flow law to one dominated by dislocation motion. It is emphasized, however, that the triple point in naturally deformed rocks is very much farther to the right than has hitherto been suspected.

ACKNOWLEDGMENTS

Various versions of this paper have been improved by critical reading by E. Cloos, H. Eugster, R. Fletcher, G. Fisher, and H. W. Green. This work was supported by the Earth Sciences Section, National Science Foundation Grant GA–10279.

REFERENCES CITED

Ashby, M. F., 1969, On interface-reaction control of Nabarro-Herring creep and sintering: Scripta Metallurgica, v. 3, p. 837–842.

Bascom, F., 1893, Structures, origin and nomenclature of the acid volcanic rocks of South Mountain: Jour. Geology, v. 1, p. 813–832.

—— 1896, The ancient volcanic rocks of South Mountain (Adams-Franklin Counties), Pennsylvania: U.S. Geol. Survey Bull. 136, 124 p.

Becke, F., 1903, Über Mineralbestand und Struktur der kristallinen Schiefer: Internat. Geol. Cong., Session 9, v. 2, Vienna 1903, p. 553–570.

Bloss, F. D., 1954, Microstylolites in a rhyolite porphyry: Jour. Sed. Petrology, v. 24, p. 252–254.

Burton, B., 1971, Influence of alumina dispersions on the diffusion-creep behavior of polycrystalline copper: Metal Science Jour., v. 5, p. 11–14.

—— 1972, Interface reaction controlled diffusional creep: A consideration of grain boundary dislocation climb sources: Materials Science and Eng., v. 10, p. 9–14.

Burton, B., and Greenwood, G. W., 1970a, Contribution of grain-boundary diffusion to creep at low stress: Metal Science Jour., v. 4, p. 215–218.

—— 1970b, The limits of the linear relation between stress and strain rate in the creep of copper and copper zinc alloys: Acta Metallurgica, v. 18, p. 1237.

—— 1971, Comments on the stress and temperature dependence of diffusion creep: Scripta Metallurgica, v. 5, p. 575–578.

Coble, R. L., 1963, A model for boundary diffusion controlled creep in polycrystalline materials: Jour. Applied Physics, v. 34, p. 1679–1682.

Cottrell, A. H., 1964, The mechanical properties of matter: New York, John Wiley & Sons, Inc., 430 p.

Dunlop, G. L., and Taplin, D.M.R., 1972, A metallographic study of superplasticity in a micrograin aluminium bronze: Jour. Materials, v. 7, p. 316–324.

Durney, D. W., 1972, Solution-transfer, an important geological deformation mechanism: Nature, v. 235, p. 315–317.

Elliott, D., 1970, The determination of flow laws in naturally deformed quartzites: Geol. Soc. America, Abs. with Programs (Ann. Mtg.), v. 2, no. 7, p. 546.

—— 1972, Deformation paths in structural geology: Geol. Soc. America Bull., v. 83, no. 9, p. 2621–2638.

Fryer, G. M., 1972, Theory of stress-induced diffusion in ionic crystals: Royal Soc. London Proc., ser. A, v. 327, p. 81–96.

Gibbs, G. B., 1965, Fluage par diffusion dans les solides polycristallines: Sci. Rev. Metalloug. Mem., v. 62, p. 781–786.

—— 1967, The role of grain boundary sliding in high temperature creep: Materials Science and Eng., v. 2, p. 269–272.

Gibbs, G. B., and Harris, J. E., 1967a, Development of the island model for grain boundaries: Materials Science and Eng., v. 2, p. 181–192.

—— 1967b, Superplasticity during creep: Jour. Inst. Metals, v. 95, p. 373–377.

—— 1969a, Diffusion at solid-solid interfaces, *in* Gifkins, R. C., ed., Interfaces: Sydney, Butterworth, 384 p.

—— 1969b, Grain boundary structure, *in* Gifkins, R. C., ed., Interfaces: Sydney, Butterworth, 384 p.

Gifkins, R. C., 1970, Transitions in creep behavior: Jour. Materials, v. 5, p. 156–165.

—— 1971, Stress and temperature dependence of diffusional creep—A reconsideration of some

recent results: Scripta Metallurgica, v. 5, p. 569–573.
Golding, H. G., and Conolly, J. R., 1962, Stylolites in volcanic rocks: Jour. Sed. Petrology, v. 32, p. 534–538.
Gordon, R. B., 1965, Diffusion creep in the Earth's mantle: Jour. Geophys. Research, v. 70, p. 2413–2418.
—— 1967, Thermally activated processes in the Earth: Creep and seismic attenuation: Royal Astron. Soc. Geophys. Jour., v. 14, p. 33–43.
Green, H. W., 1970, Diffusional flow in polycrystalline materials: Jour. Appl. Physics, v. 41, p. 3899–3902.
Greenwood, G. W., 1970, Possible effects on diffusion creep of some limitation of grain boundaries as vacancy sources or sinks: Scripta Metallurgica, v. 4, p. 171–174.
Gresens, R. L., 1966, The effect of structurally produced pressure gradients on diffusion in rocks: Jour. Geology, v. 74, p. 307–321.
Hahn, S. J., Ree, T., and Eyring, H., 1967, Mechanism for the plastic deformation of Yule Marble: Geol. Soc. America Bull., v. 78, p. 773–782.
Harrop, P. J., 1968, Self-diffusion in simple oxides: A bibliography: Jour. Materials, v. 3, p. 206–222.
Headley, T. J., Kalish, D., and Underwood, E. E., 1970, Current status of applied superplasticity, *in* Burke, J., and Weiss, V., eds., Ultrafine-grain metals: Syracuse, New York, Syracuse Univ. Press, 432 p.
Heald, M. T., 1955, Stylolites in sandstones: Jour. Geology, v. 63, p. 101–114.
Heard, H. C., and Carter, H. L., 1968, Experimentally induced "natural" intragranular flow in quartz and quartzite: Am. Jour. Sci., v. 70, p. 245–278.
Herring, C., 1950, Diffusional viscosity of a polycrystalline solid: Jour. Appl. Physics, v. 21, p. 437–445.
Holt, D. L., 1970, Superplastic, fine-grain alloys in the forming of sheet metal, *in* Burke, J., and Weiss, V., eds., Ultrafine-grain metals: Syracuse, New York, Syracuse Univ. Press, 432 p.
Kamb, W. B., 1959, Theory of preferred crystal orientation developed by crystallization under stress: Jour. Geology, v. 67, p. 153–170.
—— 1961a, The thermodynamic theory of nonhydrostatically stressed solids: Jour. Geophys. Research, v. 66, p. 259–271.
—— 1961b, Reply to discussions: Jour. Geophys. Research, v. 66, p. 3985–3988.
Karim, A., 1970, A theory of diffusional flow with apparent non-Newtonian viscosity in fine-grained materials, *in* Burke, J., and Weiss, V., eds., Ultrafine-grain metals: Syracuse, New York, Syracuse Univ. Press, 432 p.
Lee, D., 1969, Reply to: On the nature of superplastic deformation in the Mg-Al eutectic: Scripta Metallurgica, v. 3, p. 893–894.
Li, J.C.M., Oriani, R. A., Darken, L. S., 1966, The thermodynamics of stressed solids: Zeitschr. Phys. Chemie, N. F., Bd. 49, p. 271–290.
Lifshitz, I. M., 1963, On the theory of diffusion—Viscous flow of polycrystalline bodies: Soviet Physics, JETP, v. 17, p. 909–920.
Lofgren, G., 1971, Experimentally produced devitrification textures in natural rhyolitic glass: Geol. Soc. America Bull., v. 82, p. 111–124.
McKenzie, D. P., 1968, The geophysical importance of high-temperature creep, *in* Phinney, R. A., ed., History of the Earth's crust: Princeton, New Jersey, Princeton Univ. Press, 244 p.
McLean, D., 1965, The science of metamorphism in metals, *in* Pitcher, W. S., and Flinn, G. W., eds., Controls of metamorphism: Edinburgh, Oliver and Boyd, 368 p.
McLellan, A. G., 1970, Nonhydrostatic thermodynamics of chemical systems: Royal Soc. London Proc., ser. A, v. 314, p. 443–455.
Merigoux, H., 1966, Mesure du coefficient de diffusion de l'oxygine dans les feldspaths alcalius, Acad. Sci. Comptes Rendus, ser. D, v. 263, p. 1017–1018.
Mott, N. F., 1948, Slip at grain boundaries and grain growth in metals: Phys. Soc. [London] Proc., v. 60, p. 391.
Mueller, R. F., 1966, Mobility of the elements in metamorphism: Jour. Geology, v. 74, p. 565–582.
Nabarro, F.R.N., 1948, Deformation of crystals by the motion of single ions, *in* Report of a conference on the strength of solids: Phys. Soc. [London] Proc., 75 p.
Orowan, E., 1967, Seismic damping and creep in the mantle: Royal Astron. Soc. Geophys. Jour., v. 14, p. 191–218.
Paterson, M. S., 1969, The ductility of rocks, *in* Argon, A. S., Physics of strength and plasticity: Cambridge, MIT Press, 404 p.
Rainey, P. E., and Holt, D. L., 1971, An observation by transmission electron microscopy of diffusional creep in a magnesium alloy: Metallurg. Trans., v. 2, p. 3238–3239.
Raj, R., and Ashby, M. F., 1971, On grain boundary sliding and diffusional creep: Metallurg. Trans., v. 2, p. 1113–1127.
Riecke, E., 1895, Über das Gleichgewicht zwischen einem festen homogenen deformierten Körper und einer flüssigen Phase: Annalen Physic, v. 54, p. 731–738.
Riggs, K. R., and Wuttig, M., 1969, Cation grain boundary diffusion in NaCl: Jour. Appl. Physics, v. 40, p. 4682–4683.
Sander, B., 1970, An introduction to the study of fabrics of geological bodies [Trans. by Phillips, F. C., and Windsor, G.]: London, Pergamon

Press, 641 p.

Sippel, R. F., 1968, Sandstone petrology, evidence from luminescence petrology: Jour. Sed. Petrology, v. 38, p. 530–554.

Sorby, H. C., 1863, Über Kalkstein—Geschiebe mit Eindrücken: Neues Jahrb. Mineralogie, p. 801–807.

—— 1879, The anniversary address of the President: Geol. Soc. London Quart. Jour., v. 35, p. 39–95.

—— 1908, On the application of quantitative methods to the study of rocks: Geol. Soc. London Quart. Jour., v. 61, p. 171–233.

Squires, R. L., Weiner, R. T., and Phillips, M., 1963, Grain boundary denuded zones in a magnesium—½ wt% zirconium alloy: Jour. Nuclear Materials, v. 8, p. 77–80.

Stevens, R. N., 1971, Grain-boundary sliding and diffusion creep in polycrystalline solids: Philos. Mag., v. 23, p. 265–283.

Turner, F. J., and Verhoogen, J., 1960, Igneous and metamorphic petrology (2d ed.): New York, McGraw-Hill, 694 p.

Van Hise, C. R., 1896, Deformation of rocks: Jour. Geology, v. 4, p. 449–483, 593–629.

Vickers, W., and Greenfield, P., 1967, Diffusion creep in magnesium alloys: Jour. Nuclear Materials, v. 24, p. 249–260.

Voll, G., 1960, New work on petrofabrics: Geol. Jour., v. 2, p. 503–567.

Weertman, J., 1970, The creep strength of the Earth's mantle: Rev. Geophysics, v. 8, p. 145–168.

Westbrook, J. H., 1968, Analysis of nature of grain boundaries in ceramics, *in* Fulrath, R. M., and Pask, J. A., eds., Ceramic microstructures: New York, John Wiley & Sons, Inc., 231 p.

Weyl, P. K., 1959, Pressure solution and force of crystallization—A phenomenological theory: Jour. Geophys. Research, v. 64, p. 2001–2025.

Wuensch, B. J., and Vasilos, T., 1966, Origin of grain boundary diffusion in MgO: Am. Ceramic Soc. Jour., v. 49, p. 433–436.

Yang, L., Horne, G. T., and Pound, G. M., 1959, Effect of elastic strain on the electrode potential of metals, *in* Rhodin, T. N., ed., Physical metallurgy of stress corrosion fracture: New York, Interscience, 394 p.

Manuscript Received by the Society March 31, 1972

Revised Manuscript Received September 11, 1972

ERRATA

Two cited references were not printed:

Correns, C. W., and Steinborn, W., 1939, Uber die messung der sogenannten kristallisationskraft: Fortschr. Min. u. Petr., v. 23, p. 105-108.

Correns, C. W., 1949, Growth and dissolution of crystals under linear pressure: Discussions Faraday Soc., no. 5, p. 267-271.

8

Copyright ©1969 by the Geological Society of America
Reprinted from *Geol. Soc. America Bull.* **80:**1231-1264 (1969)

Principal Stress Directions from Plastic Flow in Crystals

NEVILLE L. CARTER *Department of Geology and Geophysics, Yale University, New Haven, Connecticut*
C. BARRY RALEIGH *National Center for Earthquake Research, U.S. Geological Survey, Menlo Park, California*

Abstract: Methods for determining orientations of principal stress axes in deformed rocks involve dynamic analysis of twin-gliding and of extinction bands produced by inhomogeneous translation gliding in crystals. The methods, beginning with Turner's (1953) technique for dynamic analysis of calcite twins, have been developed using as guides the results from experiments under controlled laboratory conditions. Structures induced by intragranular flow in calcite, dolomite, quartz, micas, orthopyroxenes, clinopyroxenes, olivine, and other common rock-forming materials, may now be used to derive orientations of principal stresses causing the deformation. The various methods, some new, are discussed in detail and examples of their application to tectonites are given. The usefulness of such studies is illustrated by evaluating the observed orientations of principal stresses around folds in light of new data from a theoretical analysis of large amplitude folding of viscous layers in a less viscous matrix. Other areas of research in structural geology in which these methods should prove useful have also been outlined.

CONTENTS

INTRODUCTION

Geologists have long recognized that minerals in deformed rocks display lamellar structures that have resulted from intragranular (plastic) flow; that is, twin or translation gliding. The dynamic significance of these structures was first pointed out by F. J. Turner (1948) who suggested that twin lamellae in calcite and deformation lamellae in quartz originated by plastic flow and formed in orientations of high shear stress. Later, Turner (1953) showed that the orientations of twin lamellae in calcite of three naturally deformed marbles were statistically consistent with the lamellae having been produced while under a constant stress field. Turner's hypothesis was substantiated in an extensive series of deformation experiments at high temperatures and pressure on Yule Marble carried out by D. T. Griggs, Turner, and their co-workers. These experiments established the fundamental basis for Turner's hypothesis; that is, for small strains, twin or translation gliding occurs only on those intracrystalline slip planes lying in orientations of high shear stress. Thus, statistically the orientation of deformation structures in individual grains should be related to the orientations of the principal stresses responsible for deformation of the aggregate.

Since this first application of the results from experiments to dynamic interpretations of rock fabrics, similar methods have been developed for a variety of common rock-forming materials. Most of these have been developed recently using as guidelines the results from experiments. The pertinent results are scattered throughout the literature, some in journals that are either inaccessible or not commonly read by geologists. The primary purpose of this paper is to bring together most of the pertinent information from previous studies and current unpublished data, so that structural geologists and petrologists may apply these results to tectonites in a routine fashion. Such applications are seriously needed if we are to understand the dynamic origins of geologic structures. To date, most of the discussion of the origin of such structures has been based on theoretical, geometrical, and intuitive arguments. These arguments may be verified or repudiated by studies such as those described here.

We shall restrict our discussion to structures produced during intragranular flow of common rock-forming minerals and aggregates. Understanding of the dynamics of structures produced by brittle fracture and cataclasis and of fabrics produced by recrystallization is still too limited for them to be discussed profitably, although their potential significance should be realized (Friedman, 1964). We shall treat in detail only those structures due to plastic flow and resolvable by petrographic microscopes. We shall first discuss the origin and geometry of intragranular flow in deformed crystals in general and then the methods used in dynamic interpretations of fabrics in tectonites, giving representative examples and interpretations. Finally, we shall discuss briefly an application of the results to folding in rocks and point out other areas of research for which the methods should prove useful.

ACKNOWLEDGMENTS

We are indebted to Doctors I. Borg, F. Turner, M. Friedman and R. Compton for critical review of the manuscript and many helpful suggestions. We thank Dr. H. C. Heard for use of his unpublished work on biotite crystals. Carter gratefully acknowledges the U.S. Geological Survey, Office of Earthquake Research and Crustal Studies, National Science Foundation Grant GA-1027, and Air Force Cambridge Research Laboratories Contract SAR/F19628-68-C0028 for partial support during this study.

INTRAGRANULAR FLOW OF CRYSTALS

There are two distinct mechanisms of intragranular flow induced by mechanical deformation—translation gliding and twin-gliding. Both mechanisms obey Schmid's law of critical resolved shear stress over a wide range of pressure-temperature-strain-rate conditions. Each potential deformation mechanism in a crystal will operate only when the shear stress resolved along the slip direction in the slip plane has reached a certain critical value (τ_c), which is generally a function of temperature and strain rate. The critical resolved shear stress for intragranular flow is virtually independent of the stress normal to the active slip plane.

Translation Gliding

Geometrically, the effect of translation gliding in crystals may be regarded as simple shear in which planes of atoms are displaced integral atomic distances (Fig. 1a). For each slip plane, *T*, there will be a crystallographically controlled slip direction, *t*, or directions, depending

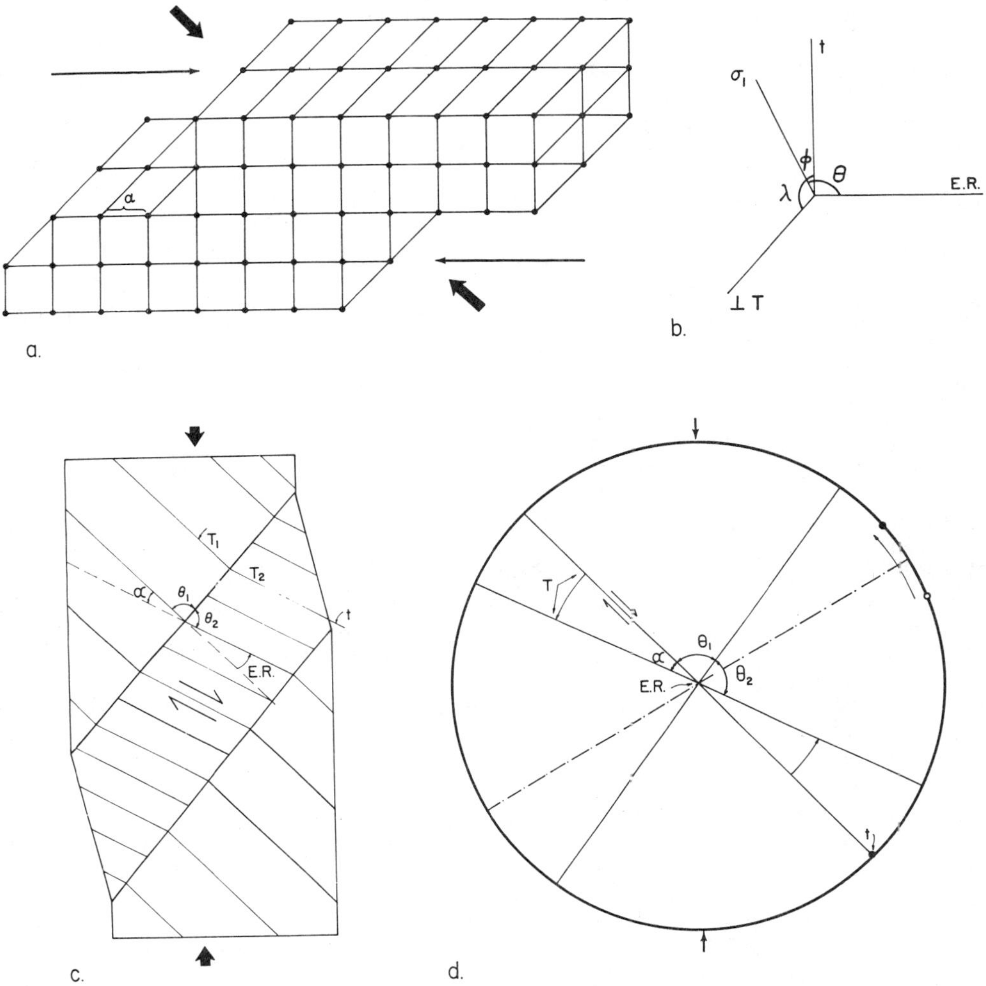

Figure 1. Translation gliding and external rotations in crystals. a. Schematic model of translation gliding by an integral number of atomic spacings (a) on the slip plane (T) in the slip direction (t) due to an externally applied force (heavy arrows). b. Conventions for describing the angular relations between σ_1 and the slip system and external rotation axis (E.R.). c. Sketch showing elements and convention for angles of external rotations in crystals. d. Equal-area projection of the relationships sketched in c.

upon the type and arrangement of atoms or ions. The slip plane T and slip direction t comprise the slip system, and most crystals possess more than one slip system. The critical resolved shear stress, τ_c, may change differently as a function of temperature and strain rate for the different slip systems in a crystal. Therefore, in a random aggregate, the active slip system will be that for which τ_c is lowest under the prevailing conditions of deformation. Under an axially symmetric system of externally applied forces, the shear stress resolved along the slip direction (τ_r) in the slip plane is given by

$$\tau_r = (\sigma_1 - \sigma_3)S_0$$

where σ_1 and σ_3 are the maximum and least principal compressive stresses, respectively, and $S_0 = \cos\lambda \cos\phi$ (Fig. 1b). The maximum resolved shear on any system is that for which $\lambda = \phi = 45°$; that is, $\tau_{max} = 1/2\,(\sigma_1 - \sigma_3)$. Slip on any system will begin when $\tau_r = \tau_c$.

Translation gliding generally is not homogeneously distributed throughout a crystal, and it is the nature and geometry of this inhomogeneous distribution that provides clues as to the orientations of principal stress axes. Slip is generally inhomogeneous because of imperfections in the crystals or the constraints placed on the crystals by neighboring grains or, in experiments on single crystals, by forcing the ends to remain coaxial. This inhomogeneous deformation gives rise to rotations in some parts of the crystals with respect to other parts and to any external coordinate frame, such as the principal stress axes (Fig. 1c). These external rotations are such that the slip plane rotates away from σ_1 and toward σ_3. The axis of external rotation (E. R.) lies in the slip plane normal to the slip direction (Figs. 1b, 1c). To locate E.R. on a projection, construct the plane bisecting the angle between the respective orientations of any crystallographic plane or direction in the rotated and unrotated parts of the crystal. All such bisectrices intersect the slip plane in the axis of external rotation (Turner, Griggs and Heard, 1954). The angle of external rotation (α) is $\pi - (\theta_1 + \theta_2)$, where θ_2 and θ_1 are the angles between the boundary separating the rotated and unrotated part and the slip plane in the rotated and unrotated part, respectively. The external rotation is related to the shear strain, s, by

$$\cot \theta_1 + \cot \theta_2 \simeq s\,.$$

Kink bands. Orowan (1942) first described kink bands in unconfined cadmium crystals compressed parallel to the basal (0001) slip plane. When the crystal bends and then becomes unstable, a region of highly localized deformation develops whose boundaries with the relatively undeformed crystal are initially nearly normal to the slip plane and to the slip direction. This localized deformation results in a physical kink in the crystal. There are other types of less common deformation bands described in the literature (Honeycombe, 1952) and the term kink band is reserved for those bands originating by localized slip in a direction initially normal to their boundaries. For kink bands in most crystals, the boundaries are nearly symmetrical with respect to the crystal structure on both sides of the boundary. In all cases, the radius of curvature of the crystal structure at the boundaries is small compared to the half-width of the band (Christie and others, 1964). The crystal structure at kink boundaries, although commonly appearing to bend abruptly to the resolution of optical microscopes, may actually bend gently over hundreds of unit cells at the much higher magnifications of electron microscopes as has been observed for kink boundaries in quartz (Christie and others, 1964).

We distinguish between four idealized types of kink bands as shown diagramatically in Figure 2. It should not be inferred that there are necessarily genetic differences between these, or that kinks of similar appearance have originated in the same way. The classification of kinking in crystals presented here is meant to provide a descriptive framework for those interested in kinking as a response to deforming stresses. For purposes of this paper it is sufficient only to relate the orientations of kink boundaries and the sense of slip within the kink bands to the orientations of the principal stresses. More detailed aspects of kinking in crystals, such as the structure of kink boundaries and preferred angles of kinking, appear to have no bearing on the subject of this paper and have been ignored. However, we emphasize that much remains to be learned about the kinking process and caution the reader against placing undue reliance on this or other classifications.

The first type of kink band (Fig. 2a) originates in crystals with a strong planar anisotropy (such as biotite or enstatite, Pl. 1, fig. a) oriented parallel to the compression axis. The crystal deforms by elastic bending about an axis in the slip plane normal to the slip direction, becomes unstable, and kinks by slip in opposite sense on each side of the east-west-trending kink boundary (Borg and Handin, 1966). Note that all three boundaries in Figure 2a are symmetrically oriented with respect to the crystal structure on both sides of the boundaries. The second type of kink band (Fig. 2b) also originates in crystals with a strong planar anisotropy oriented parallel to the compression axis, such as micas, kyanite, (Pl. 1, fig. b; Raleigh, 1965a) or gypsum (Turner and Weiss, 1965). Slip in one sense gives rise to symmetrical kink bands whose boundaries are inclined at 60° to 80° to σ_1. Interference of two such kinks originating by slip in the opposite sense (conjugate kinks) results in secondary kink boundaries oriented normal to σ_1 (Fig. 2b). Kinks of this type have also been observed in experimentally deformed phyllite (Paterson and Weiss, 1966, 1968). The actual crossing over of conjugate kinks, although common in phyllite, is quite rare in crystals, having been observed by us only in gypsum and in a few grains of experimentally deformed biotite.

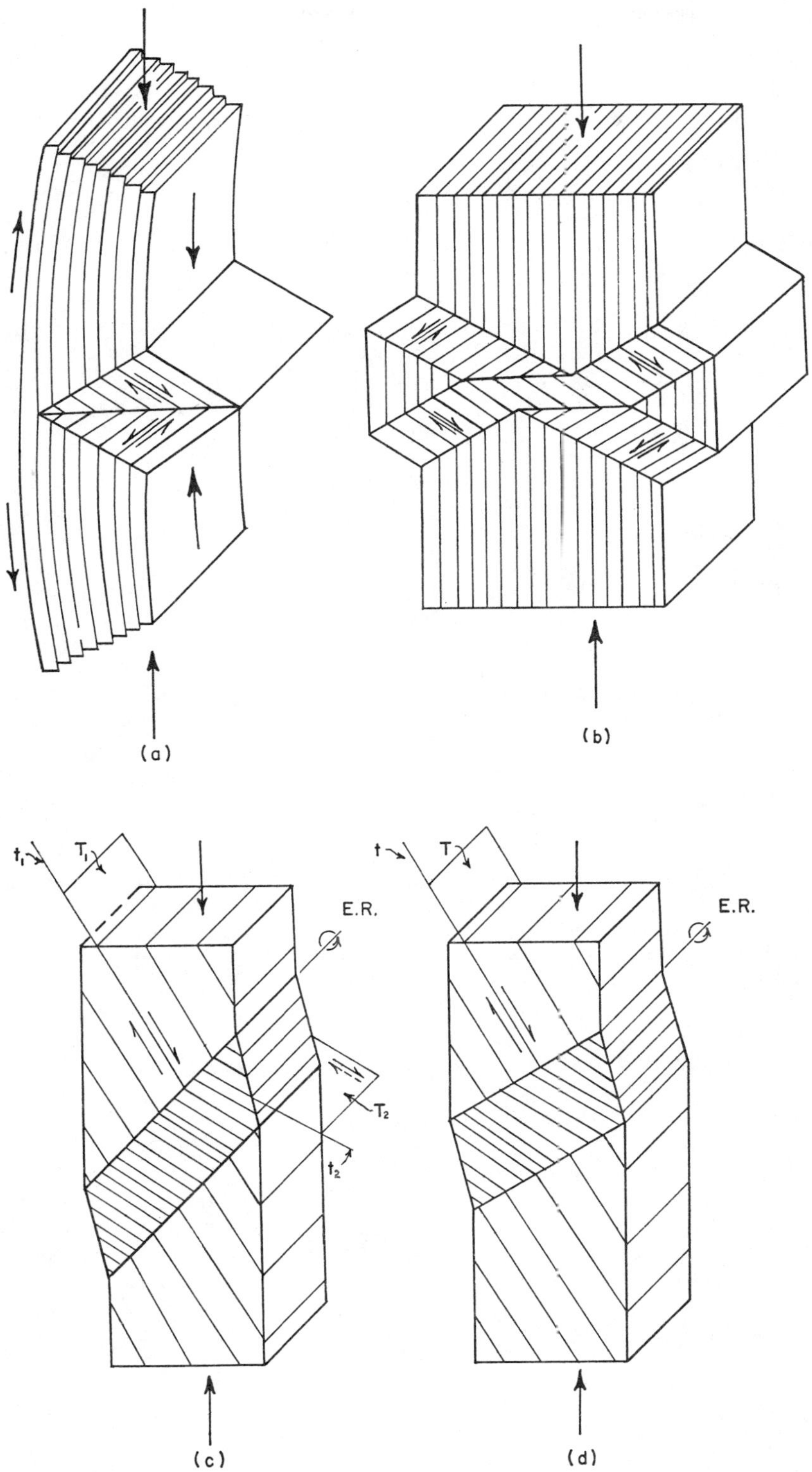

Figure 2. Schematic drawings of four types of kink bands in crystals. a. Kink bands in crystals loaded parallel to strong planar anisotropy. b. Intersecting conjugate kinks loaded parallel to a strong planar anisotropy. c. Symmetrical kink in crystal whose slip plane (T_1) is in an orientation of high shearing stress. d. Asymmetrical kink in crystal whose slip plane is in an orientation of high shearing stress.

The third and fourth types of kink band originate in crystals with relatively weak planar anisotropies when the slip plane may be in an orientation of high shear stress. The third type (Fig. 2c) is a symmetrical kink such as that observed in Plate 1, fig. c in experimentally deformed calcite (Turner and Heard, 1965). The boundaries of the fourth type are markedly asymmetrical with respect to the crystal structure on both sides (Fig. 2d); the slip plane in the relatively undeformed part is nearly normal to the kink band boundary. To our knowledge this type of kink band has been observed only in quartz (Pl. 1, fig. d; Christie and others, 1964; Carter and others, 1964).

We may visualize the geometry of kinking with the aid of Figures 1c and 2c, in which the slip plane T_1 is rotated locally to an orientation T_2 by kinking in the crystal. Ideally, where slip is limited to one system nearly normal to the band boundary the following relations must obtain (Christie and others, 1964): (1) The material within the band rotates externally with respect to the material outside the band around the axis of intersection of the slip plane and kink boundary (E.R.); this requires that the slip direction be normal to the external rotation axis as any other orientation would necessitate complex slip within the boundary. (2) At the inception of kinking, the boundary must be perpendicular to the slip plane and slip direction. (3) The atom planes parallel to the slip planes are continous across kink boundaries (except where the boundary is ruptured) so that, in the absence of elastic distortions, $\theta_1 = \theta_2$. From these relations it follows that the kink band boundary must rotate with respect to the crystal structure on one or both sides to maintain a symmetrical relationship. Rotation of the boundary need not take place if slip occurs in the same amount and in opposite sense on each side of the boundary (as for the east-west boundary in Figure 2a). Kinking with equal but oppositely directed slip on either side of the boundary is most probable in crystals oriented so that the slip plane is nearly parallel to σ_1. It is less likely for kink bands created in crystals whose slip plane is in an orientation of high shear stress (Figs. 2c, 2d; Pl. 1, figs. c and d).

The idealized kink bands sketched in Figure 2 are rarely observed in either naturally or experimentally deformed crystals. Most observations indicate that the slip planes are not exactly reflected across the boundaries of "symmetrical" kinks. For example, the upper and lower kink boundaries in the hypersthene crystal (Pl. 1, fig. a) are asymmetric, being inclined, respectively, at 83° and 88° to (100) of the host and at 72° and 74° to (100) of the kink. The central boundary, however, is very nearly symmetrical with an apical angle of 132°. Note that the upper and lower boundaries are ruptured, probably because of the large elastic distortions produced by the asymmetry, whereas the central symmetrical boundary is coherent. This kink must have originated by localized slip in opposite sense in the kinked zones on either side of the central boundary. The apical angles of kinks in kyanite (Pl. 1, fig. b) that have originated by interference of conjugate kinks also average 132°, but these kinks are asymmetrical, the average difference in inclination to the slip planes on both sides being 11°. In this case, however, the conjugate kinks themselves are very nearly symmetrical. Many more examples could be presented, but these complexities do not affect dynamic interpretations of kinking in crystals.

Undulatory Extinction. Bending of the crystal structure as in kinking need not take place about radii that are small compared with the half-width of the reoriented zone. Rather, the bending may be gradual, as shown at the top of Plate 1, fig. d, the right and left sides of Plate 1, fig. c, and Plates 3 and 4. In optically anisotropic transparent crystals, in which the gradational bands appear in thin sections as differences in extinction position in different parts of the crystals, this phenomenon is commonly called "undulatory" or "wavy" extinction and is very common in both experimentally and naturally deformed crystals. Genetically, kink bands and undulatory extinction in crystals are probably identical; they grade into one another (Pl. 1, figs. c and d; Pl. 4, figs. b and d) and they are distinguished only by the radius of curvature at the boundary.

Twin-Gliding

In the most elementary manner, the deformation due to twin-gliding may be regarded as the homogeneous simple shear of one part of a crystal structure to a mirror reflection of the undeformed part across the twin plane. In this simple model, each atom in the layer above the twin plane (Fig. 3a) has moved some fixed fraction (depending on the crystal structure) of an integral atomic translation. The distance through which atoms above the first layer move is directly proportional to their distance

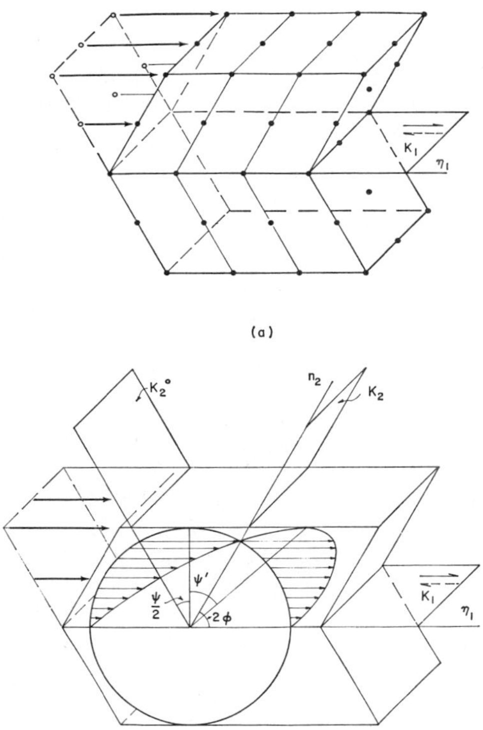

Figure 3. Schematic drawings of twin-gliding in crystals. a. Atomic planes above the twin plane (K_1) have been sheared through a constant angle to a mirror reflection of the crystal structure below K_1. b. Twin-gliding elements and conventions for angles viewed with the aid of a strain ellipsoid.

from the twin plane; thus the shear strain for a given twin law is constant. The net shear may be readily visualized with the aid of the strain ellipsoid. A sphere inscribed within a continuum before deformation (Fig. 3b) becomes a triaxial ellipsoid after twinning due to homogeneous simple shear parallel to the plane K_1 (twin plane) and the direction η_1 (twin axis). K_1 and K_2 are the circular sections of the ellipsoid, η_2 is the intersection of K_2 with the deformation plane; 2ϕ is the acute angle between the circular sections, ψ is the angle between K_2° and K_2 and ψ' is the angle of rotation of a plane initially normal to η_1. The shear strain due to twinning is given by:

$$s = \tan \psi' = 2 \cot 2\phi = 2 \tan (\psi/2) .$$

To apply this model to crystals, K_1, K_2, η_1, and η_2 must be correlated with the crystal lattice (some examples are given in the next section). Considering only deformational twins, the twin plane (K_1) is always the composition plane of the twin. This plane cannot be a symmetry plane of the crystal, and there is only one direction sense of twinning according to a given law. We note that whereas the displacements giving rise to the total net shear may be described as simple shear, the paths of individual atoms or groups of atoms are commonly much more complex (for a more complete discussion of twinning *see* Cahn, 1954, and Klassen-Neklyudova, 1964).

Twin-gliding produces a change in crystallographic orientation that generally may be readily seen in transparent anisotropic materials in thin sections between crossed polarizers (Pl. 2). The twinned bands have the appearance of sharp kink bands but, of course, originate by shear parallel to the band boundary rather than normal to it. Twin composition planes can be measured accurately and are symmetrically oriented with respect to the crystal structure of host and twin. Because of the restricted sense of shear along the twinning direction in the twin plane, important information concerning principal stress directions may be obtained from these measurements, as will be discussed in the next section.

PRINCIPAL STRESSES FROM INTRAGRANULAR FLOW IN COMMON ROCK-FORMING MINERALS

In this section we shall discuss the known mechanisms of intragranular flow of some common rock-forming minerals, emphasizing those mechanisms useful for determining principal stress directions in rocks. Wherever possible, the methods to be used for determining principal stress directions in tectonites will be justified by the results from experiments. Owing to space limitations and rapid progress in this field, we shall not generally be concerned with historical or chronological developments leading to the conclusions presented. For some materials, it will not be possible to present the results of all pertinent studies, but most of these will be referenced for the convenience of interested readers.

Let us note the limitations imposed on the techniques for determining principal stress orientations, some of which were described by Turner (1962): (1) intragranular flow in the crystals post-dates recrystallization; (2) the grains analyzed must have random or weakly preferred crystallographic orientations (this limitation may be overcome under certain circumstances as described below); and (3) techniques based on mechanical twinning and some based on kinking depend on the ability to distinguish between the undeformed and the deformed part. This distinction is not always a simple one in heavily deformed rocks unless identifiable relict crystallographic planes have been rotated passively (internally) to irrational orientations as in calcite and dolomite (Turner and Weiss, 1963, p. 339–357). For rocks in which the grains are heavily twinned (half or more of the material is in the twin orientation), these methods could lead to erroneous results and either should not be applied or the results confirmed by some other method. Thus, in general, these techniques are regarded as reliable only when applied to rocks having suffered rather small post-crystalline deformations.

Carbonates

Calcite. Calcite has been the most thoroughly studied rock-forming mineral both in experimentally and naturally deformed crystals and rocks. The extensive series of experiments on calcite crystals (Turner, Griggs and Heard, 1954; Turner and Heard, 1965; Borg and Handin, 1967) and Yule Marble[1] by Griggs, Turner and co-workers has elucidated the behavior of marble over a wide range of physical conditions. At temperatures below about 600° C calcite deforms predominantly by twin-gliding (Pl. 2, fig. a) on one or more of the $e\ \{01\bar{1}2\}$ planes parallel to the direction $[e_2{:}r_2]$(Fig. 4a). The upper layers are displaced toward the *c*-axis in the host crystal with respect to the lower layers, indicating, by definition (Turner, Griggs and Heard, 1954), a positive sense of shear. The other intragranular flow mechanisms are translation gliding on $r\ \{10\bar{1}1\}$ parallel to $[r_1{:}f_2]$ and on $f\ \{0\bar{2}\bar{2}1\}$ parallel to $[f_1{:}r_3]$ or $[f_1{:}r_2]$, both with a negative sense of shear (Fig. 4A) and twin-gliding on $\{10\bar{1}1\}$ parallel to $[r_1{:}f_2]$ with a positive sense of shear (Borg and Handin, 1967). Inasmuch as twin-gliding on e has the lowest τ_c over most of the conditions in the experiments and predominates in rocks, we shall restrict our discussion to it alone.

Turner's method (1953) for determining principal stress directions in marble involves locating the maximum and minimum principal compressive stresses most favorably oriented to produce twinning in the more strongly twinned grains. This construction is made as though S_0 is maximum (0.5) for each twinned grain. There will, therefore, be scatter among the compression and extension axes derived; however, the centers of concentrations of points should

[1] Griggs and Miller, 1951; Handin and Griggs, 1951; Turner and Chih, 1951; Griggs and others, 1951; Griggs and others, 1953; Borg and Turner, 1953; Turner and others, 1956; Heard, 1963; and Handin and others, 1960.

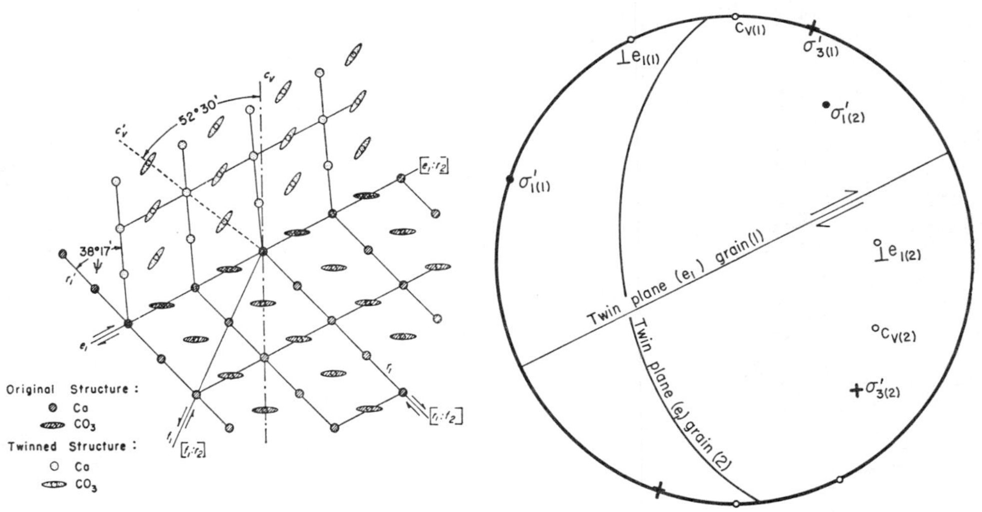

a. b.

Figure 4. Twin-gliding in calcite. a. Diagrammatic projection of the twinned calcite structure normal to the a_2 axis, showing the twinning elements (Friedman, 1964). b. Equal-area projection showing the graphic method for constructing principal stress axes best oriented to produce twin-gliding in calcite.

give the approximate orientations of σ_1 and σ_3. Inasmuch as twinning may take place on two (Pl. 2, fig. a) or all three possible twin planes, the one most profusely developed (designated e_1) is selected for the analysis. Only the c-axis in the host crystal and the normal to the twin plane need be measured; these must be separated by from 24° to 28° (26°15′ is the correct angle) for the measurements to be used in the analysis. These are plotted on an equal-area projection and σ_1'[2] constructed at 45° to e_1, away from c, in the $e_1 - c$ plane and σ_3' in the same plane, 90° from σ_1' (Fig. 4b). For simple deformations, a suitable number of such determinations (50 or more) combined from two or three mutually perpendicular thin sections from oriented specimens should give the average orientations of principal stress axes during the deformation responsible for the twinning.

To illustrate that the principal stresses derived using the Turner method agree well with known orientations of the stress, we have selected two examples from experiments: (1) a calcite-cemented sandstone specimen shortened 9.2 percent at 2 kb confining pressure and 300° C (Friedman, 1963); and (2) a specimen of Yule Marble extended 7 percent at 20° C and 10 kb confining pressure (Turner and Weiss, 1963). The *c*-axes in the calcite-cemented sandstone (Fig. 5a) have a nearly random orientation. The derived σ_1' axes (Fig. 5b) cluster about the known compression axis oriented normal to the plane of the diagram. Figure 5c shows the orientations of *c*-axes of prominently twinned grains in the Yule Marble specimen and Figure 5d the derived compression (dots) and extension (crosses) axes. The results are in excellent accord with the known principal stress orientations (*N-S* arrows represent extension axis).

The Yule Marble is strongly anisotropic with respect to the calcite crystal orientation in that there is a broad maximum of *c*-axes normal to a weak foliation. The specimen from which the data in Figures 5c and 5d were obtained was extended so that the *c*-axis maximum coincided with σ_3 and thus most of the grains were in a favorable orientation to twin (Fig. 4a). The coincidence of the unique axis of the applied stress with that of the *c*-axis fabric ensured this result. In another special case, where the compression and extension axes show orthorhombic or monoclinic symmetry and a symmetry plane happens to coincide with a symmetry plane of the *c*-axis fabric, the common symmetry plane must contain two principal stress axes (Turner and Weiss, 1963). These ideal situations are not always realized and further methods are required to analyze anisotropic rocks where the symmetries of derived compression and extension axes and of the fabric do not agree. These and other restrictions on the use of the method are discussed more fully by Turner and Weiss (1963, p. 414–421).

Applications of Turner's technique to tectonically deformed marble are far too numerous to allow a comprehensive review here.[3] We shall give just three examples in citing the studies of Friedman and Conger (1964), Nickelsen and Gross (1959), and Conel (1962), but we note that the method has been applied successfully in most of the other studies. In Friedman and Conger's study, calcite crystals whose *c*-axes are oriented radially within the elliptical wall of a gastropod shell (Fig. 6a) show markedly unequal development of twin lamellae. The grains near the top and bottom of the shell (blackened) are heavily twinned, whereas the twins are sparse (dotted) in grains along the east-west walls. The most favorable orientation of σ_1 to produce such a pattern, approximately east-west in Figure 6a, is roughly in accord with the derived compression axes (Fig. 6b). This interpretation is consistent with inferences drawn from the north-south-trending series of folds in the area.

Nickelsen and Gross (1959) studied the calcite matrix in two specimens of calcarenite of the Ordovician Conestoga Limestone, Pennsylvania. The bedding (S_1) is intersected at a small angle by a slaty cleavage along which limestone pebbles and quartz and calcite grains are flattened and elongate parallel to the *a* fabric axis (Figs. 6c through 6f). The calcite *c*-axes form an *a-c* girdle with maxima (Figs. 6c, 6d) and show over-all monoclinic symmetry, the plane of symmetry being *a-c*. Diagrams of σ_1' (dots) and σ_3' (crosses) axes (Figs. 6e, 6f) derived from twin lamellae show

[2] We shall hereafter denote the axes of maximum (σ_1) intermediate (σ_2) and least (σ_3) principal compressive stresses as σ_1', σ_2', and σ_3' when they are deduced from single grains in an aggregate or from specimens having preferred orientations.

[3] McIntyre and Turner, 1953; Gilmour and Carmen, 1954; Clark, 1954; Weiss, 1954; Crampton, 1958; Nickelsen and Gross, 1959; Hansen and Borg, 1962; Suzuki, 1962; Nissen, 1964; Friedman and Conger, 1964; Carter and Friedman, 1965; Compton, 1966, and Conel, 1962.

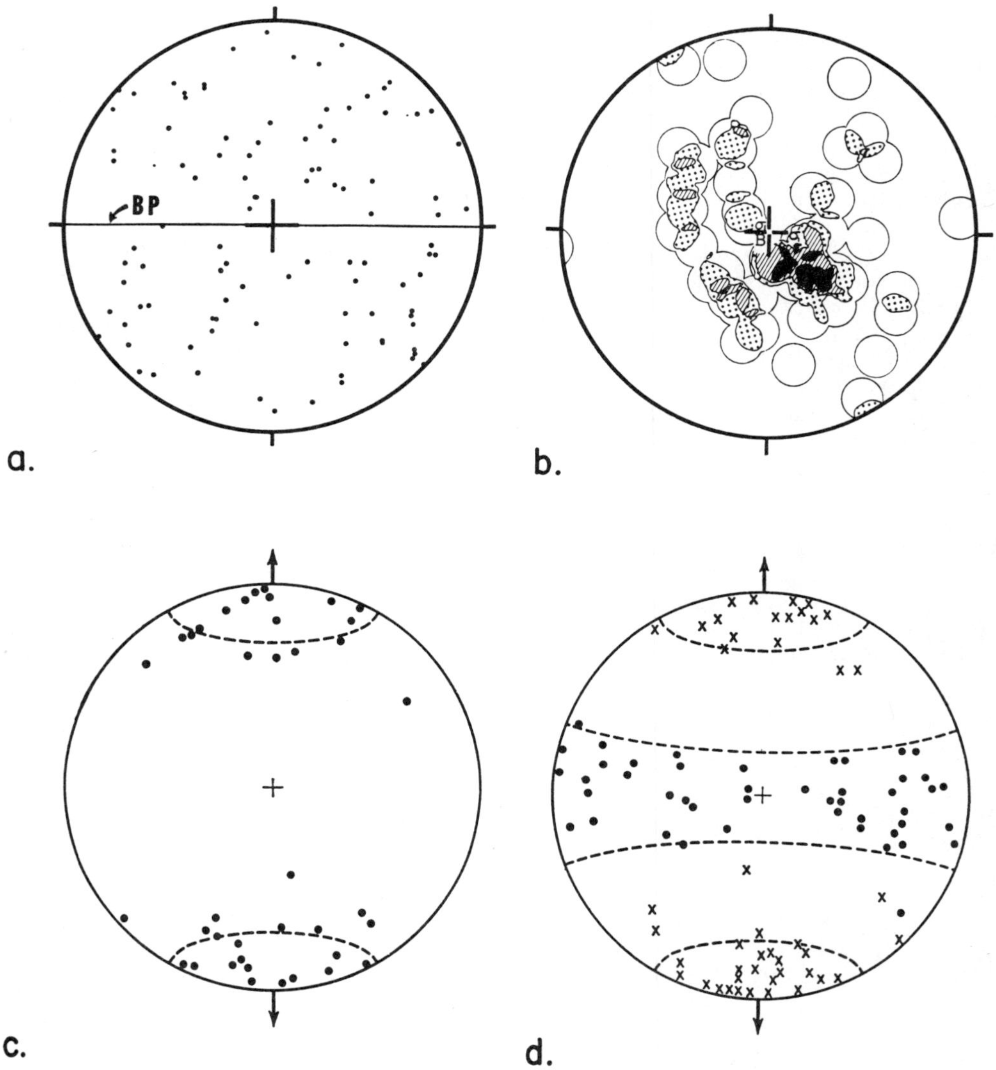

Figure 5. Principal stresses from twinning in experimentally deformed calcite. a. Calcite *c*-axes in calcite-cemented sandstone (Friedman, 1963); BP = bedding plane. b. Deduced σ_1' axes in equal-area projection agree well with known σ_1 oriented normal to plane of diagram. Fifty grains: Contours: 8, 6, 4, 2 percent per 1 percent area (Friedman, 1963). c. *c*-axes of prominently twinned calcite grains in experimentally extended Yule Marble specimen (Turner and Weiss, 1963). d. σ_3' (crosses) and σ_1' axes (dots) constructed agree well with the axially symmetric applied stress (Turner and Weiss, 1963).

▶

Figure 6. Principal stresses from natural twin-gliding of calcite. a. Gastropod shell composed of calcite with *c*-axes oriented radially. Blackened grains are those most heavily twinned suggesting an east-west orientation of σ_1 (Friedman and Conger, 1964). b. σ_1' axes deduced from twins in a. c, d. Orientation of calcite *c*-axes in two specimens of calcarenite. S_1 is the bedding plane, S_2, a slaty cleavage, a, b, and c are fabric axes. Contours for 100 *c*-axes in c: 6, 4, 2, 1 percent per 1 percent area. Contours for 71 *c*-axes in d: 9.8, 7, 4.2, 1.4 percent per 1 percent area (Nickelsen and Gross, 1959). e, f. Compression axes (dots) and extension axes (crosses) deduced from twins in the calcite. The center of concentration of compression axes is near *c* fabric axis and normal to the slaty cleavage (Nickelsen and Gross, 1959).

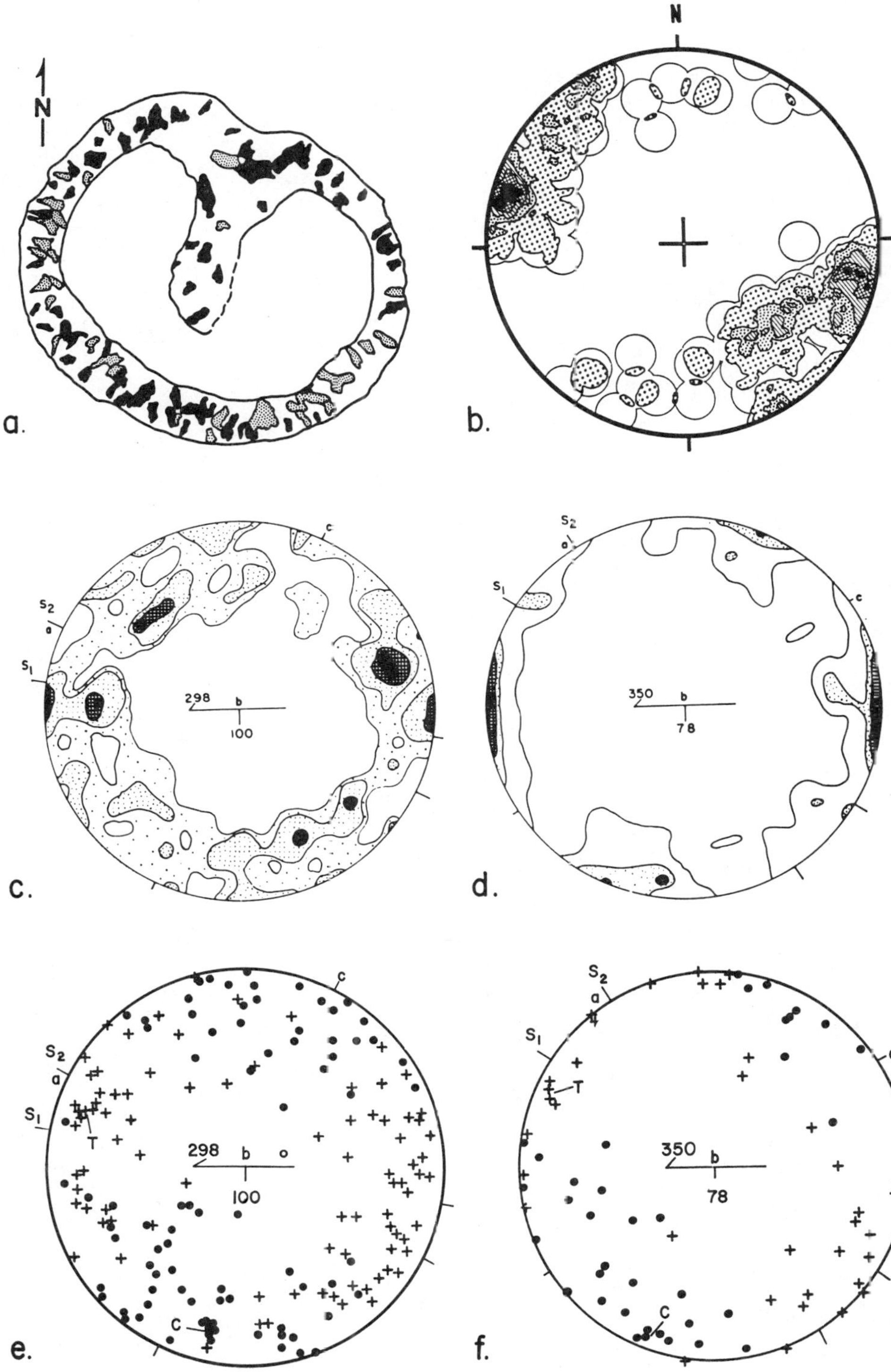
N
a.
b.
c.
d.
e.
f.
S2
a
S1
c
298
b
100
350
78
T
C

approximately orthorhombic symmetry with one symmetry plane being parallel to *a-c* allowing an interpretation of the results. It appears, therefore, that the specimens were compressed after development of the calcite *c*-axis fabric along a direction nearly parallel to the *c* fabric axis and normal to the slaty cleavage. They were simultaneously extended parallel to the *a* fabric axis in agreement with the evidence for elongations of pebbles and grains.

Conel (1962) analyzed twin lamellae in two regions of the axial portion of a folded limestone layer sandwiched between two shale layers (Fig. 7a). The calcite *c*-axes have a nearly random orientation. The composite data for the two sections from Region I and for three sections from Region II (Fig. 7a) are presented in Figures 7b and 7c, respectively. The deduced σ_1 for the specimen near the top of the trough, Region I, is parallel to bedding and σ_3 nearly normal to it. The data in Figure 7c show more scatter but suggest that σ_1 was oriented normal to the bedding and σ_3 nearly parallel to it. These results compare well with those predicted by analysis of the stress distribution in a layer of circular cross section (normal to fold axis) loaded elastically (arrows, Fig. 7a) normal to the fold axis (Conel, 1962).

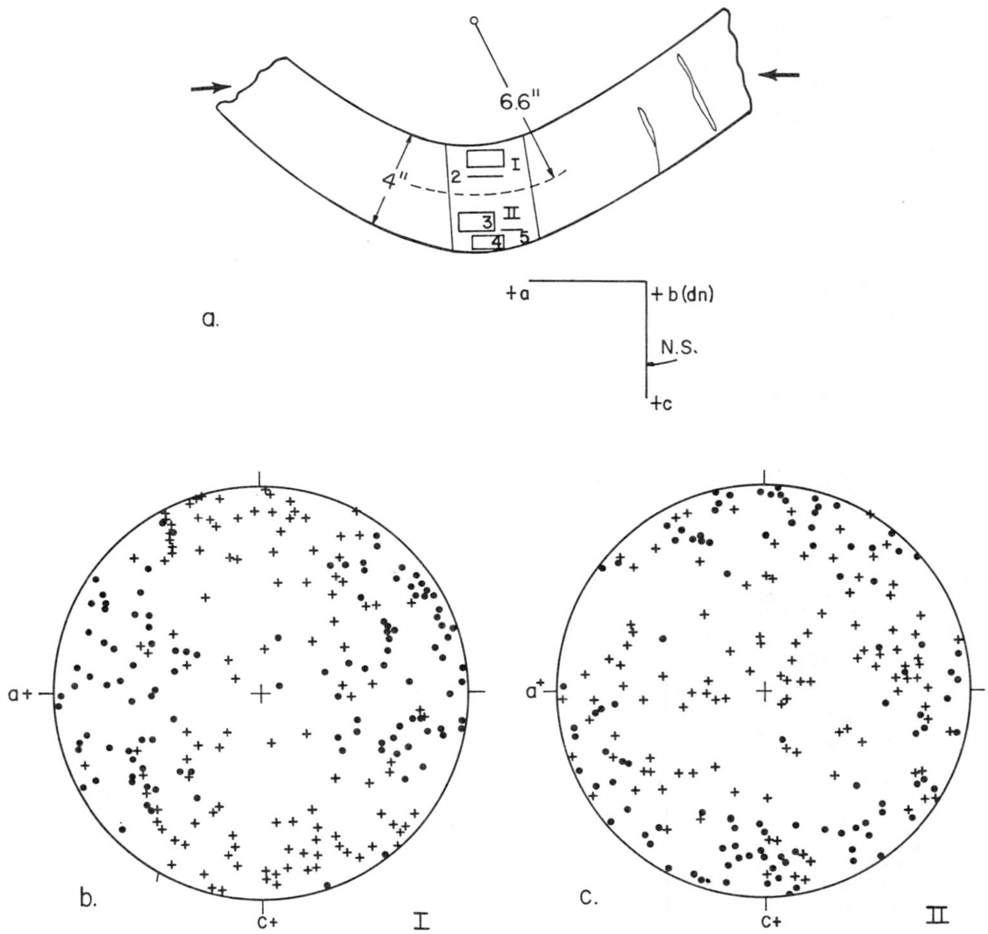

Figure 7. Dynamic analysis of calcite twin lamellae from the axial region of a small fold in limestone (Conel, 1962). a. Sketch of the fold showing orientation of thin sections in Regions I and II from which data in b and c were obtained. b, c. Compression axes (dots) and extension axes (crosses) derived from twins in Regions 1 (b) and II (c). The center of concentration of σ_1' axes in Region I is parallel to the *a* reference axis and the σ_3' axes are concentrated near the *C* reference axis spreading into a broad girdle normal to *a*. The reverse is true for Region II, but the data show considerably more scatter.

Dolomite. Dolomite has received considerably less attention than has calcite, and only three experimental studies (Handin and Fairbairn, 1953; Turner and others, 1954; Higgs and Handin, 1959) have been carried out under confining pressure. Translation gliding on the basal {0001} plane parallel to an *a*-axis was discovered by Johnson (1902) in tests conducted at atmospheric temperature and pressure and has since been confirmed by the more recent experiments. In experiments at 5 kb confining pressure (Higgs and Handin, 1959) on single crystals of dolomite favorably oriented for basal slip, large deformations are achieved at all temperatures, but τ_c *increases* with increasing temperature. At 300° to 400° C another mechanism becomes important—twin-gliding on the *f* {02$\bar{2}$1} plane parallel to [f_2:a_3] with a negative sense of shear (*see* Fig. 8a). This mechanism was first discovered by Fairbairn and Hawkes (1941) in naturally deformed dolomite. The τ_c for twin-gliding on *f* decreases with increasing temperature and is about the same as for basal slip at 400° C (Higgs and Handin, 1959). Dolomite rock is much stronger and more brittle than marble under comparable conditions (Handin and Fairbairn, 1953; Turner and others, 1954).

Twinning in dolomite may be used to determine the orientations of principal stress axes in precisely the same manner as twins in calcite. The sense of shear is opposite and the twinning elements differ ($c \wedge f = 62\frac{1}{2}°$), but the same type of analysis holds. The relevant data for determining orientations of principal stress axes from twinning are shown in Figure 8b. The compression axis is located in the *c*-*f* plane, 45° from *f* and 17½° from *c*, and the extension axis at 90° to it.

We are aware of only three published studies in which twins in dolomite have been used for dynamic interpretations (Crampton, 1958; Christie, 1958; Trommsdorff and Wenk, 1965). This lack is probably due primarily to the fact that dolomite is less common than calcite and is less often twinned. Crampton (1958) used twins in both calcite and dolomite in his study of Cambrian and Ordovician marbles in the northwest highlands of Scotland. The results for the two minerals are similar and the compression axes determined are inclined at high angles to the foliation.

Christie (1958) studied a specimen of the highly deformed Loch Ailsh dolomite rock in the mylonite zone of the Moine thrust in the northwest highlands of Scotland. Most of the grains in the specimen showed twinning and he distinguished between two types: (1) optically recognizable twins in rational crystallographic orientations, and (2) less sharply defined twins with a granular appearance occupying irrational orientations due to passive rotation by basal slip. Using the rational twins for the analysis, Christie found the compression axis to be oriented at high angles to the foliation in the Moine thrust block; he regarded the twinning as being induced in the final stage of deformation of the rocks.

A result similar to those of Crampton (1958) and Christie (1958) was obtained by Trommsdorff and Wenk (1965) in their study of the fabric of a dolomite marble from Crevola, Switzerland. The *c*-axis fabric which they regarded as having originated by recrystallization showed a strong preferred orientation normal to the foliation. Each of the 960 grains studied contained twin lamellae. The deduced compression axis was oriented nearly normal to the foliation and proved to be a late feature unassociated with a penetrative lineation.

Quartz

Until very recently, quartz, which shows evidence of extensive ductile flow in nature even in rocks of low-grade metamorphism, has been unresponsive to attempts to induce plastic flow experimentally. Large plastic flow of natural single crystals of quartz and of quartzite was achieved at 15 to 30 kb confining pressure, temperatures from 300° C to 1500° C, and strain rates of 10^{-3} to 10^{-5}/sec (Carter and others, 1964; Christie and others, 1964; Griggs and Blacic, 1964; Christie and Green, 1964). Under these conditions quartz crystals deform predominantly by slip on the basal {0001} plane parallel to an *a* {11$\bar{2}$0} axis at temperatures below about 700° C and on the *m* {10$\bar{1}$0} prism parallel to the *c* [0001] axis at temperatures above 700° C. In thin section, deformation lamellae having distinctive optical properties are developed parallel to the active slip planes. Bending of the slip planes due to inhomogeneous deformation gives rise to kink bands and undulatory extinction in zones normal to the active planes (Pl. 1, fig. d). The lamellae are most profusely developed in zones that have undergone most external rotation and hence the greatest amount of shear strain.

Zones of undulatory extinction in naturally deformed quartz (Pl. 3, figs. a and b) are most commonly oriented subparallel to the *c*-axis as in the experiments at lower temperatures. De-

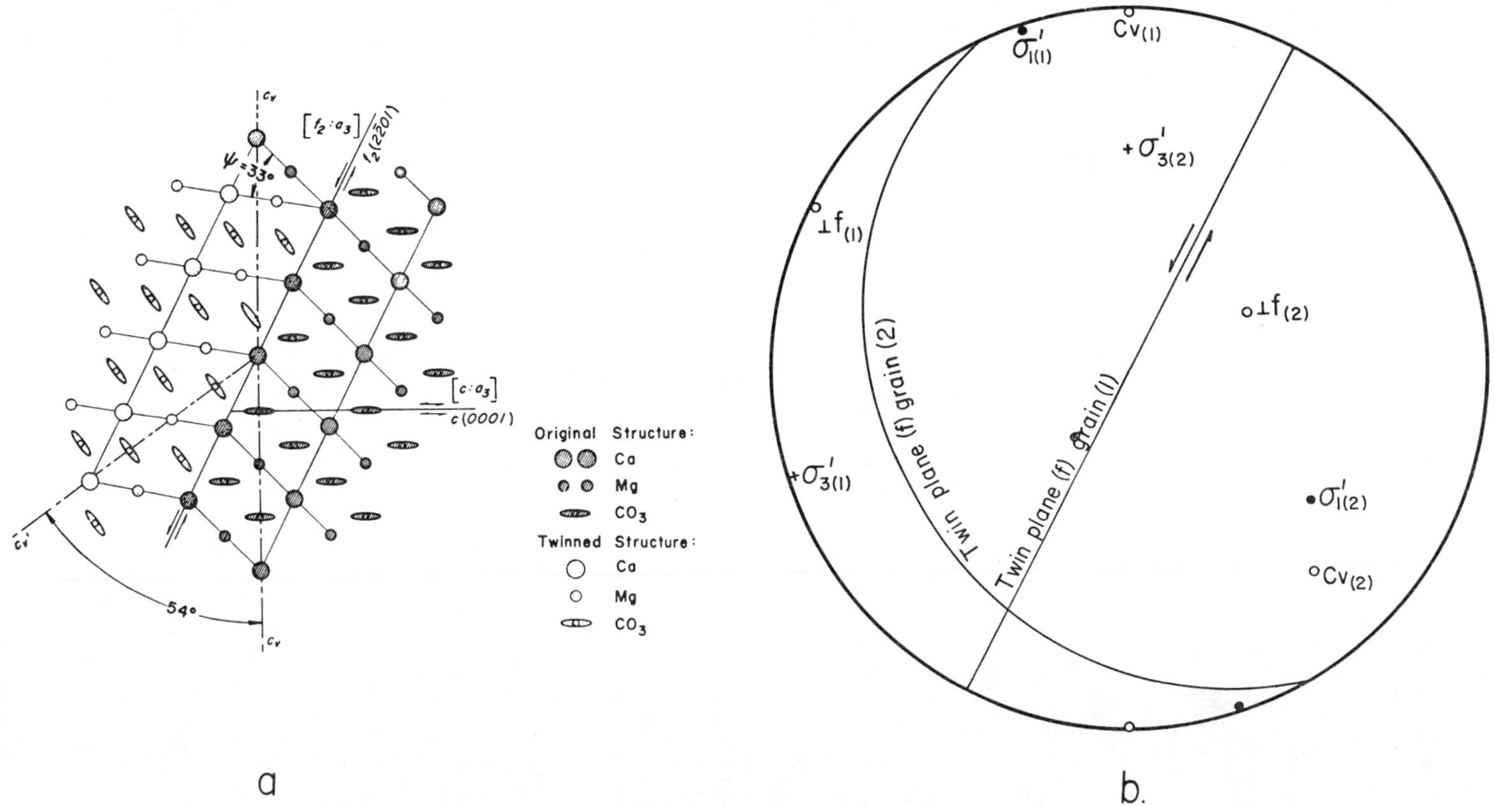

Figure 8. Twin-gliding in dolomite. a. Diagrammatic projection of the dolomite structure parallel to (11$\bar{2}$0), showing elements of twinning of $f\,\{02\bar{2}1\}$ planes and translation gliding on the basal plane (Friedman, 1964). b. Equal-area projection showing the graphic method for constructing principal stress axes best oriented to produce twin-gliding.

formation lamellae, similar to those produced experimentally, are also common features in naturally deformed quartz and frequently show unequal development in grains with undulatory extinction (Pl. 3, figs. a and b). However, natural lamellae are most commonly inclined to the basal plane at 10° to 30° (hereafter called sub-basal); basal lamellae seem to be characteristic only of quartz found near known or suspected meteorite impact sites and are presumably produced by shock (Carter, 1965). Sub-basal lamellae have recently been produced experimentally in quartz crystals and quartzite deformed dry at 6 to 10 kb confining pressure, 600° to 900° C, and strain rates of 10^{-3} to 10^{-6}/sec (Heard and Carter, 1968). The lamellae are irrational and they are generally inclined at angles less than 45° to σ_1 (Fig. 9a). In areas where the lamellae are densely developed (Pl. 3, figs. c and d) the crystal structure has been externally rotated (Fig. 9b) such that the *c*-axis in the rotated zone (c_1) lies closer to σ_1. The *c*-axes in the undeformed (c_0) and deformed parts (c_1) and poles to the lamellae are nearly co-zonal, a feature characteristic of naturally deformed quartz. Moreover, the relative orientations of these subfabric elements are fixed with respect to σ_1, suggesting a means for determining the orientations of principal stresses inducing these features.

In Figure 10, we compare data obtained from a quartzite specimen compressed at 8 kb confining pressure, 800° C, at a strain rate of 3.4×10^{-6}/sec (a, b, c; Heard and Carter, 1968) with similar data from a naturally deformed specimen (d, e, f; Carter and Friedman, 1965). The patterns of the various subfabric elements—poles to lamellae (a, d), c_0–c_1 axes in individual grains (b, e) and arrows (c, f) connecting *c*-axes (ends of arrows) with poles to lamellae (points of arrows)—are very similar. The known orientation of σ_1 for the experimentally deformed specimen is indicated by the north-south arrows ($\sigma_2 = \sigma_3$) and this leads to the interpretations of orientations of principal stress axes indicated by solid lines and 1 percent circles for the naturally deformed specimen (d, e, f). The dashed lines and circles show the principal stress orientations deduced from twinning in the calcite cement of the sandstone specimen, confirming interpretations from the quartz data.

Thus there are three methods for determining principal stress axes from intragranular flow of quartz: (1) the acute-angle method—the lamellae are inclined statistically at angles less than 45° to σ_1 (this is regarded as the least reliable criterion, as most of the lamellae in some specimens have an average inclination of about 45° to σ_1); (2) the c_0–c_1 method—planes containing c_0 and c_1 in individual grains inter-

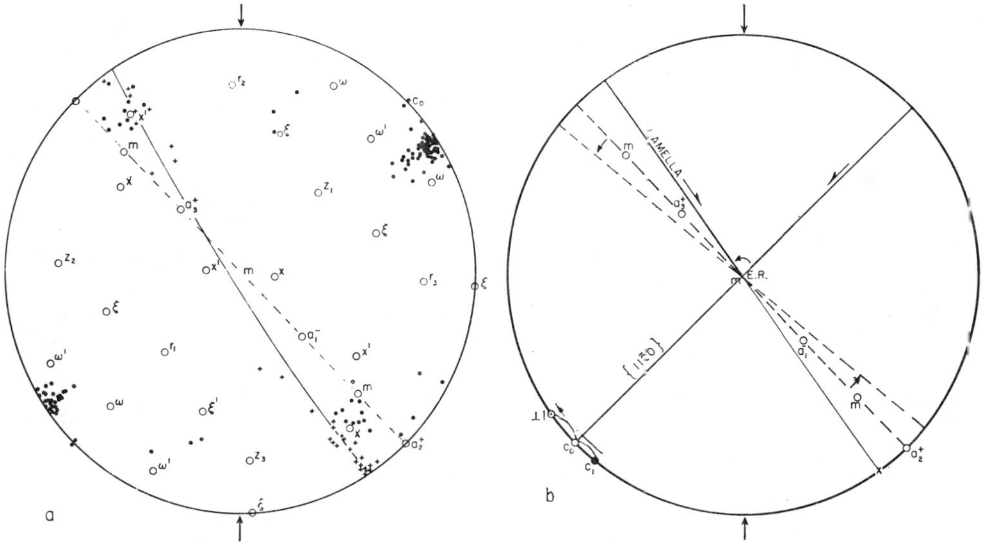

Figure 9. Experimentally deformed quartz crystals (Heard and Carter, 1968). a. Equal-area lower hemisphere projections showing the orientations of deformation lamellae with respect to the crystallography and σ_1 (N–S arrows) for six crystals. b. Relationship of lamellae, *c*-axis in host (open circle and deformed part (solid circle) to σ_1 in a representative area of one of the crystals.

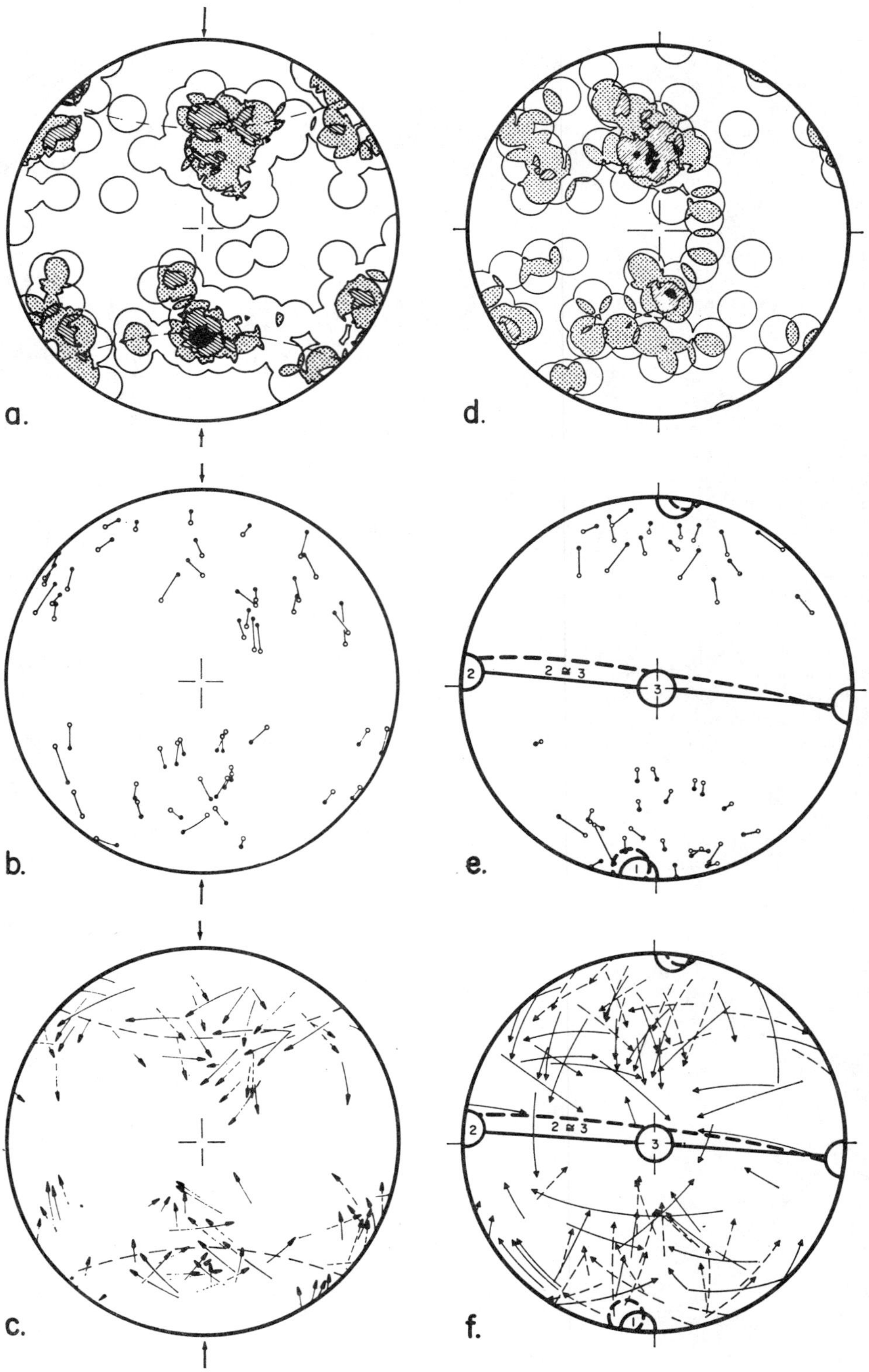
a.
d.
b.
e.
2
2 ≅ 3
3
1
c.
f.

sect σ_1 (for $\sigma_1 > \sigma_2 = \sigma_3$) or σ_3 (for $\sigma_3 < \sigma_1 = \sigma_2$) or contain σ_1 and σ_3 (for $\sigma_1 > \sigma_2 > \sigma_3$), with c_1 oriented closest to σ_1; and (3) the arrow method—the arrows point toward σ_3 (for $\sigma_3 < \sigma_1 = \sigma_2$) or toward $\sigma_2 = \sigma_3$ or contain σ_1 and σ_3 (for $\sigma_1 > \sigma_2 > \sigma_3$), with the arrow head (pole to lamellae) oriented closest to σ_3. These methods apply only to quartz containing lamellae inclined at *ca* 10° to 30° to {0001}, the orientation most commonly observed in tectonites (Carter and Friedman, 1965).

Space does not permit a review of the many studies of quartz lamellae aimed at providing dynamic interpretations.[4] Hansen and Borg (1962) correlated the orientation of deformation lamellae with principal stress axes deduced from twinning in the calcite matrix of the folded Devonian Oriskany Sandstone, Pennsylvania. They found that the quartz lamellae were inclined statistically, at angles less than 45° to σ_1 (method [1] above) and that most poles to lamellae in some of the specimens were oriented closer to σ_3 than the *c*-axis (method [3] above). In the four specimens studied, σ_2 was oriented parallel to the fold axis, σ_1 was nearly parallel to the dip direction, and σ_3 normal to the bedding (Fig. 11a).

Scott and others (1965) did a dynamic analysis based on quartz lamellae in five specimens from a small concentric fold in the Silurian Tuscarora Sandstone, New Jersey. They related *a*, *b*, and *c* fabric axes to patterns of poles of deformation lamellae and arrow diagrams by conventions set out earlier (Ingerson and Tuttle, 1945; Fig. 11b). Using methods (1) and (3) above and reasoning from probable stress trajectories during several stages of folding, they concluded that $a = \sigma_3$ and $c = \sigma_1$ (Fig. 11b). The data were therefore interpreted to indicate a compression parallel to a slightly bent layer producing the lamellae early in the folding, and then the limbs were rotated passively by strain concentrated in the hinge and in less competent surrounding layers (Fig. 11c).

Carter and Friedman (1965) used all three methods as well as twin lamellae in calcite cement to determine principal stress axes in ten specimens around an anticline in Montana (Fig. 11d). In that study, all methods gave internally consistent results. For specimens from the limbs, σ_1 is inclined at low angles to the dip and σ_3 is normal to the bedding. For specimens from the noses of the anticline, σ_1 is subparallel to the strike, σ_2 parallel to the dip (fold axis), and σ_3 normal to the bedding. These stress orientations were interpreted to indicate that the lamellae formed in response to stresses due to stable bending of the layer by predominantly vertical movements in the basement rock of the area.

Micas

Borg and Handin (1966) compressed cylindrical cores from a biotite single crystal parallel to the basal {001} plane at 500° C and 5 kb confining pressure. The initial orientation of the crystal is shown in Figures 12a and 12b. The crystals shortened near the ends of the specimen by kinking in bands whose boundaries are inclined at 60° to 90° to σ_1 (Fig. 12c). Data from a kinked region (Fig. 12c) are shown in the projection in Figure 12d; the E.R. axis, *f*, is [010], and *t* is [100]. In general, however, there is a considerable spread of the measured E.R. axes and derived slip directions as can be seen from the projection in Figures 13a and 13b showing data obtained from two crystals (Borg and Handin, 1966). The only certain system is T = {001}, t = [100], and there is a possibility of slip on the system {001} <$\bar{1}30$>. Simultaneous slip on {001}, <130>, and [010] or any combination of these could give rise to the observed spread.

Heard and Carter (unpub. data) did a few experiments at 3 kb confining pressure and

[4] Hansen and Borg, 1962; Carter and Friedman, 1965; Fairbairn, 1941; Ingerson and Tuttle, 1945; Mackie, 1947; Riley, 1947; Turner, 1948; Naha, 1959; Christie and Raleigh, 1959; Hara, 1961; Hara, 1963; Hara, 1964; Scott and others, 1965.

◀

Figure 10. Orientation of subfabric elements in experimentally (a, b, c; Heard and Carter, 1968) and naturally (d, e, f; Carter and Friedman, 1965) deformed quartzites. N–S arrows in a, b, c show the orientation of $\sigma_1(\sigma_2 = \sigma_3)$ and dashed small circles are at 45° to σ_1. Dashed lines and circles in e and f show the orientations of principal stresses deduced from calcite twins. Lower hemisphere equal-area projections. a, d. Poles to 128 sets of deformation lamellae in specimen 662 (a) and 89 sets in specimen 4 (d). Contour intervals are 7, 5, 3, 1 (a) and 6.7, 4.5, 2.2, 1.1 (d) percent per 1 percent area. b, e. *c*-axes in more (solid circles; c_1) and less (open circles: c_0) deformed parts of individual grains. c, f. Poles to lamellae (points of arrows) and *c*-axes (ends of arrows) in grains containing lamellae.

300° C on polished parallelepipeds from the same crystal. The specimens were cut so that {001} ranged in orientation from parallel to perpendicular to σ_1. In specimens whose basal plane was inclined at moderate angles to σ_1, the north-south direction in Figure 12b had the maximum S_0. Specimens whose basal plane was oriented parallel to σ_1 deformed by kinking (Pl. 4, fig. a) about an E.R. axis nearly normal to σ_1 (near F in Fig. 13a). Specimens whose basal plane was inclined to σ_1 at 30° to 55° deformed mainly by slip or sliding (we are not certain which) parallel to the "down-dip" direction, although some local kinking developed. In one specimen in which {001} was inclined at 52° to σ_1, a kink band developed with E.R. parallel to [010] and t = [100]; the sense of external rotation was correct for the applied stress. Both clockwise and counterclockwise rotations were observed for specimens compressed parallel to {001} in our experiments (Pl. 4, fig. a) and those of Handin and Borg, as is to be expected (*see* Figs. 2a and 2b).

Griggs and others (1960) did experiments on Westerly Granite at 5 kb confining pressure

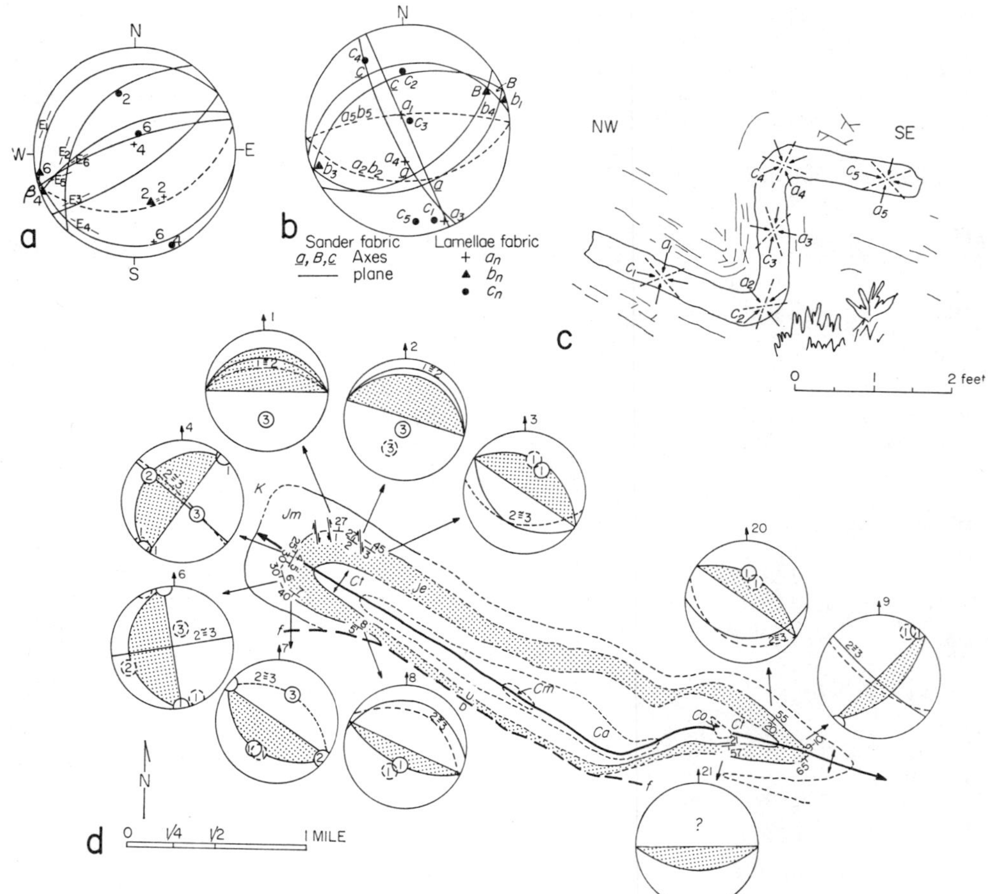

Figure 11. Principal stresses from intragranular flow in quartz. a. Orientation of planes and σ_1 (dots), σ_3 (crosses) and σ_2 (triangles) in folded calcite cemented sandstone (Hansen and Borg, 1962). b. Orientation of planes around a small fold and of fabric axes deduced from quartz lamellae orientations. The axes a_n are parallel to σ_3, b_n equals σ_2 and c_n = σ_1 (Scott and others, 1965). c. Orientations of the principal stresses around fold in quartzite (d) interpreted as passive rotation of layer. See (b) above (Scott and others, 1965). d. Orientations of principal stresses around a large fold deduced from twinning in calcite (dashed) and intragranular flow of quartz (solid). Numbers refer to subscript in σ_1, σ_2, σ_3. Stippled planes show orientation of bedding (Carter and Friedman, 1965).

and 500° C. The granite deformed by shear along diagonal zones, and biotite grains in these zones were highly kinked. The kink bands formed preferentially in grains whose {001} planes were parallel to σ_1 and the normals to the boundaries cluster about σ_1. Borg and Handin (1966) did experiments under the same conditions on Fordham gneiss, determining the intensity of kinking and orientations of kink bands in biotites as a function of the inclination of the foliation to σ_1. Most kinking was observed when the foliation, and hence most biotite {001} planes, was parallel to σ_1 (Fig. 13c). Kinking is common in the specimen in which the foliation was inclined to σ_1 at 22° with the kink boundaries nearly symmetrically disposed around the mean orientation of *S*. Kinking still occurs when the foliation is inclined at 45° to σ_1, but there is definite preferential development of kinking in grains whose {001} planes are most nearly parallel to σ_1. In those specimens, poles to kink bands are concentrated at orientations between the foliation plane and σ_1 (Fig. 13d).

It may be concluded from the experimental evidence that kinking in biotite (and pre-

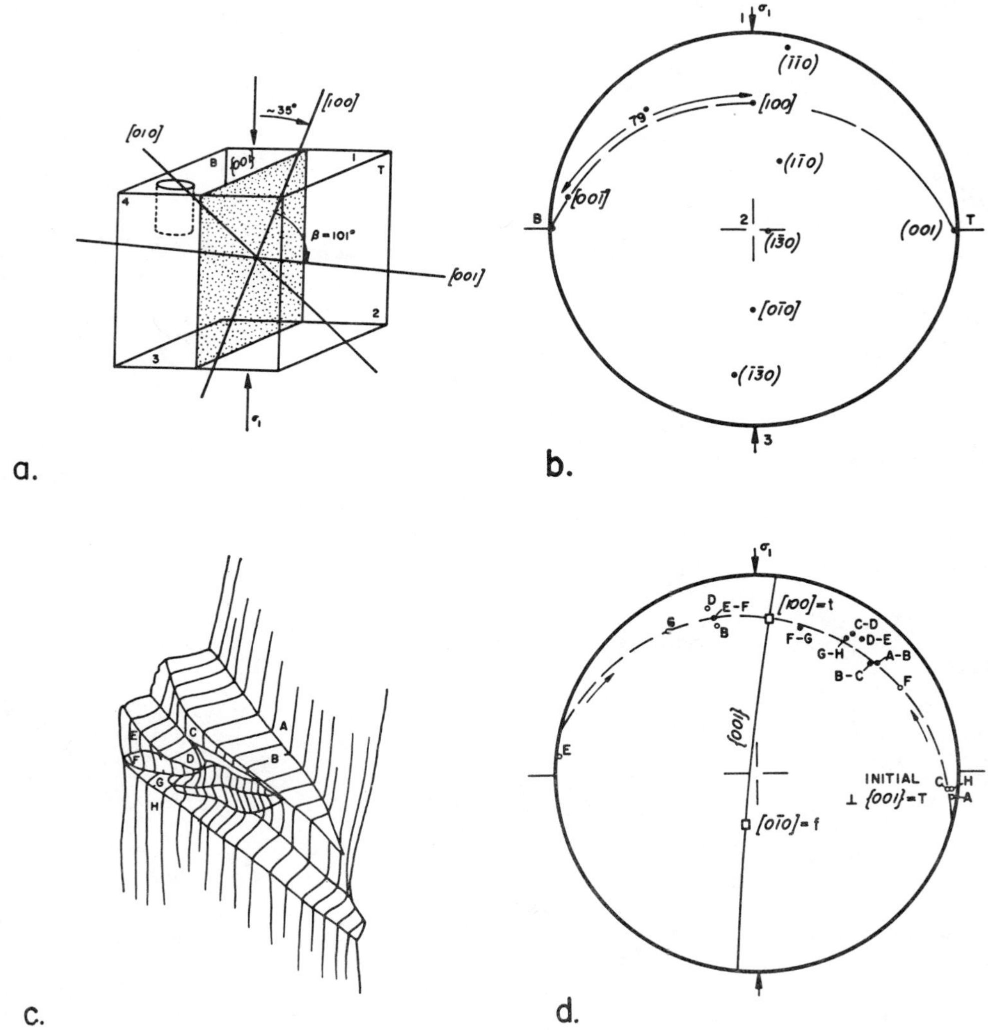

Figure 12. Kink bands in biotite experimentally deformed at 5 kb, 500° C (Borg and Handin, 1966). a. Orientation of cores and σ_1 from biotite crystal. b. Projection showing crystallographic data for biotite crystal (a). c. Sketch of kink bands in biotite crystal. d. Poles to kink boundaries (solid circles) and to slip plane (open circles) in regions designated by letters in c.

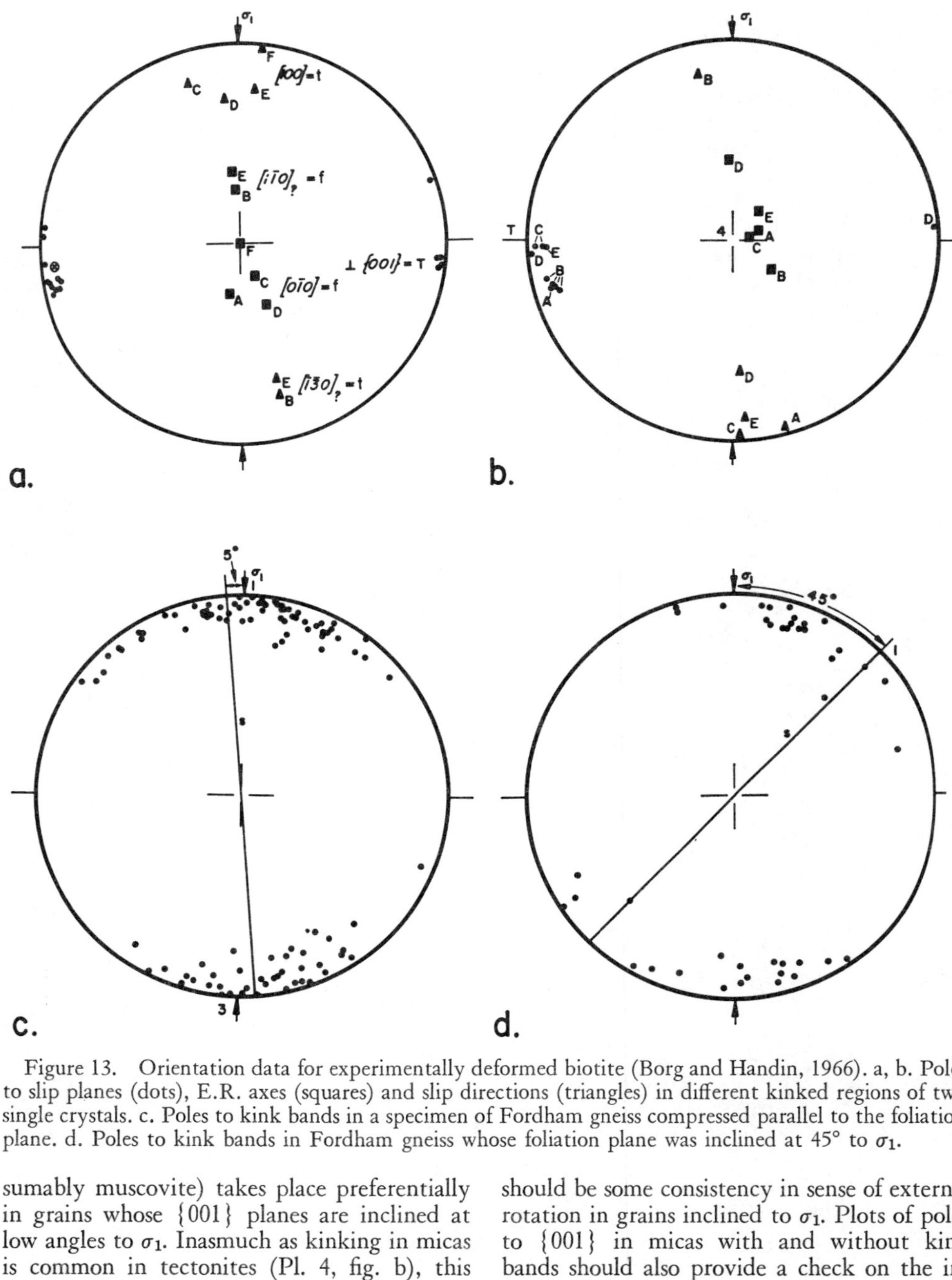

Figure 13. Orientation data for experimentally deformed biotite (Borg and Handin, 1966). a, b. Poles to slip planes (dots), E.R. axes (squares) and slip directions (triangles) in different kinked regions of two single crystals. c. Poles to kink bands in a specimen of Fordham gneiss compressed parallel to the foliation plane. d. Poles to kink bands in Fordham gneiss whose foliation plane was inclined at 45° to σ_1.

sumably muscovite) takes place preferentially in grains whose {001} planes are inclined at low angles to σ_1. Inasmuch as kinking in micas is common in tectonites (Pl. 4, fig. b), this provides a potentially powerful tool for determining principal stress directions. A plot of poles to kink boundaries should give the approximate orientation of σ_1. Grains showing slip in both senses (Figs. 2a, 2b, 4a, 4b) should be most severely restricted in orientation with {001} nearly parallel to σ_1. Moreover, there should be some consistency in sense of external rotation in grains inclined to σ_1. Plots of poles to {001} in micas with and without kink bands should also provide a check on the internal consistency of the results. Assuming homogeneous stress and that the foliation plane is also a principal plane of the stress quadric, poles to kink bands should: (1) cluster about σ_1 for $\sigma_1 > \sigma_2 = \sigma_3$; (2) form a girdle normal to σ_3 for $\sigma_3 < \sigma_1 = \sigma_2$; (3) cluster about σ_1 spreading preferentially toward σ_3 for $\sigma_1 >$

$\sigma_2 > \sigma_3$. Similar distributions should be observed for the uncommon occurrence of randomly oriented micas, but the assumption regarding the orientation of the stress quadric need not be made.

Turner (1964) did an analysis of kinking of micas in two specimens from an Innsbruck, Austria, mica schist. He was careful to point out that caution must be exercised in interpreting the data from most specimens because of the tabular habit and typically strong preferred orientation of the micas in foliated rocks. Data obtained for one of Turner's specimens are shown in Figure 14; *ab* is the foliation plane. The poles of {001} in biotite (Fig. 14a) and muscovite (Fig. 14b) show crude *ac* girdles, and there is a distinct minimum near the arbitrarily chosen reference axis, *b*. Kinked micas, however, show a more restricted range of orientations (Fig. 14c) and the orientations of grains containing conjugate kinks (*see* Fig. 2b) are still more severely restricted (Fig. 14e). The dashed great circles drawn approximately normal to the centers of concentration of {001} in two mutually perpendicular sections (Figs. 14c, 14d) intersect in X. The axis X can be equated with σ_1 for the deformation producing the kinks, assuming homogeneous stress. This orientation of σ_1 is close to that deduced from the grains showing conjugate kinks (X, Fig. 14e). Arrows drawn from the pole of {001} in the kink bands toward {001} in the host crystal show a fairly consistent change in sense approximately along the line YZ. This then leads to the interpretation of the orientation of σ_1 = X as shown and is consistent with the orientation deduced in Figures 14c, 14d, and 14e.

Pyroxenes

Orthopyroxene. Experiments on polycrystalline enstatite-bearing rocks have been conducted over a wide range of pressure-temperature conditions by many workers in recent years.[5] In all studies the crystals deform by kinking due to slip on the system {100} [001] (Pl. 1, fig. a); this seems to be the only operative slip system. An additional deformation mechanism observed in all of the experiments involves partial to total transformation of the enstatite to clinoenstatite in kink bands and in lamellae parallel to {100}. Apparently the inversion takes place by shear along {100} involving partly reconstructive processes (Borg and Handin, 1966). Clinoenstatite is not commonly observed in naturally produced kink bands, however, and will not be discussed further here.

Enstatite ($2V_Z$ = 86°) in a peridotite nodule was deformed in a series of experiments at 5 to 7 kb and 700° to 850° C by Raleigh (1965a, 1968). Measurements of the optic directions X = [010], Y = [100], and Z = [001] in 100 experimentally deformed crystals were made in areas of grains where external rotation was least (little or no conversion to clinoenstatite). Both grains with and without kink bands were measured (Figs. 15a, b, c), but those grains in which t = [001] was inclined to the section at angles greater than 50° were excluded, as possible kink bands could not have been observed. The slip plane, (100), in kinked grains of enstatite (open circles, Fig. 15a) tends to lie at angles lower than 45° to σ_1, as in the micas. Kinking will not occur, however, unless the slip direction, [001], also is inclined at less than 45° to σ_1 (Fig. 15b); hence the large number of grains that fail to kink despite the favorable orientation of (100) (Fig. 15a). Therefore, the orientation of the slip direction is the more reliable indicator of the orientation of σ_1; nearly 90 percent of the grains in which [001] is within 45° of σ_1 are kinked.

Equally reliable as an indicator of the orientation of the principal stress axes is the orientation of poles to kink boundaries (Fig. 15d). The analysis is precisely the same as that discussed for micas. An examination of naturally deformed enstatite reveals that this technique is likely to be more useful than that involving measurements of the orientation of [001]. For the latter method to be reliable, it must be possible to determine which part of the grain has an orientation nearest that prior to kinking, a requirement more easily met when working with experimentally deformed specimens than with naturally deformed material. The external rotations in natural kink bands may be sufficiently large that [001] is rotated 60° from its initial orientation parallel to σ_1. Accordingly, we suggest an analysis identical to that described for the micas but with an attempt to cross-check by measuring orientations of [001] in little deformed parts of kinked grains, if they can be recognized. Such an analysis has not yet been attempted.

[5] Borg and Handin, 1966; Raleigh, 1965a; Griggs and others, 1960; Turner and others, 1960; Riecker and Seifert, 1964; Riecker and Rooney, 1967; Raleigh, 1968.

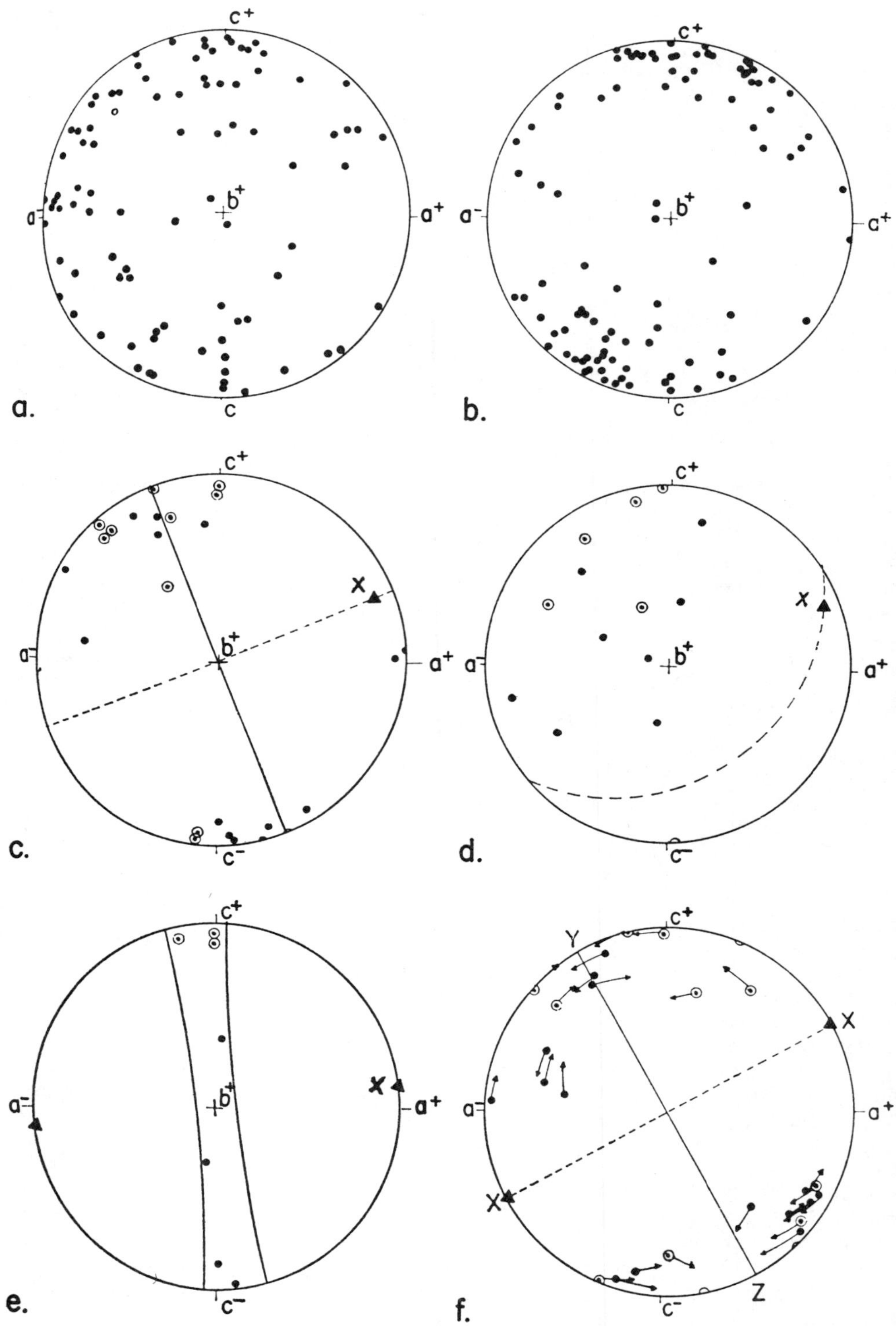
c+
a−
a+
b+
c
a.
b.
c−
X
c.
d.
e.
Y
Z
f.

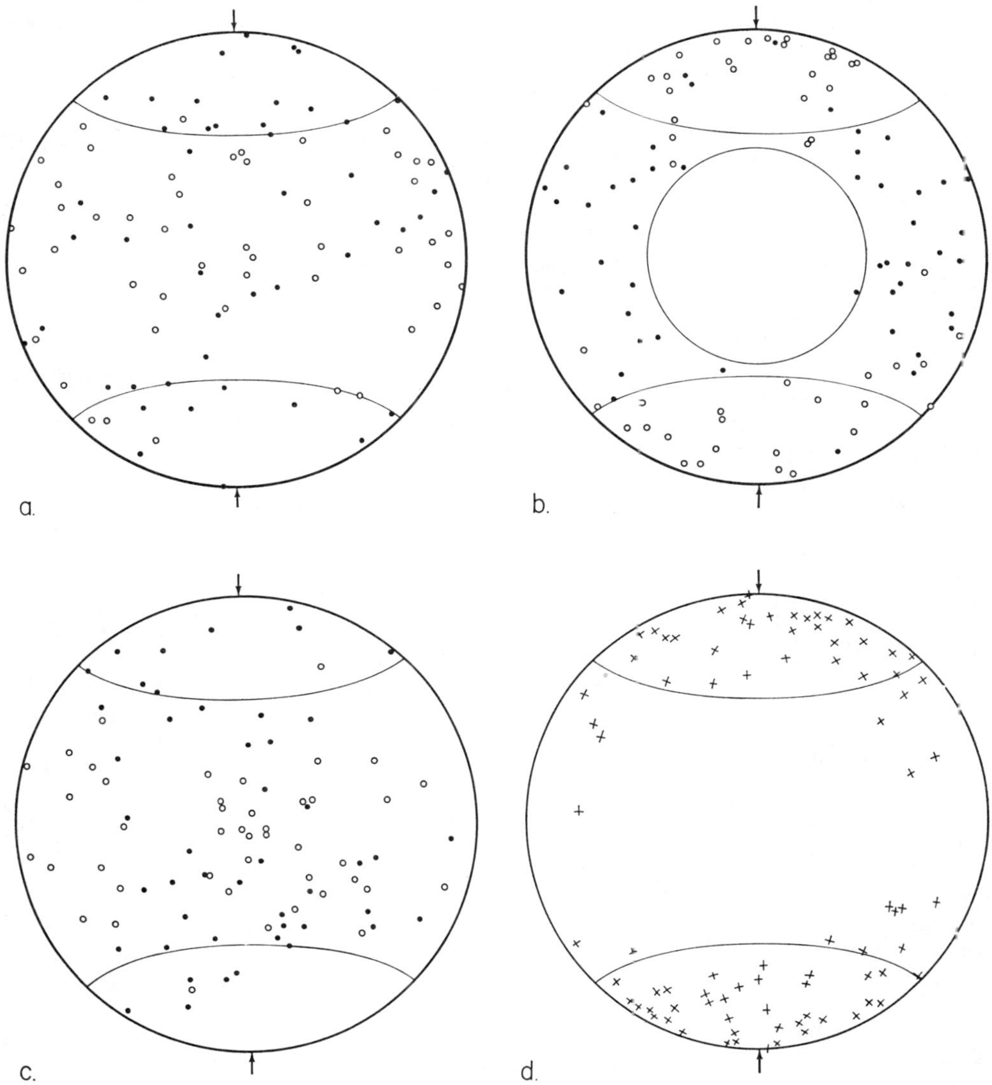

Figure 15. Kink bands in experimentally deformed enstatite (Raleigh, 1965a). Small circles about σ_1 (N–S arrows) are at 45° to σ_1. Circle in center of b encloses blind area. a. Poles to the slip plane (100) in kinked (open circles) and unkinked grains. b. The slip direction [001] in kinked (open circles) and unkinked grains. c. External rotation axes [010] in kinked (open circles) and unkinked grains. d. Poles to kink bands.

◄

Figure 14. Orientation data for kink bands in naturally deformed mica schist (Turner, 1964); *ab* is the foliation plane. a, b. Poles to {001} in all biotite (a) and muscovite (b). c, d. Poles to {001} in kinked micas from two mutually perpendicular sections. The dashed circles normal to centers of concentration intersect in X. e. Poles to {001} in grains showing conjugate kinks. f. Pole to {001} in kink band (solid circle) and in host (arrowhead in kinked grains.

Clinopyroxenes. Slip and mechanical twinning have been produced experimentally in augite and diopside (Raleigh, 1965a; Griggs and others, 1960; Raleigh and Talbot, 1967) at 5 kb confining pressure. In experiments at strain rates of 10^{-2} to 10^{-8}/sec and temperatures from 200° to 1000° C, diopside (Fig. 16a) undergoes mechanical twinning on (001) and (100) (Raleigh and Talbot, 1967). Slip on (100), [001] is the predominant translation glide mechanism and occurs in grains in which the resolved shear stress is in the reverse sense (that is, with the shear in the direction of $-c$) for twinning on (100), [001] (*see* Fig. 16a). At the lower temperatures, twins on (001) are better developed, whereas at higher temperatures (100) twins predominate. At 1000° C and a strain rate of 4.2×10^{-5}/sec, additional slip mechanisms become important (Raleigh and Talbot, 1967).

The sense and amount of shear in twinning on (100) and (001) was determined from the rotations incurred by polished surfaces in experimentally produced twins (Raleigh, 1965a; Raleigh and Talbot, 1967). The sense of shear was determined from the sense of rotation in grains in which both sets of twins were present. The twinning mechanisms are: (1) K_1, the twin-glide plane, = (100); η_1, the glide direction, = [001]; K_2 = (001); η_2 = [100]; s = 0.73; and (2) K_1 = (001); η_1 = [100]; K_2 = (100); η_2 = [001]: s = 0.73.

The common occurrence of lamellar (100) twins in naturally deformed clinopyroxenes of metamorphic rocks and peridotites (Pl. 2, fig. b) suggest the possibility of using them for dynamic analyses according to Turner's method (1953). The compression and extension axes for twinning on (100) (Fig. 16a) are oriented as shown in Figure 16b, that is, at 45° to η_1 = [001] in (010) to give a shear stress in the direction of $+c$. For the less frequent twinning on (001), σ_1' and σ_3' lie at 45° to [100] in (010) to give a maximum shear stress in the $-a$ direction (Fig. 16a). A great advantage in working with diopside with the more frequent (100) twins is that the σ_1' and σ_3' axes are approximately parallel to X and Z, respectively, obviating laborious constructions on the stereonet.

The data obtained from an experimentally deformed diopside-scapolite specimen are shown in Figures 16c and 16d (Raleigh and Talbot, 1967). Figure 16c shows the orientation of X = σ_1' in twinned grains (open circles) and X in untwinned grains (dots). The orientations of X for twinned grains are clearly restricted to within 45° of the known σ_1 (N-S arrows). The companion diagram (Fig. 16d) shows the orientation of Z = σ_3' in twinned grains (crosses) and untwinned grains; the crosses are restricted to a broad girdle normal to σ_1, reflecting the axially symmetric stress orientation.

Raleigh and Talbot (1967) tested the method by application to a naturally deformed diopside-bearing calc-silicate rock collected near a major fault in South Australia. The diopside showed well-developed lamellar twinning on (100) and associated calcite was heavily twinned. Distribution data for one of the specimens examined are shown in Figures 16e and 16f; data for the other specimen were virtually identical. The derived σ_1 is oriented nearly east-west (Fig. 16e) and the broad girdle, with maxima, of extension axes oriented nearly north-south indicates that $\sigma_2 \simeq \sigma_3$. The authors concluded that twinning in the diopside by east-west compression accompanied a severe early Paleozoic deformation responsible for the development of a series of north-south-trending thrust faults. A similar analysis of the heavily twinned calcite gave random or weak patterns, and the authors concluded that the calcite had been twinned repeatedly by subsequent second-order deformations because of the low τ_c for calcite twinning. Thus, in some instances, the high strength of the diopside (and its resistance to recrystallization) should prove useful in attempts to understand the dynamics of early, severe deformations.

Olivine

Raleigh (1965a, 1968) has produced large plastic deformation of olivine in a peridotite

---

▶

Figure 16. Principal stresses from twin-gliding in diopside (Raleigh and Talbot, 1967). a. Projection normal to [010] of crystal structure of diopside showing twinning elements. b. Graphic construction of compression and tension axes most favorably oriented to produce twinning on (100). c. Orientation of X = σ_1' in twinned (open circles) and untwinned (dots) grains in experimentally deformed specimen. N–S arrows show orientation of σ_1. d. Orientation of Z = σ_3' in twinned (crosses) and untwinned (dots) grains in same specimen as c. e. σ_1' axes for (100) twinning in a naturally deformed calc-silicate rock. 130 axes; Contours: 4, 3, 2, 1 percent per 1 percent area. f. σ_3' axes in the same specimen as e.

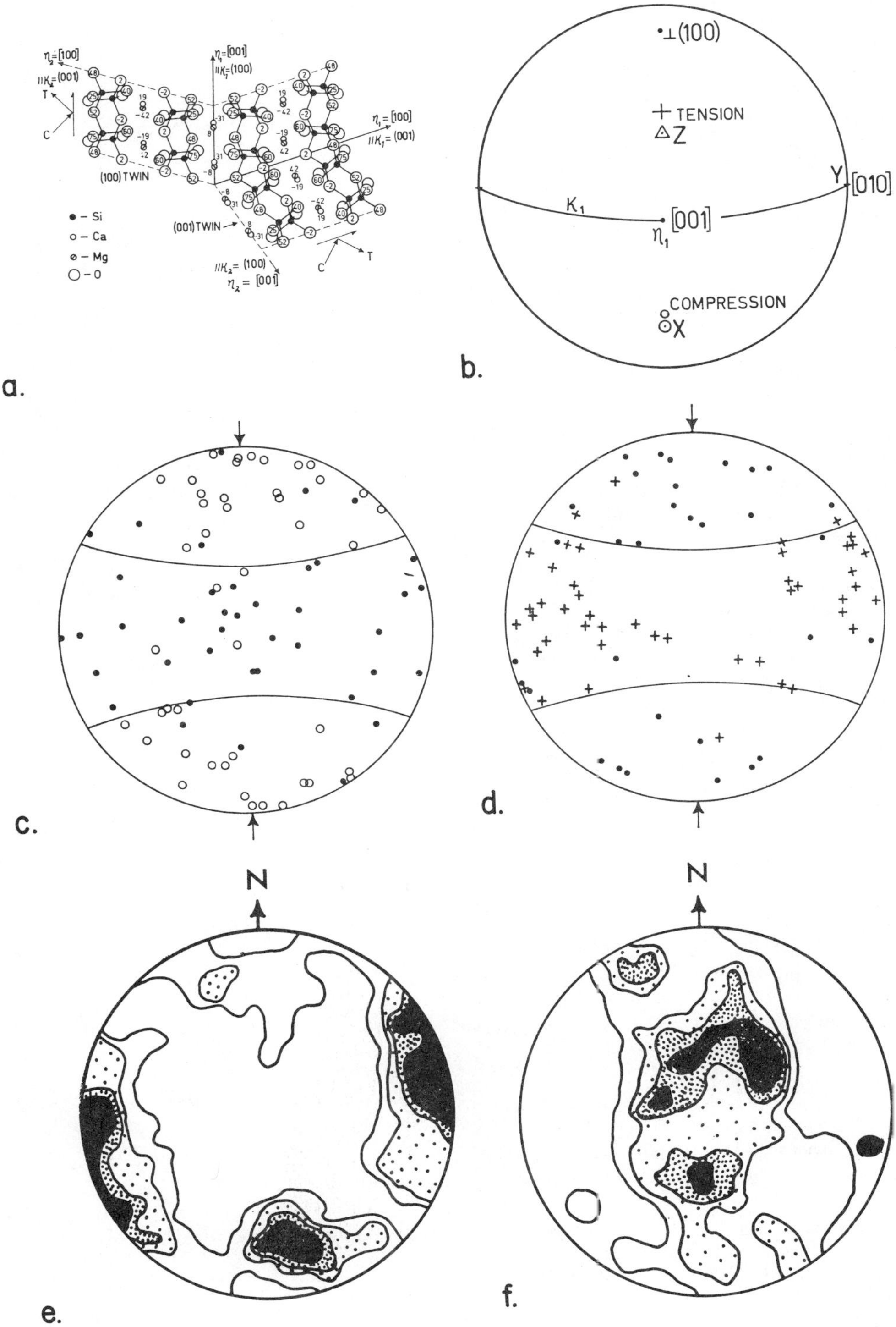

η2 = [100]
//K2 = (001)
η1 = [001]
//K1 = (100)
η1 = [100]
//K1 = (001)
(100) TWIN
(001) TWIN
//K2 = (100)
η2 = [001]
T
C
• – Si
○ – Ca
⌀ – Mg
◯ – O
a.
⊥(100)
TENSION
Z
Y
[010]
K1
[001]
η1
COMPRESSION
X
b.
c.
d.
N
N
e.
f.

deformed at 5 kb confining pressure, temperatures up to 1000° C and strain rates ranging from 10^{-2} to 10^{-6}/sec. The olivine undergoes slip on several systems; the active slip systems have been identified by a study of slip bands on polished surfaces, deformation lamellae in thin section (*see* discussion on quartz) and the geometry of kinking in the crystals. The predominant flow mechanism changes with increasing temperature and decreasing strain rate from $T = \{100\}$, $\{110\}$ $t = [001]$ through $T = \{110\}$, $t = [001]$ to slip on $T = \{0kl\}$, $t = [100]$ (Fig. 17a). The last of these systems was determined by noting that deformation lamellae have orientations ranging from (010) to (001), indicating pencil glide parallel to [100]. Kink bands in terrestrially deformed olivines (Raleigh, 1965a, 1965b, 1968) are most commonly oriented parallel to (100) with external rotation about various axes approximately in (100). Slip parallel to [100] is indicated and so only the system $\{0kl\}$ [100] will be discussed here. Kink bands parallel to (001) appear to be common only in stony meteorites (Carter and others, 1968).

In Taylor and Elam's (1926) theory of pencil glide, slip may take place on any plane in the zone of the slip direction. However, provided that strain is approximately uniform in a grain, only that plane will slip for which the resolved shear stress is greatest.[6] To test whether olivine behaves in conformity with the theory, the resolved shear stress coefficient, S_0, for the operative slip mechanism as determined for each grain was compared with the maximum possible S_0. The coefficient of resolved shear stress is

$$S_0 = \cos\phi \cos\lambda$$

where ϕ is the angle between σ_1 and the slip direction, and λ is the angle between σ_1 and the perpendicular to the slip plane (that is, deformation lamellae). Given ϕ, the slip plane for which the resolved shear stress is a maximum will be oriented so that

$$\lambda = 90 - \phi.$$

The maximum possible resolved shear stress coefficient for a given angle, ϕ, will be therefore

$$S_0 \max = \cos\phi \sin\phi.$$

Comparing S_0 for the grains measured with S_0 max (Fig. 17b) shows that the active slip plane, in accord with theory, is close to that for which the resolved shear stress is greatest regardless of the orientation of [100] (Raleigh, 1968). This observation has important consequences for the possibility of performing a dynamic analysis on deformed olivine.

To illustrate the principle underlying the technique of dynamic analysis of olivine we shall consider three orthogonal axes representing the slip direction, t, the pole of the slip plane, T, and the axis of external rotation, E.R. (Fig. 17b, *inset*). With respect to these coordinates, the direction of maximum principal compressive stress, σ_1, has the direction cosines, $\cos\phi$, $\cos\lambda$, and $\cos\theta$. From the preceding, the slip plane will be oriented such that the resolved shear stress is at its maximum, and therefore

$$\cos\lambda = \sin\phi\,.$$

But

$$\cos^2\phi + \cos^2\lambda + \cos^2\theta = 1$$

and substituting from above

$$\cos^2\phi + \sin^2\phi + \cos^2\theta = 1$$

so that

$$\cos^2\theta = 0$$

and

$$\theta = 90\,.$$

Therefore, the angle between σ_1 and the axis of external rotation should be 90° in grains bent or kinked by pencil glide. For olivine, measurements on deformed grains to determine E.R. are extremely simple. As E.R. is 90° to the slip direction, $[100] = Z$, measurements of Z in

[6] In the general case two such planes are possible although in olivine, as in silver chloride (Nye, 1949) and α-iron (Taylor and Elam, 1926), only one was observed.

▶

Figure 17. Principal stresses from experimentally deformed olivine (Raleigh, 1968). a. Slip systems in olivine as functions of temperature and strain rate. b. Orientation of the slip direction in olivine grains deformed by {0kl} [100]. pencil glide with respect to the compression axis for axially symmetric applied stress (*see* inset). c. Magnitude of the shear stress as a function of slip direction for pencil glide calculated for a ratio of $\sigma_1 : \sigma_2 : \sigma_3 = 3 : 2 : 1$. d. Orientations of external rotation axes in experimentally deformed grains (N–S arrows are parallel to σ_1). e. Orientation of [100] axes (=t) in kinked olivine grains.

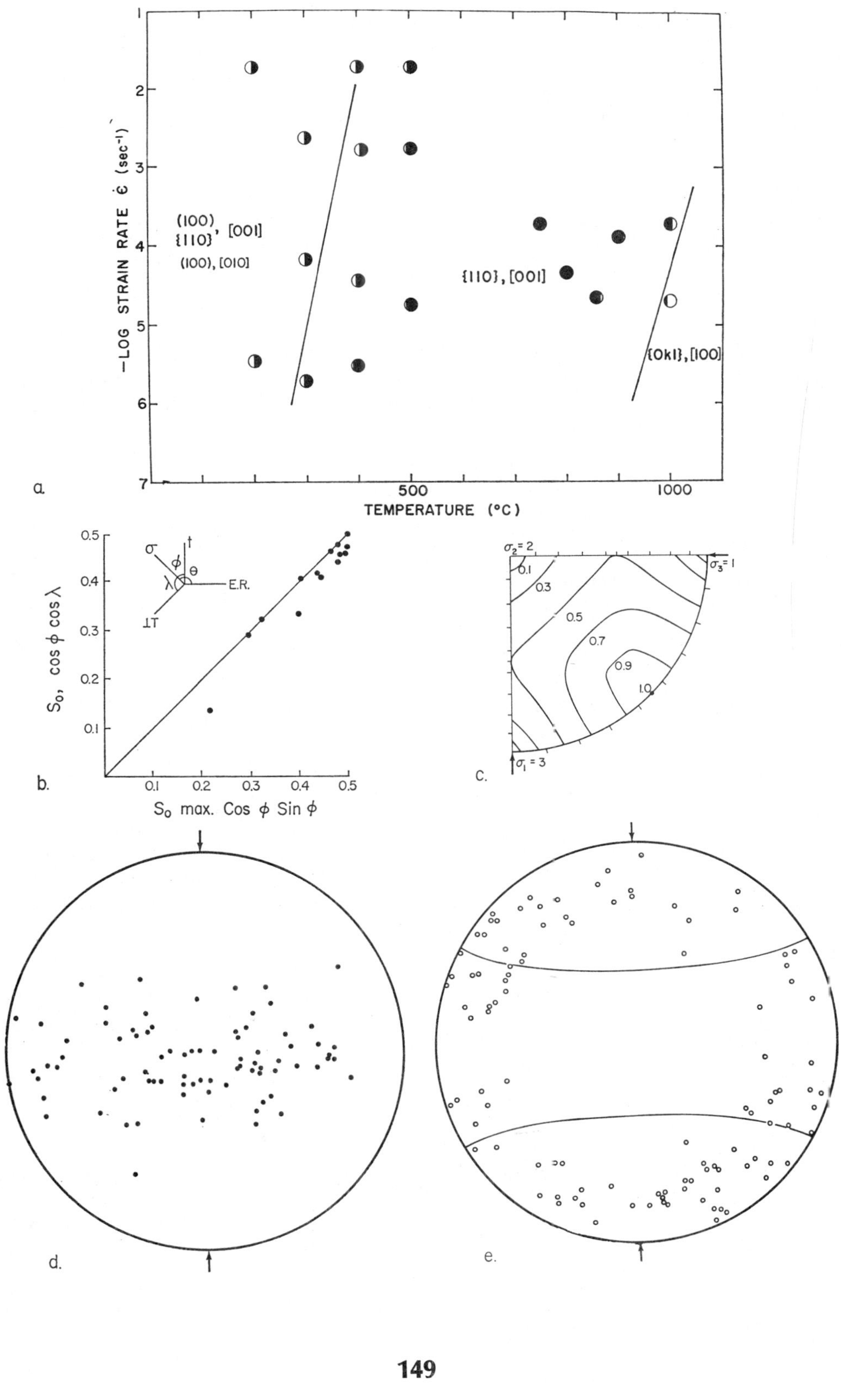

−LOG STRAIN RATE ε̇ (sec⁻¹)
TEMPERATURE (°C)
(100), {110}, [001]
(100), [010]
{110}, [001]
{0kl}, [100]
a.
S₀, cos φ cos λ
S₀ max. Cos φ Sin φ
E.R.
b.
σ₂ = 2
σ₃ = 1
σ₁ = 3
c.
d.
e.

two places across one or more kink bands when plotted on a projection will define a great circle, to which the pole is E.R. The slip direction moves along the great circle away from σ_1 in the direction of σ_3. When the axes of external rotation from a number of grains are plotted, they should lie in a girdle normal to σ_1 where $\sigma_1 > \sigma_2 = \sigma_3$.

In the experimental specimen in which pencil glide in [100] predominated, the axes of external rotation were determined from two measurements of Z in grains where there was external rotation of about 10° or greater across one or two closely spaced kink boundaries. Where there were several kink bands having smaller external rotations, several measurements of Z were taken and the mean great circle constructed through them. The normals to the great circles, E.R., are plotted in Figure 17d, where they define a girdle normal to σ_1 and parallel to $\sigma_2 = \sigma_3$. Plots of the poles to the kink bands in the specimen show no strong concentrations either parallel to or at 45° to the compression axes, and the deformation lamellae are too rare in natural specimens to be useful indicators of directions of principal stress. Therefore, the following orientation schemes are proposed:

(1) $\sigma_1 > \sigma_2 = \sigma_3$

E.R. girdle normal to σ_1; Z in more deformed sectors rotates *away from* σ_1 toward $\sigma_2 = \sigma_3$.

(2) $\sigma_1 = \sigma_2 > \sigma_3$

E.R. girdle normal to σ_3, Z in more deformed sectors rotates *toward* σ_3, away from $\sigma_1 = \sigma_2$.

The case for the general state of stress ($\sigma_1 > \sigma_2 > \sigma_3$) is somewhat more difficult, but qualitative effects can be seen by determining the variation of the shear stress with orientation in a given stress field. The shear stress τ on any plane is related to the principal stresses by

$$\tau = [\sigma_1^2 l_1^2 + \sigma_2^2 l_2^2 + \sigma_3^2 l_3^2 - (\sigma_1 l_1^2 + \sigma_2 l_2^2 + \sigma_3 l_3^2)^2]^{1/2}$$

where l_1, l_2, l_3 are the direction cosines of the normal to the plane. Taking an arbitrary ratio of the magnitudes of the principal stresses of $\sigma_1 : \sigma_2 : \sigma_3 = 3:2:1$, we may calculate the variation of τ as a function of orientation (Fig. 17c). Inasmuch as pencil glide takes place on the plane with the highest shear stress parallel to the zone axis [100], Figure 17c may be interpreted directly to show the orientations of [100] in grains that are most likely to slip. Thus those grains whose [100] axes are located, say, within the $\tau = 0.7$ contour are those most likely to deform. Consequently, a plot of the E.R. axes would give a broad maximum parallel to σ_2. This maximum would spread to a girdle parallel to the σ_2–σ_3 plane as $\sigma_2 \rightarrow \sigma_3$ or to a girdle parallel to the σ_1–σ_2 plane as $\sigma_2 \rightarrow \sigma_1$.

The foregoing discussion has assumed nearly random crystallographic orientation of the constituent grains, which is, however, rarely observed for olivine in the peridotites of orogenic belts. The effect of crystal orientations will have to be taken into account for individual analyses and is subject to the restrictions discussed for dynamic analysis of twin lamellae in calcite. In addition, an ambiguity arises in attempting to distinguish between σ_1 and σ_3 for all cases discussed because it may be difficult in naturally deformed olivines to distinguish between the relatively rotated (deformed) and unrotated parts of individual grains. This ambiguity may be resolved by plotting poles to kink-band boundaries or T = [100] in kinked grains (Fig. 17e). Although the poles and [100] show a wide range of orientations, most of them are inclined at angles less than 60° to σ_1, allowing the distinction to be made. Therefore, we propose that the orientations of the principal stress axes can be determined from distributions of external rotation axes and the identification of the axes can be made by analysis of distributions of poles to kink bands or orientations of [100].

Other Materials

Mugge and Heide (1931) and Starkey and Brown (1964) both produced albite and pericline twins by deforming anorthite at room temperature; in the latter study, most of the twins disappeared upon removal of the load. Borg and Handin (1966) conducted experiments at 5 kb confining pressure and 400° to 500° C on a labradorite crystal oriented best for both albite and pericline twinning. They succeeded in producing permanent albite twins but were uncertain of having induced pericline twinning. In one crystal, mechanically induced twinning was restricted to a narrow zone and the external rotation gave rise to a well-defined kink band (Pl. 5, fig. a). Kink bands normal to (010) similar to that produced experimentally have been observed in naturally deformed plagioclase (Pl. 5, fig. b and c; Seifert, 1965). However, the amount of albite twinning within the bands is the same as that outside indicating that the rotation must have been induced by translation gliding parallel to (010) or by fracturing (there is some rupture along the boundaries). In any event, the ubiquity of growth (primary) twins according to the albite law and the rarity of kink bands in naturally deformed plagioclase may preclude its being useful for determining principal stress orientations.

Experiments are now underway to determine the mechanical behavior and slip mechanisms in amphiboles (R.E. Riecker and T. P. Rooney, oral commun.). In our experience, however, amphiboles in tectonites rarely show any signs of plastic deformation and therefore are not likely to be useful indicators of principal stress directions. Kyanite (Pl. 1, fig. b; Raleigh, 1965a) gypsum (Pl. 5, fig. d) and brucite (Turner and Weiss, 1965) deform by kinking on a single slip plane. The analysis should be identical to that for the micas and orthopyroxenes.

APPLICATION OF THE METHODS TO NATURAL FOLDS

For an example of the usefulness of the methods discussed above, we briefly review the main results of studies concerned with determining the orientations of principal stresses around natural folds. The only methods used to date have been dynamic analyses of twin gliding in calcite and deformation lamellae in quartz. These methods have been applied to folds of many types, styles and sizes[7] and have led to internally consistent and generally similar results. For most of the studies on open folds, the deduced σ_1 lies near the bedding plane and is inclined at a few degrees to the dip direction; σ_2 is parallel to the fold axis and σ_3 is nearly normal to the bedding. With the exception of the study of Conel (1962), these results were obtained regardless of the location of the specimens within the fold. In the study of Carter and Friedman (1965), for example, all specimens were collected within the upper 20 feet of the 80 foot thick sandstone layer forming part of an anticline (*see* Figure lld). Conel (1965), on the other hand, observed a change in orientation of the principal stress axes with a change in location in accord with prediction based on elasticity theory (Fig. 7). Significantly, however, there is more scatter in the data from the "tensive" part of the fold (Region II) than from the "compressive" part (Region I), an observation relevant to the discussion below.

The orientations of principal stress axes observed for natural folds are consistent with the stress trajectories deduced theoretically for two-dimensional large-amplitude folding of a viscous layer in a less viscous matrix (Dieterich and Carter, 1969). The stress trajectories were obtained from the velocity fields determined for a series of linear incremental problems using a modification of the finite element method (Dieterich and Onat, 1969). The model regarded as most representative of natural folds had the following properties: η/η_1 (viscosity contrast = 42. L/h (wavelength/thickness) = 12, A_0 (initial amplitude = $0.1h$. The initial dip was about 2 degrees and the layer was subjected to compressive strains up to 50 increments of 2 percent average natural strain per increment. Figure 18a shows the maximum principal compressive stress (σ_1) trajectories for individual cells after 40 percent total (natural) shortening. Both the orientations of the stresses and the geometry of the fold are similar to the results and geometry of the fold studied by Conel (1962, Fig. 7) and to some folds studied by other workers.

Figure 18b is a summary plot of the variation with total strain of the magnitudes of the stresses (normalized at 1.0) at various localities within the fold (*see* inset). The areas under the curves are proportional to the total strain at some point in the fold. If we may suppose that the amount of intragranular flow in the rocks is proportional to the total strain then we should be able to predict, qualitatively, the results obtained from dynamic analyses, assuming that this simple model is applicable. This does not mean that intragranular flow is the only process governing the mechanical behavior of the material during the growth of the fold. For most of the folds, plastic flow makes only a minor contribution to the deformation, the dominant process probably being one of mass transport such as recrystallization by solid diffusion or by solution and redeposition.

All parts of the layer are subject to high compressive stresses parallel to the layer during the initial stages of folding (Fig. 18b). Therefore, if intragranular flow occurs during this stage as expected, the data obtained from all specimens should show some evidence of compression parallel to the layer as indeed do Conel's data from Region II, thus explaining

[7] Christie and Raleigh, 1959; Hansen and Borg, 1962; Conel, 1962, Hara, 1963, 1964, 1965; Scott and others, 1965; Carter and Friedman, 1965; Compton, 1966.

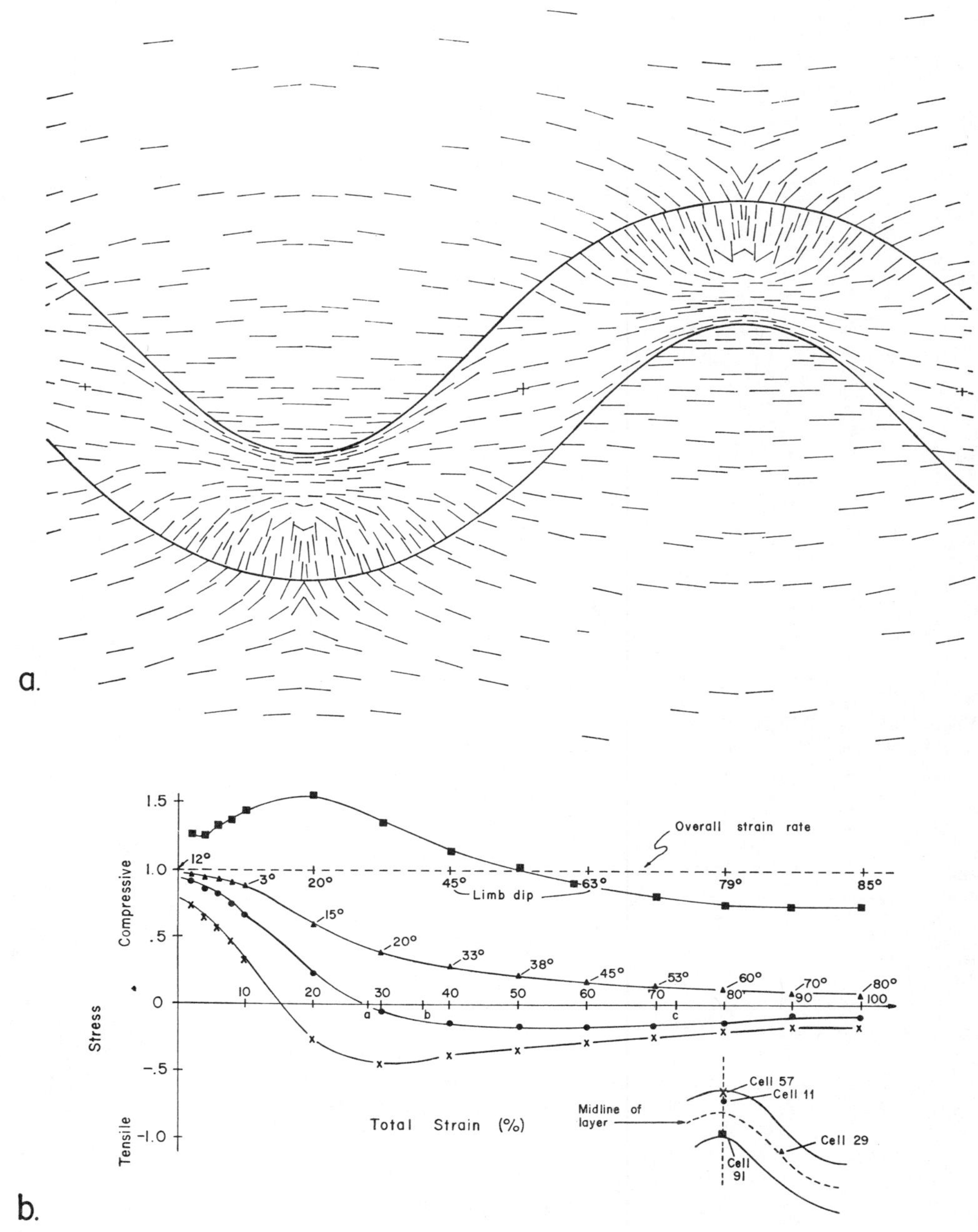

Figure 18. Theoretical stress history of folding a viscous layer in a less viscous matrix (Dieterich and Carter, 1969). a. Orientations of σ_1 after 40 percent shortening of the layer. b. Variation with total shortening (strain within matrix) of relative magnitudes of principal stresses for various localities, within fold (*see* inset). Numbers at triangles for cell 29 indicate inclination of σ_1 to dip of bed.

the scatter. The average locations of specimens analyzed by Carter and Friedman (1965) would correspond approximately to cell 11 and to cell 29 in Figure 18b and the limb dips would correspond to about 50 percent shortening. We would predict that the predominant patterns of subfabric elements should indicate σ_1 axes nearly parallel to bedding and σ_3 nearly normal to it, in agreement with the observations (Fig. 11d). For specimens obtained from the flank of the fold, the low inclinations to the dip of the derived compression axes are interpreted to indicate that most of the intragranular flow took place during early stages of folding when the stresses were high.

Thus the pertinent fabric data from natural folds are interpreted to reflect *cumulative* intragranular flow during folding controlled by the magnitude as well as the orientations of the principal stresses. These analyses and others are discussed in detail elsewhere (Dieterich and Carter, 1969), and most of the data obtained on the orientations of principal stresses from natural folds are shown to be qualitatively consistent with the viscous model discussed briefly here.

DISCUSSION AND CONCLUSIONS

Apart from the important problem of folding in rocks (on which much work remains to be done), there are many significant structural problems to which these methods might be applied. One important question concerns the transport of large thrust sheets. Do these sheets slide under the influence of gravitational forces or are they pushed from the rear? For gravity sliding the stresses are those due to body forces alone. For plates pushed from the rear the maximum principal compressive stress, at equilibrium, will have the same trend as the transport direction but will plunge toward the movement direction at an angle determined by the aspect ratio of the plate. Thus, a careful application of the methods discussed above could resolve this ambiguity. These dynamic methods could also be used to determine the direction and sense of shear along exposed faults and mylonite zones. They might also be used to infer probable orientations of concealed faults as well as the direction and sense of shear along such faults.

Another important question in structural geology is that of the origin of rock cleavage. Axial plane cleavage has been explained variously as having formed: (1) perpendicular to σ_1; (2) perpendicular to ϵ_1 (maximum principal compressive strain); or (3) at an angle to σ_1 and ϵ_1, perhaps in an orientation of maximum shearing stress or one of zero finite elongation. Of these hypotheses (2) has the strongest supporting evidence as obtained from deformed fossils and cobbles (*see* Ramsay, 1967). Chapple (1968) and Dieterich (1969) have attacked this problem theoretically using viscous models similar to that described in the last section. Dieterich found that the patterns given by directions normal to ϵ_1 correlated well with orientations of axial plane cleavage observed in natural folds and that the variations in patterns due to fanning and refraction were also correlative. Inasmuch as the stress and strain tensors do not coincide in folds (because the strain is finite and inhomogeneous) finite shear stresses exist along the newly formed cleavage planes. Dieterich (1969) showed that these shear stresses are of the correct sense and might well account for the slippage commonly observed along these planes. This conclusion is amenable to confirmation by dynamic analysis of rocks in the vicinity of suitable folds.

Of considerable importance is the interpretation of features induced by intragranular flow in ultramafic minerals of peridotites and dunites and in nodules of these materials in basalt. Do these features originate by solid flow in the upper mantle, during solid intrusion, or might they be produced during compaction of a crystal mush? If they are produced by filter pressing of accumulates, then the orientation of σ_1 derived from dynamic analyses of the structures should be normal to the layering. On the other hand, if they originate during solid flow, the orientations of the principal stress axes derived might have any inclination to the layering depending on the nature of the deformation and the crystal fabric.

The examples given above are only a few of the many significant problems amenable to investigation using the various methods of dynamic analysis. These methods are general and may be applied to any material that has undergone intragranular flow provided that the deformation mechanisms are known. The analyses are neither difficult nor excessively time consuming and most attempts to date have been successful in that the results have made possible reasonable interpretations of the structures investigated. We feel that the routine use of these methods of dynamic analysis by geologists will certainly lead to a much clearer understanding of the origin of a wide variety of structural phenomena.

REFERENCES CITED

Borg, I., and Handin, J., 1966, Experimental deformation of crystalline rocks: Tectonophysics, v. 3, p. 249–368.

——1967, Torsion of calcite single crystals: Jour. Geophys. Research, v. 72, p. 641–670.

Borg, I., and Turner, F. J., 1953, Deformation of Yule Marble, Part VI: Geol. Soc. America Bull., v. 64, p. 1343–1352.

Cahn, R. W., 1954, Twinned crystals: Advances in Physics, v. 3, p. 363–445.

Carter, N. L., 1965, Basal quartz deformation lamellae—a criterion for recognition of impactites: Am. Jour. Sci., v. 263, p. 786–806.

Carter, N. L., Christie, J. M., and Griggs, D. T., 1964, Experimental deformation and recrystallization of quartz: Jour. Geology, v. 72, p. 687–733.

Carter, N. L., and Friedman, M., 1965, Dynamic analysis of deformed quartz and calcite from the Dry Creek Ridge Anticline, Montana: Am. Jour. Sci., v. 263, p. 747–785.

Carter, N. L., Raleigh, C. B., and DeCarli, P. S., 1968, Deformation of olivine in stony meteorites: Jour. Geophys. Research (in press).

Chapple, W. M., 1968, A mathematical theory of finite-amplitude folding: Geol. Soc. America Bull., v. 79, p. 47–68.

Christie, J. M., 1958, Dynamic interpretation of the fabric of dolomite from the Moine thrust-zone in Northwest Scotland: Am. Jour. Sci., v. 256, p. 159–170.

Christie, J. M., and Green, H. W., 1964, Several new slip mechanisms in quartz: Am. Geophys. Union Trans., v. 45, p. 103.

Christie, J. M., Griggs, D. T., and Carter, N. L., 1964, Experimental evidence of basal slip in quartz: Jour. Geology, v. 72, p. 734–756.

Christie, J. M., and Raleigh, C. B., 1959, The origin of deformation lamellae in quartz: Am. Jour. Sci., v. 257, p. 385–407.

Clark, R. H., 1954, A study of calcite twinning in the Strathavon Marble, Banttshire: Geol. Mag., v. 91, p. 121–128.

Compton, R. R., 1966, Analyses of Pliocene-Pleistocene deformation and stresses in northern Santa Lucia Range, California: Geol. Soc. America Bull., v. 77, p. 1361–1380.

Conel, J. E., 1962, Studies of the development of fabrics in naturally deformed limestones: Unpub. Ph.D. dissert., California Inst. Technology.

Crampton, C. B., 1958, Structural petrology of Cambro-Ordovician limestones of northwest Highlands of Scotland: Am. Jour. Sci., v. 256, p. 145–158.

Dieterich, J. H., 1969, Origin of cleavage in folded rocks: Am. Jour. Sci., v. 267, p. 155–165.

Dieterich, J. H., and Carter, N. L., 1969, Stress-history of folding: Am. Jour. Sci., v. 267, p. 129–154.

Dieterich, J. H., and Onat, T., 1969, Slow finite deformations of viscous solids: Jour. Geophys. Research, v. 74, p. 2081–2088.

Fairbairn, H. W., 1941, Deformation lamellae in quartz from the Ajibik Formation, Michigan: Geol. Soc. America Bull., v. 52, p. 1265–1278.

Fairbairn, H. W., and Hawkes, H. E., Jr., 1941, Dolomite orientation in deformed rocks: Am. Jour. Sci., v. 239, p. 617–632.

Friedman, M., 1963, Petrofabric analysis of experimentally deformed calcite-cemented sandstones: Jour. Geology, v. 71, p. 12–37.

—— 1964, Petrofabric techniques for the determination of principal stress directions in rocks, *in* Judd, W. R., *Editor*, State of Stress in the Earth's Crust: New York, Am. Elsevier Pub. Co., Inc., 732 p.

Friedman, M., and Conger, F. B., 1964, Dynamic interpretation of calcite twin lamellae in a naturally deformed fossil: Jour. Geology, v. 72, p. 361–368.

Gilmore, P., and Carmen, M. F., 1954, Petrofabric analysis of the Loch Tay limestone from Strachar Argyll: Geol. Mag., v. 91, p. 49–60.

Griggs, D. T., and Blacic, J. D., 1964, The strength of quartz in the ductile regime: Am. Geophys. Union Trans., v. 45, p. 102–103.

Griggs, D. T., and Miller, W. B., 1951, Deformation of Yule Marble, Part I: Geol. Soc. America Bull., v. 62, p. 853–862.

Griggs, D. T., Turner, F. J., Borg, I., and Sosoka, J., 1951, Deformation of Yule Marble, Part IV: Geol. Soc. America Bull., v. 62, p. 1315–1406.

Griggs, D. T., Turner, F. J., Borg, I., and Sosoka, J., 1953, Deformation of Yule Marble, Part V: Geol. Soc. America Bull., v. 64, p. 1327–1342.

Griggs, D. T., Turner, F. J., and Heard, H. C., 1960, Deformation of rocks at 500° to 800° C: Geol. Soc. America Mem. 79, 382 p.

Handin, J. W., and Fairbairn, H. W., 1953, Experimental deformation of Hasmark dolomite: Geol. Soc. America Bull., v. 64, p. 1429–1430.

Handin, J. W., and Griggs, D. T., 1951, Deformation of Yule Marble, Part II: Geol. Soc. America Bull., v. 62, p. 863–886.

Handin, J. W., Higgs, D. V., and O'Brien, J. K., 1960, Torsion of Yule Marble under confining pressure: Geol. Soc. America Mem. 79, p. 245–274.

Hansen, E. C., and Borg, I. Y., 1962, The dynamic significance of deformation lamellae in quartz of a calcite-cemented sandstone: Am. Jour. Sci., v. 260, p. 321–336.

Hara, I., 1961, Dynamic interpretation of the simple type of calcite and quartz fabrics in the naturally deformed calcite-quartz vein: Hiroshima Univ. Jour. Sci., Ser. C., v. 4, p. 35–53.

—— 1963, Petrofabric analysis of a drag fold: Hiroshima Univ. Geol. Rept. 12, p. 463–492.

—— 1964, The quartz lamella fabrics in a concentric fold: Hiroshima Univ. Jour. Sci., Ser. C, p. 369–394.

—— 1965, Quartz fabrics in a kink band: Hiroshima Univ. Jour. Sci., Ser. C, v. 4, p. 1–12.

Heard, H. C., 1963, Effect of large changes in strain rate in the experimental deformation of Yule Marble: Jour. Geology, v. 71, p. 162–195.

Heard, H. C., and Carter, N. L., 1968, Experimentally induced "natural" intragranular flow in quartz and quartzite: Am. Jour. Sci., v. 266, p. 1–42.

Higgs, D. V., and Handin, J. W., 1959, Experimental deformation of dolomite single crystals: Geol. Soc. America Bull., v. 70, p. 245–278.

Honeycombe, R. W. K., 1952, Inhomogeneities in the plastic deformation of metal crystals, 1–11; Jour. Inst. Metals, v. 80, p. 45–56.

Ingerson, E., and Tuttle, O. F., 1945, Relations of lamellae and crystallography of quartz and fabric directions of some deformed rocks: Am. Geophys. Union Trans., v. 26, p. 95–105.

Johnson, A., 1902, Biegungen und Translationen: Neues Jahrb. Mineralogie Abh., Teil II, p. 133–153.

Klassen-Neklyudova, M. V., 1964, Mechanical twinning of crystals: New York, Consultants Bureau, 213 p.

Mackie, J. B., 1947, Petrofabric analyses of two quartz-piemontite schists from northwest Otago: Royal Soc. New Zealand Trans., v. 86, p. 362–368.

McIntyre, D. B., and Turner, F. J., 1953, Petrofabric analysis of marbles from mid-Strathspey and Strathavon: Geol. Mag., v. 90, p. 225–240.

Mugge, O., and Heide, F., 1931, Einfache Schiebungen am Anorthit: Neues Jahrb. Mineralogie Abh., I, v. 64, p. 163–170.

Naha, K., 1959, Time of formation and kinematic significance of deformation lamellae in quartz: Jour. Geology, v. 67, p. 920–924.

Nickelsen, R. P., and Gross, G. W., 1959, Petrofabric study of Conestoga limestone from Hanover, Pennsylvania: Am. Jour. Sci., v. 257, p. 276–286.

Nissen, H. U., 1964, Dynamic and kinematic analysis of crinoid stems in a quartz graywacke: Jour. Geology, v. 72, p. 346–360.

Nye, J. F., 1949, Plastic deformation of silver chloride; 1. Internal stresses and the glide mechanism: Royal Soc. London, A, v. 198, p. 190–204.

Orowan, E., 1942, A type of plastic deformation new in metals: Nature, v. 149, p. 643.

Paterson, M. S., and Weiss, L. E., 1966, Experimental deformation and folding in phyllite: Geol. Soc. America Bull., v. 77, p. 343–373.

Paterson, M. S., and Weiss, L. E., 1968, Folding and boudinage of quartz-rich layers in experimentally deformed phyllite: Geol. Soc. America Bull., v. 79, p. 795–812.

Raleigh, C. B., 1965a, Glide mechanisms in experimentally deformed minerals: Science, v. 150, p. 739–741.

—— 1965b, Structure and petrology of an Alpine Peridotite on Cypress Island, Washington, U.S.A.: Beitr. Mineralogie u. Petrographie, v. II, p. 719–741.

—— 1968, Mechanisms of plastic deformation of olivine: Jour. Geophys. Research, v. 73, p. 5391–5406.

Raleigh, C. B., and Talbot, J. L., 1967, Mechanical twinning in naturally and experimentally deformed diopside: Am. Jour. Sci., v. 265, p. 151–165.

Ramsay, J. G., 1967, Folding and fracturing of rocks: McGraw Hill Book Co., New York, 568 p.

Riecker, R. E., and Rooney, T. P., 1967, Deformation and polymorphism of enstatite under shear stress: Geol. Soc. America Bull., v. 78, p. 1045–1054.

Riecker, R. E., and Seifert, K. E., 1964, Shear deformation of upper-mantle mineral analogs; tests to 50 kilobars at 27° C: Jour. Geophys. Research, v. 69, p. 3901–3911.

Riley, N. A., 1947, Structural petrology of the Baraboo quartzite: Jour. Geology, v. 55, p. 453–475.

Scott, W. H., Hansen, E., and Twiss, R. J., 1965, Stress analysis of quartz deformation lamellae in a minor fold: Am. Jour. Sci., v. 263, p. 729–746.

Seifert, K. E., 1965, Deformation bands in albite: Am. Mineralogist, v. 50, p. 1469–1472.

Starkey, J., and Brown, W. L., 1964, Kunstlicher erzeugung mechanischer zwillinge in Anorthit: Zeitschr. Kristallographie, v. 120, p. 388–392.

Suzuki, T., 1962, Petrofabric studies of crystalline limestone from the southern marginal belt of the Sambagama metamorphic zone in Central Shikoka, Japan: Hiroshima Univ. Jour. Sci., Ser. C, v. 4, p. 131–162.

Taylor, G. I., and Elam, C. F., 1962, The distortion of iron crystals: Roy. Soc. London, Proc. A, v. 112, p. 337–361.

Trommsdorff, V., and Wenk, H. R., 1965, Die regelung des dolomites von Crevola (Simplongruppe): Schweizer Mineralog. Petrog. u. Mitt., v. 45, p. 551–570.

Turner, F. J., 1948, Note on the tectonic significance of deformation lamellae in quartz and calcite: Am. Geophys. Union Trans., v. 29, p. 556–569.

—— 1953, Nature and dynamic interpretation of deformation in calcite of three marbles: Am. Jour. Sci., v. 251, p. 276–298.

—— 1962, "Compression" and "tension" axes deduced from $\{01\bar{1}2\}$ twinning in calcite: Jour. Geophys. Research, v. 67, p. 1660.

—— 1964, Analysis of kinks in micas of an Innsbruck mica schist: Neues Jahrb. Mineralogie Monatsh., 9–11, p. 347–356.

Turner, F. J., and Chih, C. S., 1951, Deformation of Yule Marble, Part III: Geol. Soc. America Bull., v. 62, p. 887–906.

Turner, F. J., Griggs, D. T., Clark, R. H., and Dixon, R. H., 1956, Deformation of Yule Marble, Part VII: Geol. Soc. America Bull., v. 67, p. 1259–1294.

Turner, F. J., Griggs, D. T., and Heard, H. C., 1954, Experimental deformation of calcite crystals: Geol. Soc. America Bull., v. 65, p. 883–934.

Turner, F. J., Griggs, D. T., Heard, H. C., and Weiss, L. E., 1954, Plastic deformation of dolomite rock at 380° C: Am. Jour. Sci., v. 252, p. 477–488.

Turner, F. J., and Heard, H. C., 1965, Deformation of calcite crystals at different strain rates, *in* Heard and others, Studies of heterogeneous strain in experimentally deformed calcite, marble, and phyllite: California Univ. Pubs. Geol. Sci., v. 46, p. 103–126.

Turner, F. J., Heard, H. C., and Griggs, D. T., 1960, Experimental deformation of enstatite and accompanying inversion to clinoenstatite: Internat. Geol. Cong. 21, Part 18, p. 399–408.

Turner, F. J., and Weiss, L. E., 1963, Structural analysis of metamorphic tectonites: McGraw Hill Book Co., New York, 545 p.

Turner, F. J., and Weiss, L. E., 1965, Deformational kinks in Brucite and Gypsum: Natl. Acad. Sci. Proc., v. 54, p. 359–364.

Weiss, L. E., 1954, A study of tectonic style: California Univ. Pubs. Geol. Sci., v. 30, p. 1–102.

MANUSCRIPT RECEIVED BY THE SOCIETY AUGUST 22, 1968
REVISED MANUSCRIPT RECEIVED JANUARY 2, 1969
PUBLICATION AUTHORIZED BY THE DIRECTOR, U.S. GEOLOGICAL SURVEY

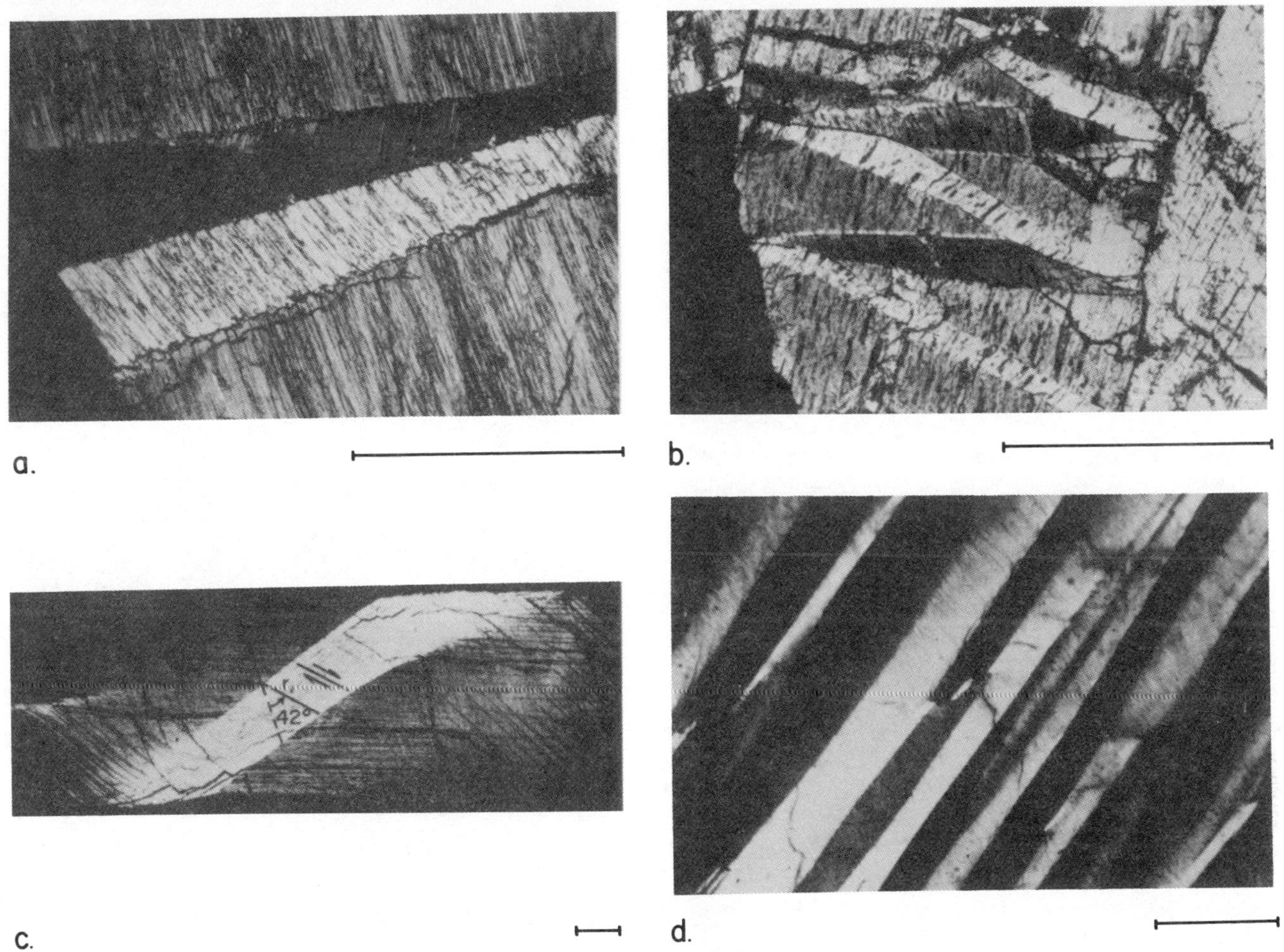

Thin sections between crossed polarizers. Scale lines beneath a, b, c represent 1 mm; beneath d, 0.1 mm. (a) Symmetrical kink in enstatite. (b) Conjugate kinks in kyanite. (c) Symmetrical kink in calcite by slip on r_1 (Turner and Heard, 1965). (d) Asymmetrical kinks in quartz. Note that the kink bands grade into undulatory extinction near the top of the photo (Carter and others, 1964).

KINK BAND IN CRYSTALS

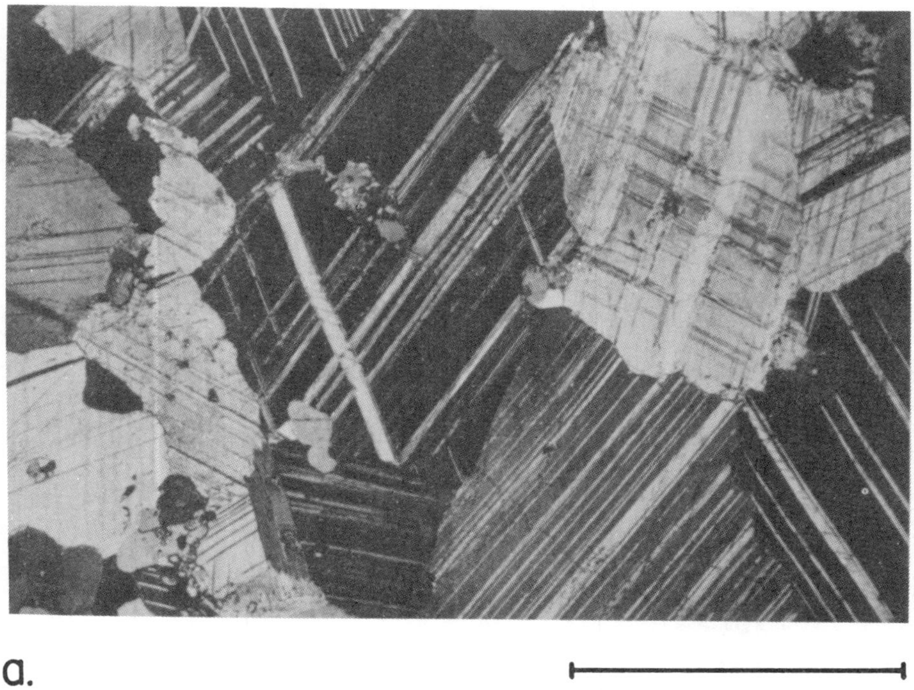

a.

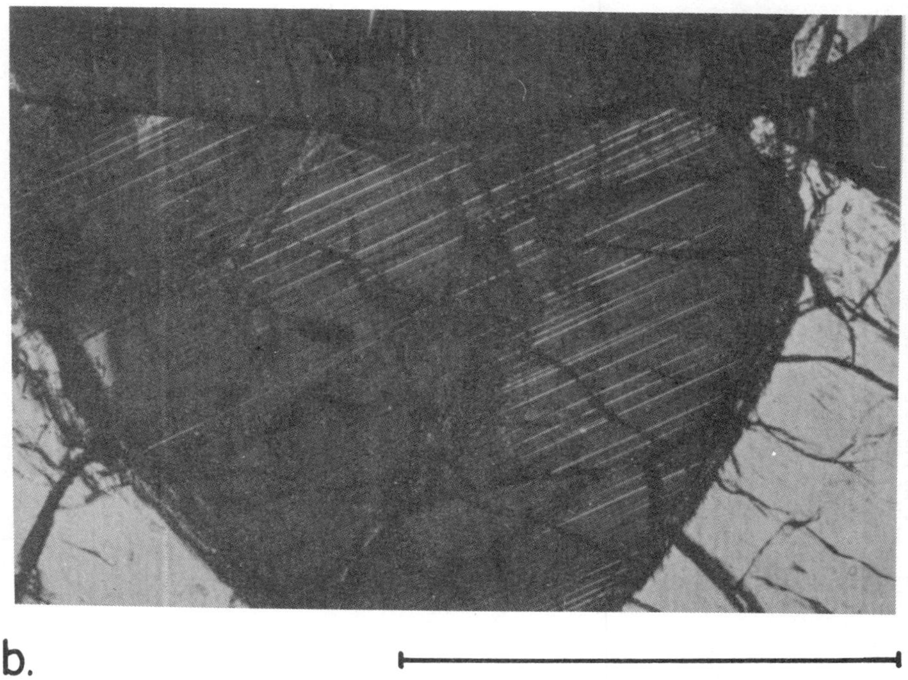

b.

Thin sections beneath crossed polarizers. Scale lines beneath photos represent 1 mm. (a) Twinned bands in naturally deformed calcite. (b) Thin twinned bands (white linear features) in experimentally deformed diopside (Raleigh and Talbot, 1967).

TWIN-GLIDING IN CRYSTALS

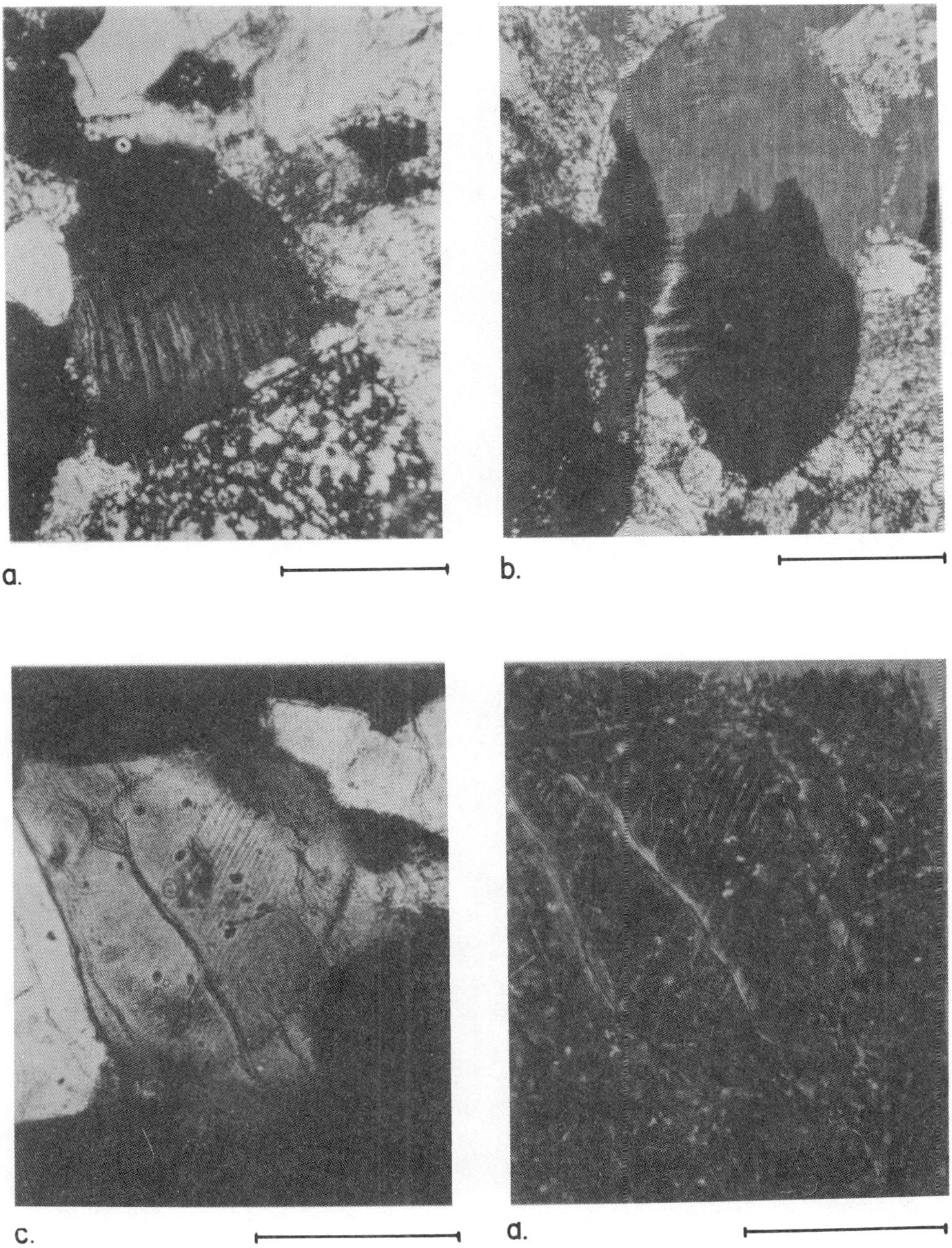

Thin sections between crossed polarizers. Scale lines beneath photomicrographs represent 0.1 mm. (a) Deformation lamellae (N–S-trending linear features) normal to zone of undulatory extinction (E–W light region) in naturally deformed quartz grain (Carter and Friedman, 1965). (b) Deformation lamellae (E–W-trending features) normal to zone of undulatory extinction (N–W-trending light bands) in naturally deformed quartz (Carter and Friedman, 1965). (c and d) Deformation lamellae in quartz grain experimentally deformed at 800° C, 10 kb, and a strain rate of 3.9×10^{-6}/sec in bright field illumination (c) and phase contact (d). The lamellae (N–E-trending linear features) are normal to zone of undulatory extinction in (c) and are more prominent and asymmetrical in (d). (Heard and Carter, 1968.)

DEFORMATION LAMELLAE AND UNDULATORY EXTINCTION IN QUARTZ

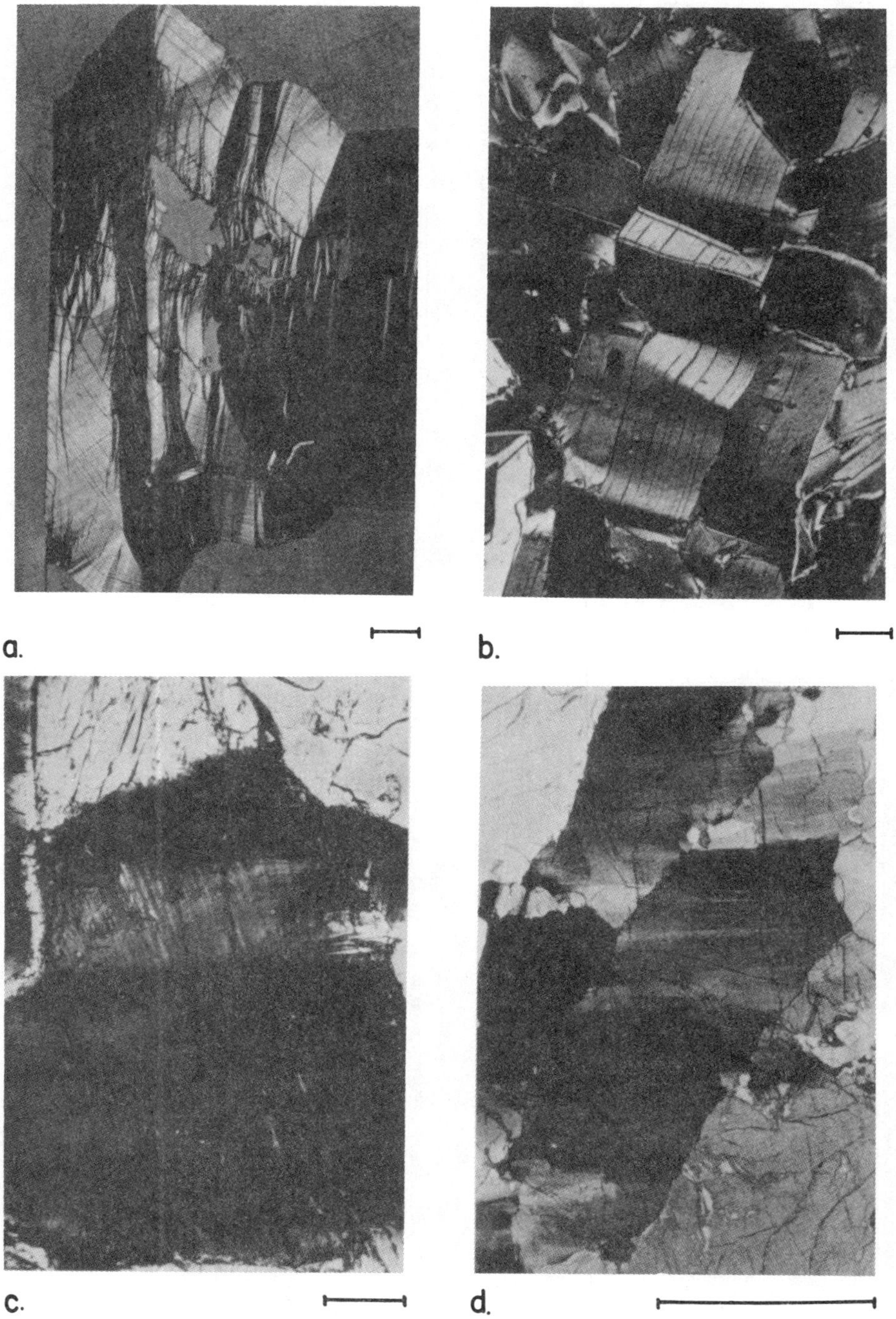

Thin sections between crossed polarizers. Scale lines beneath b and c represent 0.1 mm, beneath a and d 1.0 mm. (a) Kink bands in biotite crystal compressed parallel to the basal plane at 3 kb and 300° C. (b) Kink bands in naturally deformed biotite. (c) Deformation lamellae (N–S-trending linear features) normal to zones of undulatory extinction in experimentally deformed olivine grain (Raleigh, 1968). (d) Kink bands in naturally deformed olivine.

DEFORMATIONAL STRUCTURES IN BIOTITE AND OLIVINE

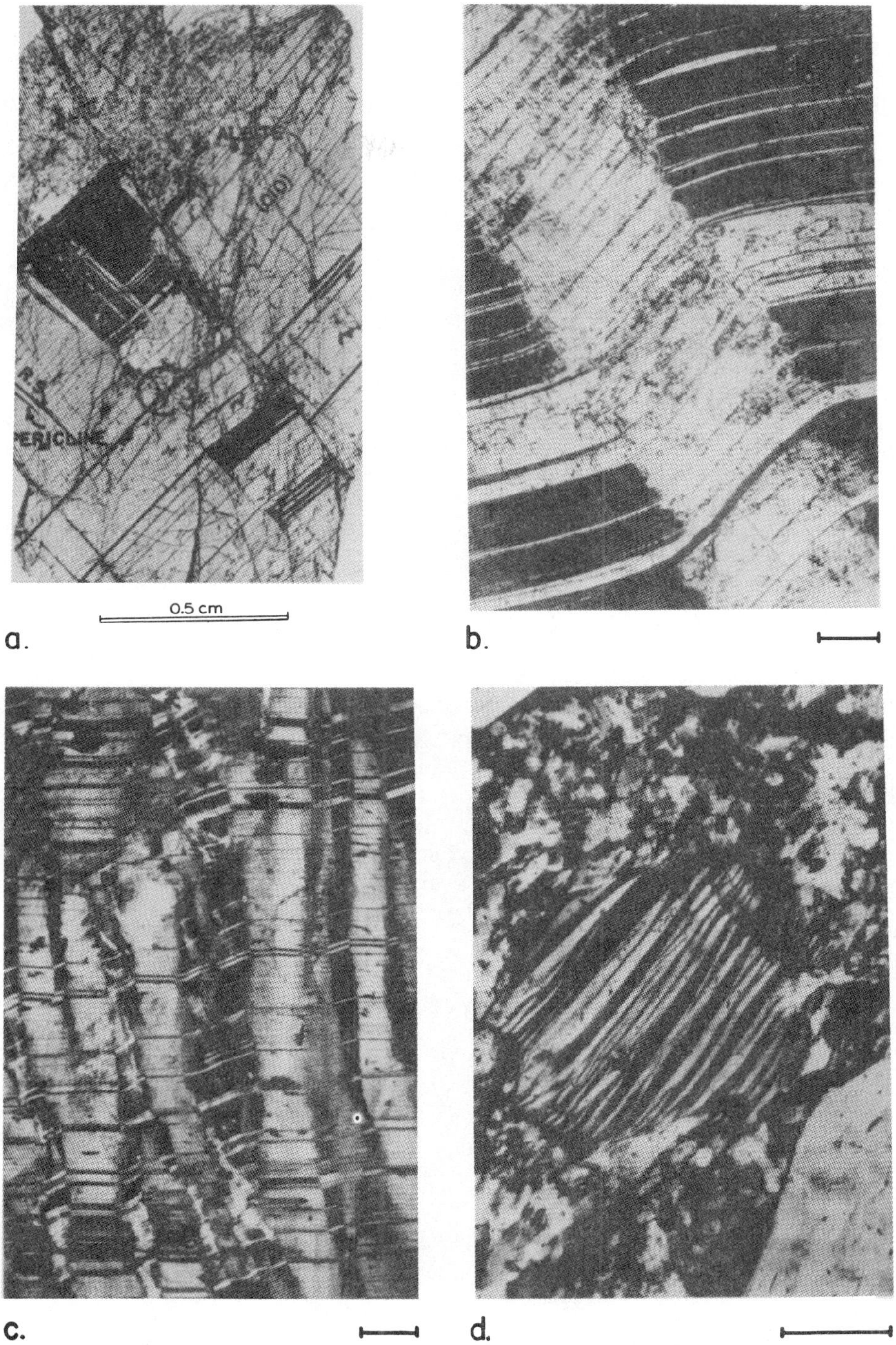

Thin sections beneath crossed polarizers. Scale lines beneath b and d represent 0.1 mm, beneath c, 1 mm. (a) Kink band produced experimentally by twin-gliding according to the albite twin law (Borg and Handin, 1966). (b and c) Kink bands in naturally deformed plagioclase. Note that the density of albite twins remains constant on both sides of kink bands (Seifert, 1965). (d) Kink bands in gypsum.

KINK BANDS IN PLAGIOCLASE AND GYPSUM

9

Copyright ©1963 by the Geological Society of Australia
Reprinted from *Geol. Soc. Australia Jour.* **10:**193–208 (1963)

THE CHRONOLOGICAL ANALYSIS OF CRYSTALLIZATION AND DEFORMATION OF SOME TASMANIAN PRECAMBRIAN ROCKS

By ALAN SPRY

(WITH 2 PLATES AND 3 TEXT-FIGURES)

(*Received October 5th, 1962; read at Hobart, December 20th, 1962.*)

ABSTRACT

The analysis of the time-relations between different minerals, between a mineral and some aspect of deformation, and between different aspects of deformation, enables the complex history of some regionally metamorphosed rocks to be recognized. Two separate periods of deformation occurred during the regional metamorphism of Precambrian schists, phyllites, quartzites and amphibolites from Frenchman's Cap, the Mersey-Forth area and Port Davey, Tasmania.

INTRODUCTION

Although deformation and crystallization may be regarded as essentially simultaneous in regional metamorphism, the study of certain metamorphic provinces shows that their effects may be independent. Major metamorphic events (the Hercynian in the Pyrenees, Zwart, 1960a and 1960b; the Caledonian in Scotland, Rast, 1959) have been shown to be complex and to consist of a number of periods of deformation. Crystallization and deformation are not necessarily simultaneous and may begin and end at different times. The metamorphic history of a rock is essentially the chronological relation of crystallization to deformation (Read, 1948); in some rocks the relation is simple but many regionally metamorphosed rocks contain evidence of several periods of alteration. The time element is of great importance in any metamorphic study but is commonly overlooked.

The technique of chronological analysis of deformation and crystallization has not been applied to Australian metamorphic rocks but a brief survey indicates that it would be a fruitful line of investigation in the metamorphic complexes of Broken Hill (New South Wales), the Kanmantoo Group (South Australia), the Brisbane Polymetamorphics (Queensland), etc.

The older Precambrian metamorphosed rocks of Tasmania have been studied by this method and will be used as an example of the use of chronological analysis.

CHRONOLOGICAL ANALYSIS OF CRYSTALLIZATION AND DEFORMATION

Chronological analysis consists of a study of the relations between the various effects of crystallization (recrystallization of existing minerals or the growth of new minerals) and deformation (strain of existing minerals, production of new S-surfaces, folding of existing S-surfaces).

The following three time-relations are examined: 1. The crystallization of a mineral with respect to some evidence of strain; in most cases to a foliation but also to a lineation, a fold or to a pattern of preferred orientation shown by petrofabric analysis. 2. The relative time of crystallization of one mineral with respect to another. 3. The relations between different phases of deformation, i.e. between different foliations or folds.

I. Time-Relations of Mineral Crystallization to some Evidence of Deformation

The crystallization of a mineral may be referred to as pre-, syn-, or post-tectonic according to whether it crystallized before, during or after a given phase of deformation. The terms pre-, para- and post-crystallization deformation have similar meanings but different emphasis.

The following terminology is used (Sander, 1950; Turner, 1948; Rast, 1958; Zwart, 1960, etc.).

The term S-surface is non-genetic and refers to any recognizable surface within a rock. Surfaces are numbered in order of formation. Lines of inclusions (trails, trends) can be seen within some porphyroblasts under the microscope. The S-surface within the crystal is called S_i (S-internal) and that in the body of the rock outside of the crystal is called S_e (S-external). *Helicitic* structure consists of curved trends within a crystal which has enveloped a previously folded S-surface. *Rotational* (snowball, pinwheel) structure is a symmetrical spiral pattern of S_i formed by the simultaneous crystallization and rotation of a crystal.

Pretectonic Crystallization

During regional metamorphism, porphyroblasts (e.g. andalusite, staurolite, cordierite, etc.) may grow due to elevation of temperature before the onset of appreciable movements (Zwart, 1960, p. 176). Such porphyroblasts may contain regularly distributed inclusions which may be either (a) related to the crystal structure, e.g. the cruciform patterns in chiastolite (Harker, 1932, pp. 43, 44) or some trends in albite (Ingerson, 1938); or (b) related to bedding or foliation remnants, i.e. are trails of tiny particles of iron-ore, graphite or mica.

Fig. 1. Textural criteria for the recognition of the time-relations between crystallization and deformation.

Pretectonic: (a) Undulose extinction and deformation lamellae in quartz; (b) and (c) "pressure fringes" around pretectonic magnetite or pyrite; (d) crinkled biotite; (e) "rodded" garnet; (f) bent and fractured twins in plagioclase; (g) later micas wrapping around a fractured garnet crystal; (h) large amphibole partially replaced by small amphibole crystals.

Syntectonic: (i) Snowball structure in garnet (web-like variety); (j) snowball structure in albite, note S_i concordant with S_e; (k) differentially crenulated S_i, after Zwart, 1960a.

Post-tectonic: (l) Helicitic structure in feldspar, S_i concordant with S_e; (m) randomly oriented micas oblique to foliation; (n) late micas which have crystallized mimetically along a previously folded foliation; (o) late albite porphyroblast; (p) massive post-tectonic rim on garnet with syntectonic core; (q) randomly oriented chlorite replacing garnet.

Multistage Deformation: (r) Helicitic structure in albite, S_i discordant with S_e; (s) a late generation of mica along S_2 which is oblique to S_1, the foliation marked by early mica; (t) late amphibole crystallized around garnet takes on a bluish tint adjacent to the garnet; (u) helicitic structure in albite, S_i is S_1 and the crystal is wrapped around by S_2; (v) snowball structure in garnet has S_1 as S_i and the crystal is wrapped around by S_2; (w) folded relics of S_1 between planar S_2.

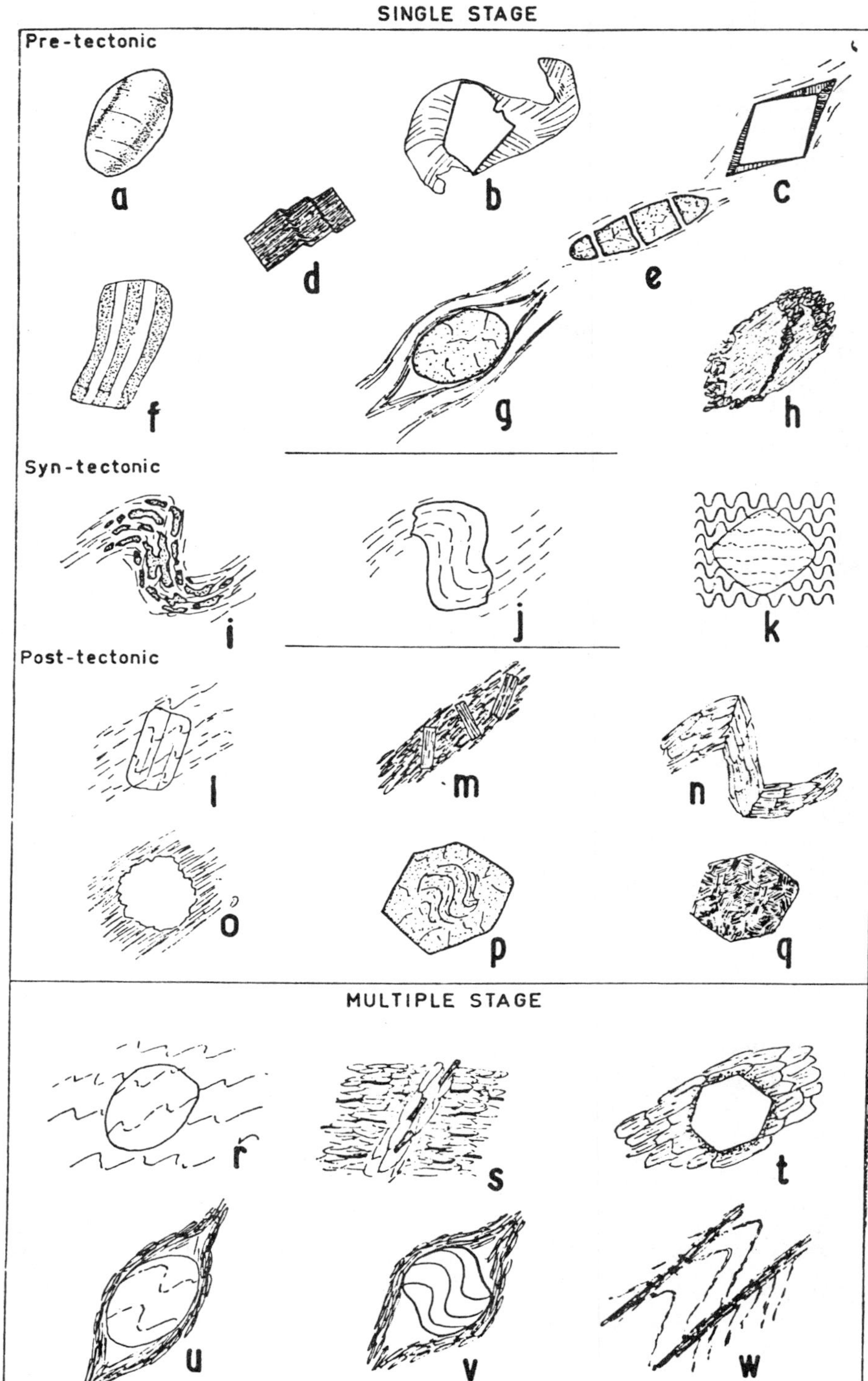
SINGLE STAGE
Pre-tectonic
a
b
c
d
e
f
g
h
Syn-tectonic
i
j
k
Post-tectonic
l
m
n
o
p
q
MULTIPLE STAGE
r
s
t
u
v
w

Fig. 1.

Most pretectonic crystals are obliterated by later crystallization and those crystals which persist are commonly modified. Quartz becomes undulose and develops deformation lamellae, crystals are bent or fractured ("rodded" garnets of Bailey, 1923), twins are produced in calcite or plagioclase, twin-planes or cleavages are fractured or warped (see Fig. 1). Pretectonic crystals are most easily recognized by the effect they have on the later foliation which wraps around them. Platy crystals, such as mica, whose parallelism produces a schistosity, will wrap around the crystal partly because existing crystals become aligned along the flow layers (which themselves curve around the crystal) and partly because new crystals grow in this direction. The bowing of the schistosity is not necessarily due to flattening perpendicular to schistosity. Rast (1958, p. 422) described "spindles" of mica which were formed by syntectonic crystallization of mica around pretectonic garnet.

This structure has been interpreted differently by Krige (1916, p. 580), Hietanen (Cloos and Hietanen, 1941) and Zwart (1960, p. 215) who considered that such crystals may be post-tectonic and that the micas wrapping around the porphyroblast have been pushed aside as it grew. The evidence contrary to this view is as follows:

(a) Post-tectonic crystals of albite, cordierite, tourmaline, andalusite, chloritoid, etc., which contain S_i concordant with S_e do not push adjoining micas aside. The micas abut the porphyroblast without disturbance of the foliation.

(b) The foliation wraps around demonstrably pretectonic bodies such as pebbles in sheared conglomerate, relict feldspar crystals in sheared granite or fossils in slate.

"Pressure fringes", "shadows" or "tails" or quartz, albite or mica are common in the vicinity of early porphyroblasts, e.g. magnetite or pyrite in slate. The minerals within the shadows are syntectonic and explanations have been advanced by Mugge (1930), Pabst (1931) and Fairbairn (1949).

Syntectonic Crystallization

Although crystallization during deformation is probably one of the most important processes by which metamorphic rocks are deformed, it leaves little evidence by which it may be recognized. Apart from minerals in pressure shadows, the only crystals which can be said to be certainly syntectonic are those which show "snowball" structure. This structure is found in such minerals as garnet, albite, staurolite, kyanite and chloritoid and has been discussed in detail by Spry (in press). Earlier papers of importance include those of Krige (1916), Schmidt (1918), Becke (1924) and Mugge (1930). This structure appears in thin sections as curved trails of inclusions forming a symmetrical S-shape (or its reverse). The crystal envolops S_e as it grows, but rotates simultaneously so that the S_i are rotated progressively away from parallelism with S_e.

A structural analysis (Spry, in press) of the spiral S_i allows the growth law of the crystal to be determined and may give some indication of the amount of strain in the rock.

Post-tectonic Crystallization

Crystallization after movements have ceased is extensive in some rocks. Post-tectonic minerals, such as feldspar, mica, quartz, kyanite, tourmaline, chloritoid, calcite, etc., are normally randomly oriented but may have a preferred orientation due to mimetic growth along an existing foliation.

Post-tectonic porphyroblasts are commonly idioblastic and grow across the foliation which buts against them without deflection. The most characteristic feature of post-tectonic crystals (particularly albite) is helicitic structure (Teall, 1897; Cunningham-Craig, 1909; Niggli, 1912; Bailey, 1923; Turner, 1933; Hutton, 1940; Turner and Hutton, 1941; Cloos and Hietanen, 1941; Reynolds, 1942; Rast, 1958; Zwart, 1960a, 1960b; Jones, 1961). A porphyroblast may not expel all of the unwanted material in the space it ultimately occupies. Some porphyroblasts envelop considerable quantities of inclusions which preserve their original orientation. Straight or folded S-surfaces (bedding or foliation) can be traced continuously through such porphyroblasts, i.e. S_e are continuous with S_i. The orientation pattern of inclusions (e.g. quartz in albite) as shown by petrofabric analysis is homogeneous with that of the matrix (Turner and Hutton, 1941).

Helicitic structure must have a folded form to be clearly recognizable; a simple S shape which may represent two small helicitic folds, may be confused with a syntectonic rotational structure. The symmetry and position of the structure in the porphyroblast is important in this regard. Clough (1897) and others interpreted the Cowal albites as post-tectonic but Bailey (1923) regarded them as syntectonic.

It is possible that most of the evidence for syntectonic crystallization is destroyed by post-tectonic recrystallization because crystals or crystal-aggregates with high energy (due to lattice strain or fineness of grain) tend to be replaced by lower energy crystals. A foliation produced by layers of small, parallel lenticular mica flakes may be accentuated by mimetic growth of coarse flakes along the schistosity or may be disrupted by large oblique flakes. Sahama (1935, 1936) and Harris and Rast (1958) have suggested that quartz fabrics tend to be disorientated in the post-tectonic phase, perhaps by annealing.

Decrease of temperature or increase of water pressure may cause post-tectonic diaphthoresis and produce pseudomorphs of early minerals by non-oriented aggregates of later minerals, e.g. chloritization of garnet, pinitization of aluminium silicates. Watson (1948) and Rast (1958) recognized the post-tectonic formation of sillimanite and kyanite in cross-cutting veins.

II. Time-Relations of Minerals to Each Other

In addition to the chronological information given by the textures discussed above, it is possible to use the relations of one mineral to another to establish the sequence of crystallization. Inclusions of one mineral in another may have crystallized first and then have been enveloped, or may have crystallized essentially simultaneously by exsolution, or may have crystallized later by replacement.

If it can be shown that the host and the inclusions do not have an exsolution or replacement relationship, it is probable that the host is younger than the inclusions. This is particularly so where the inclusions consist of a variety of minerals, are idioblastic or have a preferred orientation unrelated to the crystal lattice of the host. This principle has been accepted in the interpretation of helicitic structures and in more complex examples by Cloos and Hietanen (1941, p. 114).

Replacement textures are difficult to interpret (Edwards, 1947; Read, 1949) in that it is not clear in many examples which mineral is replacing which. The order may be clear at an early stage when the original mineral is still predominant and the replacer is localized along cracks or cleavages but it is less clear when the replacer is predominant, even if the original mineral remains as similarly oriented relics (e.g. replacement perthite). In some examples it is possible to recognize the nature of the original mineral after complete replacement by the shape of the pseudomorph but, although a six-sided pseudomorph composed of interlocking chlorite flakes may be assumed to have been a rhombdodecahedron of garnet, the origin of web-like aggregates of staurolite or garnet after mica (Harker, 1938, fig. 2; Rast, 1958, plate XIIA) is less certain.

III. Time-Relations Between Different Deformation Phases

Attempts to apply the principles derived from the simple examples discussed so far are not always easy because there are anomalies which indicate that the structures of some rocks reflect more than one phase of deformation. The complex history of a rock may be recognized from mesoscopic structures, e.g. the presence of multiple foliations, lineations or fold trends, but detailed understanding of structural evolution is only possible by the complete integration of the microscopic fabric with the mesoscopic and macroscopic structures (Fig. 3).

It is difficult to determine the order in which the multiple foliations of some rocks were formed but in general the oldest S-surface is the most strongly folded and disrupted. During the first phase of deformation S_0 (bedding) is folded and S_1 is formed as a planar foliation; during the next phase, S_0 is again folded (becoming very complex), S_1 is folded, and S_2 is produced as a planar foliation.

Multiple-phase metamorphism produces departures from the simple pre-, syn-, and post-tectonic textures described earlier. Minerals formed at an early stage may persist through all later stages, in fact the only evidence for the existence of an early foliation may be the helicitic structure preserved in porphyroblasts. All early-formed minerals are pretectonic to later phases thus an albite crystal which envelops folded S_1 but which is wrapped around by S_2 is post-tectonic to F_1 and pretectonic to F_2. Zwart (1960b) introduced the useful term *interkinematic* for such crystals.

S_i is continuous with S_e in simple syntectonic or post-tectonic crystals but later deformation may rotate the porphyroblast or produce a completely new S_e in a different direction. Clough (1897), Turner and Hutton (1941) and Zwart (1960) have interpreted discordant S_i and S_e as indicating that the porphyroblast has received "a later twist".

Multiple-stage deformation is indicated by the following characteristics of snowball garnet in schists from the Raglan Range (Western Tasmania): (i) Discordant S_i and S_e. (ii) Rotational structure in the core of the garnets but massive rims (syntectonic followed by post-tectonic growth). (iii) Non-parallel axes of rotation of the garnets. (iv) Differences between the fabric of the quartz inclusions in the garnets and that of the surrounding matrix. (v) Triclinic symmetry of the quartz-fabric of the rock.

CHRONOLOGICAL ANALYSIS OF SOME TASMANIAN PRECAMBRIAN ROCKS

The procedure in chronological analysis is to: (1) recognize each genetically separate S-surface; (2) attribute each one to a separate phase of deformation; (3) determine their order of formation; (4) relate the crystallization of each mineral to the others and then to each S-surface and hence to each deformational phase; (5) relate each microscopic S-surface to a macroscopically visible foliation and hence to large-scale structural features.

1. *Examples of Some Relations Between Foliations*

Some Tasmanian Precambrian rocks (Spry, 1957; Spry, 1958; Spry, 1962; Spry and Zimmerman, 1949; Gee, 1962) contain at least three recognizable foliations (Fig. 3). Two of these (S_1 and S_2) are attributed to two movements (F_1 and F_2) during the regional metamorphism of the Precambrian Frenchman Orogeny and are accompanied by the growth of metamorphic minerals. A group of possibly unrelated movements, referred to here as F_3, may include tectonic events ranging from Precambrian to Devonian times. Only slight mineral growth occurred during F_3 which is thus separable from the regional metamorphism and is not discussed in detail here. As the emphasis is on chronology, all structures produced at the one time will be given the same number, i.e. bedding is S_0, the foliation produced during F_1 is S_1 and the lineation L_1; the structures produced during F_2 are S_2 and L_2 and those during F_3 are S_3 and L_3.

The foliations are recognized and separated by using a combination of field and microscope criteria. The macroscopic characteristics are described first.

Bedding (S_0) is recognizable with certainty only in massive quartzites of the Mary and Fisher Groups where ripple marks and cross-bedding are visible. The contacts between the quartzite and phyllite or schist layers are probably strongly transposed bedding. The lamination in many schists and phyllites is not S_0 but S_2. S_1 is a foliation marked by parallel micas and is best developed in Franklin, Joyce and Howell Group quartzites where it is parallel, or almost parallel, to S_0. The dominant foliation in the mica schists and many of the phyllites is S_2 which is parallel to the axial surface of isoclinal folds of S_0 and S_1. S_2 is weakly developed in most quartzites but has been accentuated by mimetic growth in the schists. The latest structure is a fracture cleavage (S_3) which has been formed along the axial surfaces of chevron folds or kinks or along the appressed limbs of small asymmetrical folds.

S_0, S_1, and S_2 are commonly parallel in these rocks because of isoclinal folding and consequently are difficult to distinguish macroscopically. Microscopic criteria allow them to be separated, particularly in rocks in which all three occur.

The appearance of S_0, S_1 and S_2 is shown (Plate 2, No. 1) in a Mary Group phyllite. An alternation of layers richer in quartz or in sericite is the oldest structure; it is parallel to bedding in adjacent quartzite and is probably S_0. Parallel to it is S_1 which is a foliation formed by parallel mica flakes. This foliation has been crumpled and a strain-slip cleavage (S_2) has developed obliquely to it. The new foliation is emphasized by the growth of quartz and

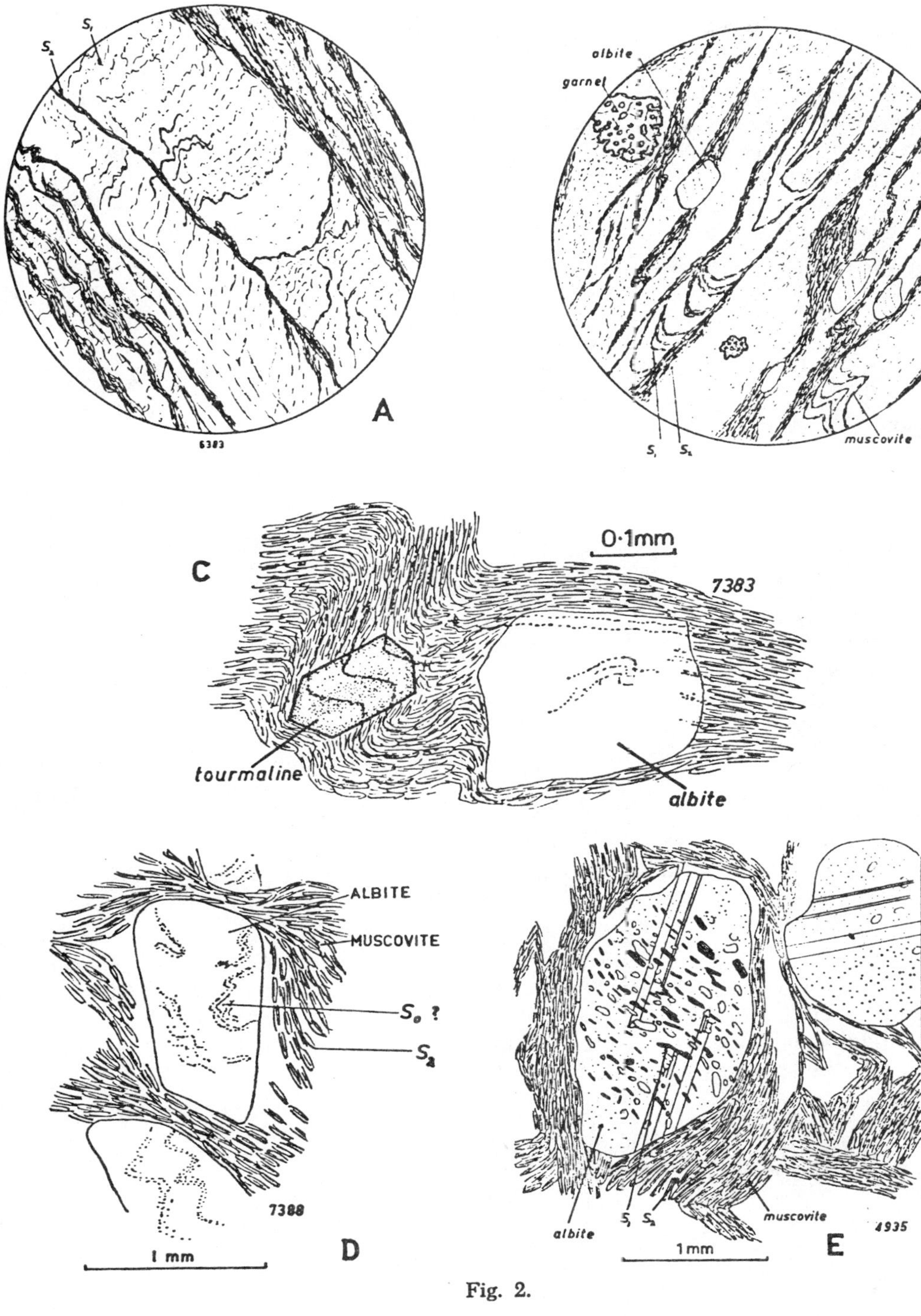

S_2
S_1
A
6383
albite
garnet
S_1
S_2
muscovite
B
C
0·1mm
7383
tourmaline
albite
ALBITE
MUSCOVITE
S_0 ?
S_2
7388
1 mm
D
albite
S_1
S_2
muscovite
4935
1mm
E

Fig. 2.

mica and becomes the dominant foliation in most schists and phyllites where it is marked by a pronounced layering with twisted relics and fold cores of S_1 between the layers (Plate 2, No. 2).

The development of S_2 has been studied by Gee (1962) in the transition zone between Mary Group phyllite and Franklin Group schist. S_1 is strongly developed in the phyllite both microscopically and macroscopically and S_2 becomes increasingly strong as the schist is approached until S_1 is not recognizable macroscopically.

Many phyllites do not contain S_0 but have S_1, S_2 and S_3, the latter forming as a sporadic fracture cleavage. The only mineral growth along S_3 is a little dolomite in some Scotchfire Group calc-schists and some quartz, chlorite and biotite in some Howell Group schists.

2. *Examples of Some Relations Between Minerals*

The replacement of one mineral by another (e.g. chlorite replacing garnet or biotite) is a simple example of a pair of minerals of different age. More important in these rocks, however, is the occurrence of two generations of one mineral.

Mortar texture is common in the massive quartzites of the Mary and Fisher Groups (Plate 2, No. 5). Large undulose quartz grains with deformation lamellae represent an older generation (F_1) and these have been broken down into an aggregate of small, elongate, unstrained grains of a second generation (F_2). Petrofabric analysis, to be presented elsewhere, shows that the two generations of quartz have different fabrics. Quartz-diagrams from the large, older grains have orthorhombic symmetry whereas those from the small, younger grains have monoclinic symmetry.

The texture of amphibolites within Franklin Group schists is similar. Large, old amphibole crystals are broken and bent and are partially replaced around the edges and along cracks by small, young, fresh amphibole.

The time-relations of the garnet and the small actinolite crystals in the amphibolites can also be determined. The actinolite is normally pale green but takes on a bluish tint immediately adjacent to garnet. This implies that the garnet was formed first and that slight compositional changes took place in the actinolite when it was recrystallized next to the garnet. The bluish colour is probably due to slight accession of ferrous iron.

The difficulty of interpreting the meaning of inclusions of one mineral within another was discussed earlier but some relations seem clear. In some schists, small idioblastic garnet crystals occur as inclusions within albite but not elsewhere in the rock (Plate 1, No. 3). The conclusion that the garnet is older than the albite seems reliable.

Fig. 2. (a) Folded relics of S_1 between planar S_2 in Mary Group phyllite, Frenchman's Cap, x 20; (b) folded relics of S_1 between slightly curved S_2 in Franklin Group garnet schist, Frenchman's Cap, S_i in albite is S_1, x 20; (c) helicitic structure in tourmaline and albite, Dove Schist, upper Mersey River, S_i concordant with S_e; (d) helicitic structure in albite of Howell Group albite schist, upper Mersey River, S_2 wraps around the albite which is pretectonic to S_2, the S_i in the albite may be S_0; (e) helicitic structure in albite of schist from Port Davey, S_2 wraps around the crystal, S_1 is enclosed as S_i, S_i and S_e are discordant, the crystal is post-tectonic to F_1 but pretectonic to F_2.

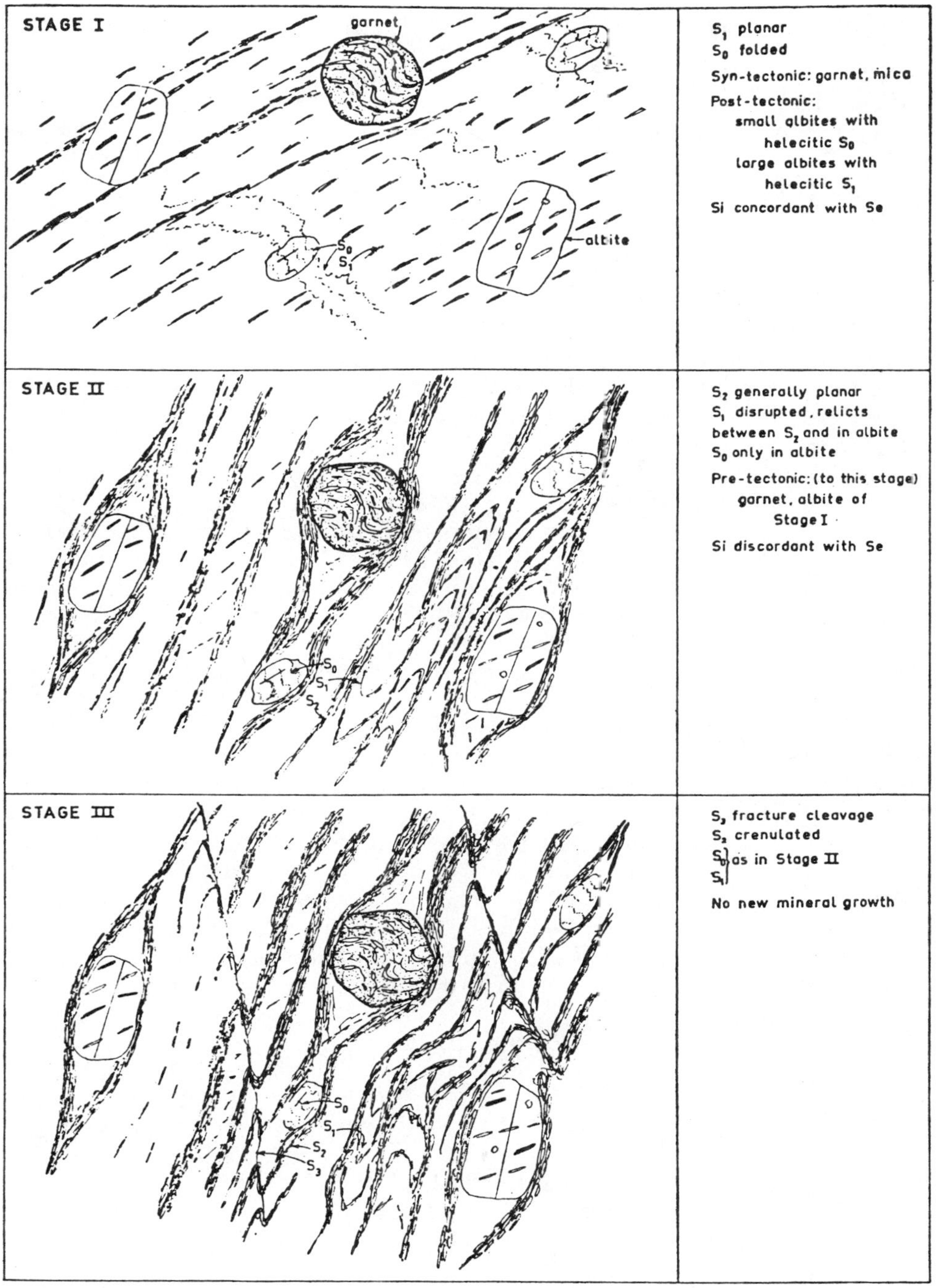

Fig. 3. Evolution of the microfabric of the Franklin Group (Frenchman's Cap) garnet-albite schists.

3. *Examples of Some Relations Between Minerals and Foliations*

The time-relations between various minerals and foliations give a great deal of information about the history of a rock. Using the criteria given earlier, certain minerals can be shown to be pre-, syn-, or post-tectonic with respect to each foliation. Thus a mineral which is syntectonic to S_1 (i.e. F_1) is pretectonic to S_2 (F_2).

Garnet is important in the Franklin, Dove and Howell Group mica schists. Some crystals have a snowball centre (syntectonic) with a massive, idioblastic rim (post-tectonic) (Plate 1, No. 2). Micas outlining S_2 wrap around the garnet which is thus pretectonic to S_2. Many garnets have a regular system of cracks, some are fragmented and many have pressure fringes of quartz and mica (Plate 1, No. 1).

Albite is perhaps the most useful of all minerals in chronological analysis. It occurs in several forms in Franklin, Dove and Howell schists. The smaller crystals are untwinned and contain lines of inclusions of tiny opaque granules which may represent S_0 (Plate 1, No. 4). In some crystals the S_i outline folds as a helicitic (post-tectonic) structure but in others the S_i show the symmetrical spiral form of snowball structure (syntectonic). The large albite crystals reach 1 cm. across and contain simple or multiple twins, e.g. Plate 1, No. 3. Some crystals in the Howell Group contain helicitic structure with contorted S_0 (?), Fig. 2d, but most contain fairly straight lines of coarse quartz, mica and garnet inclusions (S_1), Fig. 2e. These S_i are discordant with S_e, i.e. the micas of S_2 wrap around the albite crystals and many albite crystals are shattered, thus the albite is interpreted as being post-tectonic to S_1 and pretectonic to S_2. The albite in Plate 2, No. 4 is post-tectonic to the foliation. Some garnet inclusions in the albite have snowball structure and are thus syntectonic to S_1, as is also the elongate quartz and mica which forms the S_i. Quartz and mica also outline S_2 and are probably thus syntectonic to F_2 also. Coarse muscovite was also formed post-tectonically to S_2; it buts against old albite crystals and is not deformed by folds of S_2. Mimetic growth of muscovite along S_2 is strong in some rocks, particularly the coarse schists of the lower Forth River.

Two varieties of chlorite are found in some Franklin Group schists. A pale green variety (magnesian prochlorite?) forms syntectonically along S_2 and a green penninite? occurs as randomly oriented flakes replacing garnet (post-tectonic to F_2).

COMPLETE ANALYSIS OF SELECTED SPECIMENS OF TASMANIAN PRECAMBIAN ROCKS

The full interpretation of the history of multiply deformed rocks requires the integration of all field and microscope data. Examples of the treatment of some schists, quartzites and amphibolites are given below.

Howell Group schists from Walter's Marsh on the Mersey River contain an average of 50 p.c. quartz, 25 p.c. muscovite, 20 p.c. albite, 5 p.c. garnet, 4 p.c. chlorite with accessory rutile, apatite and tourmaline.

A single foliation is visible macroscopically. It is marked by parallel mica flakes or a lamination and is parallel to interbedded quartzites. In thin section this foliation is marked by layers alternately richer in quartz or in mica, both of which are elongate parallel to the foliation. This foliation is not bedding but is S_2 and is younger than another foliation S_1 which is visible as contorted relicts between the layers of S_2 and also within porphyroblasts. A younger foliation (S_3) occurs as sporadic fractures at about 15° to S_2.

Micas of S_2 wrap around small garnets with snowball structure which are probably syntectonic to S_1. Some garnets are replaced by randomly oriented biotite or chlorite which is thus post-tectonic to S_2.

Albite forms large porphyroblasts with helicitic structure. Some S_1 consist of contorted trails of tiny opaque granules (S_0?) but most are of gently curved lines of elongate quartz, parallel muscovite flakes (S_1) and small idioblastic garnets and tourmalines. S_2 wraps around these porphyroblasts which are thus post-tectonic to S_1 and pretectonic to S_2.

Garnet and tourmaline inclusions in albite are probably post-tectonic to S_1 and a little older than the albite. Some tourmaline is post-tectonic to S_2?; it is randomly oriented and contains S_2? as helicitic structure.

Biotite occurs as small flakes in albite (syntectonic to S_1), as medium-sized flakes along S_2 (syntectonic to S_2?) and as large randomly oriented flakes along S_3 (post-tectonic to S_3).

Chlorite occurs as envelopes to the garnet where it is drawn out along S_2 (syntectonic to S_2) but some is randomly oriented along S_3 (post-tectonic to S_3).

Quartz is elongate along S_1 in the layers between S_2 or within the albites, thus is partly syntectonic to S_1. Most is elongate along S_2 and is thus probably syntectonic to S_2. A little occurs in coarse aggregates along S_3 and is post-tectonic to S_3.

The occurrence of muscovite is generally similar to that of quartz. It is elongate along S_1 (syntectonic to S_1) and also along S_2 (syntectonic to S_2).

The history of the Howell Group schists is as follows:

		F_1		F_2		F_3	
quartz		——	- - - -	——	- - - -		——
muscovite		——		——	——		——
garnet		——	——				
biotite		——	——		——		
albite		——	——		——		
chlorite			——	——	——		——
tourmaline					——		
Deformation stage	pre-	syn- tectonic	post-	syn- tectonic	post-	syn- tectonic	post-
S formed		S_1		S_2		S_3	
S folded		S_0		S_0, S_1		S_0, S_1, S_2	

Analysis of quartzites is much more difficult because of the lack of useful minerals such as albite and garnet, and the apparent readiness of quartz to recrystallize and obliterate earlier structures.

The presence of two generations of quartz in Mary Group quartzites with mortar texture has already been outlined and it is reasonable to correlate these with F_1 and F_2.

The medium-grade quartzites of the Franklin, Joyce and Howell Groups consist of quartz and a little muscovite with accessory tourmaline, rutile, apatite and zircon. Garnet occurs in a few. Macroscopically they contain thin layers of schist and have a colour banding, probably representing S_0, with a single strong foliation which is parallel, or almost parallel, to this direction. This foliation is marked by parallel muscovite flakes. Both S-surfaces are folded isoclinally and a second tectonic foliation which is produced parallel to the axial surfaces of these folds in both quartzites and schists is S_2. The older foliation is then S_1. Difficulty in recognizing the identity of these S-surfaces macroscopically is caused by their parallelism on the limbs of the isoclinal folds.

In thin section most quartzites appear to have a single foliation marked by elongate quartz and mica (Plate 2, No. 6). This foliation is due to coincidence between S_1 and S_2. The mica is largely syntectonic to S_1 which was folded by S_2 folds but the quartz in such folds is not "unrollable" and was recrystallized during F_2 (Gee, 1962).

Garnet is present in a few quartzites as web-like crystals with snowball structure. S_i are discordant with S_e and thus these are probably syntectonic to F_1.

The history of the Franklin Group quartzites is as follows:

		F_1		F_2		F_3	
quartz				—	—		
muscovite		—					
garnet		—	—				
Deformation stage	pre-	syn-tectonic	post-	syn-tectonic	post-	syn-tectonic	post-
S formed		S_1		S_2			
S folded		S_0		S_0, S_1		S_0, S_1, S_2	

The history of the amphibolites is similar to that of the schists and quartzites. Those in the Franklin Group consist of actinolite, garnet, biotite, chlorite, albite, quartz, zoisite and epidote, with accessory sphene, rutile, magnetite, zircon and calcite.

The rocks consist of small parallel actinolite prisms producing a foliation, which is parallel to S_2 in the enclosing schists, wrapping around large actinolite and garnet crystals which are shattered. The garnet and large actinolite crystals are pretectonic to the foliation which, by analogy with the massive quartzites, is

S_2. The small actinolite crystals are partially syntectonic to S_2 but much of the amphibole is post-tectonic and poorly oriented.

Some garnet has a snowballed centre and massive idioblastic rim and thus is partly syntectonic and partly post-tectonic to F_1. Biotite forms large, rusty-brown, twisted flakes which wrap around the garnet and are parallel to S_2. These are probably old (F_1) relicts which are pretectonic to S_2 but which have been rotated bodily into parallelism with it.

Albite is not common in the amphibolites. A few large, fractured and sericitized crystals are older than all other minerals and may be original phenocrysts. Some albite contains S_1 as a helicitic structure and is post-tectonic to S_1 but pretectonic to S_2. Small, irregularly-shaped, fresh albite crystals with S_i concordant with S_e (S_2) are post-tectonic to S_2.

Garnet amphibolites on the lower Forth River contain no evidence of S_1 and only slight evidence of S_2. Post-tectonic crystallization has destroyed most of the preferred orientation of hornblende.

The history of the normal (Franklin Group) garnet amphibolites is as follows:

		F_1		F_2		F_3	
actinolite			———	———	———		
garnet		———	———				
biotite		——	———				
albite		?	———	?	———		
quartz			———	———			
calcite					———		
chlorite					———		
Deformation stage	pre-	syn-tectonic	post-tectonic	syn-tectonic	post-tectonic	syn-tectonic	post-tectonic
S produced		S_1?		S_2			
S folded				S_1?			

CONCLUSIONS

The chronology of the development of foliations and the crystallization of metamorphic minerals can be determined from microscopic data for certain of the Tasmanian Precambrian rocks. The history of adjacent schists, phyllites, quartzites and amphibolites is similar and the general history of one region, e.g. Frenchman's Cap, is essentially similar to that of other areas, e.g. Mersey-Forth, Ulverstone, Port Davey, etc.

The Frenchman Orogeny can be divided into two phases, F_1 and F_2. The difference in metamorphic grade between the chlorite zone (Mary, Fincham, Scotchfire and Fisher Groups) on one hand, and the garnet zone (Franklin, Joyce and Howell Groups) on the other, was produced during F_1. The grade of metamorphism during F_2 was equivalent to chlorite zone in both assemblages.

REFERENCES

BAILEY, E. B., 1923: Metamorphism of the South-West Highlands. *Geol. Mag., 60*, pp. 317-331.

BECKE, F., 1924: Struktur und Klüftung. *Fortschr. Min. Krist. Pet., 9*, pp. 185-220.

CLOOS, E., and HIETANEN, A., 1941: Geology of the Martic Overthrust and Glenarm Series. *Spec. Pap. geol. Soc. Am., 35.*

CLOUGH, C. T., 1897: In W. Gunn *et al.*: The Geology of Cowal. *Mem. geol. Surv. Scot.*

CUNNINGHAM-CRAIG, E. H., 1904: Metamorphism in the Loch Lomond District. *Quart. J. geol. Soc. Lond., 60*, pp. 10-29.

EDWARDS, E. B., 1947: *Textures of the Ore Minerals.* Aust. Inst. Min. Met.

FAIRBAIRN, H. W., 1949: *Structural Petrology of Deformed Rocks.* Cambridge.

GEE, D., 1962: Structure and Petrology of the Raglan Range. Unpub. Thesis, Univ. of Tasm.

HARKER, A., 1938: *Metamorphism.* London.

HARRIS, A. L., and RAST, N., 1960: The Evolution of Quartz Fabrics in the Metamorphic Rocks of Central Perthshire. *Trans. Edin. geol. Soc., 18* (1), pp. 51-78.

HUTTON, C. O., 1940: Metamorphism in the Lake Wakatipu Region, Western Otago. *Geol. Mem. N.Z. Dep. Sci. Ind. Res., 5.*

INGERSON, E., 1938: Albite Trends in some Rocks of the Piedmont. *Amer. J. Sci., 35*, p. 127.

JONES, K. A., 1961: Origin of Albite Porphyroblasts in Rocks of the Ben-More-Am Binnein Area, Western Perthshire. *Geol. Mag., 98*, pp. 41-55.

KRIGE, L. J., 1916: Petrographishe Untersuchungen im Val Piora und Umgebung. *Ecl. geol. Helv., 14*, p. 519.

MUGGE, O., 1930: Bewegungen von Porphyroblasten. *Neues Jb. Min. Geol. Paläont., 61A*, p. 469.

NIGGLI, P., 1912: Die Chloritoidschiefer und die S-sedimentare Zon am Nordostrand. *Beitr. Geol. Karte Schweize., 36*, 1931.

RAST, N., 1958: Metamorphic History of the Schichallion Complex. *Trans. roy. Soc. Edinb., 62*, 2, pp. 413-432.

READ, H. H., 1949: A Contemplation of Time in Plutonism. *Quart. J. geol. Soc. Lond., 105*, pp. 101-56.

REYNOLDS, D. R., 1942: The Albite-Schists of Antrim and Their Petrogenetic Relationship to Caledonian Orogenesis. *Proc. roy. Irish Acad., 48*, 13, p. 43.

SAHAMA, T. G., 1938: Zur Regelung der Gesteine im Kristallin von Liverpoollund in Ost-Grönland. *Medd. Grønland, 95*, p. 6.

———, 1936: Die Regelung von Quartz und Glimmer in den Geisteinen, den Finnische-Laplandischen Granulitformation. *Bull. Comm. géol. Finl., 113.*

SANDER, B., 1950: *Einführung in die Gefügekunde der Geologischen Körper.* Springer, Vienna.

SCHMIDT, W., 1918: Bewegungspuren in Porphyroblasten Krystalliner Schiefer. *S.B. Akad. Wiss. Wien. Mat-Nat. Kl. Abt.*, 1, *127*, p. 193.

SPRY, A., 1957: Precambrian Rocks of Tasmania, Part II, Mt. Mary Area. *Pap. roy. Soc. Tasm., 91*, pp. 95-108.

———, 1958: Precambrian Rocks of Tasmania, Part III, Mersey-Forth Area. *Pap. roy. Soc. Tasm., 92*, pp. 117-137.

———, 1962a: The Precambrian in "The Geology of Tasmania". *J. geol. Soc. Aust.*, 9 (2).

———, 1962b: Origin and Significance of Snowball Structure in Garnet. *In press.*

———, and ZIMMERMAN, D., 1959: Precambrian Rocks of Tasmania, Part IV, Mt. Mullens Area. *Pap. roy. Soc. Tasm., 93*, pp. 1-9.

TEALL, J. J., 1897: In W. GUNN ET AL.: The Geology of Cowal. *Mem. geol. Surv. Scot.*

TURNER, F. J., 1933: The Metamorphic and Intrusive Rocks of Southern Westland. *Trans. N.Z. Inst., 63*, pp. 178-284.

———, 1948: Mineralogical and Structural Evolution of the Metamorphic Rocks. *Mem. geol. Soc. Amer., 30.*

TURNER, F. J., and HUTTON, C. O., 1941: Some Porphyroblastic Albite Schists from Waikauaiti River, Otago. *Trans. roy. Soc. N.Z., 71*, pp. 223-240.

WATSON, J., 1948: Late Sillimanite in the Migmatites of Kildonan, Sutherland. *Geol. Mag.*, *85*, pp. 149-162.

ZWART, H. J., 1960a: Relations between Folding and Metamorphism in the Central Pyrenees. *Geol. Mijnb.*, *39e*, pp. 163-180.

———, 1960b: Chronological Succession of Folding and Metamorphism in the Central Pyrenees. *Geol. Rdsch.*, *50*, pp. 203-18.

EXPLANATION OF PLATES

PLATE 1

No. 1. Garnet porphyroblast in Franklin schist No. 6986. Garnet is pretectonic to S_2 which wraps around it. Typical regular set of fractures.

No. 2. Garnet from schist No. 6290, Franklin Group, Raglan Range. Snowball (syntectonic) core and massive (post-tectonic) rim.

No. 3. Large albite porphyroblast with S_1 contained as S_i and micas of S_2 wrapping around. Franklin Group schist No. 6746.

No. 4. Small albite with helicitic structure. S_0? enclosed as S_i and S_2 micas wrapping around. Franklin Group schist No. 6986.

No. 5. Randomly oriented, post-tectonic chloritoid in Lachlan Conglomerate.

No. 6. Chevron folds in schist No. 9371 from the Whyte Schist. The strong foliation is probably S_1 and the fracture cleavage along the folds is probably S_2.

PLATE 2

No. 1. Mary Group phyllite with S_0 (bedding), S_1 (foliation parallel to bedding), and S_2 (strain-slip cleavage oblique to S_0 and S_1).

No. 2. Folded relics of S_1 between planar S_2 in Dove Schist, No. 7382.

No. 3. Multiple foliations in Mary Group phyllite (No. 8130). The contorted sericite flakes outline S_1 and the moderately curved dark lines are probably S_3. S_0 and S_2 are not visible.

No. 4. Dove schist with post-tectonic albite and tourmaline crystals containing helicitic structure.

No. 5. Mortar texture in massive quartzite from the Mary Group, No. 6783.

No. 6. Elongate quartz in Franklin Group quartzite, No. 6283.

A. Spry,
Department of Geology,
University of Tasmania,
Hobart.

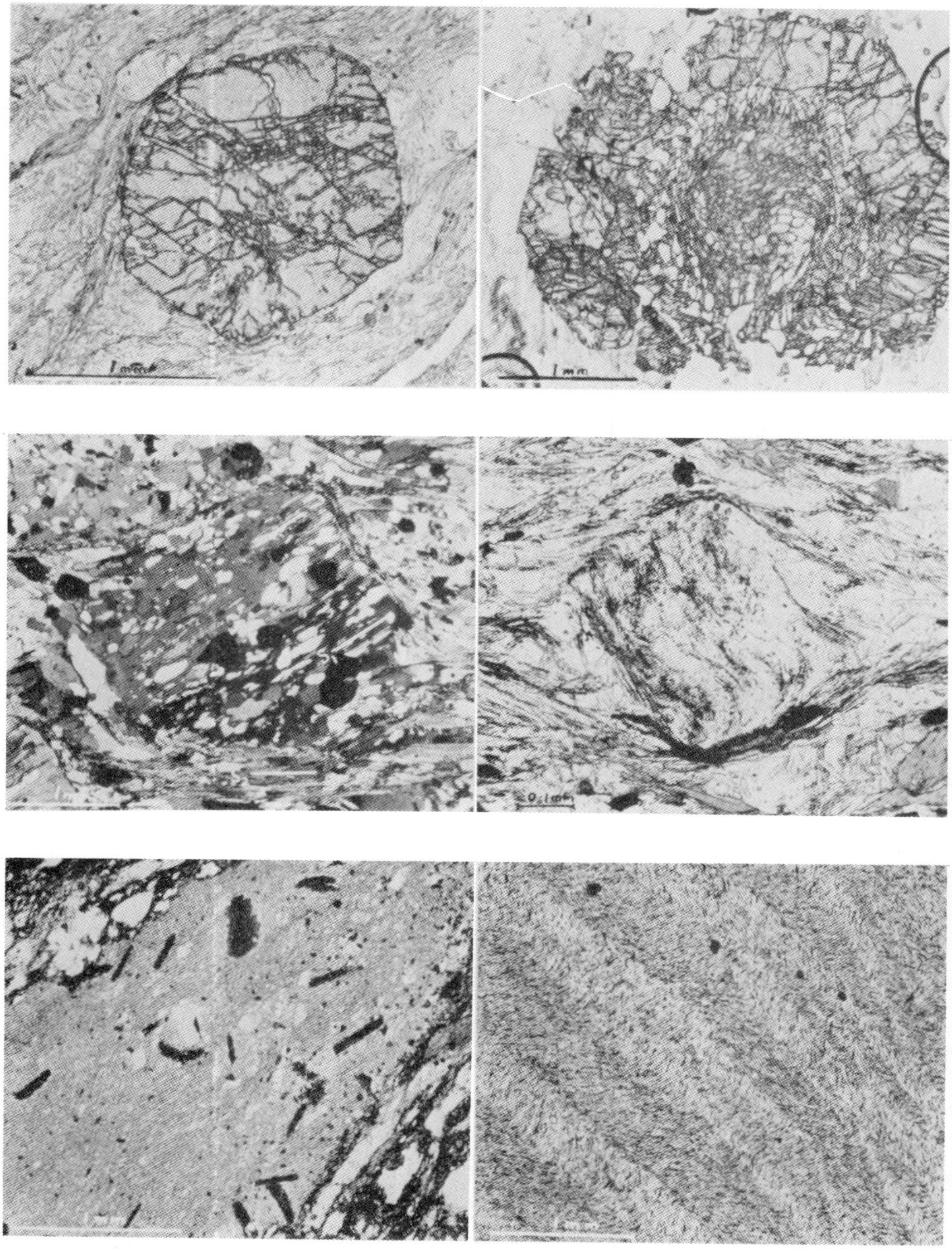

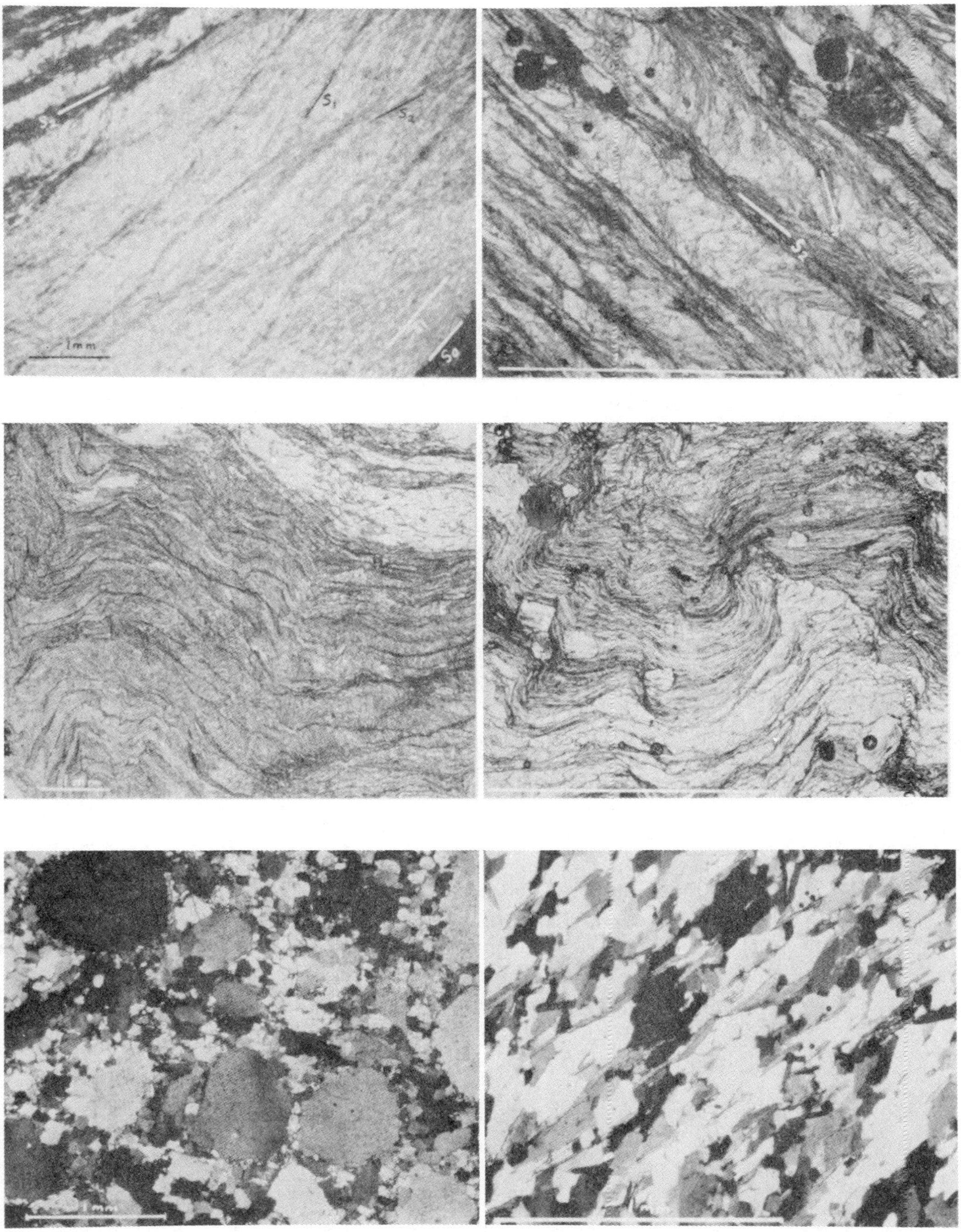
S2
S1
S2
1mm
S0
S2
1 mm

10

Copyright ©1975 by Cambridge University Press
Reprinted by permission from *Geol. Mag.* **112**:467–480 (1975)

Textural patterns at porphyroblast margins and their use in determining the time relations of deformation and crystallization

C. C. FERGUSON & B. HARTE

(Plates 1–2)

Summary. Textural patterns characteristic of supposed pre-, post- and to a lesser extent syn-tectonic porphyroblasts are reviewed. Difficulties of interpretation arise where an *Si* (internal schistosity) is absent, and *Se* (external schistosity) behaviour adjacent to porphyroblast margins is the only available criterion; examples may be found showing truncation of *Se* by pre-*Se* porphyroblasts. Truncation and/or deflection textures are not by themselves sufficient criteria for determining the time relations of particular phases of deformation and crystallization.

1. Introduction

Thin-section examination of the textures and microstructures of metamorphic rocks is a well established technique used to elucidate the local time relations of crystallization and deformation. Establishing such relations is clearly of considerable importance, especially if the complex nature of orogenesis is to be understood; though several workers have pointed out that in correlating minor structures from outcrop to outcrop, the time equivalence of similar fabrics or folds is assumed but is rarely capable of being firmly established (Offler & Fleming, 1968; Park, 1969; Pitcher & Berger, 1972).

The textural patterns used to decide the age of a mineral relative to a deformation episode have been outlined by Zwart (1960*a*, *b*, 1962, 1963), Spry (1963, 1969), and others. In essence the criteria are simple and straightforward, but in practice they may be ambiguous. This is partly because complete sets of key textural criteria are by no means ubiquitously developed in metamorphic rocks – a good example of this is the lack of a clear inclusion trail (*Si*) in many porphyroblasts. In addition, the deformation and crystallization history of many orogenic belts is seen to be highly complex in detail, and the matrix texture and structure, which forms the reference framework for determining the time relations of many porphyroblasts, may bear the imprint of many events, which may not always be readily distinguishable. A matrix whose essential character and *s*-surface (*Se*) is the product of a single combined deformation and crystallization event, as figured, for example, in many of the ideal textural relations of Zwart (1960*a*, *b*, 1962, 1963) may be a relatively rare phenomenon.

Distinguishable deformation events may produce cleavages of similar orientation, and an apparently simple set of planar structures (schistosity and foliation – the term is used to describe a layering of metamorphic origin) may be the product of a multiple event (see, for example, Roberts, 1974, pp. 100–4). In addition, such *s*-surfaces (*Se*) may be modified by later phases of deformation, or by late-stage annealing recrystallization and grain growth (Rast, 1965, p. 88). The pre-, syn- and post-tectonic terminology introduced by Sander & Pernt (1923) therefore embodies a potentially misleading simplicity, and the precise tectonic event or resultant structure referred to must be carefully established. Syntectonic porphyroblast textures have been actively re-examined in recent years (e.g. Powell & Treagus, 1970; Rosenfeld, 1970; Wilson, 1971; Elliott, 1972); this paper is concerned with some of the criteria commonly used in designating crystals as pre-tectonic or post-tectonic, and in particular the disposition of *Se* adjacent to porphyroblast margins.

2. Pre- and post-tectonic porphyroblasts – some current textural criteria

Where porphyroblasts contain a well-developed *Si*, interpretation of the textural relations often appears relatively straightforward. But where an *Si* is absent, and interpretation depends on the behaviour of *Se* adjacent to the porphyroblast, uncertainties become apparent. The arrangement of *Se* adjacent to a porphyroblast is dependent on a number of factors:

(i) The manner of growth of the porphyroblast, and its interaction with the matrix during growth.

(ii) The mechanism of formation and history of modification of *Se*.

(iii) The timing of porphyroblast development with respect to the history of *Se*.

(iv) The mineral species of the porphyroblast and its competence relative to that of the matrix during deformation.

The mechanics and mechanisms pertinent to the above are often uncertain. Our knowledge is largely dependent on observation of the rocks themselves, and an understanding of one factor is dependent upon a knowledge of the others, with the consequent danger of circular argument. A good example of the problems involved is seen in a recent paper by Misch (1971) and the correspondence following its publication. Misch (1971) considered that he had unequivocal evidence relevant to factor (i) because he had established factor (iii), a fact strenuously disputed by others (Shelley, 1972; Spry, 1972; Ferguson & Harvey, 1972).

2.a. Pre-tectonic porphyroblasts

Crystals formed prior to an episode of deformation may be recognized by evidence of mechanical strain (undulose extinction, deformation twinning, disruption, etc.; see, for example, Spry (1969), p. 252). However, the absence of obvious mechanical strain, particularly in porphyroblasts, is not taken as sufficient evidence of a post-tectonic origin, and indeed, where slight strain occurs it may not be clear which episode of deformation was the cause. *The absence of strain in many porphyroblasts present during deformation is important*; it is well documented for syn-tectonic porphyroblasts (Zwart, 1960*a*, *b*; Spry, 1969, p. 254), and is implicit in many descriptions of pre-tectonic porphyroblasts though Zwart (1960*a*, p. 218, 1960*b*, p. 179) is one of the few authors who explicitly notes it.

In the absence of obvious mechanical strain other criteria are used. According to Zwart (1963), porphyroblasts formed prior to a given *s*-surface 'show *Se* curving completely around the crystal without any connection due to the plastic flow of the schist around the pre-existing crystal' (Fig. 1A). Similarly, Spry (1963) states that pre-foliation crystals 'are most easily recognized by the effect they have on the later foliation which wraps around them'. 'Foliation' is used here to include schistosity and other structures and has a broader meaning than that given in section 1 – cf. Spry (1969, p. 207). The interpretation of such crystals is clearest when they display an internal *s*-surface (*Si*) which is quite inconsistent with *Se* (Fig. 1B).

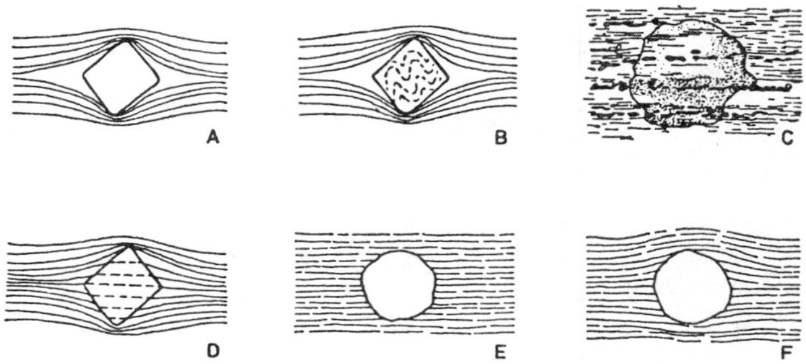

Figure 1. Idealized textural patterns discussed in text (A) after Zwart (1963, fig. 11B), (B) same but with *Si* added by present authors, (C) after Fettes (1971, fig. 2*b*), (D) after Zwart (1963, fig. 11D), (E) after Misch (1971, fig. 1A), (F) after Misch (1971, fig. 1B).

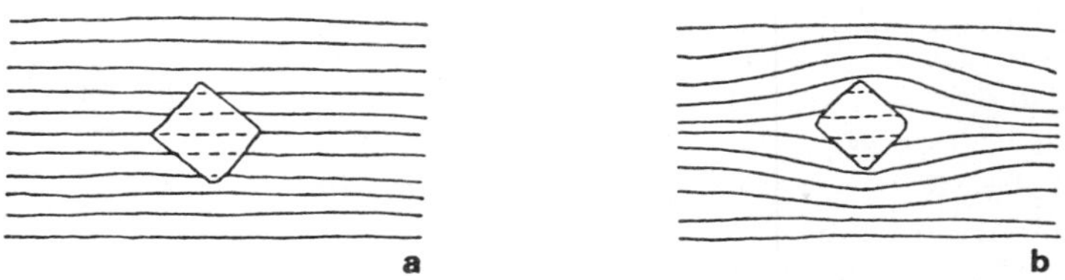

Figure 2. (a) Porphyroblast, with *Si*, grown subsequent to the formation of the matrix schistosity. (b) Modification of (*a*) by flattening deformation.

In the absence of an *Si*, and with a multiple deformation history, the deflection of *Se* alone is not necessarily unambiguous. Thus in the history depicted in Figure 2, a porphyroblast has grown after the formation of *Se*, and the rock has later undergone flattening without any general recrystallization of *Se*. With an *Si* present and showing similar features to *Se* as in Figure 2, the actual history is the most likely interpretation, but if the *Si* is absent the conclusion might be reached that the porphyroblast formed altogether prior to *Se* (compare Figs 2B and 1A). Similar situations are illustrated by Ramsay (1962, figs 15, 16), wherein porphyroblasts are post-tectonic with respect to the initial formation of matrix schistosity but are syn- or pre-tectonic with regard to rotation of the matrix schistosity during flattening. In all these cases interpretation of the textures might be further complicated if a later-stage annealing and essentially mimetic matrix recrystallization occurred, as may be common in moderate to high-grade rocks.

2.b. Post-tectonic porphyroblasts

Post-*Se* porphyroblasts are most surely recognized when *Si* and *Se* are continuous and incorporate elements of the same nature (Fig. 1C). Such textures provide clear evidence of constant volume replacement during porphyroblast growth (Zwart, 1962, p. 42; Rast, 1965, pp. 92–3).

Whether porphyroblasts can sometimes force aside and cause deflection of the matrix *Se* by their growth is a question of some importance in establishing time relations of crystallization and deformation. Recent general opinion (for example, Zwart, 1962, p. 42; Rast, 1958, 1965; Spry, 1969, pp. 150–1) has been against such deflection being produced by porphyroblast growth. However, Misch (1971, p. 246) recognises porphyroblasts which have grown by 'pure chemical replacement of matrix fabric without any distension' (Fig. 1E), and others in which a replacement pattern is 'combined with minor or major distension' of *Se* (Fig. 1F). The latter pattern involves, according to Misch, variable amounts of pushing-out of pre-existing *Se* by the growing porphyroblast and provides 'unequivocal evidence for such pushing out because the coexisting replacement relationship shows that the porphyroblast post-dates *s* of the matrix which it truncates' and he claims (p. 247) that 'included remnants of schistose matrix as *Si* . . . is not a necessary condition' for identifying this pattern.

The insufficiency of pattern Figure 1F as a criterion for crystals which have, by growth, both replaced and displaced their matrix has been pointed out by Shelley (1972), Spry (1972) and Ferguson & Harvey (1972). Zwart (1962, p. 42) and Spry (1972) have noted that the problem of matrix distortion as a result of porphyroblast growth is probably insoluble with data from the rocks themselves – a viewpoint which is of further interest because it suggests a general acceptance amongst geologists (e.g. Zwart, 1962, p. 42; Rast, 1965, p. 72; Spry, 1972; Ghosh & Sengupta, 1973, p. 174) that some strain may often occur not readily correlated with an identifiable set of fold or cleavage structures.

Misch's paper (1971) also attracts attention with respect to the establishment or otherwise of another textural 'principle'; namely that sharp cross-cutting or truncation of *Se*, as seen in figures 1E and 1F, demonstrates replacement and post-*Se* porphyroblast growth even in the absence of an *Si*. The critics of Misch's displacement criterion did not dispute this evidence for replacement, though Spry (1972) and Ferguson & Harvey (1972) pointed out that the pattern of combined truncation and deflection (Fig. 1F) might be formed syn-*Se* (an interpretation following Zwart (1962), fig. 1).

Use of the sharp truncation of *Se* by a porphyroblast, in the absence of an *Si*, as evidence for the post-*Se* age of the porphyroblast has been accepted explicitly and implicitly by other authors. Bastin (1950, p. 37) adopted the criterion, noting that 'If they [pyrite porphyroblasts] had antedated the development of the schistosity they should show fracturing and other evidence of deformation, and the schistosity of the rock should wrap around them'. We have already seen that mechanical strain is not always exhibited by pre-tectonic porphyroblasts, and other descriptions of pyrite porphyroblasts raise further doubts concerning Bastin's simple textural interpretation. Particularly intriguing are the descriptions of Frankel (1957, especially pp. 400–1), wherein pyrite crystals typically show both of Bastin's criteria, namely lack of both obvious internal deformation and deflection of *Se*, but these are associated with pressure shadows (equivalent to the pressure fringes of Spry (1969), p. 240), and locally abraded margins produced by deformation. Pabst's descriptions (1931, p. 68) also imply the coexistence of pressure shadows around pyrite together with the sharp transection of *Se* by pyrite and its pressure shadow. Pabst (1931, p. 57) also notes that pressure shadows are not always developed around pyrite porphyroblasts.

Rast (1958, p. 422) adopts the truncation criterion stating that garnets 'are post-tectonic their edges cutting abruptly across the mica flakes'. The criteria for replacement proposed by Zwart in a series of influential papers (Zwart, 1960*a*, *b*, 1962, 1963) also implicitly include truncation in the absence of an *Si* as a sufficient criterion for post-tectonic growth. For example, Zwart (1963, p. 335) stated:

> '. . . most, if not all crystals grow by chemical replacement only without disturbing the surrounding rock . . . In many cases *Si* runs straight through the crystal and is continuous into *Se*; in other cases porphyroblasts contain no *Si* but the pattern of *Se* is similar. Both cases are indicative of pure chemical replacement, and the absence or presence of *Si* is of no concern . . .'

and further on (p. 337)

> 'However, when the s-plane [external schistosity] is used for renewed movements the [post-*Se*] crystal may be rotated or the matrix may be flattened or crumpled as shown in Fig. 11 D . . .'

Zwart's fig. 11 D is reproduced here as Figure 1 D and shows truncation combined with deflection of *Se*. Excluding *Si*, the arrangement is identical to the pattern B (reproduced as Fig. 1 F) of Misch (1971, p. 246), and it is interesting to note that Zwart's figures (e.g. 1962, fig. 1) often show a strong element of truncation about pre- and syn-tectonic porphyroblasts. Indeed in his depiction (1962, and also 1960*a*, *b*) of porphyroblasts formed pre- and syn-microfolding of *Se*, he anomalously shows no modification of the arrangement of *Se* adjacent to the porphyroblast.

Clearly, textures in which porphyroblasts (without *Si*) simultaneously truncate and deflect *Se* may be interpreted as pre-, syn-, or post-*Se* according to some uncertain estimate of the degree of truncation or deflection or the influence of deformation subsequent to *Se* and porphyroblast formation. Presumably because of these ambiguities, many authors have used truncation as evidence of post-tectonic growth only in the absence of deflection of *Se* close to the porphyroblast (e.g. Rast, 1965; Sturt & Harris, 1961; Roberts, 1968; Spry, 1969). In such instances the implication is that the porphyroblasts are not only post-*Se*, but also later than any flattening, slip, or other strain, which has affected *Se*. We shall return to a discussion of whether these porphyroblasts are necessarily post-*Se* in a later section; but it seems that at the most, deflection and truncation textures are only likely to show whether any or no deformation has occurred around the porphyroblasts since their growth. To correlate this presence or absence with particular phases of deformation in a long history of deformation and crystallization is extremely difficult.

None the less the use of matrix truncation, with or without deflection, as a criterion for post-*Se* porphyroblast growth has largely remained unquestioned, and few workers have indicated contrary views. Sengupta (1963, p. 92) has outlined some hypothetical textural relations and possible interpretations; where a 'foliation is undisturbed around porphyroblasts' with 'lines of mica ending abruptly against them', he includes as a theoretically possible interpretation that the porphyroblasts could be pre-foliation with 'development of foliation controlled purely by blastetrix'. However, he seems to discount this as a practical possibility (1963, p. 95). Elliott (1972, pp. 2635–6), in a discussion of coaxial deformation paths, suggests that porphyroblasts with post-tectonic features, namely *Si* parallel to *Se* but presumably also with truncation, may be syn-tectonic, though the detailed *Si* and *Se* characteristics are obscure. In terms of published interpretations of *observed* textural relations, the authors know of only 3 cases where the use of truncation as a criterion for post-foliation growth has been disputed. Johnson (1962, p. 48) described andalusite porphyroblasts from the Banffshire Dalradian which truncate the S_3 foliation but which themselves appear to be deformed by the F_3 movements with which S_3 is associated. This, together with the absence of any clear-cut fabric within the porphyroblasts (although included biotites are comparable in size, and frequency to those in the foliated groundmass). suggested to Johnson that the andalusites antedate F_3 in spite of their 'truncation' of S_3. Harte & Johnson (1969, p. 66) noted staurolite porphyroblasts which 'truncate' the coarse-grained regional schistosity but also have fine-grained *Si* orientated at large angles to *Se*; they considered, in spite of the truncation, that the staurolites antedate the matrix schistosity. Finally, Galwey & Jones (1962) observed garnets intersecting, without deflecting, matrix phyllosilicates occurring close to garnets with associated augen structures; they considered that these differences might be explained by different nucleation and growth environments of garnet and this would not necessarily indicate post- or syn-tectonic growth.

3. An example of truncation texture adjacent to pre-*Se* porphyroblasts

In view of the uncertainties, it is worth while describing in detail some textures in which extensive matrix truncation, combined with only minor *Se* deflection, can be demonstrated with reasonable certainty to occur adjacent to pre-*Se* porphyroblasts in a moderately high-grade rock.

The example studied, a quartz–plagioclase–muscovite–biotite–garnet–staurolite schist from Glen Lethnot, Angus, is referred to by Harte & Johnson (1969), and background information with respect to the metamorphic and structural history of the area may be obtained from their paper. The matrix of the rock is dominantly composed of felts of muscovite (0.7–1.5 mm thick) separated by broader foliae of quartz and muscovite showing a polygonal or elongate-polygonal texture. Apparent muscovite dimensions in thin sections are typically 0.3 × 0.04 mm although long dimensions of 0.5–0.6 mm are common in the mica welts. Matrix quartz grains have diameters typically ranging from 0.1 to 0.3 mm; quartz grain size is, however, apparently related to mica abundance, and, in mica-deficient foliae, may exceed 0.5 mm. The porphyroblasts are garnet (up to 2 mm diameter), rare plagioclase (maximum 2.5 × 1.5 mm) and abundant staurolite (maximum 2.5 × 0.75 mm). Small staurolite crystals may be enclosed within garnet and, less commonly, the reverse situation is observed. Larger staurolite porphyroblasts are also sometimes partially enclosed within garnet and one example has been noted of a small staurolite enclosed within garnet which is itself almost completely enclosed within a large staurolite.

The textural relations between the porphyroblasts, *Se* and *Si* have been

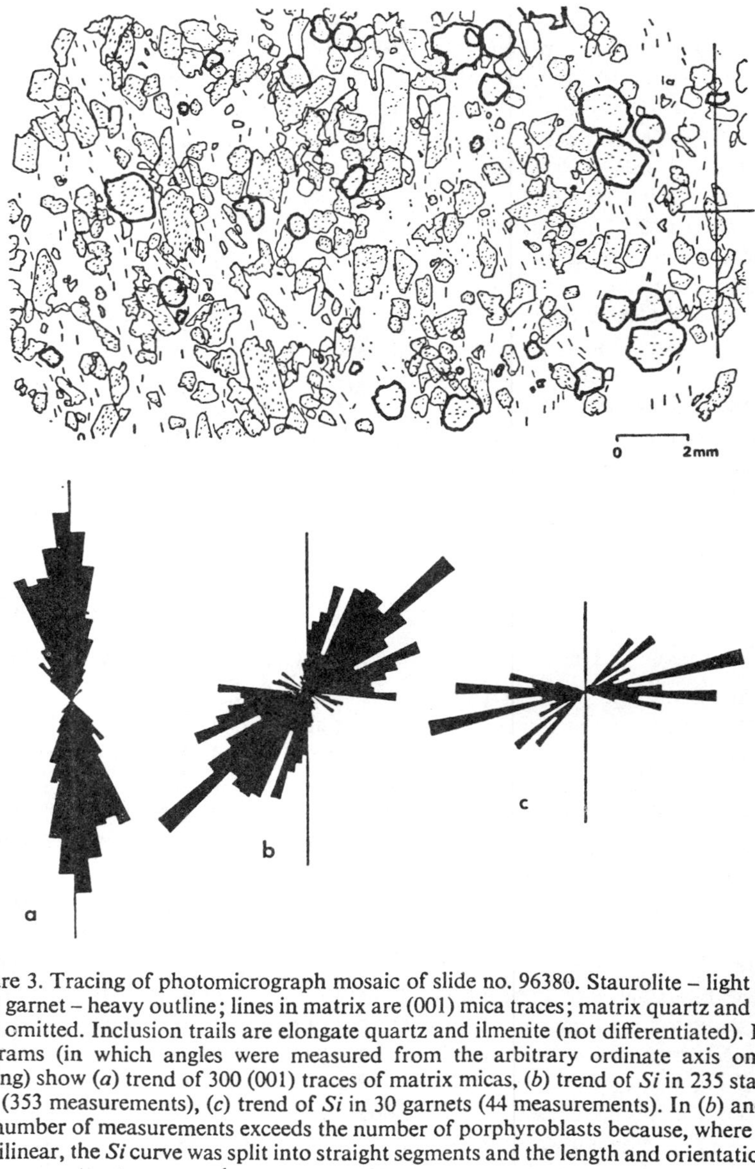

Figure 3. Tracing of photomicrograph mosaic of slide no. 96380. Staurolite – light outline; garnet – heavy outline; lines in matrix are (001) mica traces; matrix quartz and feldspar omitted. Inclusion trails are elongate quartz and ilmenite (not differentiated). Rose diagrams (in which angles were measured from the arbitrary ordinate axis on the tracing) show (*a*) trend of 300 (001) traces of matrix micas, (*b*) trend of *Si* in 235 staurolites (353 measurements), (*c*) trend of *Si* in 30 garnets (44 measurements). In (*b*) and (*c*) the number of measurements exceeds the number of porphyroblasts because, where *Si* is curvilinear, the *Si* curve was split into straight segments and the length and orientation of each segment was measured.

analysed in detail by preparing a mosaic of 72 photomicrographs covering one thin section (number 96380 in Department of Mineralogy and Petrology Harker Collection, University of Cambridge); essential features were carefully traced and the tracing (original size 100 × 60 cm) is reproduced in Figure 3. From Figure 3 and Plate 1 it can be seen that the porphyroblasts clearly and consistently 'truncate' the matrix foliation although there is usually some gentle 'bulging' of *Se* in the vicinity of the porphyroblasts. This porphyroblast-*Se* relationship is similar to those shown diagrammatically in Figure 1 D and 1 F or intermediate between these and Figure 1 E.

The inclusions within garnet and staurolite porphyroblasts are mainly elongate quartz grains (typically 45 × 10 μm) and plates of ilmenite, showing a strong preferred dimensional orientation (Pls 1, 2*a*). Although orientated arrangements of inclusions of particular shapes may arise in several ways, including crystallographic control of inclusions by the host, exsolution, etc., we believe that the inclusions and the structure they define are relicts of an *s*-surface partially incorporated within the porphyroblasts during growth because they show the following features:

(i) Inclusions form continuous trails extending from margin to margin of the porphyroblasts. In thin section, the trails are linear or, especially in larger porphyroblasts, slightly curvilinear.

(ii) In staurolite porphyroblasts, the inclusion trails show some tendency to be parallel or subparallel to the traces of (010) or (110) but also occur at a wide range of angles to these traces. Interpenetrant staurolite twins, usually with (231) twin plane, are common and inclusion trails invariably cross the composition plane without change in orientation.

(iii) Where staurolite is included within garnet (or vice versa) the inclusion trails pass from one to the other without deflection.

That the porphyroblasts and their included *Si* antedate *Se* is shown by the following combination of observations:

(i) *Si* is not continuous with *Se* but makes a variable angle with it (Fig. 3 and Pl. 1).

(ii) Quartz inclusions are much smaller than matrix quartz grains, suggesting that the matrix overgrown by the porphyroblasts was finer grained than the present matrix. The distance between adjacent *Si* grains and differences in their optical orientations demonstrates that this is so, even allowing for some dissolution of the quartz grains as they were incorporated into the porphyroblasts; also, *Si* quartz grains are markedly elongate and show a pronounced dimensional orientation (Pl. 2*a*) whilst *Se* quartz is equidimensional or only slightly elongate.

An attempt further to determine differences or similarities between *Si* and *Se*, by measuring quartz c-axes on the universal stage, proved difficult; this was partly due to the effects of relative rotations of the porphyroblasts since their growth and partly to the small size of *Si* quartz grains. The former was overcome by measuring the c-axis of 100 *Si* quartz grains in a cluster of 3 porphyroblasts whose margins impinged on each other and within which the *Si* trend in thin section was continuous. It is felt that if rotation of these porphyroblasts has occurred they have probably rotated as a cluster, and not relative to each other. Due to the small size of *Si* grains it was not possible to measure the c-axis of every quartz inclusion within the porphyroblasts. The larger included grains were, on the whole, less elongate than smaller (unmeasurable) grains, and occasionally had rather irregular shapes. Owing to their much larger size nearly every matrix quartz grain was measurable, and 100 grains were measured in the vicinity of the porphyroblast cluster. The data are reproduced in Figure 4. The first three eigenvalues (λ_1, λ_2, λ_3) for the *Si* quartz grains indicate a uniform distribution ($\lambda_1 \simeq \lambda_2 \simeq \lambda_3$) whilst a weak bimodal distribution is indicated for the matrix quartz grains ($\lambda_1 \neq \lambda_2 \simeq \lambda_3$) – see Mardia (1972, p. 225). However, the

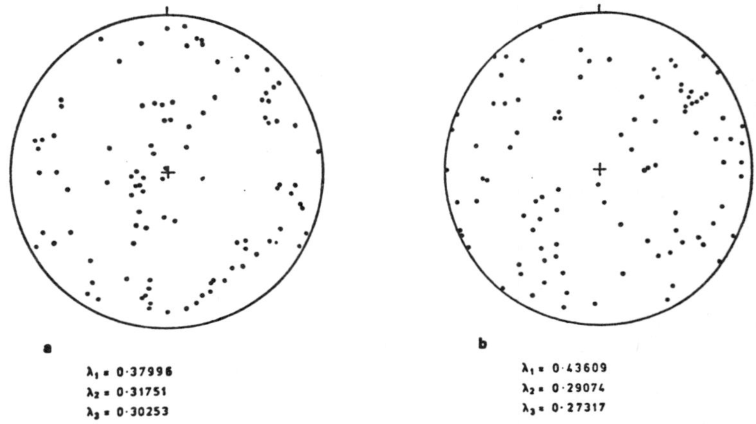

Figure 4. Stereographic plots (equal area net, lower hemisphere) of (*a*) c-axes of 100 quartz grains occurring as inclusions in a cluster of three porphyroblasts whose margins impinge on each other, (*b*) c-axes of 100 quartz grains in the matrix in the immediate vicinity of the porphyroblast cluster. λ_1, λ_2, λ_3 are the first three eigenvalues of each data set.

difference in the spherical mean directions of the two data sets, although large, is not statistically significant at the 95 % confidence level. Thus, while some differences between *Si* and *Se* quartz fabrics may exist, it is obvious that little emphasis can be placed on their interpretation. (For details of statistical measures and tests used, see Mardia, 1972.)

(iii) The distribution of inclusions within the porphyroblasts suggests that the matrix overgrown by the porphyroblasts showed either a more homogeneous distribution of minerals than the present matrix, or, at least, that any inhomogeneity or foliation was on a different scale to the present one. Thus, where quartz or ilmenite rich foliae in the present matrix abut against porphyroblasts they are not reflected in the abundance of porphyroblast inclusions, irrespective of porphyroblast-matrix rotation.

(iv) *Si* orientations within the abundant staurolites are approximately continuous from staurolite to staurolite (Fig. 3) and have a mean direction of 47° (circular 'standard deviation', 17°); garnet *Si* trails also show a preferred orientation with mean direction 78° ('standard deviation', 10°). This difference is statistically significant at 95 % confidence level (Watson and Williams' two-sample test). Similar amounts of relative rotation between the matrix and porphyroblasts of a given mineral species are therefore indicated during the deformation (D_3 of Harte & Johnson, 1969) principally responsible for the present nature and orientation of *Se*. These observations show some similarity to the features described and discussed by Ramsay (1962, pp. 323–4), except for the differences in angular rotation between staurolite and garnet, which suggest relative rotation of the porphyroblasts to one another within the matrix with angular displacements varying as a function of shape (cf. Zwart, 1960*a*, *b*; Chadwick, 1968; Harte & Johnson, 1969). Preliminary measurements (on 172 staurolite porphyroblasts) support the latter, in that crystals with length/width ratios of 2.0 or greater have a mean $Si \wedge Se$ angle of 60° ('standard deviation', 11°) while in crystals with length/width of less than 2.0 the angle is 43° ('standard deviation', 13°). Again, this difference is statistically significant at 95 % confidence level.

(v) Although garnet inclusion trails are often oriented at a higher angle to *Se* than those of staurolites it is notable that when these minerals wholly or partially include one another, their *Si*'s show a continuity of nature and orientation. This suggests that garnet and staurolite overlapped in their times of growth, and, together with the previous observations, that there was one period of pre-*Se* growth.

4. Truncation, without deflection, of *Se* at porphyroblast margins as a criterion of post-tectonic porphyroblast growth

We conclude from the previous section that truncation combined with some deflection of *Se* against a porphyroblast margin does not, by itself, exclude the possibility of pre-*Se* porphyroblast growth. Doubts also arise, from descriptions of pyrite porphyroblasts (see above, section 2.b) as to whether truncation, even in the complete absence of deflection, necessarily indicates post-*Se* growth; or, stated another way, whether complete truncation necessarily indicates replacement of the coexisting matrix *Se*.

As a criterion of post-tectonic growth, matrix truncation is often combined with the absence of 'deflected' or 'eyed' structures (equivalent to the pressure shadows of Spry, 1969, pp. 246–7) which result from the deformation of a less competent matrix about a pre-existing porphyroblast. Such usage depends on the assumptions:

(i) deflection of matrix is always produced by deformation,

(ii) truncation can only originate through the replacement of *Se* by a porphyroblast.

Particularly strong evidence against the truncation criterion is provided by recent descriptions (Williams, 1972*a*, *b*) of low-grade metamorphic rocks showing clastic grains sharply transecting the dominant preferred orientation of phyllosilicates; similar observations have been made by us in low-grade metagreywackes from other localities (Pl. 2*b*).

Situations where ambiguity of identification of truncation may occur are also readily found; for example, the matrix may undergo an essentially mimetic recrystallization, and although the porphyroblasts remain later than the initial formation of any schistose structure they truncate, they now antedate the minerals defining that structure.

Barring such ambiguities of identification, and some less plausible mechanisms, two major categories of situation are possible in which truncation is directly misleading – they are discussed in the following sections.

4.a. Varying amounts of strain within a deformed rock body

Common experience shows that the amount of strain occurring during the deformation of a body of rock can vary on both major and minor scales throughout the body (Cloos, 1947; Ramsay & Graham, 1970; Williams, 1972*a*; Pabst, 1931; Stephansson, 1974). If some porphyroblast-bearing parts of the rock body largely escape strain during a period of deformation, the *Se*-porphyroblast margin textural pattern will also escape unmodified.

For example, consider a rock showing a foliation in which mica felts lacking quartz alternate with layers containing abundant quartz as well as mica, and with all micas showing a preferred orientation parallel to the foliation. Porphyroblasts grow by replacement within this rock, which is then subjected to further deformation. If the permanent strain is controlled by pressure solution of quartz – a process recently emphasized by several authors (Durney, 1972; Williams, 1972*a*; Stephansson, 1974) – porphyroblasts wholly contained in mica felts may retain truncation, while those in quartz-rich foliae develop deflection, against *Se*. Using only *Se* to porphyroblast margin relationships the resultant textures might be taken to indicate two periods of porphyroblast growth. Such behaviour, though it cannot be proved, may be shown by rock 96380, described in section 3, and other rocks in the Glens Lethnot and Esk area. Within the latter rocks, small porphyroblasts, usually of staurolite, sometimes occur within thick mica felts and show no deflection of *Se*. A distinct *Si* is usually absent, but any inclusions present are typically of fine grain size, and this, coupled with the general evidence for only one phase of porphyroblast growth (see above, section 3, and Harte & Johnson, 1969), suggests that truncation may have been retained in the manner described above while the micas have additionally undergone a largely mimetic recrystallization.

4.b. Stress- and interface-controlled preferred orientation

The orientation of slaty cleavage has often been shown to be normal to the minimum axis of the finite strain ellipsoid (Ramsay, 1967, p. 180; Siddans, 1972) and this same relationship has been found to apply to schistosity formed in a shear zone (Ramsay & Graham, 1970). However, strain has by no means been demonstrated as the only major control of preferred orientation associated with cleavage and schistosity in rocks. Indeed, in fairly homogeneous rocks which have suffered only one deformation of relatively simple geometry, it is hardly possible to distinguish between strain control and stress control (see Elliott (1972), p. 2635).

Strömgard (1973) has modelled the stress variation around an equidimensional rigid inclusion (or porphyroblast) and shows (p. 18) that the stress trajectories are only slightly modified adjacent to the inclusion. Although the deformation geometry and rheological properties used in the model are very simple, this work suggests that in a situation where preferred orientation is stress controlled (see

Flinn (1965) for a general discussion of possible mechanisms) *Se* would terminate abruptly against the porphyroblast.

Examples of the growth of elongate crystals (fibres) in preferred directions controlled by stress or kinematics were shown by Durney & Ramsay (1973) with reference to syn-tectonic crystal growth in veins and pressure shadows (equivalent to pressure fringes of Spry (1969), p. 240). The growth of these pressure shadows is a result of the tendency of the matrix to pull away from the porphyroblast or other body in the direction of maximum elongation, and the fibres usually appear to grow parallel to the least principal stress, while the material for them is probably derived by pressure-solution. In other cases, especially adjacent to euhedral pyrite crystals, Durney & Ramsay (1973) found elongate crystals orientated perpendicular to the pyrite crystal faces (see also Mügge, 1930; Pabst, 1931; Frankel, 1957). Wickham (1973) described phenomena in which tremolite (?) fibres formed 'bridges' between quartz grains in a carbonate rock as the quartz grains moved apart during deformation. He argued (1973, pp. 35–6) that the fibres grew parallel to the displacement vector but that their orientation was probably stress controlled. Similar features, but with sericite rather than tremolite, appear to be shown by Figure 2 of Mukhopadhyay (1973).

In all cases the fibrous crystals are truncated abruptly at the margin of the porphyroblast or other relatively competent body, and can extend parallel to the general matrix cleavage. Fibres and pressure fringes of tremolite or sericite/muscovite are reported (Frankel, 1957; Williams, 1972*b*; Wickham, 1973; Mukhopadhyay, 1973), but most commonly they are of quartz and calcite and form an entity distinct from the matrix *Se*. Despite the latter, the structures have important implications for textural interpretations, and increased solubility and diffusion coefficients (e.g. at higher temperature) may promote the formation of fringes with more complex mineral assemblages similar to those forming *Se*. Elliott (1972, pp. 2635–6) also suggested that the quartz fibres of pressure fringes might become incorporated within a porphyroblast as an *Si* with the formation of misleading textural patterns.

Of particular importance, in demonstrating the relevance of the features of pressure fringes and shadows to pervasive cleavage structures, is Williams's account (1972*b*) of 'mica beards'. In the latter, phyllosilicates with preferred orientation commonly truncate against, and may even penetrate, detrital quartz grains. The control (Williams, 1972*b*, p. 375) seems to be dominantly one of preferential growth normal to the surface of detrital quartz grains, possibly in combination with stress control. The preferred orientation of micas in these pressure shadows or mica beards is parallel to rock cleavage and in many cases is the dominant element defining that cleavage.

5. Conclusions

We believe that the localized truncation of *Se* at porphyroblast margins, with or without deflection, is not by itself a *sufficient* criterion for direct replacement, or, therefore, syn- or post-*Se* growth. Furthermore, we consider that *any* interpretation of porphyroblast age relationships, which is based solely on *Se* behaviour at porphyroblast margins in rocks affected by repeated deformation, is extremely hazardous.

Acknowledgements. We thank Drs P. K. Harvey, R. F. Cheeney and R. C. Fletcher, who have each improved the manuscript at some stage of its development, Mrs B. A. Walden, who prepared most of the figures, and Mr D. Jones, who helped with the photography.

References

Bastin, E. S. 1950. Interpretation of ore textures. *Mem. Geol. Soc. Am.* **45.**

Chadwick, B. 1968. Deformation and metamorphism in the Lukmanier region, Central Switzerland. *Bull. geol. Soc. Am.* **79,** 1123–50.

Cloos, E. 1947. Oolite deformation in the South Mountain fold, Maryland. *Bull. geol. Soc. Am.* **58,** 843–918.

Durney, D. W. 1972. Solution-transfer, an important geological deformation mechanism. *Nature, Lond.* **235,** 315–16.

Durney, D. W. & Ramsay, J. G. 1973. Incremental strains measured by syntectonic crystal growths. *In* De Jong, K. A. & Scholten, R. (Eds): *Gravity and Tectonics,* 67–96. John Wiley & Sons Inc., New York.

Elliott, D. 1972. Deformation paths in structural geology. *Bull. geol. Soc. Am.* **83,** 2621–38.

Ferguson, C. C. & Harvey, P. K. 1972. Porphyroblasts and 'crystallisation force': some textural criteria: discussion. *Bull. geol. Soc. Am.* **83,** 919–20.

Fettes, D. J. 1971. Relation of cleavage and metamorphism in the Macduff Slates. *Scott. J. Geol.* **7,** 248–53.

Flinn, D. 1965. Deformation in metamorphism. *In* Pitcher, W. S. & Flinn, G. W. (Eds): *Controls of Metamorphism,* 46–72. Oliver & Boyd, Edinburgh.

Frankel, J. J. 1957. Abraded pyrite crystals from the Witwatersrand gold mines. *Mineralog. Mag.* **31,** 392–401.

Galwey, A. K. & Jones, K. A. 1962. Inclusions in garnets. *Nature, Lond.* **193,** 471–2.

Ghosh, S. K. & Sengupta, S. 1973. Compression and simple shear of test models with rigid and deformable inclusions. *Tectonophysics* **17,** 133–75.

Harte, B. & Johnson, M. R. W. 1969. Metamorphic history of Dalradian rocks in Glens Clova, Esk and Lethnot, Angus, Scotland. *Scott. J. Geol.* **5,** 54–80.

Johnson, M. R. W. 1962. Relations of movement and metamorphism in Banffshire. *Trans. Edin. geol. Soc.* **19,** 29–64.

Mardia, K. V. 1972. *Statistics of Directional Data.* 357 pp. Academic Press, London.

Misch, P. 1971. Porphyroblasts and 'crystallisation force': some textural criteria. *Bull. geol. Soc. Am.* **82,** 245–52.

Mügge, O. 1930. Bewegungen von Porphyroblasten in Phylliten und ihre Messung. *Neues Jb. Abh.* **61,** 469–510.

Mukhopadhyay, D. 1973. Strain measurements from deformed quartz grains in the slaty rocks from the Ardennes and the northern Eifel. *Tectonophysics* **16,** 279–96.

Offler, R. & Fleming, P. D. 1968. A synthesis of folding and metamorphism in the Mt. Lofty Ranges, South Australia. *J. geol. Soc. Aust.* **15** (2), 245–66.

Pabst, A. 1931. 'Pressure-shadows' and the measurement of the orientation of minerals in rocks. *Am. Miner.* **16,** 55–70.

Park, R. G. 1969. Structural correlation in metamorphic belts. *Tectonophysics* **7,** 323–38.

Pitcher, W. S. & Berger, A. R. 1972. *The Geology of Donegal.* 435 pp. John Wiley & Sons Inc., New York.

Powell, D. & Treagus, J. E. 1970. Rotational fabrics in metamorphic minerals. *Mineral. Mag.* **37,** 801–14.

Ramsay, J. G. 1962. The geometry and mechanics of formation of 'similar' type folds. *J. Geol.* **70,** 309–27.

Ramsay, J. G. 1967. *Folding and Fracturing of Rocks.* 568 pp. McGraw-Hill, New York.

Ramsay, J. G. & Graham, R. H. 1970. Strain variation in shear belts. *Can. J. Earth Sci.* **7,** 786–813.

Rast, N. 1958. Metamorphic history of the Schichallion Complex, Perthshire. *Trans. R. Soc. Edin.* **63,** 413–31.

Rast, N. 1965. Nucleation and growth of metamorphic minerals. *In* Pitcher, W. S. & Flinn, G. W. (Eds): *Controls of Metamorphism*, 73–102 Oliver & Boyd, Edinburgh.

Roberts, D. 1968. The structural and metamorphic history of the Langstrand-Finfjord area, Soroy, Northern Norway. *Norg. geol. Unders.* **253**, 1–160.

Roberts, J. L. 1974. The structure of the Dalradian rocks in the S.W. Highlands of Scotland. *Jl geol. Soc. Lond.* **130**, 93–124.

Rosenfeld, J. L. 1970. Rotated garnets in metamorphic rocks. *Spec. Pap. geol. Soc. Am.* **129**, 1–105.

Sander, B. & Pernt, M. 1923. Zur petrographisch-tektonischen Analyse: I. *Jb. Geol. Bundesanst. Wien* **73**, 183–253.

Sengupta, A. 1963. Preliminary results of the study of porphyroblasts from Sakoli metamorphites, Bhandara district. *J. geol. Soc. India* **4**, 88–96.

Shelley, D. 1972. Porphyroblasts and 'crystallisation force': some textural criteria: discussion. *Bull. geol. Soc. Am.* **83**, 919–20.

Siddans, A. W. B. 1972. Slaty cleavage – a review of research since 1815. *Earth-Sci. Rev.* **8**, 205–32.

Spry, A. 1963. The chronological analysis of crystallisation and deformation of some Tasmanian Precambrian rocks. *J. geol. Soc. Aust.* **10**, 193–208.

Spry, A. 1969. *Metamorphic Textures.* 350 pp. Pergamon Press, London.

Spry, A. 1972. Porphyroblasts and 'crystallisation force': some textural criteria: discussion. *Bull. geol. Soc. Am.* **83**, 1201–2.

Stephansson, O. 1974. Stress-induced diffusion during folding. *Tectonophysics* **22**, 233–51.

Strömgard, K.-E. 1973. Stress distribution during formation of boudinage and pressure shadows. *Tectonophysics* **16**, 215–48.

Sturt, B. A. & Harris, A. L. 1961. The metamorphic history of the Loch Tummel area, central Perthshire, Scotland. *Lpool Manchr geol. J.* **2**, 689–711.

Wickham, J. S. 1973. An estimate of strain increments in a naturally deformed carbonate rock. *Am. J. Sci.* **273**, 23–47.

Williams, P. F. 1972(*a*). Development of metamorphic layering and cleavage in low grade metamorphic rock at Bermagui, Australia. *Am. J. Sci.* **272**, 1–47.

Williams, P. F. 1972(*b*). 'Pressure shadow' structures in foliated rocks from Bermagui, New South Wales. *J. geol. Soc. Aust.* **18**, 371–7.

Wilson, M. R. 1971. On syntectonic porphyroblast growth. *Tectonophysics* **11**, 239–60.

Zwart, H. J. 1960(*a*). The chronological succession of folding and metamorphism in the Central Pyrenees. *Geol. Rdsch.* **50**, 203–18.

Zwart, H. J. 1960(*b*). Relations between folding and metamorphism in the Central Pyrenees, and their chronological succession. *Geol. Mijnb.* **39**, 163–80.

Zwart, H. J. 1962. On the determination of polymetamorphic mineral associations and its application to the Bosost area (Central Pyrenees). *Geol. Rdsch.* **52**, 38–65.

Zwart, H. J. 1963. Metamorphic history of the Central Pyrenees. II. Valle de Aran. *Leidse Geol. Meded.* **28**, 321–76.

C. C. F.
Department of Geology
University of Nottingham
Nottingham, NG7 2RD

B. H.
Grant Institute of Geology
University of Edinburgh
Edinburgh
Scotland

EXPLANATION OF PLATES

Plate 1(*a, b*). Truncation of the matrix schistosity at porphyroblast margins in rock 96380. The lack of continuity between *Si* and *Se* is particularly clear in (*a*). Scale bars represent 0.5 mm.

Plate 2(*a*). Elongate nature of quartz inclusions and their preferred orientation in staurolite in rock 96380. (*b*) Truncation of the rock cleavage, defined by phyllosilicates, at the margin of a clastic quartz grain, Portaskaig Boulder Bed, Islay, Scotland. Scale bars represent 0.1 mm.

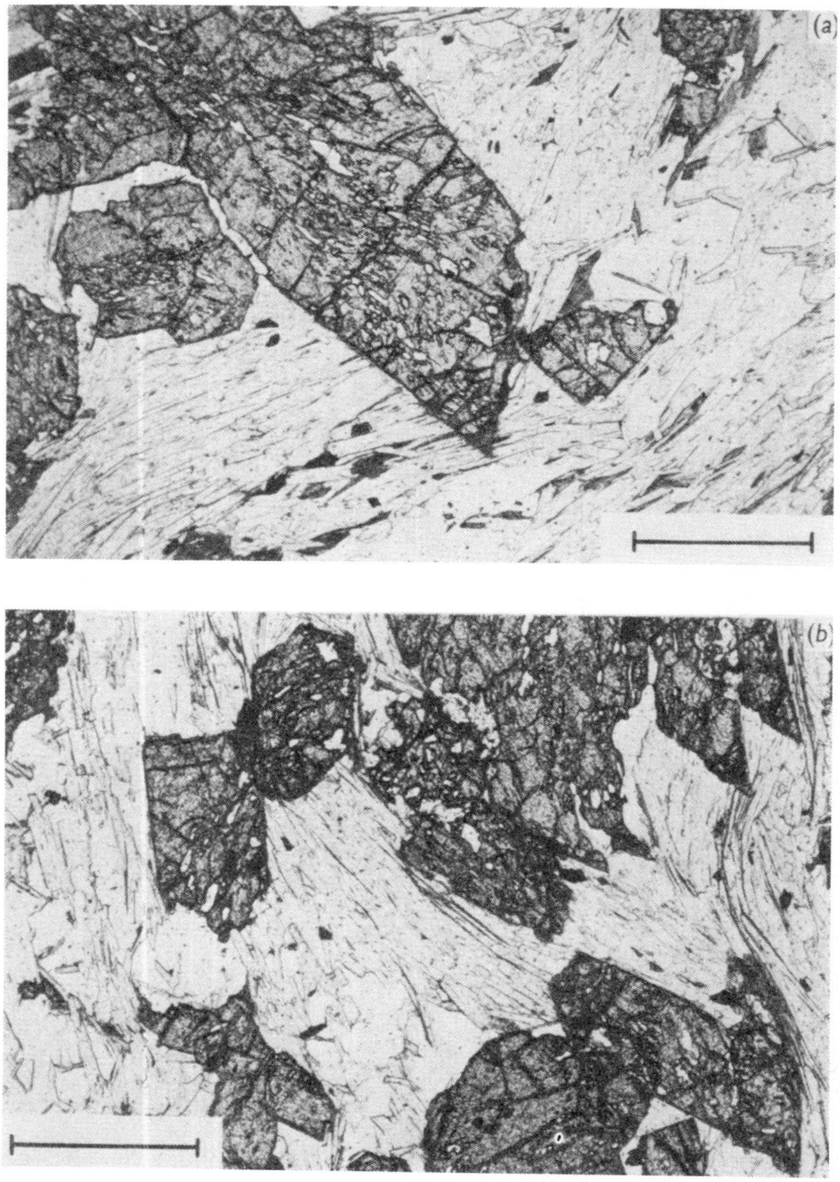

Plate 1. Truncation of the matrix schistosity.

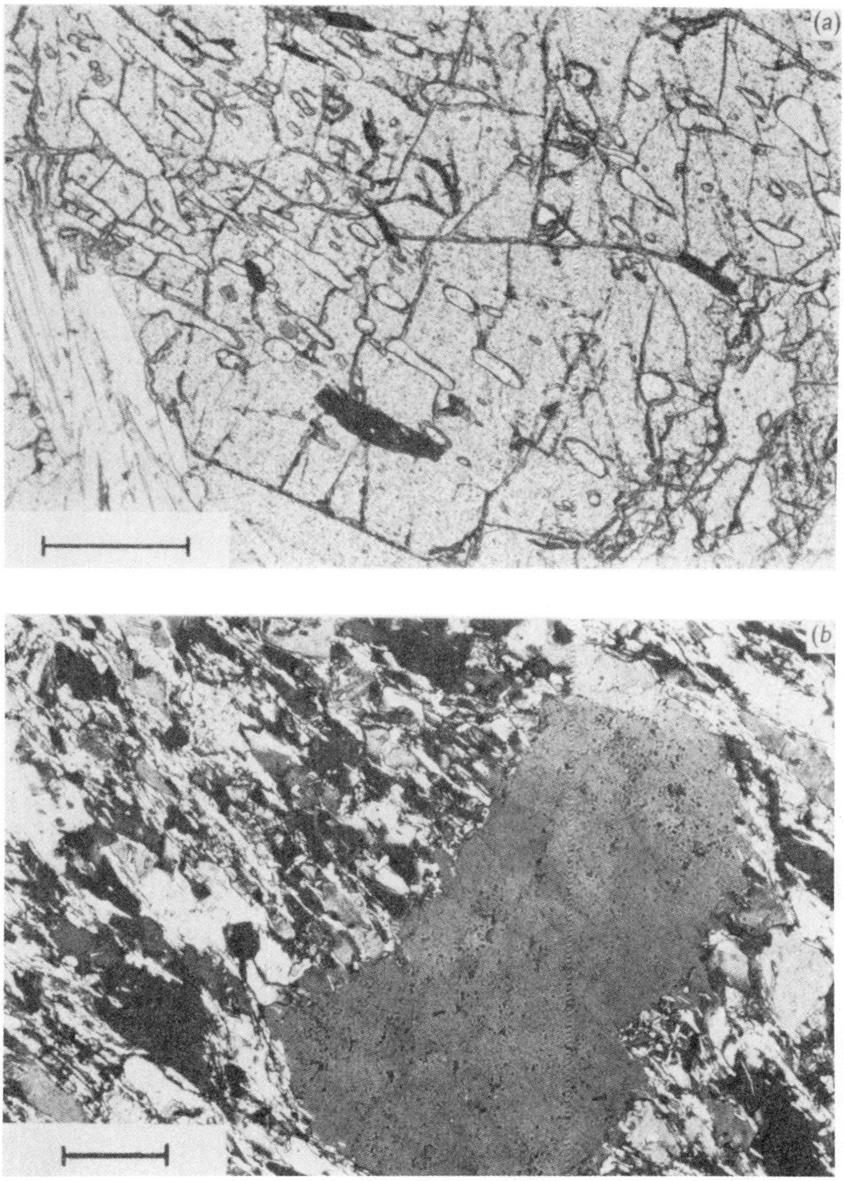

Plate 2. Elongate nature of quartz inclusions and truncation of rock cleavage.

11

Copyright ©1969 by the American Journal of Science

Reprinted from pages 43, 44-49, and 62-63 of *Am. Jour. Sci.* **267**:43-63 (1969)

PARACRYSTALLINE MICROBOUDINAGE OF ZONED GRAINS AND OTHER CRITERIA FOR SYNKINEMATIC GROWTH OF METAMORPHIC MINERALS

PETER MISCH

Department of Geology, University of Washington, Seattle, Washington 98105

[*Editor's Note:* In the original, material precedes this excerpt.]

2. PARACRYSTALLINE MICROBOUDINAGE

2A. Crystal boudinage during growth recorded by zoning, and implications regarding crystallization schistosity.[1]—Snowball porphyroblasts and certain microfolds provide evidence for synkinematic recrystallization at various stages after formation of the original schistosity. Evidence for recrystallization synchronous with deformation on the original schistosity is more elusive. Yet, if it can be found, such evidence has a critical bearing on the origin of crystallization schistosity—which may be attributed either to synkinematic or postkinematic recrystallization.

If homogeneous in composition, the aligned but individually undeformed crystals of a given phase rarely yield any clue as to whether they grew while deformation parallel to *s*[2] and/or stretching along linea-

[1] Crystallization schistosity = *Kristallisationsschieferung* (Becke, 1913, p. 37, 51) is used in a purely descriptive sense here, for schistose (or gneissose) rocks characterized by disregarded orientation of elongate crystals which, however, individually lack mechanical deformation in terms of lattice distortion.

[2] "*s*" is used here as defined by Sander (1930; compare, Knopf and Ingerson, 1938), except that bedding is not included as in Sander's original usage. Thus, *s* stands for metamorphically produced planar schistosity and/or foliation, whether it parallels or transects inherited bedding, if any. Following Sander, coexisting *s* planes are referred to, from older to younger, as s_1, s_2, et cetera.

tion[3] were active, or whether they represent static annealing after deformation (mimetic recrystallization = *Abbildungskristallisation,* Sander, 1930, p. 172). If, however, the aligned crystals are zoned, they contain a record of their growth history which might be checked against the record of deformation, provided such a record can be found in the crystals. In this manner, it may be demonstrated whether crystallization schistosity represents synkinematic or postkinematic recrystallization.

Evidence of this type was found in zoned sodic amphiboles in blueschist layers of the Shuksan Greenschist of the Northern Cascades of Washington (Misch, 1959, 1961, 1966). The mineralogy of these amphiboles (briefly outlined by Misch, 1959) need not be discussed here except for those features which provide a relative time sequence in terms of amphibole growth history.

The dominant sodic amphibole is *crossite,* arbitrarily subdivided into *"crossite 1" through "crossite 4"* with decreasing Fe^{3+}/Al ratio, "crossite 1" attaining the Fe^{3+}-content (but not the optic orientation!) of riebeckite. Crossite is followed by *"uniaxial" blue amphibole* in which Fe^{3+}/Al is close to 1 and by still relatively Fe^{3+}-rich *glaucophanes,* highly aluminous glaucophane not being stable in the particular subfacies of a broadly defined blueschist facies developed in this area (Misch, 1959, 1966). Further, there is *"soda-actinolite",* a solid solution of actinolite and glaucophane or relatively Fe^{3+}-poor crossite; and *"crossactinolite",* a solid solution of actinolite and Fe^{3+}-richer crossite, which shares the normal-symmetric orientation of crossite.

Most of the sodic amphiboles are zoned; relations between contiguous zones are gradational, as a rule. *"Normal" zoning* involves decreasing Fe^{3+}/Al; crossite 1→2→3→4→ uniaxial → glaucophane, with part or all of this series present in a given rock, depending on bulk composition. A variant of "normal" zoning involves increasing actinolite admixture; crossite → crossactinolite or soda-actinolite, glaucophane → soda-actinolite; also, crossite → crossactinolite → soda-actinolite, and soda-actinolite → actinolite. Though more rarely preserved, *"reverse" zoning* also occurs, both in terms of Fe^{3+}/Al ratio and of actinolite admixture. Where found, "reverse" invariably predates "normal" zoning ("reverse" cores, "normal" rims). This zoning sequence is consistent throughout the area studied, that is, for several dozen miles, and it may be interpreted in terms of gradually changing physical conditions (briefly discussed by Misch, 1959). Thus, the zoned amphiboles provide a regionally valid relative time sequence. Further, the "amphibole history" of the blueschists can be correlated with their overall metamorphic history. For instance, in the northern parts of the area, the reverse →

[3] Lineation may represent *a* (the direction of "tectonic transport") or, more commonly, *b*—as defined by Sander (1930). Though representing stretching at many places, *b* lineation may form in other ways also, as by distributed rotational motion like microfolding or "rolling of pencils", either of which commonly, but not necessarily, is accompanied by stretching along *b*. Where *b* is referred to in this paper, it has been identified on the basis of being parallel to axes of minor folds.

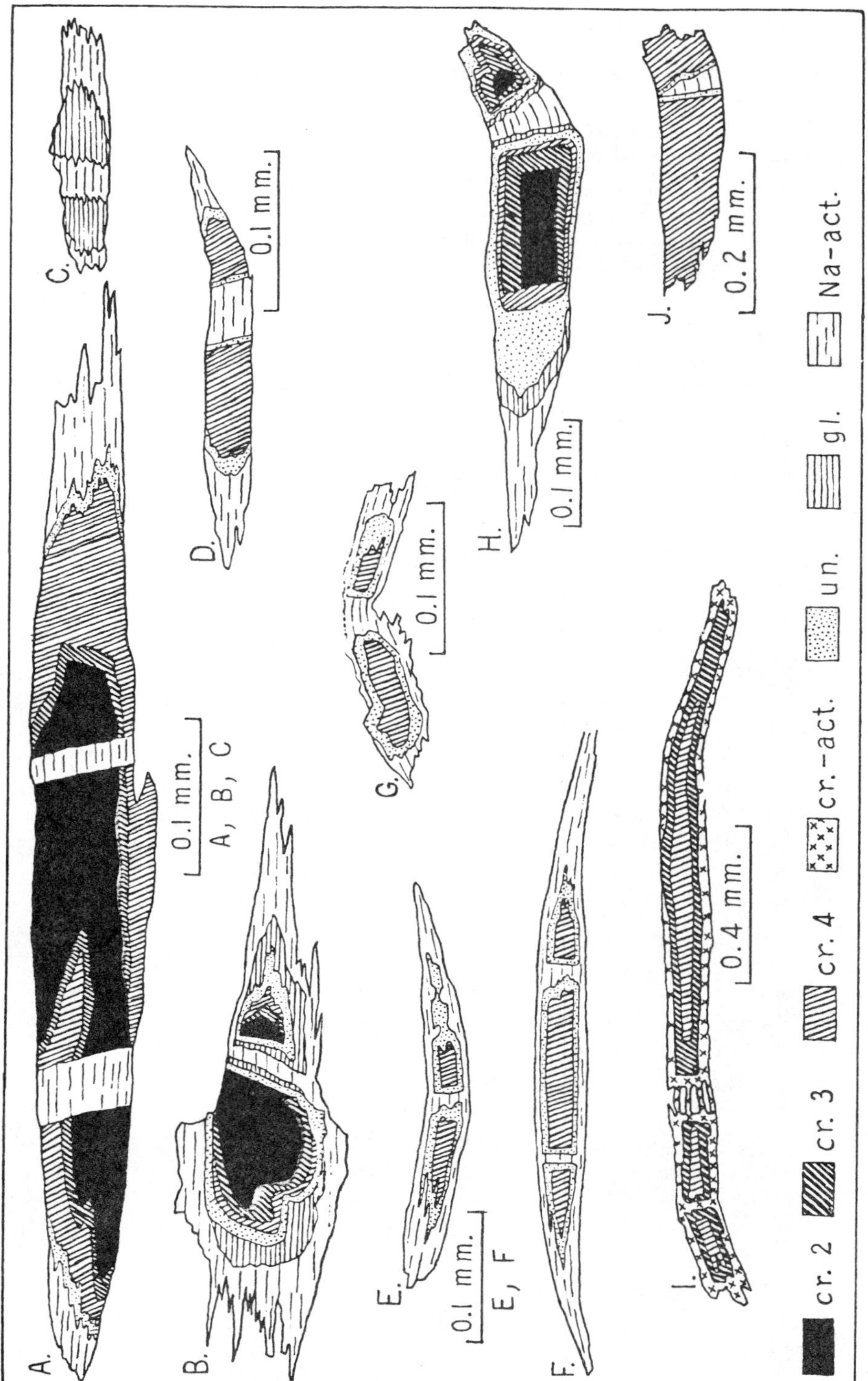

A.
B.
C.
D.
E.
F.
G.
H.
I.
J.
0.1 mm.
A, B, C
0.1 mm.
E, F
0.1 mm.
0.1 mm.
0.1 mm.
0.4 mm.
0.2 mm.
cr. 2
cr. 3
cr. 4
cr.-act.
un.
gl.
Na-act.

normal zoning sequence seen in the sodic amphiboles is matched by the sequence, prochlorite → greenish biotite → prochlorite. The metamorphism recorded by the amphiboles and associated minerals was preceded by a hypothetical pre-blueschist-facies stage[4] and followed by minor retrogression chiefly characterized by late- and, dominantly, postkinematic stilpnomelane.

In addition to providing a consistent relative time sequence, the zoned amphiboles in the Shuksan blueschists contain a record of deformation, in terms of "stretching" of growing crystals parallel to the schistosity. This involves elongation on the *s* plane (commonly with preferred orientation of amphibole (100) parellel to *s*) and, in rocks having *b* lineation, maximum stretching along *b* (with amphibole tending to align itself parallel to *b*).

The results of "stretching" of growing zoned amphibole crystals are illustrated by the examples shown on figure 1. The process visualized is pictured on figure 2. "Stretching" was initiated by rupture across the prism axis and involved pulling-apart of the fragments. However, the rupture was healed as soon as it occurred, and continued pulling-apart of the original fragments was compensated by continuous filling of the developing gap. Thus, successive amphibole zones were deposited not only on the outer surface of the crystal but also grew normal to the walls of the opening rupture. The continuity of the amphibole structure was maintained throughout. Where this pattern is developed, obviously the width of the actual opening rarely, if ever, exceeded submicroscopic dimensions. Being a site of reduced pressure, the extensional rupture attracted diffusing amphibole constituents and favored their precipitation.

The pattern described proves that amphibole crystallization was exactly synchronous with deformation that involved elongation parallel to the schistosity. If the amphiboles were homogeneous instead of zoned, no visible record would remain of what actually happened. Undoubtedly, this applies to the generally (optically) unzoned actinolites which charac-

---

Fig. 1. Paracrystalline microboudinage of aligned grains of zoned soda-amphiboles in fine-grained blueschists from 8 localities in the metabasaltic Shuksan Greenschist between Skagit River and Mount Shuksan, Northern Cascades, Washington. cr. 2 = "crossite 2"; cr. 3 = "crossite 3"; cr. 4 = "crossite 4"; cr.-act = "crossactinolite"; un. = "uniaxial" blue amphibole (in crystal J, "uniaxial" zone has actinolite admixture, is green-blue); gl. = glaucophane; Na-act. = soda-actinolite. Crystal A from soda-actinolite-bearing crossite schist (no. Sh 114). B and C from soda-actinolite-glaucophane-crossite schist with minor soda-pyroxene (Sh 115). D from soda-actinolite-bearing crossite schist (Sh 125). E and F from glaucophane-soda-actinolite-crossite schist (Sh 26). G from soda-actinolite-bearing crossite schist (Sh 131). H from glaucophane- and soda-actinolite-bearing crossite schist (Sh 83). I from crossactinolite-bearing crossite schist (Sh 78). J from soda-actinolite- and crossactinolite-bearing crossite schist (Sh 77). All rocks carry pistacite, albite, and subordinate prochlorite; quartz and phengitic mica occur in part of the rocks.

[4] This is suggested by such metamorphically produced compositional banding as predates the blueschist assemblage and, locally, is transected by the schistosity associated with that assemblage.

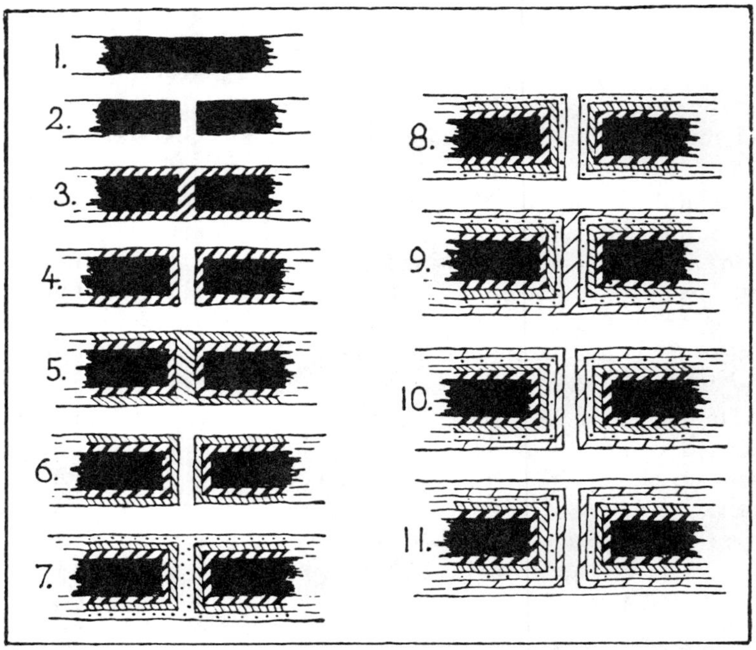

Fig. 2. Deformation-growth sequence involved in paracrystalline microboudinage of zoned crystals. A single site of rupture across a prism is shown. To allow graphic representation, stretching and accretion have to be treated as alternating finite steps. Actually, both occur simultaneously, the steps, as a rule, being infinitesimal.

terize the predominant rock types of the Shuksan Greenschist[5]. For it would seem highly unlikely that overall deformation-crystallization history in the actinolite-rich greenschists was different from that in the blueschist intercalations, their structures being identical and their mineralogical differences merely reflecting rock-compositional variation.

The pattern discussed above is modified if the rate of pulling-apart of segments of an amphibole prism exceeds the rate of amphibole crystallization in the gap, due to accelerated deformation and/or insufficient diffusion of amphibole constituents to the site of rupture. Instead of solid filling, only delicate amphibole bridges form across the gap (fig. 1-I). The remainder of the opening space is filled by precipitation of other phases (for example, albite, quartz, et cetera) that diffuse to the site of reduced pressure. This bridge pattern is very rare in the Shuksan blueschists but was found to predominate at a Californian locality (fig. 5). Such bridge patterns would be diagnostic in an unzoned, homogeneous crystal also.

In some of the zoned amphiboles, paracrystalline stretching was accompanied by rotational movement between segments, resulting in bent or kinked prisms (fig. 1-G, H, J; by contrast, crystals D, E, F, I show ordinary postcrystalline bending).

[5] Owing to lower Fe^{3+}/Fe^{2+} ratios in these rocks (Misch, 1959).

The regionally consistent amphibole zoning sequence in the Shuksan blueschists and regional distribution of crystals showing paracrystalline microboudinage make it possible to determine when initial rupture and subsequent stretching took place in individual crystals, in different crystals and rocks from the same locality, and in samples from different localities. For the few examples depicted on figure 1, crystallization-deformation history is plotted on figure 3. Regional integration of all available data indicates: (1) initial rupture occurred throughout the "amphibole history" of the rocks, including the early stage recorded by reverse zoning; (2) once initial rupture had occurred, stretching usually continued at the ruptured site during the remainder of the growth history of the crystal. It follows that all the blueschist metamorphism was synkinematic.

[*Editor's Note:* Material has been omitted at this point.]

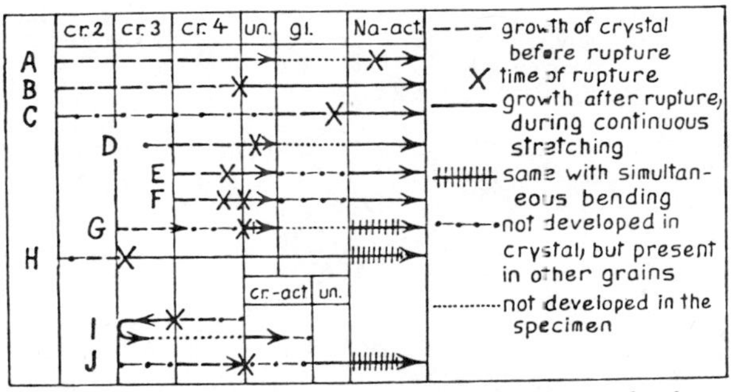

Fig. 3. Interpreted history of zoned soda-amphibole crystals shown on figure 1. Arrows point from cores to rims.

REFERENCES

Backlund, H., 1918, Petrogenetische Studien an Taimyrgesteinen: Geol. Fören. Stockholm Förh., v. 40, p. 101-203.

Becke, F., 1913, Über Mineralbestand und Struktur der kristallinen Schiefer: Akad. Wiss. Wien Denkschr., v. 75, p. 1-53.

Flett, J. S., 1912, Petrological contributions, *in*, Geology of Ben Wyvis, Carn Chuinneag, Inchbae, and the surrounding country: Scotland Geol. Survey Mem., 93, 189 p.

Grubenmann, Ulrich, and Niggli, P., 1924, Die Gesteinsmetamorphose: Berlin, Gebrüder Borntraeger, 539 p.

Harker, Alfred, 1931, Metamorphism, 2d ed.: London, Methuen & Co., 362 p.

Knopf, E. B., and Ingerson, F. E., 1938, Structural petrology: Geol. Soc. America Mem. 6, 270 p.

Krige, L. J., 1918, Petrographische Untersuchungen im Val Piora und Umgebung: Ecologae Geol. Helvetiae, v. 14, p. 519-654.

Misch, P. O. 1949, Metasomatic granitization of batholithic dimensions; pt. 3: Am. Jour. Sci., v. 247, p. 673-705.

——— 1959, Sodic amphiboles and metamorphic facies in Mount Shuksan belt, Northern Cascades, Washington [abs.]: Geol. Soc. America Bull., v. 70, p. 1736-1737.

——— 1961, New criteria for synkinematic growth of metamorphic minerals [abs.]: Geol. Soc. America Spec. Paper 68, p. 44-45.

——— 1966, Tectonic evolution of the Northern Cascades of Washington State, *in* Gunning, H. C., ed., Tectonic history and mineral deposits of the Western Cordillera: Canadian Inst. Mining Metallurgy Spec. v. 8, p. 101-148.

Niggli, P., 1912, Die Chloritoidschiefer des nordöstlichen Gotthardmassives: Geol. Karte Schweizer Beitr., neue folge, Lfg. 36.

Page, B. M., ms, 1939, The geology of the southeast quarter of Chinsaukum quadrangle: Ph.D. dissert. Stanford Univ.

Sander, Bruno, 1930, Gefügekunde der Gesteine: Vienna, J. Springer, 352 p.

Schmidt, W., 1918, Bewegungsspuren in Porphyroblasten kristalliner Schiefer: Akad. Wiss. Wien Sitzungsber., Abt. 1, v. 127, p. 293-310.

Spry, Alan, 1963, The origin and significance of snowball structure in garnet: Jour. Petrology, v. 4, p. 211-222.

Turner, F. J., and Weiss, L. E., 1963, Structural analysis of metamorphic tectonites: New York, McGraw Hill, 545 p.

ERRATA

Footnote 1, line 3 should read: "by preferred orientation . . . ."

Page, B. M., reference above, the title should read: "The geology of the southeast quarter of Chiwaukum . . . ."

Part III

THE GEOMETRY OF STRAIN

Editor's Comments on Papers 12 Through 20

12 **FLINN**
On Folding During Three-Dimensional Progressive Deformation

13 **CLOOS**
Excerpts from *Oölite Deformation in the South Mountain Fold, Maryland*

14 **DUNNET**
Excerpt from *A Technique of Finite Strain Analysis Using Elliptical Particles*

15 **ROBIN**
Excerpt from *Determination of Geologic Strain Using Randomly Oriented Strain Markers of Any Shape*

16 **HANNA and FRY**
A Comparison of Methods of Strain Determination in Rocks from Southwest Dyfed (Pembrokeshire) and Adjacent Areas

17 **ELLIOTT**
Deformation Paths in Structural Geology

18 **MITRA**
A Quantitative Study of Deformation Mechanisms and Finite Strain in Quartzites

19 **RAMSAY**
Shear Zone Geometry: A Review

20 **SCHWERDTNER**
Determination of Stratigraphic Thickness of Strained Units

The geometry of strain may be defined as the shape and orientation of the strain ellipsoid. The strain ellipsoid is the form taken by an originally spherical volume of deformed rock that underwent homogeneous strain. Particularly clear descriptions of strain ellipsoids are given by Means (1976, pp. 130-150) and Hobbs et al. (1976, pp. 25-40).

It is impossible in the space available to present papers that address all aspects of strain geometry, and papers reprinted here were chosen primarily to demonstrate the basic principles and to present some of the most useful methods of strain analysis. For the most part, papers that apply strain geometry to solving geologic problems are excluded, although a brief discussion of some applications is given at the end of this introduction.

Flinn (Paper 12) presents an analysis of the way in which lines and planes rotate as a result of progressive strain and discusses how this relates to the flattening of folds. Flinn's "deformation plot" has become one of the most-used methods of presenting the results of strain analyses, though Ramsay (1967, p. 329) and Flinn (1978) suggest an improvement using logarithmic strain values for the coordinates of the graph (K = log *a*/log *b*) rather than axial ratios of the strain ellipsoid. An alternative method for representing strain graphically was presented by Hsu (1966) and is discussed by several geologists, including Hossack (1968), Burns and Spry (1969), and Owens (1974*a*).

Paper 13 by Cloos is an acknowledged classic in the field of strain analysis. In it, Cloos discusses many features relevant to strain studies, such as relative strength of strain markers and their matrix, the effects of original shape and orientation of strain markers, and the importance of understanding deformation textures. He also outlines many of the field and laboratory procedures for carrying out strain analyses. Cloos elaborated on some of these points in a later book on microtectonics (1971).

One of Cloos's more important observations is that originally ellipsoidal objects will take on a range of shapes and orientations as a result of deformation, depending on their original shape and orientation. Dunnet (Paper 14) took up this point and developed one of the most useful methods of analyzing strain in many rocks with initial random fabric (the Rf/ϕ method). The Rf/ϕ technique is based on the fact that on any outcrop or cut surface there is a relationship between the final shape (Rf) of a deformation ellipse and its orientation (angle ϕ). Dunnet and Siddans (1971) modified the Rf/ϕ method to analyze strain in rocks that initially had non-random fabrics. Lisle (1977) and Peach and Lisle (1979) suggested a modification of the Rf/ϕ method that improves its accuracy. In order to obtain a three-dimensional strain ellipsoid, the above Rf/ϕ methods require that measurements be made on two, or preferably three, planes cut along predetermined principal strain directions. Siddans (1980), however, expanded their use to measurements made in any three orthogonal planes, based on initially random fabrics.

Unfortunately, as noted by De Paor (1980), application of the Rf/ϕ technique runs into problems where there is a marked ductility contrast between the objects used as strain markers and their matrix (a problem shared by most methods of strain analysis). Where this is so, the grains chosen for analysis will most likely be more competent than their matrix and thus will have undergone less strain than the rock as a whole. Added to this is the problem of cohesion between the grains making up the rock. For example, an elongate grain of greater competence than its matrix or the strength of which is greater than the intergranular cohesion will rotate during deformation, upsetting the Rf/ϕ relationships on which the method is based. Also, large errors in strain estimation can occur where the initial fabric of the rock is even slightly different from that assumed in the strain analysis (Seymour and Boulter, 1979).

Robin (Paper 15) presents a simple, easy-to-use method of determining the shape of the strain ellipsoid for markers with any initial shape, provided they originally had random orientation. Schnorr and Schwerdtner (1981) devised an empirical test of the precision of the Robin method and found that, under suitable circumstances, it can yield results reproducible within 10% (a figure probably better than that for other strain methods). This method does not require that measurements be made in the principal planes of strain, although it is subject to the same limitations discussed above for Rf/ϕ methods.

Hanna and Fry (Paper 16) compare several methods of determining the two-dimensional state of strain on plane surfaces, including a new graphical method devised by one of the authors (Fry, 1979). This method is based on the spacing between objects in a deformed rock and is particularly useful for analyzing either the strain undergone by the matrix surrounding randomly scattered rigid objects, or the total strain if objects and matrix are deformed uniformly. If combined with certain other methods (such as the Rf/ϕ technique), it may also be used for obtaining some idea of any possible differential strain between objects and matrix.

Numerous other methods have been devised for analyzing strain, though some are useful only to a limited range of rocks and none is applicable to all types of rocks: Elliott (1970), Talbot (1970), McLeish (1971), Durney and Ramsey (1973), Owens (1974*b*), Matthews et al. (1974), Groshong (1974), Borradaile (1974), Lisle (1974), Schwerdtner et al. (1974), Lipshie et al. (1976), Hirsinger (1976), Paulis (1976), Ramberg (1976), Shimamoto and Ikeda (1976), Stauffer and Burnett (1979), Miller and Oertel (1979), Harvey and Laxton (1980). Grunsky et al. (1980) described a computer-graphics technique for determining the shapes and orientations of deformed grains. Methods invented before 1967 are summarized by Ramsay (1967, pp. 142-254). Milton (1980) and Gendzwill and Stauffer (1981) discussed methods of obtaining the shape and orientation of the strain ellipsoid from measurements of strain ellipses on any three surfaces, thus requiring no pre-estimations of the principal planes of strain; for these methods, however, the strain ellipses must be determined by some other means, such as the Rf/ϕ technique. Gay (1968*a*, 1968*b*, 1968*c*) and Ramberg (1975) made important contributions to our understanding of the strain, displacement, and rotation of particles in a rock undergoing deformation. Flinn (1978, 1979*a*, 1979*b*) summarized the computation of strains using matrix equations; this is the most powerful method of handling calculations related to three-dimensional deformation and has been used by numerous workers for the past decade.

The methods of strain analysis described by the above authors are useful for estimating the total strain undergone by a rock or by certain components of that rock. This, however, tells us nothing about the kinematic path or the deformational history which must be reconstructed if we are to develop realistic concepts of the development of folds, faults, intrusions, and foliation. An important paper by Elliott (Paper 17) goes a long way in describing how we can use various aspects of the fabric of deformed rocks to determine the types of deformation a rock has undergone and the relative orientation of strain axes during various stages of deformation. Mitra (Paper 18) expands on aspects of Elliott's paper and clearly shows how, in suitable rocks, the total strain resulting from one period of deformation can be partitioned into components due to different strain mechanisms.

A key paper which applies modern theory in strain geometry to ductile shear is that by Ramsay and Graham (1970). Since its publication, the concept of ductile shear zones has developed considerably; some aspects of the state of the art are summarized by Ramsay in Paper 19. Papers by Escher and Watterson (1974), Coward (1976), and Cobbold (1977) are recommended reading.

The final paper (20) reprinted in this benchmark edition is by Schwerdtner, who describes a method of obtaining the original thickness of a deformed layer once the strain geometry is known (e.g., the original thickness of a formation, as required for stratigraphic studies in an area where the formation has been deformed).

Strain concepts have many applications in geology. A complete discussion of them is well beyond the scope of one "benchmark" book, however the following brief list of papers may be useful.

The mechanics of folding has received considerable attention in the literature (e.g., Biot, 1961; Biot et al., 1961; Ramberg, 1964; Chapple, 1968; Dieterich and Carter, 1969; Hudleston and Stephansson, 1973; Shimamoto and Hara, 1976), and much of this work has involved mathematical and experimental modeling of stress, viscosity, strain rate, and strain geometry. Valuable though this work is, it is primarily two dimensional (in the fold profile). Future work should include variations in strain along fold axes and should consider folds developed where the direction of maximum elongation is oblique to the fold axis (e.g., Treagus and Treagus, 1981). Groshong (1975) makes a valuable contribution to the study of rock fabrics and the strain history of rocks within natural folds, a type of approach that promises to be useful for understanding folding processes (also see Pfiffner, 1980).

Little has been written about the strain geometry of intrusion, though papers by Fletcher (1972), Dixon (1975), Schwerdtner and Troeng (1978), Schwerdtner et al. (1978), Morgan (1980), and Davidson (1980) make important contributions to understanding the strain related to diapirism, the development of mantled gneiss domes, and perhaps plumes or undations in the Earth's mantle. Other studies of diapiric processes have been mainly concerned with viscosity, density, and driving force (e.g., Ramberg, 1967).

Some words of caution are in order for anyone wishing to attempt strain analysis. First, there is the problem of interpreting a regional strain pattern where the strain data is obtained from essentially a planar surface (topography). The resulting lack of knowledge of the strain geometry at higher and lower structural levels leads to ambiguous strain patterns which could, for example, be the result of either crustal shortening or diapirism (Schwerdtner, 1976, 1977). All methods of strain analysis, moreover, require the acceptance of a number of assumptions about the original state of the rock (e.g., the shape and orientation of grains, the relative competence of the rock's components, and the intergranular cohesion). Apparently less well understood is that deformational mechanism is of equal importance for the application of most of the methods (Elliott, Paper 17; Mitra, Paper 18; and Gay, 1968*c*). For example, a method based on grain shape and orientation that assumes the post-strain geometry to result only from the ductile change of shape of the grains is likely to be in error because it ignores, among other factors, the effects of intergranular movements (Stauffer, Paper 5), rotation of the strain markers (Gay, 1968*b*), and pressure solution (Elliott, Paper 7). In fact, it is probably a rare rock which satisfies all the necessary assumptions for any method of strain analysis to yield numbers that provide an accurate measure of strain. The numbers obtained should be considered to be estimates at best and should be used only for comparative purposes: for example, for comparing the approximate amounts of grain elongation and degrees of alignment between equivalent rocks or between the components of a single sample (Stauffer, 1967). Also, methods based on grain shape and orientation, especially when applied to heterogeneous rocks such as conglomerates or granites, should not be called *strain analysis* at all (though the alignments are due to strain) but rather should be given some different name. Perhaps *shape alignment analysis* would serve.

REFERENCES

Biot, M. A., 1961, Theory of Folding of Stratified Viscoelastic Media and Its Implications in Tectonics and Orogenesis, *Geol. Soc. America Bull.* **72:**1595–1620.

Biot, M. A., H. Ode, and W. L. Roever, 1961, Experimental Verification of the Theory of Folding of Stratified Viscoelastic Media, *Geol. Soc. America Bull.* **72:**1621–1632.

Borradaile, G. J., 1974, Bulk Finite Tectonic Strain Estimates from the Deformation of Neptunian Dykes, *Tectonophysics* **22:**127–139.

Burns, K. L., and A. H. Spry, 1969, Analysis of Shape of Deformed Pebbles, *Tectonophysics* **7:**177–196.

Chapple, W. M., 1968, A Mathematical Theory of Finite-Amplitude Folding, *Geol. Soc. America Bull.* **79:**47–68.

Cloos, E., 1971, *Microtectonics along the Western Edge of the Blue Ridge, Maryland and Virginia,* Johns Hopkins Univ. Press, Baltimore, 234p.

Cobbold, P. R., 1977, Description and Origin of Banded Deformation Structures. I. Regional Strain, Local Pertubations, and Deformation Bands, *Canadian Jour. Earth Sci.* **14:**1721–1731.

Coward, M. P., 1976, Strain Within Ductile Shear Zones, *Tectonophysics* **34:**181–197.

Davidson, D. M., Jr., 1980, Emplacement and Deformation of the Archean Saganaga Batholith, Vermillion District, Northeastern Minnesota, *Tectonophysics* **66:**179–195.

De Paor, D. G., 1980, Some Limitations of the Rf/ϕ Technique of Strain Analysis, *Tectonophysics* **64:**T29–T31.

Dieterich, J. H., and N. L. Carter, 1969, Stress-History of Folding, *Am. Jour. Sci.* **267:**129–154.

Dixon, J. M., 1975, Finite Strain and Progressive Deformation in Models of Diapiric Structures, *Tectonophysics* **28:**89–124.

Dunnet, D., and A. W. B. Siddans, 1971, Non-random Sedimentary Fabrics and Their Modifications by Strain, *Tectonophysics* **12:**307–325.

Durney, D. W., and J. G. Ramsay, 1973, Incremental Strains Measured by Syntectonic Crystal Growths, in *Gravity and Tectonics,* K. A. DeJong and R. Scholten, eds., Wiley, New York, pp. 67–96.

Elliott, D., 1970, Determination of Finite Strain and Initial Shape from Deformed Elliptical Objects, *Geol. Soc. America Bull.* **81:**2221–2236.

Escher, A., and J. Watterson, 1974, Stretching Fabrics, Folds and Crustal Shortening, *Tectonophysics* **22:**223–231.

Fletcher, G. R., 1972, Application of a Mathematical Model to the Emplacement of Mantled Gneiss Domes, *Am. Jour. Sci.* **272:**197–216.

Flinn, D., 1978, Construction and Computation of Three-Dimensional Progressive Deformations, *Geol. Soc. (London) Jour.* **135:**291–305.

Flinn, D., 1979*a*, The Computations of Deformations, *Geol. Jour.* **14:**87–98.

Flinn, D., 1979*b*, The Deformation Matrix and the Deformation Ellipsoid, *Jour. Structural Geol.* **1:**299–308.

Fry, N., 1979, Random Point Distributions and Strain Measurement in Rocks, *Tectonophysics* **60:**89–105.

Gay, N. C., 1968*a*, The Motion of Rigid Particles Embedded in a Viscous Fluid during Pure Shear Deformation of the Fluid, *Tectonophysics* **5:**81–88.

Gay, N. C., 1968*b*, Pure Shear and Simple Shear Deformation of Inhomogeneous Viscous Fluids. 1. Theory, *Tectonophysics* **5:**211–234.

Gay, N. C., 1968*c*, Pure Shear and Simple Shear Deformation of Inhomogeneous Viscous Fluids. 2. The Determination of the Total Finite Strain in a Rock from Objects such as Deformed Pebbles, *Tectonophysics* **5:**295–302.

Gendzwill, D. J., and M. R. Stauffer, 1981, Analysis of Triaxial Ellipsoids, Their Shapes, Plane Sections, and Plane Projections, *Internat. Assoc. Math. Geol. Jour.* **13:**135–152.

Groshong, R. H., Jr., 1974, Experimental Test of Least-Squares Strain Gage Calculation Using Twinned Calcite, *Geol. Soc. America Bull.* **85:**1855–1864.

Groshong, R. H., Jr., 1975, Strain, Fractures, and Pressure Solution in Natural Single-Layer Folds, *Geol. Soc. America Bull.* **86:**1363–1376.

Grunsky, E. C., P. Y. E. Robin, and W. M. Schwerdtner, 1980, Orientation of Feldspar Porphyroclasts in Mylonite Samples from the Southern Churchill Province, Canadian Shield, *Tectonophysics* **66:**213–224.

Harvey, P. K., and R. R. Laxton, 1980, The Estimation of Finite Strain from the Orientation Distribution of Passively Deformed Linear Markers: Eigenvalue Relationships, *Tectonophysics* **70:**285–307.

Hirsinger, V., 1976, Numerical Strain Analysis Using Polar Coordinate Transformations, *Internat. Assoc. Math. Geol. Jour.* **8:**183–202.

Hobbs, B. E., W. D. Means, and P. F. Williams, 1976, *An Outline of Structural Geology,* Wiley, New York, 571p.

Hossack, J. R., 1968, Pebble Deformation and Thrusting in the Bygdin Area (Southern Norway), *Tectonophysics* **5:**315–339.

Hsu, T. C., 1966, The Characteristics of Coaxial and Non-coaxial Strain Paths, *Jour. Strain Anal.* **1:**216–222.

Hudleston, P. J., and O. Stephansson, 1973, Layer Shortening and Fold-Shape Development in the Buckling of Single Layers, *Tectonophysics* **17:**299–321.

Lipshie, S. R., G. Oertel, and J. M. Christie, 1976, Measurement of Preferred Orientation of Phyllosilicates in Schists, *Tectonophysics* **34:**91–99.

Lisle, R. J., 1974, Deformed Lineations as Finite-Strain Structures, *Tectonophysics* **21:**165–179.

Lisle, R. J., 1977, Clastic Grain Shape and Orientation in Relation to Cleavage from the Aberystwyth Grits, Wales, *Tectonophysics* **39:**381–396.

McLeish, A. J., 1971, Strain Analysis of Deformed Pipe Rock in the Moine Thrust Zone, Northwest Scotland, *Tectonophysics* **12:**469–503.

Matthews, P. E., R. A. B. Bond, and J. J. Van den Berg, 1974, An Algebraic Method of Strain Analysis Using Elliptical Markers, *Tectonophysics* **24:**31–67.

Means, W. D., 1976, *Stress and Strain: Basic Concepts of Continuum Mechanics for Geologists,* Springer-Verlag, New York, 339p.

Miller, D. M., and G. Oertel, 1979, Strain Determination from the Measurement of Pebble Shapes; a Modification, *Tectonophysics* **55:**T11–T13

Milton, N. J., 1980, Determination of the Strain Ellipsoid from Measurements on any Three Sections, *Tectonophysics* **64:**T19–T27.

Morgan, J., 1980, Deformation Due to the Distension of Cylindrical Igneous Contacts: a Kinematic Model, *Tectonophysics* **66:**167–178.

Owens, W. H., 1974*a*, Representation of Finite Strain State by Three-Axis Planar Diagrams, *Geol. Soc. America Bull.* **85:**307–310.

Owens, W. H., 1974*b*, Mathematical Model Studies on Factors Affecting the Magnetic Anisotrophy of Deformed Rocks, *Tectonophysics* **24:**115–131.

Paulis, R. V., 1976, A Method of Strain Using Initially Orthogonal Planar and Linear Markers, *Tectonophysics* **34:**T29–T36.

Peach, C. J., and R. J. Lisle, 1979, A Fortran IV Program for the Analysis of Tectonic Strain Using Deformed Elliptical Markers, *Comput. Geosci.* **5:**325–334.

Pfiffner, O. A., 1980, Strain Analysis in Folds (Infrahelvetic Complex, Central Alps), *Tectonophysics* **61:**337–362.

Ramberg, H., 1964, Selective Buckling of Composite Layers with Contrasted Rheological Properties; A Theory for Simultaneous Formation of Several Orders of Folds, *Tectonophysics* **1:**307–341.

Ramberg, H., 1967, *Gravity, Deformation and the Earth's Crust,* Academic, London, 214p.

Ramberg, H., 1975, Particle Paths, Displacement and Progressive Strain Applicable to Rocks, *Tectonophysics* **28:**1–37.

Ramberg, H., 1976, The Strain in a Sheet Intersecting the Strain Ellipsoid at any Angle, *Soc. Geol. France Bull.* **18:**1417–1422.

Ramsay, J. G., 1967, *Folding and Fracturing of Rocks,* McGraw-Hill, New York, 568p.

Ramsay, J. G., and R. H. Graham, 1970, Strain Variation in Shear Belts, *Canadian Jour. Earth Sci.* **7:**786–813.

Schnorr, P., and W. M. Schwerdtner, 1981, An Empirical Test of Sample Size and Precision of Robin's Method of Strain Analysis, *Tectonophysics* **73:**T1–T8.

Schwerdtner, W. M., 1976, A Principal Difficulty of Proving Crustal Shortening in Precambrian Shields, *Tectonophysics* **30:**T19–T23.

Schwerdtner, W. M., 1977, Geometric Interpretation of Regional Strain Analyses, *Tectonophysics* **39:**515–531.

Schwerdtner, W. M., and B. Troeng, 1978, Strain Distribution within Arcuate Diapiric Ridges of Silicone Putty, *Tectonophysics* **50:**13–28.

Schwerdtner, W. M., D. H. Waddington, and G. Stollery, 1974, Polycrystalline Pseudomorphs as Natural Gauges of Incremental Strain, *Neues Jahrb. Mineralogie Monatsh.* **¾:**174–182.

Schwerdtner, W. M., R. H. Sutcliffe, and B. Troeng, 1978, Patterns of Total Strain in the Crestal Region of Immature Diapirs, *Canadian Jour. Earth Sci.* **15:**1437–1447.

Seymour, D. B., and C. A. Boulter, 1979, Tests of Computerized Strain Analysis Methods by the Analysis of Simulated Deformation of Natural Unstrained Sedimentary Fabrics, *Tectonophysics* **58:**221–235.

Shimamoto, T., and I. Hara, 1976, Geometry and Strain Distribution of Single-Layer Folds, *Tectonophysics* **30:**1–34.

Shimamoto, T., and Y. Ikeda, 1976, A Simple Algebraic Method for Strain Analysis from Deformed Ellipsoidal Objects; 1 Basic Theory, *Tectonophysics* **36:**315–337.

Siddans, A. W. B., 1980, Analysis of Three-Dimensional, Homogeneous, Finite Strain Using Ellipsoidal Objects, *Tectonophysics* **64:**1–16.

Stauffer, M. R., 1967, Tectonic Strain in some Volcanic Sedimentary, and Intrusive Rocks near Canberra, Australia: A Comparative Study of Deformation Fabrics, *New Zealand Jour. Geology and Geophysics* **10:**1079–1108.

Stauffer, M. R., and A. I. Burnett, 1979, Down-plunge Viewing: A Rapid Method for Estimating the Strain Ellipsoid for Large Clasts in Deformed Rocks, *Canadian Jour. Earth Sci.* **16:**290–304.

Talbot, C. J., 1970, The Minimum Strain Ellipsoid Using Deformed Quartz Veins, *Tectonophysics* **9:**47–76.

Treagus, J. E., and S. H. Treagus, 1981, Folds and the Strain Ellipsoid: A General Model, *Jour. Structural Geol.* **3:**1–18.

12

Copyright ©1962 by the Geological Society of London

Reprinted by permission from *Geol. Soc. London Quart. Jour.* **118**:385–428 (1962)

ON FOLDING DURING THREE-DIMENSIONAL PROGRESSIVE DEFORMATION

BY DEREK FLINN, PH.D. D.I.C. A.R.C.S. F.G.S.

Submitted 13 June 1961 ; revised manuscript received 20 March 1962 ; read 1 November 1961

CONTENTS

SUMMARY

This paper is concerned with the geometry of homogeneous strain in rocks. Equations relating rotation of planes and lines to the deformation of the rock containing them are derived with the aid of the deformation ellipsoid. The Fresnel construction is borrowed from optical mineralogy to determine from the deformation ellipsoid the principal axes of compression and extension in any plane, and an expression is derived for determining the lengths of these axes. Further equations are derived which define the surfaces of no infinitesimal strain and no finite strain for all possible ellipsoids, and it is shown that these surfaces may be used for rapidly determining whether the principal axes in any plane have suffered shortening or extension.

With these tools the rather complex geometry of three-dimensional homogeneous strain is examined in detail. It is shown, for instance, that during deformation planes and lines develop preferred orientations reflecting the symmetry of the deformation, and pre-existing folds rotate bodily in space and either open or close or both during the deformation. In order to draw further conclusions of general interest the condition of homogeneous strain is relaxed in order to consider the deformation of layered rocks with competence differences between the layers. It is argued that during the deformation of such rocks folds and boudinage form parallel to the principal directions of strain in the layers. The conditions that determine whether folds or boudinage are formed are examined and the subsequent development of the structures after their generation is followed. These methods of analysis are used in the construction of models of superimposed fold-systems which are shown to be similar to some recently described field examples.

It is argued that the tectonic axial cross must be replaced by the deformation ellipsoid. The structures treated in the paper are usually derived by shear-plane hypotheses, but it is considered that such hypotheses are superfluous.

I. INTRODUCTION

IT is the task of the structural geologist to determine, from a study of structures in rocks, the stress-system that gave rise to those structures. This can only be done by first determining the strain or deformation the rocks have suffered. Yet there is a tendency in structural geology to make assumptions about the nature of this strain. In particular it is often implicitly assumed, on insufficient evidence, that strain is essentially two-dimensional.

The normal procedure in structural geology is to deduce the strain in an area from the structures occurring there. In this paper an attempt is make to work in the reverse direction. Strain systems are postulated and the structures to be expected are deduced. This is done for a large number of strain systems in order to find rules, not involving the assumption of two-dimensional strain, for the interpretation of patterns of structures in terms of causal strain. The type of deformation considered is flow, either plastic or viscous, beyond the elastic limit of the rocks and of the type commonly called irrotational finite homogeneous strain (Jaeger 1956, p. 23). The deformation is considered to take place at constant volume. The rocks deformed may be structurally isotropic or anisotropic, but in the first part of the paper are assumed to be homogeneous. Later it is argued that this assumption can be dropped in certain situations; conclusions of more general interest are then drawn. The study is restricted to irrotational strain and to deformation at constant volume only in order to limit the length of the paper. The methods employed may be extended to cover both.

Finite homogeneous strain is conveniently represented by the deformation ellipsoid, that is the ellipsoid resulting from the deformation of an originally spherical portion of the rock being deformed. The discussion that follows is largely concerned with the properties of the deformation ellipsoid and the interrelationships of all the possible shapes it may assume. All possible deformation ellipsoids will be considered in order to cover all possible states of homogeneous strain.

It is sometimes possible from a study of the rocks to determine directly the shape and orientation of the deformation ellipsoid for those rocks (Cloos 1947; Flinn 1956; Voll 1960, pp. 536–42). In other instances an estimate is possible (Flinn 1958, p. 124); the purpose of this paper is to increase the number of instances in which such an estimate can be made.

II. The Geometry of the Deformation Ellipsoid

The most convenient method of studying geometrical relationships among ellipsoids is to plot them, according to the ratios of their axes, on rectangular co-ordinates (Flinn 1956, fig. 4, p. 487). The major axes of the deformation ellipsoid will be designated $Z > Y > X$ in this paper in order to conform to the practice of optical mineralogists in describing the optical indicatrix (Wahlstrom 1951, p. 143), and so avoid confusion when some of the methods of optical mineralogy are applied to the deformation ellipsoid. The ratios $a = Z/Y$ and $b = Y/X$ give the most convenient plot, which will be called the *deformation plot* (Fig. 1). Both ordinate and abscissa have an origin of unity and the same scale. Each possible ellipsoid is represented by a single point, and each point on the plot represents a possible ellipsoid.

When a rock deforms, an original sphere within it changes *pro-*

Fig. 1.—(opposite) Deformation plot. $Z > Y > X$-deformation ellipsoid axes. V_z is the angle between Z and the pole to a circular section.

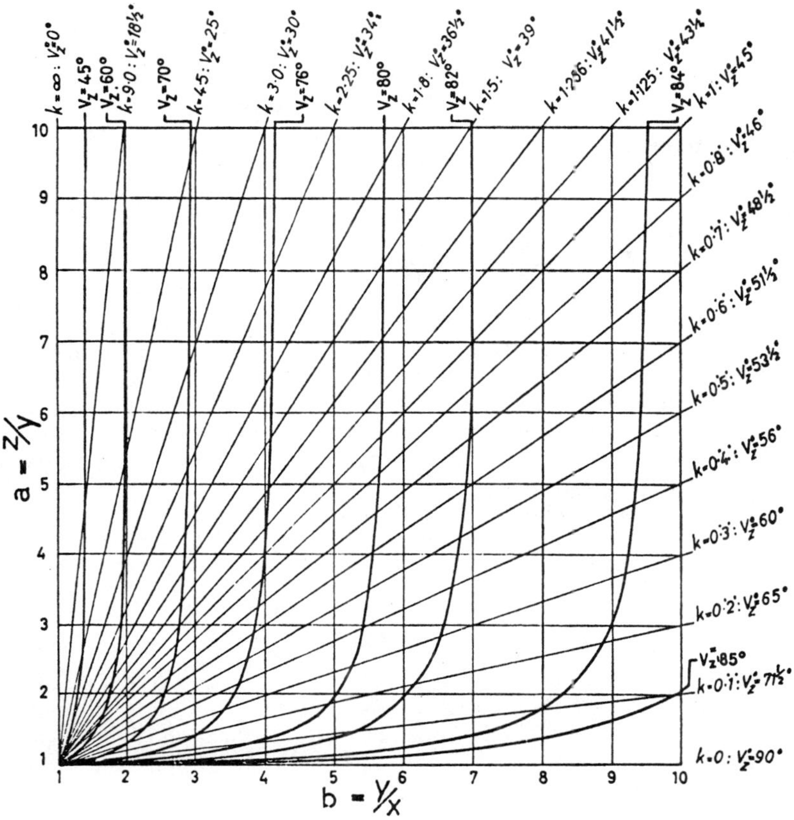
a = Z/Y
b = Y/X
k = ∞ : V°z = 0°
Vz = 45°
Vz = 60°
k = 9·0 : V°z = 18½°
Vz = 70°
k = 4·5 : V°z = 25°
k = 3·0 : V°z = 30°
Vz = 76°
k = 2·25 : V°z = 34°
Vz = 80°
k = 1·8 : V°z = 36½°
Vz = 82°
k = 1·5 : V°z = 39°
k = 1·286 : V°z = 41½°
k = 1·125 : V°z = 43½°
Vz = 84°
k = 1 : V°z = 45°
k = 0·8 : V°z = 46°
k = 0·7 : V°z = 48½°
k = 0·6 : V°z = 51½°
k = 0·5 : V°z = 53½°
k = 0·4 : V°z = 56°
k = 0·3 : V°z = 60°
k = 0·2 : V°z = 65°
Vz = 85°
k = 0·1 : V°z = 71½°
k = 0 : V°z = 90°

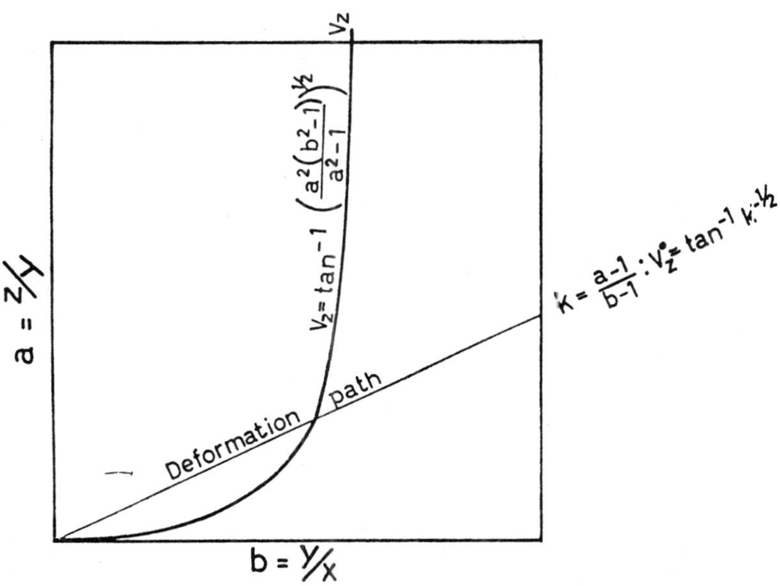
a = Z/Y
b = Y/X
Vz
Vz = tan⁻¹ (a²(b²−1)/(a²−1))^½
Deformation path
k = (a−1)/(b−1) : V°z = tan⁻¹ k^−½

gressively through a series of ellipsoids varying continuously in shape until the deformation ceases. Such deformation will be called *progressive deformation*. The locus of the projections of these related ellipsoids on the deformation plot will be called the *deformation path*. Thus any given stress-system is related, not to a single deformation ellipsoid, but to a series of ellipsoids or, alternatively, to a deformation path. If the three principal stresses do not vary relatively to one another during deformation and the stress-strain relationship is linear (e.g. viscous flow), the deformation path will be a straight line radiating from the origin ($a = b = 1$) of the deformation plot. Curved deformation paths are to be expected when the stress-strain relationship is non-linear (as, for example, in plastic flow), and when the deformation history is considered for longer than a single episode. It is convenient to restrict the discussion to straight deformation paths. This restriction is made solely to limit the length of the paper; curved deformation paths can be treated by the methods presented below, but the constructions and calculations are more complicated.

In effect, instead of considering all possible deformation ellipsoids, we now consider all possible deformation paths. The deformation path will be called k, where

$$k = \frac{a-1}{b-1} = \frac{X(Z-Y)}{Y(Y-X)}.$$

for straight deformation paths. From this it is clear that $0 < k < \infty$. $k = 0$ represents the path connecting all oblate (or, in mineralogical terms, negative) uniaxial ellipsoids, and $k = \infty$ all prolate (or positive) uniaxial ellipsoids; $k = 1$ (the diagonal of the plot) connects all ellipsoids in which the intermediate axis remains equal to the diameter of the original sphere. Such ellipsoids can be produced both by simple shear and by pure shear and have been much studied by structural geologists for this reason. In this paper deformation along the path $k = 1$ is often used for examples, because this path lies along the boundary between constriction-type deformation ($1 < k < \infty$) and flattening-type deformation ($0 < k < 1$).

The angle 2V in the deformation ellipsoid. An important property of an ellipsoid is the angle contained between its two circular sections. This angle, always measured over the X axis, will be called $2V_z$ in conformity with mineralogical practice. It is convenient to mark lines of equal V_z on the deformation plot. The values can be computed from the standard formula for V_z (Wahlstrom 1951, p. 145):

$$\tan^2 V_z = \frac{1/X^2 - 1/Y^2}{1/Y^2 - 1/Z^2}.$$

However, a more convenient relationship for our purpose is obtained if a and b are substituted for Z/Y and Y/X respectively, giving

$$\tan^2 V_z = \frac{a^2(b^2-1)}{a^2-1}. \qquad (1)$$

As $b^2 < \tan^2 V_z + 1$, all curves of V_z bend asymptotically towards the ordinate direction. This may be confirmed from the deformation plot (Fig. 1), on which are plotted curves of equal V_z. Examination of the plot also shows that for most possible ellipsoids the angle $2V_z$ is very large. In fact, positive ellipsoids occur only very close to the ordinate.

If k is substituted for $(a - 1)/(b - 1)$ in (1) we have

$$\tan^2 V_z = \frac{a^2(a + 2k - 1)}{k^2(a + 1)}, \tag{2}$$

which gives the variation in V_z along any straight deformation path. When $k = 1$ (i.e. $a = b$) $\tan V_z = a = b$ (equivalent to equation (10), Ramberg 1959, p. 107, and to equation (9), Jaeger 1956, p. 34); also $\tan^2 V_z$ will be at a minimum when a is at a minimum. The minimum value of a is 1, and when $a = 1$, $\tan^2 V^0_z = 1/k$ (V^0_z being the initial value of V_z). Therefore, for a given deformation path k, V_z has a minimum value of $\tan^{-1} k^{-1/2}$ at the origin and increases progressively with increasing deformation according to equation (2). Initial values of V_z (V^0_z) for various values of k have been computed and are given in Fig. 1.

(ii) *Principal directions in the deformation ellipsoid.* It has been assumed that the rock which is deformed is structurally homogeneous, but it may nevertheless be layered, folded, and lineated, etc., as indicated by the presence of a pattern of structurally inactive bands, surfaces, or lines formed, for instance, by non-uniform colour distribution. Elements of such surfaces small enough to be considered planar, moved parallel to themselves to the centre of the ellipsoid and extended, will cut its surface along ellipses that in special cases are circles. The major and minor axes of these ellipses are the directions of maximum compression and/or elongation in the planes. Were it not for the assumption that the rocks are structurally homogeneous, these axes would be directions of possible boudinage or fold-axes. Geometrically they correspond to the extinction directions of mineralogy, and in conformity with mineralogical practice they will be called *principal directions*. The best method of determining the position of principal directions is the Fresnel construction (Burri 1950, pp. 53 and 286).

A fold already present in the rock is represented in the deformation ellipsoid by a bundle of planes intersecting in a common line, the fold-axis (β). The loci on the projection sphere of principal directions for such a bundle cutting the ellipsoid at any angle are in general three curves. In mineralogy these curves are called extinction curves, but here they will be called *principal direction curves.*

Extinction curves have been investigated in some detail in recent years by Joel & Garaycochea (1957, pp. 399–406). There are three curves: a large equatorial curve, which in general is not a great circle, and two smaller polar curves, which in general are not small circles. The equatorial curve intersects the circular sections of the ellipsoid at Y and contains either Z or X. The polar curves contain the axis of the bundle of planes (the A_4 rotation axis of the universal stage in optical mineralogy) and either X or Z. If the equatorial curve is a great circle, either the rotation axis is parallel to an ellipsoid axis, or it lies in a circular section, or the ellipsoid is uniaxial; in these cases the polar curves are small circles. If the curves have a plane of symmetry the rotation axis lies in a symmetry plane of the ellipsoid.

(iii) *Standardized volume for the deformation ellipsoid.* It will occasionally be necessary to determine the actual lengths of principal directions in the ellipsoid. This can be done quite easily from the equation for the ellipsoid, but the value obtained is of restricted use unless the volume of the ellipsoid is standardized. It is most convenient to use an ellipsoid of volume $4/3\pi$ whose axes, expressed in terms of a and b, become

$$Z = a^{2/3}b^{1/3}, \quad Y = a^{-1/3}b^{1/3}, \quad X = a^{-1/3}b^{-2/3}. \tag{3}$$

In the past it has been usual to report the shapes of deformed objects, such as pebbles, in the form $Z/X : Y/X : X/X$ (e.g. Flinn 1956). This method has little to recommend it. The dimensions should each be divided by $(ZYX)^{1/3}$ so that the pebbles are all reduced to a volume of $4/3\pi$ and an original diameter of unity. A comparison of the recalculated Y with 1 then shows at a glance whether k is greater than, equal to, or less than unity, that is, whether the deformation involves flattening or constriction.

(iv) *The length of lines in any direction in the standardized deformation ellipsoid.* The equation for the ellipsoid expressed in polar coordinates (Burri 1950, p. 43) is

$$\frac{\cos^2 \psi_1}{X^2} + \frac{\cos^2 \psi_2}{Y^2} + \frac{\cos^2 \psi_3}{Z^2} = \frac{1}{n^2}, \tag{4}$$

where $Z > Y > X$ are the lengths of the symmetry axes of the ellipsoid, and n is the length of the line through the centre which lies at angle ψ_1 to X, ψ_2 to Y, and ψ_3 to Z. Substituting the axial lengths of the standard deformation ellipsoid (3) and reforming the equation gives

$$n^2 = \frac{a^{-2/3}b^{2/3}}{a^{-2}\cos^2 \psi_3 + \cos^2 \psi_2 + b^2 \cos^2 \psi_1}. \tag{5}$$

In solving this equation it occasionally is convenient to make use of the relationship (Burri 1950, p. 58):

$$\cos^2 \psi_1 + \cos^2 \psi_2 + \cos^2 \psi_3 = 1 \tag{6}$$

(v) *The line of no finite deformation.* Although equation (5) is convenient for calculating the actual lengths of individual directions in the ellipsoid, it is not convenient for quickly determining the state of elongation or shortening of many different directions. This is better done by determining which directions have their original lengths, by computing the line of intersection of the deformation ellipsoid with its parent undeformed sphere. This line of intersection will be called the *line of no finite deformation.* It is not independent of volume change. For $k = 0$ and $k = \infty$ the line is a small circle round X and Z respectively; for $k = 1$ it is two intersecting great circles, the circular sections of the ellipsoid (Ramberg 1959, p. 107). For all ellipsoids on other deformation paths the line is neither a great circle nor a small circle.

The equation for the line may be derived from (5) by putting $n = 1$:

$$b^2 \cos^2 \psi_1 + \cos^2 \psi_2 + a^{-2} \cos^2 \psi_3 = a^{-2/3}b^{2/3}. \tag{7}$$

This equation is for constant-volume deformation. If the volume changes during deformation, n is made equal to the diameter of a sphere of the appropriate final volume. In practice, if the points of intersection of the line with two symmetry planes of the ellipsoid are known, it is often possible to sketch in the rest of the line sufficiently accurately. With the help of equation (6) it is possible to derive from (7) three equations for the intersection of the line of no finite deformation with the three symmetry planes of the deformation ellipsoid:

$$Z\text{–}Y \text{ plane}; \ 1 < k < \infty, \ \cos^2 \psi_3 = \frac{a^{-2/3}b^{2/3} - 1}{a^{-2} - 1}, \tag{8a}$$

$$Y\text{–}X \text{ plane}; \ 0 < k < 1, \ \cos^2 \psi_2 = \frac{a^{-2/3}b^{2/3} - b^2}{1 - b^2}, \tag{8b}$$

$$Z\text{–}X \text{ plane}; \ 0 < k < \infty, \ \cos^2 \psi_3 = \frac{a^{-2/3}b^{2/3} - b^2}{a^{-2} - b^2}. \tag{8c}$$

These have been solved for integer values of a and b up to 10, and the results are presented in Table I (overleaf).

(vi) *Line of no infinitesimal deformation.* For determining the state of elongation and shortening of various directions in a deformation ellipsoid, the *line of no infinitesimal deformation* is as important as the line of no finite deformation (Ramberg 1959). It is the line of intersection of an infinitesimally deformed ellipsoid with a sphere of the same volume. It is thus independent of volume change during deformation. This fact may be used in deriving a general equation for the line. Ramberg's (1959) equations are of use only for determining intersections of the line with symmetry planes of the ellipsoids corresponding to $k = 0$, 1, and ∞. Equation (4) may be re-written to describe an infinitesimally deformed ellipsoid:

$$\frac{\cos^2 \psi_1}{(n + dX)^2} + \frac{\cos^2 \psi_2}{(n + dY)^2} + \frac{\cos^2 \psi_3}{(n + dZ)^2} = \frac{1}{n^2}.$$

This can be reduced with the aid of equation (6) and the usual approximations to

$$dX \cos^2 \psi_1 + dY \cos^2 \psi_2 + dZ \cos^2 \psi_3 = 0, \tag{9}$$

Since the volume is constant,

$$(n + dX)(n + dY)(n + dZ) = n^3,$$

which reduces to

$$dX + dY + dZ = 0. \tag{10}$$

Table I.—Line of No Finite Deformation.

	$b = 1$			$b = 2$			$b = 3$			$b = 4$			$b = 5$		
a	x	y	z	x	y	z	x	y	z	x	y	z	x	y	z
1	—	—	—	26½	—	26½	21½	—	21½	18½	—	18½	16½	—	16½
2	45½	45½	—	26	90	0	20½	—	11½	17	—	11½	15	—	11
3	40	40	—	24	59	—	19½	90	0	15½	—	7	13	—	7½
4	36½	36½	—	22½	51	—	17	63	—	14	90	0	12	—	4½
5	34	34	—	21	46½	—	16	56½	—	13	67½	—	11½	90	0
6	32	32	—	20	43	—	15	52	—	12½	60½	—	11	70	—
7	30½	30½	—	19	40½	—	14½	48½	—	12	55½	—	10	63	—
8	29	29	—	18	38½	—	14	45½	—	11½	52	—	9½	58½	—
9	28	28	—	17½	37	—	13½	43½	—	11	49½	—	9½	55	—
10	27	27	—	17	35½	—	13	42	—	10	47	—	9	52½	—

	$b = 6$			$b = 7$			$b = 8$			$b = 9$			$b = 10$		
a	x	y	z	x	y	z	x	y	z	x	y	z	x	y	z
1	15	—	15	13½	—	13½	12½	—	12½	11½	—	11½	11	—	11
2	13	—	10	12	—	9½	11	—	9	10	—	8½	9½	—	8
3	11½	—	7½	10½	—	7	10	—	7	9	—	6½	8½	—	6½
4	10½	—	5½	9½	—	5½	9	—	5½	8½	—	5½	7½	—	5½
5	10	—	3½	9	—	4½	8½	—	4½	7½	—	4½	7	—	4½
6	9½	90	0	8½	—	3	8	—	3½	7	—	4	6½	—	4
7	9	71½	—	8	90	0	7½	—	2	7	—	3	6½	—	3
8	8½	65	—	8	73	—	7	90	—	6½	—	2	6	—	2½
9	8½	60½	—	7½	67	—	7	74	—	6½	90	0	6	—	1½
10	8	57½	—	7½	62½	—	6½	68	—	6½	75	—	6	90	0

$a = Z/Y$ and $b = Y/X$, where $Z > Y > X$ are major axes of the deformation ellipsoid.

$x = \psi_3$ in Z–X plane—see equation (8c)
$y = \psi_3$ in Z–Y plane—see equation (8a)
$z = \psi_2$ in Y–X plane—see equation (8b)

Combining (9) and (10) we obtain

$$\frac{dY}{dX} = \frac{(\cos^2 \psi_1 - \cos^2 \psi_3)}{(\cos^2 \psi_3 - \cos^2 \psi_2)}. \tag{11}$$

Now

$$k = \frac{a - 1}{b - 1} = \frac{Z/Y - 1}{Y/X - 1}.$$

This may be rewritten

$$k = \frac{\dfrac{n + dZ}{n + dY} - 1}{\dfrac{n + dY}{n + dX} - 1},$$

which reduces to

$$k = \frac{dZ - dY}{dY - dX}. \tag{12}$$

Substituting (10) in (12 we have

$$k = \frac{dX + 2dY}{dX - dY}$$

or

$$\frac{dY}{dX} = \frac{k - 1}{k + 2}. \tag{13}$$

From (11) and (13)

$$\frac{k - 1}{k + 2} = \frac{(\cos^2 \psi_1 - \cos^2 \psi_3)}{(\cos^2 \psi_3 - \cos^2 \psi_2)}. \tag{14}$$

Substitution of (6) in (14) gives

$$\cos^2 \psi_1 = \frac{3k \cos^2 \psi_3 - k + 1}{3}, \tag{15}$$

from which the complete lines of no infinitesimal deformation for all ellipsoids may be calculated. The line depends entirely on k; it does not change as the ellipsoid changes shape along a straight deformation path. From equations (15) and (6), if two values of $\cos^2 \psi$ are each equal to 1/3, the third must also equal 1/3. All lines of no infinitesimal deformation therefore pass through the point $\psi_1 = \psi_2 = \psi_3 = 54° 44'$. As with the line of no finite deformation, it is often sufficient in practice to calculate the points of intersection of the line with symmetry planes of the ellipsoid. Equations (15) and (6) give:

$$Z\text{–}Y \text{ plane};\ 1 < k < \infty,\ \cos^2 \psi_3 = \frac{k - 1}{3k}, \tag{16a}$$

$$Y\text{–}X \text{ plane};\ 0 < k < 1,\ \cos^2 \psi_1 = \frac{k + 2}{3}, \tag{16b}$$

$$Z\text{–}X \text{ plane};\ 0 < k < \infty,\ \cos^2 \psi_2 = \frac{k + 2}{3(k + 1)}. \tag{16c}$$

Equations (16a, b, and c) have been solved for various values of k; the results are presented in Table II (overleaf). Ramberg's results (Ramberg 1959, pp. 105, 118, 120) are reproduced.

TABLE II.—LINE OF NO INFINITESIMAL DEFORMATION.

k	x	y	z
∞	54° 44′	54° 44′	—
9·0	52° 44′	57° 3′	—
4·5	51° 7′	59° 23′	—
3·0	49° 46′	61° 52′	—
2·25	48° 38′	64° 31′	—
1·8	47° 44′	67° 22′	—
1·5	46° 55′	70° 32′	—
1·286	46° 9′	74° 12′	—
1·125	45° 34′	78° 52′	—
1·000	45° 00′	90° 00′	00° 00′
0··8·	44° 26′	—	11° 8′
0··7·	43° 51′	—	15° 48′
0··6·	43° 5′	—	19° 28′
0··5·	42° 16′	—	22° 38′
0··4·	41° 22′	—	25° 29′
0··3·	40° 14′	—	28° 8′
0··2·	38° 53′	—	30° 37′
0··1·	37° 16′	—	32° 57′
0·0	35° 16′	—	35° 16′

$x = \psi_3$ in Z–X plane—see equation (16c)
$y = \psi_3$ in Z–Y plane—see equation (16a)
$z = \psi_2$ in Y–X plane—see equation (16b)

$$k = \frac{X(Z-Y)}{Y(Y-X)}$$

where $Z > Y > X$ are the major axes of the deformation ellipsoid.

(vii) *Rotation of planes and directions during deformation.* During deformation along a straight deformation path the line of no infinitesimal deformation remains fixed, the line of no finite deformation changes shape and size so as to enclose a smaller area and, as shown by Ramberg (1959, p. 106), a structural direction rotates and may cross both lines. The paths traced out by the poles of structural directions or planes on the projection net during deformation will be called *structural movement paths*. The equations for these paths can be derived quite easily. They are independent of volume change.

Consider first the rotation of a plane. Let the intersection point of the plane and its normal lie at the centre of the ellipsoid. Let the normal to the plane be also the diagonal of a parallelepiped whose edges are parallel to the symmetry axes of the ellipsoid. The magnitude of the diagonal and, therefore, of the parallelepiped is fixed by its intersection with the surface of the ellipsoid. Consider deformation between two states indicated by superscripts (·) and (×), where state(·) is nearer the origin of the deformation plot than state (×). Let $d^{\cdot}e^{\cdot}f^{\cdot}$ and $d^{\times}e^{\times}f^{\times}$ be the dimensions of the two parallelepipeds (· and × respectively) measured along their edges so that $d^{\cdot}$ and $d^{\times}$ are parallel to X, $e^{\cdot}$ and $e^{\times}$ are parallel to Y, and $f^{\cdot}$ and $f^{\times}$ are parallel to Z. Then the following relationships hold :

$$\frac{d^{\cdot}}{X^{\cdot}} = \frac{d^{\times}}{X^{\times}} \quad (17a), \qquad \frac{e^{\cdot}}{Y^{\cdot}} = \frac{e^{\times}}{Y^{\times}} \quad (17b), \qquad \frac{f^{\cdot}}{Z^{\cdot}} = \frac{f^{\times}}{Z^{\times}} \quad (17c)$$

Now consider the direction angle ψ_1, i.e. the angle between X and the diagonal of the parallelepiped:

$$\tan \psi_1^{\cdot} = \frac{[(e^{\cdot})^2 + (f^{\cdot})^2]^{1/2}}{d^{\cdot}}, \tag{18a}$$

$$\tan \psi_1^{\times} = \frac{[(e^{\times})^2 + (f^{\times})^2]^{1/2}}{d^{\times}}. \tag{18b}$$

By substituting for $d^{\cdot}$ (17a) in (18a)

$$\tan \psi_1^{\cdot} = \frac{[(e^{\cdot})^2 + (f^{\cdot})^2]^{1/2} X^{\times}}{d^{\times} X^{\cdot}}. \tag{19}$$

From (19) and (18b)

$$\tan \psi_1^{\cdot} \cdot \frac{X^{\cdot}}{[(e^{\cdot})^2 + (f^{\cdot})^2]^{1/2} X^{\times}} = \frac{\tan \psi_1^{\times}}{[(e^{\times})^2 + (f^{\times})^2]^{1/2}}, \tag{20}$$

whence, with (17b) and (17c)

$$\tan \psi_1^{\times} = \tan \psi_1^{\cdot} \cdot \frac{X^{\cdot}}{X^{\times}} \cdot \frac{\left[\left(\frac{Y^{\times}}{Y^{\cdot}}\right)^2 \cos^2 \psi_2^{\cdot} + \left(\frac{Z^{\times}}{Z^{\cdot}}\right)^2 \cos^2 \psi_3^{\cdot}\right]^{1/2}}{(\cos^2 \psi_2^{\cdot} + \cos^2 \psi_3^{\cdot})^{1/2}}. \tag{21}$$

Finally, substituting $X = a^{1/3}b^{2/3}$, $Y = a^{1/3}b^{-1/3}$, $Z = a^{-2/3}b^{-1/3}$ (these are the axes of the reciprocal ellipsoid of the standardized deformation ellipsoid [equation (3)] since the pole of the plane is being moved) into equation (21):

$$\tan \psi_1^{\times} = \tan \psi_1^{\cdot} \cdot \frac{b^{\cdot}}{b^{\times}} \cdot \frac{\left[\left(\frac{a^{\cdot}}{a^{\times}} \cos \psi_3^{\cdot}\right)^2 + \cos^2 \psi_2^{\cdot}\right]^{1/2}}{(\cos^2 \psi_3^{\cdot} + \cos^2 \psi_2^{\cdot})^{1/2}}. \tag{22a}$$

By a similar method, but using ψ_2 and ψ_3 in (18) instead of ψ_1 we obtain

$$\tan \psi_2^{\times} = \tan \psi_2^{\cdot} \cdot \frac{\left[\left(\frac{a^{\cdot}}{a^{\times}} \cos \psi_3^{\cdot}\right)^2 + \left(\frac{b^{\times}}{b^{\cdot}} \cos \psi_1^{\cdot}\right)^2\right]^{1/2}}{(\cos^2 \psi_3^{\cdot} + \cos^2 \psi_1^{\cdot})^{1/2}}, \tag{22b}$$

$$\tan \psi_3^{\times} = \tan \psi_3^{\cdot} \cdot \frac{a^{\times}}{a^{\cdot}} \cdot \frac{\left[\left(\frac{b^{\times}}{b^{\cdot}} \cos \psi_1^{\cdot}\right)^2 + \cos^2 \psi_2^{\cdot}\right]^{1/2}}{(\cos^2 \psi_1^{\cdot} + \cos^2 \psi_2^{\cdot}}. \tag{22c}$$

The equations (22a, b, and c) give the movement paths of poles of planes during deformation. By a similar argument, but using the deformation ellipsoid instead of the reciprocal ellipsoid, a set of equations can be obtained defining the movement paths of poles of directions, such as lineations, during progressive deformation:

$$\tan \psi_1^{\times} = \tan \psi_1^{\cdot} \cdot \frac{b^{\times}}{b^{\cdot}} \cdot \frac{\left[\left(\frac{a^{\times}}{a^{\cdot}} \cos \psi_3^{\cdot}\right)^2 + \cos^2 \psi_2^{\cdot}\right]^{1/2}}{(\cos^2 \psi_3^{\cdot} + \cos^2 \psi_2^{\cdot})^{1/2}}, \tag{23a}$$

$$\tan \psi_2^{\times} = \tan \psi_2^{\cdot} \cdot \frac{\left[\left(\frac{a^{\times}}{a^{\cdot}} \cos \psi_3^{\cdot}\right)^2 + \left(\frac{b^{\cdot}}{b^{\times}} \cos \psi_1^{\cdot}\right)^2\right]^{1/2}}{(\cos^2 \psi_3^{\cdot} + \cos^2 \psi_1^{\cdot})^{1/2}}, \tag{23b}$$

$$\tan \psi_3^{\times} = \tan \psi_3^{\cdot} . \frac{a^{\cdot}}{a^{\times}} . \frac{\left[\left(\frac{b^{\cdot}}{b^{\times}} \cos \psi_1^{\cdot}\right)^2 + \cos^2 \psi_2^{\cdot}\right]^{1/2}}{(\cos^2 \psi_1^{\cdot} + \cos^2 \psi_2^{\cdot})^{1/2}}. \quad (23c)$$

By the same methods, but with the elimination of $d^{\cdot}e^{\cdot}f^{\cdot}$ instead of $d^{\times}e^{\times}f^{\times}$ after (20), sets of equations for the reversal of deformation may be obtained :

$$\tan \psi_1^{\cdot} = \tan \psi_1^{\times} . \frac{b^{\times}}{b^{\cdot}} . \frac{\left[\left(\frac{a^{\times}}{a^{\cdot}} \cos \psi_3^{\times}\right)^2 + \cos^2 \psi_2^{\times}\right]^{1/2}}{(\cos^2 \psi_3^{\times} + \cos^2 \psi_2^{\times})^{1/2}}, \quad (24a$$

$$\tan \psi_2^{\cdot} = \tan \psi_2^{\times} . \frac{\left[\left(\frac{b^{\cdot}}{b^{\times}} \cos \psi_1^{\times}\right)^2 + \left(\frac{a^{\times}}{a^{\cdot}} \cos \psi_3^{\times}\right)^2\right]^{1/2}}{(\cos^2 \psi_1^{\times} + \cos^2 \psi_3^{\times})^{1/2}}, \quad (24b)$$

$$\tan \psi_3^{\cdot} = \tan \psi_3^{\times} . \frac{a^{\cdot}}{a^{\times}} . \frac{\left[\left(\frac{b^{\cdot}}{b^{\times}} \cos \psi_1^{\times}\right)^2 + \cos^2 \psi_2^{\times}\right]^{1/2}}{(\cos^2 \psi_1^{\times} + \cos^2 \psi_2^{\times})^{1/2}}. \quad (24c)$$

Equations (24a, b, and c) reverse the movement of poles of planes along their movement paths ('unroll' folds) ; the following equations (25a, b, and c) reverse the movement of poles of directions :

$$\tan \psi_1^{\cdot} = \tan \psi_1^{\times} . \frac{b^{\cdot}}{b^{\times}} . \frac{\left[\left(\frac{a^{\cdot}}{a^{\times}} \cos \psi_3^{\times}\right)^2 + \cos^2 \psi_2^{\times}\right]^{1/2}}{(\cos^2 \psi_3^{\times} + \cos^2 \psi_2^{\times})^{1/2}}, \quad (25a)$$

$$\tan \psi_2^{\cdot} = \tan \psi_2^{\times} . \frac{\left[\left(\frac{a^{\cdot}}{a^{\times}} \cos \psi_3^{\times}\right)^2 + \left(\frac{b^{\times}}{b^{\cdot}} \cos \psi_1^{\times}\right)^2\right]^{1/2}}{(\cos^2 \psi_3^{\times} + \cos^2 \psi_1^{\times})^{1/2}}, \quad (25b)$$

$$\tan \psi_3^{\cdot} = \tan \psi_3^{\times} . \frac{a^{\times}}{a^{\cdot}} . \frac{\left[\left(\frac{b^{\times}}{b^{\cdot}} \cos \psi_1^{\times}\right)^2 + \cos^2 \psi_2^{\times}\right]^{1/2}}{(\cos^2 \psi_1^{\times} + \cos^2 \psi_2^{\times})^{1/2}}. \quad (25c)$$

A study of these equations reveals several of the more important properties of structural movement paths. From equation (22) it is clear that

$$\tan \psi_1^{\times} < \tan \psi_1^{\cdot},\ \tan \psi_3^{\times} \geqslant \tan \psi_3^{\cdot}, \text{ and } \tan \psi_2^{\times} \gtrless \tan \psi_2^{\cdot}.$$

In progressive deformation the pole of a plane therefore always moves away from the longest axis Z and towards the shortest axis X of the deformation ellipsoid, but not necessarily by the shortest route. On the other hand it can be seen from equations (23) that the poles of structural directions move away from X towards Z. The structural movement paths are identical for these two types of pole (cf. equations (23) and (24)) but they move in opposite directions along them.

If a pole lies in a symmetry plane of the ellipsoid one of its direction angles is 90°, and since $\tan 90° = \infty$ and $\cos 90° = 0$, one of the three equations defining its structural movement path falls out and the other two simplify greatly and provide two complementary angles. Poles lying in symmetry planes of the ellipsoid therefore remain in those planes during deformation and no pole can leave the quadrant of the ellipsoid in which it originally lay; there is thus no need to consider the signs of the trigonometrical functions in the computations.

Fig. 2 shows structural movement paths of poles of planes for $k = 0$, 1, and ∞, the three extreme types. Movement paths for $k = 1$ lie between those for $k = 0$ and $k = \infty$ and show maximum deviation from both. Movement paths for all other values of k lie between these; their general aspect can be visualized with the help of Fig. 2. For deformation paths other than $k = 0$ and $k = \infty$, each point inside the traces of the symmetry planes of the deformation ellipsoid is the origin of a unique structural movement path.

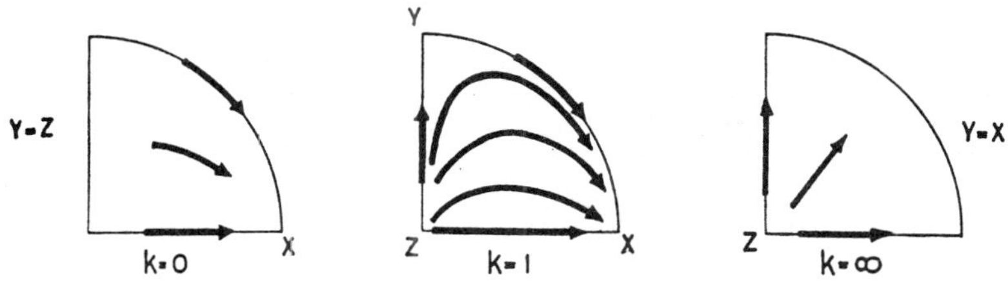

FIG. 2.—Movement paths of poles of planes for $k = 0$, 1, and ∞. For movement paths of poles of lines (structural directions) reverse the sense of movement.

It is never necessary to solve more than two of the three equations in a set for each computation of a point on a structural movement path, but the equations as they are presented are awkward to solve. Equations (22) and (23) may be simplified by putting $a^{\cdot}$ and $b^{\cdot}$ equal to 1, thus making each computation from the undeformed state. A further simplification is however possible. This depends partly on the fact mentioned above (p. 396) that a pole lying in a symmetry plane remains in the symmetry plane during deformation, and partly on a property of the equations that has not yet been mentioned; that the three equations in each group give a unique solution for each computation because the planes rotate without folding (a well-known property of finite homogeneous strain).

The simplified method of solving the equations is as follows: on the stereo-plot (Fig. 3) project the pole of the line to be deformed ($p^{\cdot}$) on to two of the symmetry planes of the ellipsoid by drawing through the pole two great circles from the poles of the two symmetry planes to cut the traces of these planes at $m^{\cdot}$ and $n^{\cdot}$. The projections ($m^{\cdot}$ and $n^{\cdot}$) of the pole on the symmetry planes are then moved by the appropriate simplified formulae (a simple multiplication or division of the tangent of an angle for each projection of the pole—see Fig. 3) just as if they were actual poles, to $m^{\times}$ and $n^{\times}$. Then a reversal of the projection procedure gives the new position of the pole being deformed, $p^{\times}$. The justification for this procedure lies in the fact that the structural direction to be rotated may be represented by the line of intersection of any two planes. Here planes containing a symmetry axis of the deformation ellipsoid are chosen so that one of the direction angles is equal to 90°, its cosine is equal to zero, and a great simplification of the equations results. The intersection of the two planes after deformation gives the position of the rotated structural direction.

Both the constructions and the computations are simpler than those involved in the direct rotation of the pole. Furthermore, if the deformation of a series of poles lying on a great circle has to be carried out, the pole of this great circle is first moved to its new position by the method given immediately above. Then, since all the poles will still lie on it after deformation, it is only necessary to make one further simple calculation for each pole not lying in a symmetry plane and none for those that do.

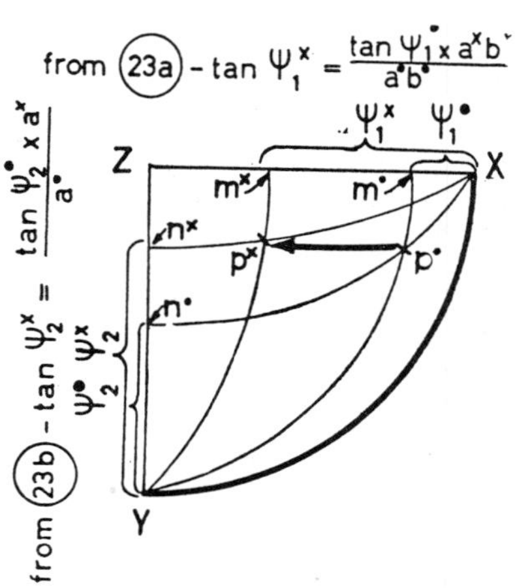

FIG. 3.—Computation of the rotation of a pole of a line (structural direction) during deformation from ($a^{\cdot}$, $b^{\cdot}$) to ($a^{\times}$, $b^{\times}$), where $a = Z/Y$ and $b = Y/X$. The computation of the rotation of a pole of a plane is carried out in a similar manner but with equations simplified from (22a) and (22b).

(viii) *Discussion.* The equations presented above are of two types. Equations (22), (23), (24), and (25) describe the rotation in three dimensions during homogeneous strain of structural lines and planes, that is of physically recognizable or material lines and planes. Equations (15) and (7) define lines that have no material existence and are not physically recognizable in the rock. The material lines and planes in general rotate through the non-material lines during the course of the deformation, although the non-material lines may also be changing in attitude and form. Failure to recognize that the material and non-material structures rotate at different rates has in the past led to confusion in the literature. For example, Jaeger (1956, p. 30, equation [37]) gives an equation which purports to give the initial position of the line of no finite deformation (a non-material line) for two-dimensional finite homogeneous strain ($k = 1$), whereas in fact it gives the initial position of the material line which at the end of the deformation happened to coincide with the non-material line of no finite deformation. The initial position of the line of no finite deformation is, of course, on the line of no infinitesimal deformation. Likewise Jaeger (1956, p. 38 and p. 29, equation [34]) gives equations which purport to give the final position of the non-material line of no infinitesimal deformation for two-dimensional finite homogeneous strain ($k = 1$). It is shown above that for $k = 1$ the line of no infinitesimal deformation is stationary at 45° to Z, and Jaeger's two equations really give the final position of the material line which, at the start of the deformation, happened to coincide with the non-material line of no infinitesimal deformation.

III. The Geometry of Deformation : Descriptive

(a) Homogeneous Rocks

The rotation of planes and lines in homogeneous rocks during finite homogeneous strain can be determined for all deformation paths with the aid of equations (22), (23), (24), and (25). Two subjects of particular interest to structural geologists that can be studied with the aid of these equations are (i) patterns of preferred orientation shown by originally randomly oriented planes and lines after deformation and (ii) changes caused in structures already present before deformation.

(i) *Deformation of random orientations of planes and lines.* The development of preferred orientation among randomly oriented planes during deformation can be studied with the help of the equations (22a, b, and c). It is easiest to consider $k = 0$ and $k = \infty$, since for $k = 0$ equation (22a) reduces to

$$\tan \psi_1{}^{\times} = \frac{\tan \psi_1{}^{\bullet} b^{\bullet}}{b^{\times}}$$

and for $k = \infty$ (22c) reduces to

$$\tan \psi_3{}^{\times} = \frac{\tan \psi_3{}^{\bullet} a^{\times}}{a^{\bullet}}$$

and all structural movement paths are great circles (symmetry planes) radiating from the unique symmetry axis of the ellipsoid. These simplified equations are equivalent to those used by Sander in a kinematic theory of *Regelung* according to grain form (Sander 1950, p. 112). Fig. 4 shows how rapidly the poles of planes become concentrated and large areas of the net become denuded during the deformation of an originally uniform distribution shown plotted along the vertical radial line. To the left of this line the progressive movement of the poles towards the X–Y plane during deformation of $k = \infty$ type is shown up to $a = 10$. To the right of the vertical line the migration of poles towards the X direction during deformation of $k = 0$ type is shown up to $b = 10$. At $b = 10$ and $k = 0$, half the poles have become concentrated within a radius of 10° from X, and at $b = 3$, 9 per cent are, giving, in terms of concentration intensity alone, a highly significant concentration (Flinn 1958). For $k = \infty$ the concentrations developed are not nearly so intense as the same number of poles are uniformly distributed in the X–Y

plane. At $a = 10$ the concentration level is only about 7 per cent, which considered alone is barely significant.

Since the structural movement paths are different for different deformation paths (Fig. 2) a different distribution or preferred orientation of poles will be developed on each deformation path. For an original uniform distribution of poles of planes and for $k = \infty$, a uniform girdle of poles develops with the girdle axis parallel to Z. For $k = 0$ a point concentration develops at X. For $k = 1$ a girdle develops with its axis parallel to Z and a decreasing concentration of poles from X towards Y. As k changes in value from ∞ to 0 the expected preferred orientation changes gradually

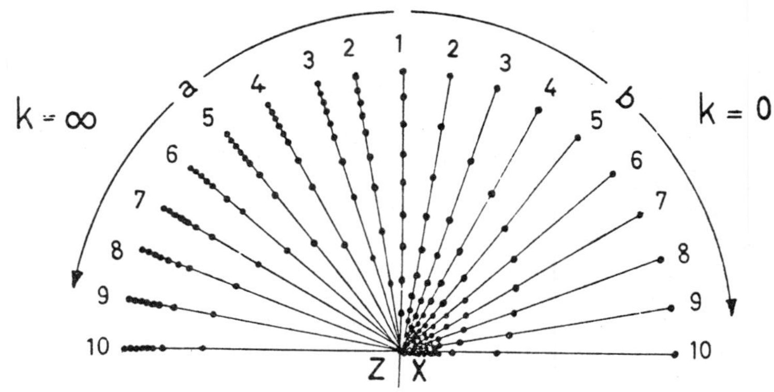

FIG. 4.—The migration of poles of planes caused by progressive deformation along deformation paths $k = 0$ and $k = \infty$. For explanation see text (p. 399).

from a uniform girdle in the X–Y plane to a point concentration at X. The variation in pattern of concentration of poles of planes predicted above is that shown by poles of (001) cleavage of mica and of [0001] axes of quartz in the Funzie conglomerate if the pebbles are considered as deformation ellipsoids (Flinn 1956). It is clear that the symmetries of the pole distributions are the same as the symmetries of the deformations that gave rise to them.

The preferred orientations developing among randomly oriented lines during finite homogeneous strain may be determined with the aid of equations (23a, b, and c). This will not be done here because the process is similar to that described above for planes. The results can easily be predicted from a knowledge of the preferred orientation of planes developed during the same deformation, since the poles of the lines and the poles of the planes move in opposite directions along the same movement paths.

(ii) *Deformation of patterns of planes and of lines.* If instead of being randomly oriented the planes or lines are arranged in a pattern, then after deformation the pattern will have changed. An interesting pattern to study is a fold. In order to study the effects of finite homogeneous strain on a fold, the fold has to be plotted on the stereo-net as a bundle of planes intersecting in a line, the fold-axis β.

During deformation the distribution of the planes in the bundle will in general change. If $k = \infty$, β is near the $X-Y$ plane, and the planes are uniformly distributed, deformation will give rise to a flattening of the bundle towards the plane containing the structural movement path of β (Fig. 12). This is due to the fact that the planes rotate about the fixed lines of their intersections with the $X-Y$ plane; in these circumstances the lines of intersection are closely spaced. If at the start of the deformation the planes are already closely grouped about the structural movement path, they will become even more so during deformation. If, on the other hand, at the start the planes are concentrated in a sub-parallel arrangement normal to the movement path, then their intersections with the $X-Y$ plane are widely spaced, and during deformation as the planes rotate about these intersections the bundle will open out to a more uniform distribution. A limit is set to both the closing-up and opening-out of the bundle by the distribution of the planes at the beginning of the deformation.

For $k = 0$ the behaviour of the bundle is the reverse of that for $k = \infty$, and the final distribution of the planes at infinite deformation, whatever their original distribution, is parallel to the $Z-Y$ plane of the deformation ellipsoid. If β lies near X at the start of the deformation and the planes are uniformly distributed about β then the deformation causes them to close up towards a position normal to the β structural movement path and parallel to the $Z-Y$ plane (Fig. 8). This occurs because they are rotating about the fixed lines of their intersection with the $Z-Y$ plane. If the planes are closed up sub-parallel to the structural movement path at the start of the deformation, then during the deformation they open out to a uniform distribution and then close up towards a position normal to the movement path.

The deformation of a bundle of planes for $0 < k < \infty$ may be predicted from the behaviour of bundles being deformed along paths corresponding to $k = 0$ and $k = \infty$ for the following reason. A study of the movement paths of poles of planes in Fig. 2 shows that for $k = 1$ the movement paths start at Z parallel to those for $k = \infty$, diverge more and more from them until half-way between X and Z, and then correspond more and more closely to the movement paths for $k = 0$, to which they finally become parallel at X. At $k = 1$ these movement paths show maximum deviation from those for $k = 0$ and $k = \infty$ at the great circle $Y–Z\,45°X$. As k increases, this line of maximum deviation moves towards X (at $k = \infty$ it coincides with the $X-Y$ plane), and as k decreases it moves towards Z (at $k = 0$ it coincides with the $Z-Y$ plane). It is clear, therefore, that a bundle of planes lying with β near X at the start of a deformation will change during the deformation first in the manner described above (p. 401) for such bundles deforming along $k = \infty$; when the β-axis has passed the line of maximum deviation the bundle will deform as described above (p. 401) for a similar bundle deforming along $k = 0$. The relative amounts of behaviour of $k = 0$ and $k = \infty$ type depend on the value of k, the position of β at the start of the deformation, and the amount of deformation.

For $k = 1$ with the β-axis starting near X and ending near Z, there are the following possibilities : (1) A bundle of uniformly distributed planes closes up twards parallelism with the β-axis structural movement path, then opens again to uniform distribution, and finally closes towards parallelism with the Z–Y plane. (2) A bundle initially closed up normal to the β-axis movement path opens out and then closes up again the same way towards parallelism with the Z–Y plane. (3) A bundle closed up parallel to the movement path closes up still more, then opens out to uniform distribution, and finally closes up towards parallelism with the Z–Y plane. A bundle with its β-axis starting near Y follows only the $k = 0$ type of behaviour (Figs. 13 and 14).

(b) Inhomogeneous Rocks

(i) *Discussion of folding.* Under the assumption of homogeneous deformation made at the beginning of the paper, no folds or boudinage can be generated. The equations (22a, b, and c) defining the movement paths of poles of planes give a single solution, so that the pole of a plane does not multiply itself as happens when folding takes place. All that can happen is the opening or further closing of a fold already in existence, and the rotation of planes and structures already in existence. Field-evidence makes it clear, however, that folds are generated during deformation, and as this seems to happen particularly in rocks that are clearly inhomogeneous, the assumption of homogeneous strain employed above must be re-examined.

According to the equations given above, if a layered rock is deformed with a component of compression parallel to the layering and if the layering has no structural significance, no folds will be generated, but folds already present will develop further at the same time as the layers thicken or thin in response to the compression. It is possible to construct folds of this type graphically (Fig. 15) ; they have the characteristics of similar (parallel) folds (Hills 1953, p. 84 ; de Sitter 1956, p. 182) as the planes defining the folds will be parallel after deformation if they were before deformation. This folding is the result of flow of the rock—original spheres have flowed into ellipsoids—so that the folds must be 'flow folds'. 'Flow folds', however, are usually said to result from internal movements that are not regularly orientated with respect to the stress-directions, which seems to imply turbulent flow (de Sitter 1956, p. 182).

The folds that are expected to develop during deformation of the type investigated above are thus similar folds. In order to develop during deformation they have to be already present in the rocks ; they cannot be generated by this type of deformation.

Ramberg (1959, 1960) has shown that where layers of greater and lesser competence alternate in a rock, compression parallel to the layering causes greater flow in the less competent than in the more competent layers. Friction at the interfaces may then cause the more competent layers to buckle. In this way folds are generated in layered rocks with competence differences between the layers. These folds will be generated as concentric folds (Hills 1953, p. 81) but once formed will develop further, partly as concentric folds and partly as similar folds. Which type of folding dominates depends on the competence difference between the layers. If the difference in competency is great a more competent layer folds in a concentric manner, forcing the less competent rock on either side to accommodate itself to it up to a distance of about half or three-quarters of the original wavelength (Ramberg 1960, p. 43). This appears to be the type of folding modelled by Ramberg (1959), for his ptygmatic folds are of concentric type showing very little or no change of thickness of layers as measured normal to their surfaces.

As the difference in competency between the layers decreases, increasing flow deformation takes place in the more competent layers. On the deformation plot, the projection point of the deformation ellipsoid for the more competent layers moves out from the origin along the deformation path towards the projection point of the deformation ellipsoid for the incompetent layers. The folding is a combination of concentric and similar types, the component of concentric folding decreasing as the difference in competency decreases. The original wavelength of the folds is related to both the difference in competence and to the thickness of the layers (Ramberg 1959, p. 148). Disharmonic folding may occur where the more competent layers are separated by more than a wavelength, as the influence of the more competent layers extends to only half a wavelength.

A study of the rotation of planes and lines during the deformation of layered rocks is complicated by the fact that rotation is only partly due to finite homogeneous strain. It is also due to the buckling or concentric folding of the layers, and this component cannot be determined by the equations presented here. No method of dealing with concentric folding is presented in this paper. Instead it is assumed that the competence differences between the layers, while being sufficient to allow the generation of new folds, are insufficient to modify those folds during their subsequent development.

(ii) *Principal directions.* When the lines of no finite deformation and no infinitesimal deformation for a particular deformation ellipsoid are plotted on an equal-area net (see Fig. 6), they divide the net into three different areas. All directions whose poles lie inside the line of no finite deformation have been elongated, while all directions lying outside it have been shortened. The line of no infinitesimal deformation divides the area outside the line of no finite deformation into two. Directions lying between the two lines have at first been shortened and later elongated but insufficiently to restore their original lengths. Directions outside the line of infinitesimal deformation have all been shortened. The three areas will be called the *area of elongation*, the *area of reduced shortening*, and the *area of shortening*.

During progressive deformation the poles of structural directions move along movement paths that bring them continuously nearer the longest axis of the ellipsoid. In general, movement paths cross both the lines of no finite and no infinitesimal deformation, but only in a direction resulting in elongation. Thus, in the course of a single deformation a direction may first suffer shortening until the line of no infinitesimal deformation is crossed, then elongation until at the line of no finite deformation it has regained its initial length, and finally absolute elongation (de Sitter 1958, p. 281 ; Ramberg 1959, p. 107). At any stage during deformation all directions lying in the area of reduced shortening have previously lain in the area of shortening, since at the start of the deformation the lines of no infinitesimal and no finite deformation coincided. Fig. 5 shows changes in lengths of structural directions of different original attitudes after deformation of different amounts along the deformation paths $k = 0$ and $k = \infty$; $k = 0$ gives the maximum shortening possible and $k = \infty$ the maximum elongation possible on any deformation path and for any value of k.

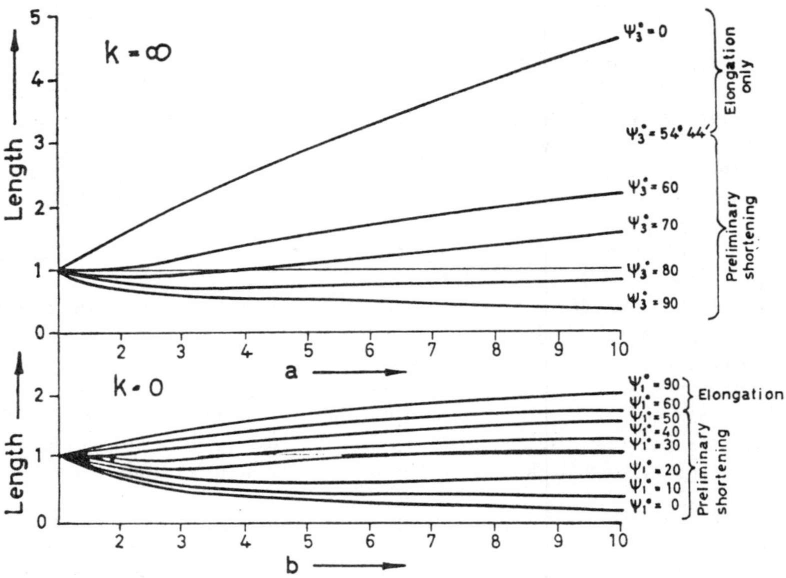

FIG. 5.—Shortening and elongation of lines of different original orientation during progressive deformation, for $k = 0$ and $k = \infty$. $\psi_3{}^0$ is the original angle between the line and Z, $\psi_1{}^0$ is the original angle between the line and X.

If one principal direction in a plane is a direction of shortening, then the other principal direction in that plane is a potential fold-axis. On the other hand, if one principal direction is a direction of extension then the other is a potential boudinage or pinch-and-swell axis (Ramberg 1955). Whether or not folds or boudinage develop at these principal directions depends partly on the properties of the rock and partly on factors which are analysed in the following pages.

(iii) *Principal directions and planes.* If the cyclographic projection of any plane is drawn on a plot of the lines of no infinitesimal and no finite deformation for a particular ellipsoid, it is possible to see at a glance the state of elongation or shortening of the principal directions in that plane. To fix the principal directions accurately the Fresnel construction is necessary, but to determine qualitatively the state of elongation or shortening of the principal directions it is necessary only to see which areas the trace of the plane crosses.

Systematic examination shows that if the trace of the plane lies entirely in the area of shortening (only possible for $1 < k < \infty$) both principal directions will be shortened, giving two simultaneous fold-axes at right-angles in the same plane. Such fold-systems—*double folds*—have the appearance of interfering wave-systems. Ramberg (1959, p. 130 and pl. 6, fig. B) has produced such folds experimentally for a plane lying normal to Z in a deformation ellipsoid of type $k = \infty$.

If the plane lies entirely in the area of elongation (only possible for $0 < k < 1$) both principal directions will be elongated, giving two simultaneous boudinage axes in the same plane at right-angles.

In general, a plane crosses more than one area and the X' principal direction is a direction of shortening, making the Z' principal direction a fold-axis, while the Z' principal direction is one of elongation, making the X' principal direction a boudinage axis ($X < X' < Y < Z' < Z$ in mineralogical nomenclature). When there is no change of length parallel to one principal direction the other is neither a fold-axis nor a boudinage axis and will be called a *direction of no structure*. The types of pairs of principal directions that can occur are shown in Table III.

TABLE III.

Value of k	Pairs of potential structures
$0 < k < 1$	Boudinage axis normal to boudinage axis (double boudinage axes) Boudinage axis normal to fold-axis Boudinage axis normal to no structure (rare)
$k = 1$	Boudinage axis normal to fold-axis Boudinage axis normal to no structure (rare) Fold-axis normal to no structure (rare) No structure normal to no structure (very rare)
$1 < k < \infty$	Fold-axis normal to fold-axis (double fold-axis) Fold-axis normal to boudinage-axis Fold-axis normal to no structure (rare)

When all possible structures are plotted on a net, it is seen (Fig. 6) that the distributions of principal directions shown in Table IV (p. 407) occur for the various deformation paths.

These facts and Fig. 6 agree with the conclusions drawn by Ramberg (1959) for the particular deformation paths $k = 0, 1, \infty$, and for specially orientated planes. Ramberg's (1959) work on ptygmatic folds and Fig. 6 make it tempting to conclude that by collecting a large amount of data from a ptygmatically folded area it would be possible to determine both the shape and orientation of the deformation ellipsoid for the deformation from the pattern of distribution of the plotted data. This is, however, improbable. Deformation must be considered as a progressive process. If the ptygmatic veins before folding occupied all possible attitudes, after even a weak deformation they would have been rotated into a high

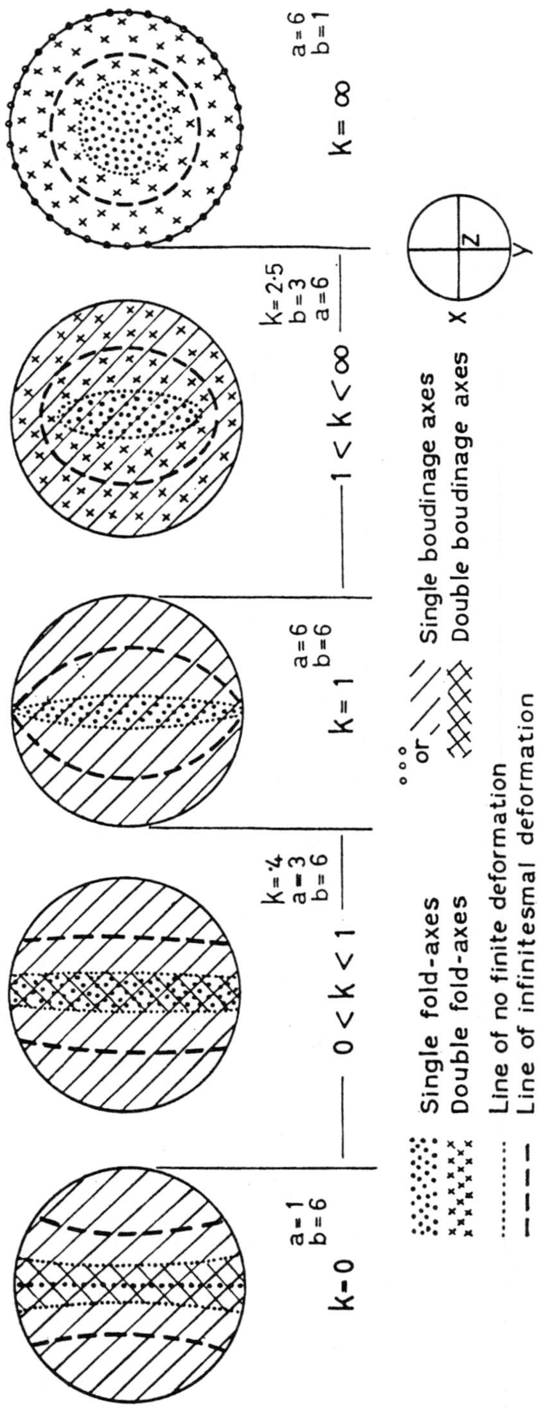

FIG. 6.—The distribution of possible fold-axes and boudinage axes for all values of k, shown in relation to the lines of no finite and no infinitesimal deformation.

TABLE IV.

Value of k	Areas of {shortening / reduced shortening}	Area of elongation
$k = 0$	Boudinage axes	Double boudinage axes Fold-axes in Z–Y plane
$0 < k < 1$	Boudinage axes	Double boudinage axes Fold-axes
$k = 1$	Boudinage axes	Fold-axes
$1 < k < \infty$	Boudinage axes Double fold-axes	Fold-axes
$k = \infty$	Boudinage axes in X–Y plane Double fold-axes	Fold-axes

degree of preferred orientation, so that field examination would produce data covering only a small part of the net. Nevertheless, the frequency of the folds and the distribution of their axes might give some indication of whether the value of k is greater or less than unity.

The behaviour and structural significance of principal directions during progressive deformation is most easily studied for deformation along deformation paths corresponding to $k = 0$ and $k = \infty$. Equations (22a, b, and c) and (23a, b, and c) show that poles of planes and of structural directions lying in symmetry planes must remain in those planes during deformation. In uniaxial ellipsoids all planes containing the unique symmetry axis are symmetry planes. During deformation along $k = 0$ and $k = \infty$ each plane therefore rotates about the fixed line of its intersection with the unique symmetry plane of the ellipsoid (for $k = 0$ the Z–Y plane and for $k = \infty$ the X–Y plane). The rotation axis is one of the two principal directions of the plane; the other principal direction rotates along the great circle whose pole is the rotation axis.

For $k = 0$ (Fig. 8) with a plane originally steeply inclined to Z–Y, the Z' principal direction that is also the axis of rotation starts as the axis of a fold, and during rotation of the plane changes to an axis of unfolding and then to an axis of boudinage. Folds formed on such axes will here be called *ephemeral folds* (p. 424–5). The other principal direction X' starts as a boudinage axis near X and rotates along a great circle towards the Z–Y plane, remaining a boudinage axis. For $k = \infty$ (Fig. 12) with a plane steeply inclined to Z, the axis of rotation is the X' principal direction and starts as a fold-axis, changing during progressive deformation to an axis of unfolding and then to a boudinage axis. The associated Z' principal direction starts as a fold-axis near the X–Y plane and continues as a fold-axis as it moves along a great circle whose pole is the X' principal direction towards the Z axis (Fig. 12).

(iv) *Principal directions and folds.* A fold already in existence at the commencement of a deformation may be treated as a bundle of planes. During deformation each plane of the bundle behaves individually exactly as described immediately above. Collectively the principal directions in these planes form three loci, the three principal direction curves, along which the lengths of the principal directions change systematically.

Fig. 8. For $k = 0$, the equatorial principal direction curve coincides with the trace of the Z–Y plane, and the two polar curves are small circles passing through X and the axis of the bundle, β. During deformation β moves along a great circle from X towards the Z–Y plane. At the same time the polar curves become progressively larger until at infinite deformation, with the β axis lying in the Z–Y plane, they each enclose a quarter of the surface area of the ellipsoid, and simultaneously contract again to small circles of radius nil at β, while the equatorial curve reforms in the plane normal to β. During their expansion the principal direction polar curves are loci of boudinage axes (X' principal directions) which tend to be concentrated on those parts of the polar curves nearest to the Z–Y plane. Each plane of the bundle rotates about its Z' ($= Z = Y$) principal direction (the line of intersection with the Z–Y plane) and each such principal direction passes through that part of the sequence—axis of folding, axis of unfolding, and boudinage axis (indicated in Figs. 8 to 14 by F, U–F, and B respectively)—required by the original attitude of the plane.

Fig. 12. For $k = \infty$, the X–Y plane is the plane of the equatorial principal direction curve and the polar curves pass through Z and the bundle axis. The Z polar principal direction curves start as large ones and decrease in size with progressive deformation until at infinite deformation they have a radius of nil. They are the loci of fold-axes (Z' principal directions) moving radially along great circles towards Z and tending to become increasingly concentrated on those parts of the curves nearest to Z. The equatorial principal direction curve is, as for $k = 0$, the locus of boudinage axes ($X' = X = Y$ principal direction) some of which may have passed through a stage of ephemeral folding.

Consideration of the rotation of planes and of the structural significance of principal directions for $0 < k < \infty$ is complicated by the fact that in general the axes of rotation of planes are fixed neither relative to the deformation ellipsoid nor relative to the planes themselves; the movement paths of principal directions and the movement paths of structures generated at those principal directions diverge, and neither principal directions nor structures remain stationary. (For $k = 0$ and $k = \infty$, one principal direction in each plane is always stationary, and the structural and principal direction movement paths of the other coincide.) The principal direction curves are neither great circles nor small circles except when β lies in either a circular section or a symmetry axis of the ellipsoid. Finally, complication is caused by the change-over from an X polar principal direction curve to a Z polar principal direction curve during deformation.

The traces of the circular sections divide the surface of the deformation ellipsoid into two pairs of diametrically opposing areas, one pair containing Z and another containing X. If β lies in the area

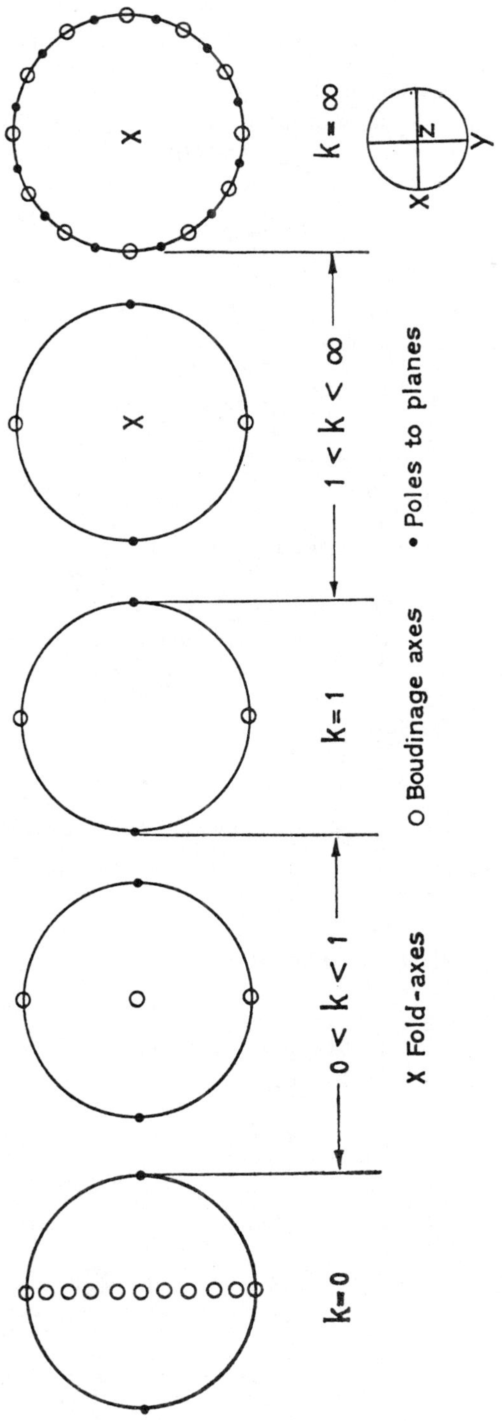

FIG. 7.—The distribution at infinite deformation of poles of planes and of fold-axes and boudinage axes, for all values of *k*.

containing Z then it lies on a Z polar principal direction curve, but if it lies in the area containing X then it lies on an X polar principal direction curve. Since $2V_z$ always increases with increasing deformation (equation [2]), the traces of the circular sections must approach the Z–Y plane, and the area containing Z always decreases. The bundle axis, β, also approaches Z (equation [23]). During deform-

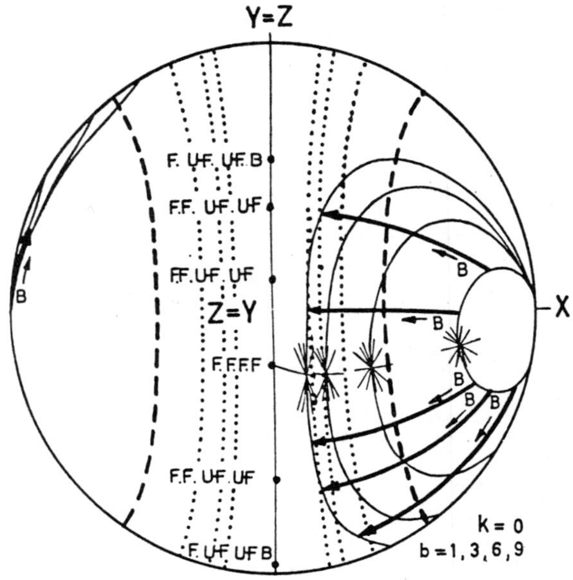

Fig. 8.

Figs. 8–14.—The course of deformation of earlier folds (bundles of planes) during deformation along various straight deformation paths (k), displayed by plotting several stages in each deformation. For each figure the stages are given in order as values of either a or b, e.g. a = 1, 3, 9.

- - - - -	line of no infinitesimal deformation ;
....	line of no finite deformation ;
————	principal direction curves ;
———	movement paths of principal directions ;
✱	bundle axis (fold-axis) with traces of planes ;

F	axis of folding	These structures are generated at the principal directions. Once a structure is formed it migrates along structural movement paths which, in general, diverge from the movement paths of principal directions. Where these symbols appear in a row at a stationary principal direction they are to be read from left to right and indicate the sequence of structures generated at that principal direction in each of the stages plotted.
U–F	axis of unfolding	
B	axis of boudinage	
O	axis of no structure	

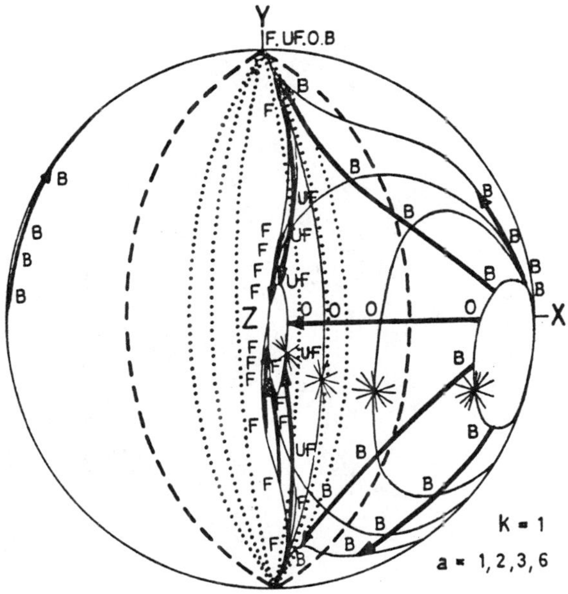

Fig. 9.

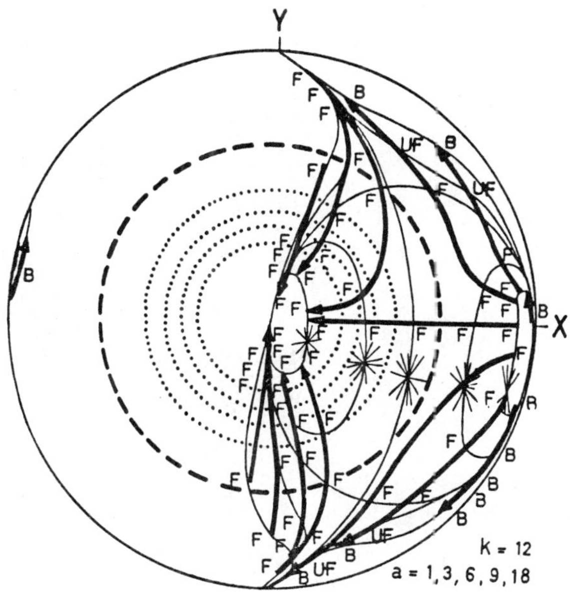

Fig. 10.

(for explanation see facing page)

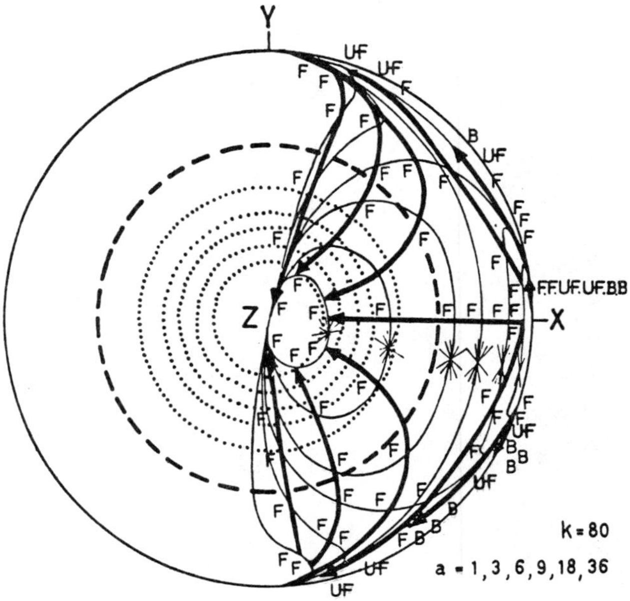

Fig. 11.

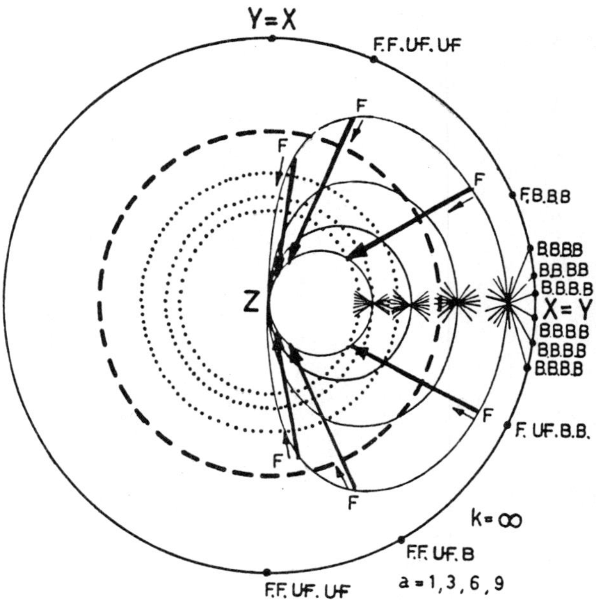

Fig. 12.

(*for explanation see p.* 410)

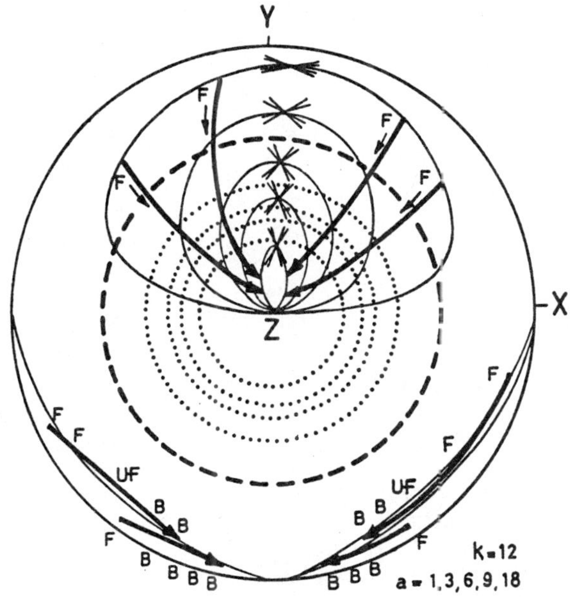

FIG. 13.

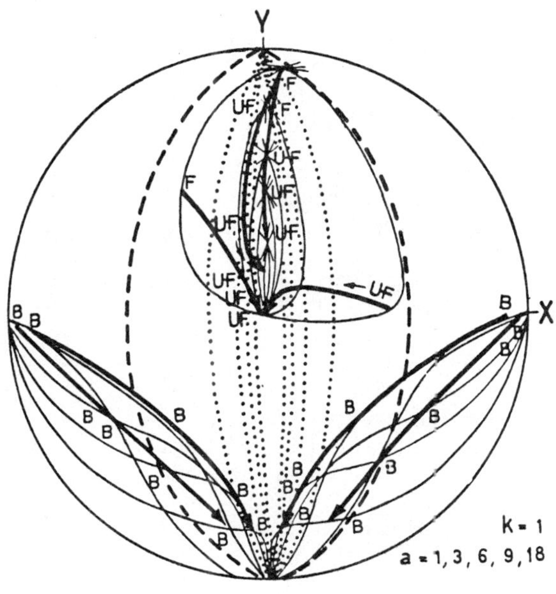

FIG. 14.

(*for explanation see p.* 410)

ation the direction angle between X and the projection of β on the Z–X plane (projected by the great circle from Y through β) changes as follows:

$$\tan^2 \psi_1{}^{\times} = \tan^2 \psi_1 \cdot a^2b^2 \qquad \text{from (23a)}$$

because $\psi_2 = 90°$ and $\cos 90° = 0$.

Since $k = (a - 1)/(b - 1)$ (p. 388) and $\tan^2 V^0{}_z = 1/k$ (p. 389), from equation (1)

$$\tan^2 V^{\times}{}_z = \tan^2 V^0{}_z \frac{a^2(b+1)}{(a+1)}$$

By definition, a and $b \geqslant 1$; therefore

$$a^2b^2 \geqslant \frac{a^2(b+1)}{a+1}.$$

During deformation, β therefore always approaches Z faster than the traces of the circular sections. With sufficient deformation a bundle axis originally lying on an X polar curve will cross the circular section and come to lie on a Z polar curve (unless it originally lay exactly 90° from Z). A change in the reverse direction, from a Z polar curve to an X polar curve, is not possible.

Over most of the deformation plot, $2V^0{}_z$ is so large and the circular sections are so near Z at the start of the deformation that no change-over is likely except in the case of extreme deformation or a bundle axis lying very near Z at the start of the deformation. A change-over is chiefly to be expected at high values of k (say $k > 9$), that is, during strong constriction.

Figs. 8 to 12 show the process of a change-over from an X polar principal direction curve to a Z polar principal direction curve and how the process varies with varying k towards the special cases of $k = 0$ and $k = \infty$, where no change-over is possible. They also show, together with Figs. 13 and 14, the shapes of movement paths of principal directions and the types of structures generated along them.

There are two types of principal direction movement path, which are associated with the two types of principal direction X' and Z'. The X' movement paths originate on the X polar principal direction curves and move towards Y, transferring to an X equatorial principal direction curve on the way. The Z' principal direction movement paths originate on a Z equatorial principal direction curve and move towards Z, transferring on the way to a Z polar principal direction curve. The starting and finishing points for a principal direction on its movement path are controlled by the original position of the plane, the amount of deformation, and the value of k. In Figs. 8 to 14 these three variables have been chosen in order best to illustrate the courses of the principal direction movement paths; the examples are not to be considered as typical deformations. It may be noted that neither X' nor Z' principal direction movement paths cross a symmetry plane of the ellipsoid, and only those coincident with symmetry planes can cross circular sections, in which special case they change from X' to Z' principal direction movement paths.

The nature of the structures generated along an X' or Z' principal direction movement path is determined by the state of elongation or shortening along the associated Z' or X' principal direction movement path respectively. The nature of the structures is most easily determined by superimposing lines of no finite and no infinitesimal deformation over the plot of the movement paths; this has been done in Figs. 8 to 14. It may be seen that along both X' and Z' principal direction movement paths the sequence *axis of folding–axis of unfolding–axis of boudinage* may occur as deformation increases. Since an axis of unfolding generated anywhere other than parallel to a previously formed axis of folding can only give rise to boudinage, many of the axes of unfolding in the diagrams are really boudinage axes.

For $0 < k < 1$, boudinage axes are generated along X' principal direction movement paths. As k increases above 1, fold-axes are generated along an increasing length of the X' principal direction movement paths starting from the X end. For $1 < k < \infty$ fold-axes are generated along Z' principal direction paths, but if $k < 1$ boudinage axes are generated instead of fold-axes along increasing lengths of the paths measured from Z.

For $0 < k < 1$ the X polar principal direction curve is the locus of generation of boudinage axes. At $k = 1$ the X' principal direction coinciding with the intersection of the polar principal direction curve and the Z–X symmetry plane is a direction of no structure because there is no change of dimension parallel to Y (Fig. 9). When $k > 1$ it becomes a fold-axis, and as k increases an increasing length of the polar principal direction curve on either side of the symmetry plane generates fold-axes. After the change-over, when the X polar principal direction curve has become an X equatorial principal direction curve, boudinage axes are generated along it instead of fold-axes. For $1 < k < \infty$ the Z polar principal direction curve is the locus of generation of fold-axes. At $k = 1$ the Z' principal direction coinciding with the intersection of the principal direction curve and the Z–X symmetry plane is a direction of no structure. If $k < 1$ this direction becomes a boudinage axis and as k decreases an increasing arc of the X polar principal direction curve on either side of the symmetry plane generates boudinage axes.

Whether or not a newly generated structure develops depends on the relation between its structural movement path and the principal direction movement path. For $k = 0$ and $k = \infty$ these two paths coincide in direction and distance of movement, so that the newly generated structures remain at the principal direction and continue to develop. In such cases the principal direction curves are also loci of actual structures as well as of principal directions. For $0 < k < \infty$ the two types of movement path diverge. To avoid overcrowding Figs. 8 to 14 structural movement paths have been omitted, but they can be visualized with the aid of Fig. 2.

At $k = 0$ the structural and principal direction movement paths are coincident and approach the nearest point on the Z–Y plane by the shortest route. As k increases above 0 the two movement path types diverge near the Z–Y plane owing to the fact that the X' principal direction movement paths turn towards Y and the structural movement path turns towards Z. As k increases further the divergence commences nearer and nearer to X, the origin of the

X' principal direction movement path. As a result, for $0 < k < 1$ boudinage is generated and developed along the part of the X' principal direction movement path nearest to X and then carried away by a divergent structural movement path towards Z. Although boudinage is continuously generated along the remainder of the X' principal direction movement path it is just as continuously removed by divergent structural movement paths and so has no chance to develop. As k increases above 1, fold-axes are generated and developed along the parts of the X' principal direction movement paths nearest to X, and owing to increasing divergence there is increasingly less chance for boudinage to develop.

At $k = \infty$, Z' principal direction movement paths and associated structural movement paths coincide. As k decreases they diverge, and this divergence takes place first at the beginning of the paths near Y. It is therefore the later-generated Z' structures (fold-axes or boudinage axes) that develop and the earlier ones (fold-axes) that never develop owing to continuous removal by divergent structural movement paths.

This tendency for the suppression of the development of folds in the early stages of deformation and of boudinage in the later stages operates in favour of fold development for $k > 1$ and boudinage development for $k < 1$.

Fig. 7 is an aid to visualizing the deformation. It shows the final destinations of fold-axes and boudinage axes for deformation along all deformation paths. All fold-axes and boudinage axes must lie in planes. At infinite deformation for all values of $k < \infty$ the poles of all planes are coincident with X and for $k = \infty$ they form a girdle (not necessarily uniform) in the X–Y plane. For $k \leqslant 1$ all fold-axes formed during the deformation are destined to become first axes of unfolding and then boudinage axes, since at infinite deformation no fold-axes remain, there being only boudinage axes parallel to Z and Y. (For $k = 0$ they are scattered, but not necessarily uniformly, in the Z–Y plane). For $k > 1$ fold-axes formed during the deformation and not unfolded are concentrated at Z, while boudinage axes are concentrated at Y (for $k = \infty$ they are scattered, but not necessarily uniformly, in the X–Y plane). Folds present before the deformation, if not unfolded, are isoclinally folded and rotated until the axes are parallel to Z.

IV. The Geometry of Deformation : Geological Interpretation

Of particular interest at the present time is the problem of superimposed folding, which for the purposes of this paper is best classified as follows :

Type I : small folds superimposed on earlier larger ones
Type II : large folds superimposed on earlier smaller ones
Type III : both generations of fold of the same order of size

(i) *Type III* is the most complicated to study, as each attitude of the plane in the first fold has to be individually treated as for type II. If it is necessary to take into account the component of concentric folding during the development of the folds the problem is even more complicated. As this paper is not primarily concerned with superimposed folding the matter will not be taken further.

(ii) *Type I* is easier to deal with, as the new superimposed folds may be treated as lineations or micro-folds which are so small relative to the earlier fold that their effect on the attitude of the earlier fold may be ignored. Figs. 8 to 14 are examples of this type of superimposed folding and could represent, for example, a schistosity in a uniform pelite folded first into a relatively large fold and later deformed again. The schistosity behaves as alternating very thin layers of material of different competence so that during the second deformation folds of small wavelength are generated. Boudinage would not, however, be expected in such rocks. The early large fold is represented in Figs. 8 to 14 by the bundle of planes and its axis β by the bundle axis.

It is necessary here to distinguish between active and passive folding. *Active* folding or unfolding takes place because the fold-axis is parallel or nearly parallel to a principal direction. *Passive* folding or unfolding takes place when the fold-axis is not connected with a principal direction, for example the deformation of an earlier-formed fold. A fold that is unfolding will be called an *opening fold*, and a fold that is closing will be called a *closing fold*. The passive deformation of the earlier-formed fold may be determined from the behaviour of the bundle of planes outlined above. It is not, however, possible to reconstruct a fold from its stereographic plot alone; information concerning the axial plane is necessary to make geological sense of the bundles of planes, There are three significant positions for the axial plane: tangential to the structural movement path of β, normal to it, and oblique to it. The sequences of passive deformation of earlier formed folds are shown in Table V (p. 419).

At the same time as the early fold is being passively deformed, new folds (here only micro-folds are considered) are being generated along the principal direction curves and either actively developed and then rotated towards Z or not developed at all. After generation and development along coincident or nearly coincident principal direction movement paths and structural movement paths, the structural movement path may in many instances diverge from the principal direction movement path only very obliquely at first. With a difference in competency between the layers it seems likely that as long as the fold-axis remains nearly parallel to the principal direction the fold will continue to develop actively, and no further structure will be generated at the principal direction. However, if the competence difference is not too great the fold-axis may pass out of the field of influence of the principal direction; the further development of the fold will then be passive, and a new fold may be generated and developed at the principal direction. In this manner in the course of a single deformation, especially if $k > 1$ and for Z' structures, several folds may develop one after the other with obliquely inclined axes on the same surface.

For $k = \infty$, the Z' structures lie along a small circle passing through Z and the early fold-axis β (Fig. 12), and the X' structures lie in the X–Y plane. When the axial plane of the early fold is parallel to the β structural movement path (the fold being further compressed) the maximum divergence between the Z' lineation and β occurs on the limbs and the minimum on the hinge. On the hinge the divergence should reach nil, and if the limbs lie parallel to the axial plane the divergence registered on them will be the maximum possible and equal to the angle between Z and β. If, however, the axial plane is normal to the movement path this maximum divergence occurs on the hinge and the minimum on the limbs. An oblique axial plane causes asymmetrical distribution of the deviation on the fold. Lineations due to X' micro-folds may also be present and are generated normal to the Z' lineations on the same plane, but during deformation they become increasingly restricted to planes nearly normal to the β structural movement path, being unfolded or replaced by boudinage on other planes.

For $k = 0$, the Z' structures lie along the great circle containing Y and Z and the X' structures on a small circle passing through β and X (Fig. 8). With the axial plane parallel to the β structural movement path and the fold passively opening during deformation towards complete unfolding, the maximum divergence of Z' lineations occurs on the hinge and the minimum on the limbs: the minimum possible, on limbs parallel to the axial plane, is $(90° - \beta \wedge X)$. If the axial plane is normal to the β structural movement path and the fold is closing towards isoclinal, the minimum divergence occurs on the hinge $(90° - \beta \wedge X)$, and the maximum on the limbs. In both cases the greater the divergence between the lineation and β the more likely is the micro-fold lineation to be replaced by a parallel boudinage. X' structures are invariably boudinage. An oblique attitude of the axial plane to the β structural movement path gives rise to an asymmetrical distribution of the divergence over the fold, but with increasing deformation the rotation of the axial plane causes the distribution of the divergence to approximate more and more closely to that for an axial plane normal to the β structural movement path.

The possible angular relationships between the new lineation and the old fold-axis are summarized in Table VI.

Thus, for $k = 0$ and $k = \infty$ a minimum divergence between the older fold-axes and the younger Z' lineation in the hinge region means that the older fold is being passively closed on its old axis and a maximum divergence on the hinge means that it is being passively opened. The field appearance of a late lineation superimposed on an earlier fold by a deformation of the type considered here may be seen in Fig. 16a and b (which has, however, been constructed to represent the later folding of an earlier formed lineation [p. 422]).

For $0 < k < \infty$ it is much less easy to describe the angular relationship between early fold-axes and later lineations. This is because the principal direction curves are not loci of structures; structures tend to be carried away on divergent structural movement paths. However, up to quite high values of k, say $k = 12$, the combined effects of the shapes of the principal direction curves and of the structural movement paths lead to a concentration of Z' lineations in a belt containing Z and approximating to a great circle. Over most of the deformation plot the behaviour therefore approximates to that for $k = 0$, especially in the earlier stages of deformation before change-over from X to Z polar principal direction curves takes place. But the positions of maximum and minimum deviation of the Z' lineations from the earlier fold-axis will in general be asymmetrically distributed over the fold; and the folds themselves, owing to the rotation of the axial planes, will mostly close up towards isoclinal parallel to Z–Y.

TABLE V.

Value of k	Axial plane tangential to β structural movement path	Axial plane normal to β structural movement path	Axial plane oblique to β structural movement path
$k = \infty$	Some folding	Some unfolding	Some folding or unfolding depending on degree of obliquity
$k = 0$	Unfolding—possibly complete	Folding—possibly to isoclinal with axial plane parallel to Z–Y	
$0 < k < \infty$: β in X–Z plane near X	Fold closes up to some extent, then opens up, possibly completely	Fold opens to some extent, then closes, possibly to isoclinal parallel to Z–Y	Sequence of opening and closing according to degree of obliquity always ending, if deformation goes far enough, as isoclinal parallel to Z–Y
$0 < k < \infty$: β oblique to X–Z plane near X	Fold closes to some extent, then opens to some extent, then closes, possibly to isoclinal parallel to Z–Y	Fold opens to some extent, then closes, possibly to isoclinal parallel to Z–Y	

TABLE VI.

Value of k	(1) Axial plane of early fold parallel to β structural movement path	(2) Axial plane of early fold normal to β structural movement path	(3) Axial plane of early fold oblique to β structural movement path
$k = \infty$	$Z' \wedge \beta$ maximum on limbs; minimum on hinge; fold passively closing (see Fig. 16a and b, lineation $A^{\times}$)	$Z' \wedge \beta$ maximum on hinge; minimum on limbs; fold passively opening (see Fig. 16a and b, lineation $B^{\times}$)	Asymmetrical distribution tending to (1) or (2) depending on degree of obliquity
$k = 0$	$Z' \wedge \beta$ maximum on hinge; minimum on limbs; fold passively opening (see Fig. 16a and b, lineation $B^{\times}$)	$Z' \wedge \beta$ maximum on limbs; minimum on hinge; fold passively closing (see Fig. 16a and b, lineation $A^{\times}$)	Asymmetrical distribution tending to (2) with increasing deformation

(iii) Repeated folding of *type II*, that is large younger folds superimposed on earlier smaller ones, may be simulated by constructions very similar to those shown in Figs. 8 to 14. The following example probably represents a common situation (Fig. 16c). For ease of computation k is taken as 1 and the lineations $A^{\cdot}$ and $B^{\cdot}$ are given an initial dispersion along arcs of great circles passing through Z. However, a wide variation of k and initial dispersion of the lineations along great circles not passing through Z or even along small circles makes very little difference to the final result. In order to observe the maximum effect, suppose that the early folds are repre-

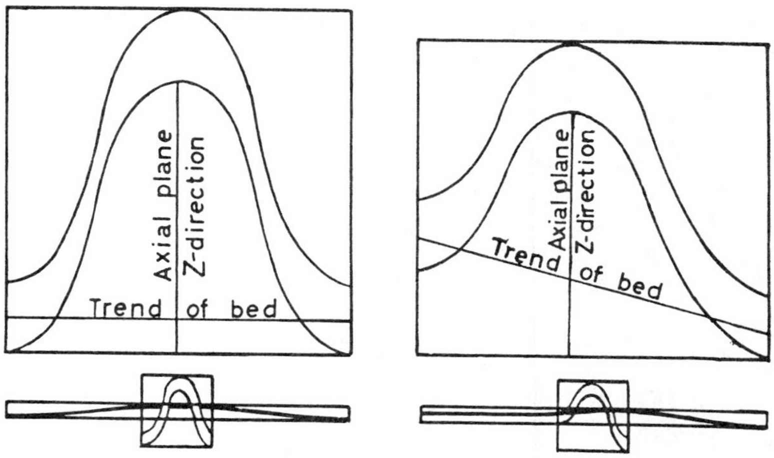

FIG. 15.—Similar folding during finite homogeneous strain. The small-scale versions show how pairs of slightly curved parallel lines become strongly folded by homogeneous strain. The folds are shown in detail in the larger versions and are seen to be of 'similar' type. $k = 1$, $a = 5$. Section in Z–X plane.

sented by lineations $A^{\cdot}$ or $B^{\cdot}$ on almost planar surfaces which lie nearly normal to Z. For any value of k, a potential fold-axis arises parallel to a Z' principal direction nearly parallel to Y. In the presence of a competence difference between the layering a fold will be generated and developed at this principal direction. The fold develops rapidly as the axis is nearly normal to X. The new fold-axis and the axial plane rotate towards Z and towards the Z–Y plane at the same time. The old lineations on each segment of the surface being folded rotate along appropriate structural movement paths towards Z and their dispersion after a given amount of deformation is shown in Fig. 16c by the arcs $A^{\times}$ and $B^{\times}$ respectively. These structural movement paths can only be great circles if they coincide with a symmetry plane of the ellipsoid, or $k = 0$ or ∞, or if there is no component of concentric folding. This should be compared with Weiss's (1959, p. 100, fig. 76) conclusion for shear folding that all such movement paths are great circles.

At any given moment in the deformation the old lineation is

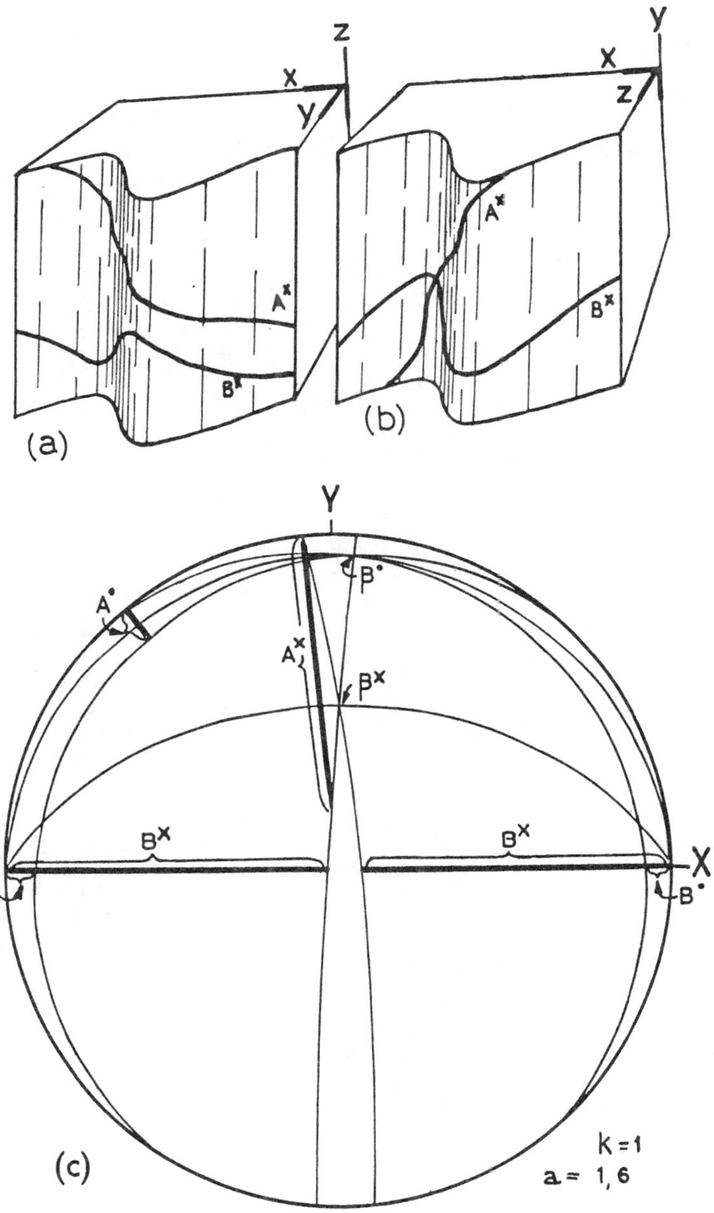

Fig. 16.—Early lineation folded by a younger fold.

(a), (b) The appearance of the resultant fold-systems for different atitudes of the deformation ellipsoid to the earth's surface.

(c) Stereo-plot. The early lineations—$A^{\cdot}$ and $B^{\cdot}$—lie in a plane nearly normal to Z. Later folding takes place about fold-axis β and disperses the early lineations along the lines $A^{\times}$ and $B^{\times}$ respectively. The folding is dominantly similar, but a component of concentric folding is present and is represented by a slight folding of the plane about β in the undeformed state.

dispersed along a great circle only under the following circumstances :

(1) When the new fold-axis is parallel to Y.

(2) When there is no component of concentric folding, and the lineation was already dispersed along an arc of a great circle before the deformation started. The smaller the component of concentric folding, and the larger the initial angle between the new fold-axis and the old lineation, the less will the dispersed lineation depart from a great circle distribution. The departure will rarely be great enough to be recognized from field data with their natural scatter.

The locus of the dispersed lineation does not pass through Z, unless it or the plane containing it did so before the new folding started or infinite deformation has occurred. The later deformation has the effect of rotating the locus of the dispersed lineation nearer to Z. Ramsay (1960, p. 89) finds that for shear folding the locus of early lineations always passes through *a* (the *a* direction of the *abc* tectonic axial cross). It should be noted that Ramsay does not orient the *abc* tectonic axial cross according to the rules laid down by Sander (1948, p. 68).

In superimposed folding of type II the angular relation between the lineations and the fold-axis is more variable than in type I, and there would be little point in trying to describe the relations in detail ; each case needs to be considered individually, taking k, the original position of the plane, and the original position of the lineation into account. In general, an old lineation is deflected as it passes over the hinge of a new fold, in much the same way as a new lineation is deflected as it passes over the hinge of an old fold. If the folding contains a component of concentric folding this also has to be taken into account. As the component of concentric folding increases the dispersed lineation increasingly conforms to a small circle about the new fold-axis (Sander 1948, p. 171).

Fig. 16a and b show some possible geological relationships shown in Fig. 16c for various orientations of the deformation ellipsoid relative to the horizontal. Note the similarity between the results of this deformation (Fig. 16a and b) and Ramsay 1960, fig. 4, 5, and 6. Also of interest in this connexion is Clifford 1959, fig. 11. In Fig. 16a and b the old lineation is distributed symmetrically over the new fold and Z lies in the axial plane, because at the start of the second deformation the surface about to be folded lay nearly normal to Z, that is it had a rather special orientation. If the surface had had a more general orientation Z would not have lain in the axial plane and the dispersion of the lineation would not have been arranged symmetrically over the fold.

The deflexion of a new lineation superimposed on an older fold appears to be very similar to that of an old lineation folded by a later folding ; Fig. 16a and b therefore serves to illustrate both cases. A distinction between these two cases cannot be made on purely geometrical grounds from a stereo-projection of a field occurrence, except under specially favourable circumstances of deformation and orientation.

The attitudes of the axial planes of new folds generated during the deformation cannot be determined from the equations presented here, since folds cannot be generated in the conditions under which the equations are valid. The generation of the fold and, therefore, the initial position of the axial plane depend on a component of concentric folding. Once a fold and its axial plane have been generated, they both come under the influence of the similar component in the folding and the axial plane behaves like any other plane in that its pole approaches X, but not necessarily by the shortest route.

Typical ptygmatic folds as modelled by Ramberg seem to be disharmonic concentric folds. Ramberg (1959, p. 125) notes that in his experiments the axial planes were not always parallel to the longest axis of the deformation ellipsoid. In his photographs the axial planes appear to be generated normal to the trend of the layer and to rotate slightly towards the longest axis of the deformation ellipsoid during deformation. It is not, however, safe to predict the course of rock deformation from simple shear experiments.

It is possible to learn something of the behaviour of axial planes during similar folding by the graphical construction of the process. From Fig. 15 it can be seen that upright folds develop in layers originally trending parallel to the shortening direction, and overturned folds develop in layers originally inclined to the shortening direction. Goguel (1948, p. 317) concludes that overturning is due to the inclination of layering to stress-axes. In the absence of evidence to the contrary, overturning is best considered as a sign of original asymmetrical orientation of layering to the strain-axes, rather than of rotation of the stress-axes during deformation.

When fold-axes do not lie in the Z–Y plane, the axial planes cannot be perpendicular to the maximum strain (X). Weiss's (1959, p. 87, fig. 5) construction of new fold-axes superimposed on earlier folds does not apply to folding due to homogeneous strain. He draws the great circle whose pole is the direction of maximum pressure and considers the intersections of this great circle with the traces of the planes in the early fold to be the axes of the new superimposed folds, because he considers this great circle to be the trace of the axial planes of the new folds. It has been shown in this paper that new fold-axes generated parallel to principal directions lie in the Z–Y plane (the plane normal to the direction of greatest compression) only if the old fold-axis is parallel to X or if $k = 0$.

V. General Conclusions and Discussion

The effects of three-dimensional finite homogeneous strain have been studied. It has been shown that some commonly occurring geological structures could be the result of finite homogeneous strain, and that when such structures are found in the field it may be possible to determine from them the orientation and shape of the associated deformation ellipsoid.

(i) *Preferred orientation.* If a homogeneous mass of rock containing randomly oriented planes, flakes, or rods (of no structural significance) formed, for example, by an unequal distribution of a pigment, suffers homogeneous strain, then the planes, flakes, or rods develop a preferred orientation. The preferred orientation changes in character as the strain changes and the two have the same symmetry. The preferred orientations predicted above are similar to preferred orientations commonly found in nature, and for two-dimensional strain they are similar to preferred orientations predicted for flakes and rods and experimentally produced (Sander 1948). If the planes, flakes, or rods have a preferred orientation before deformation, then after the deformation their distribution will be even more restricted than if they had been randomly distributed. However, this is probably not the mechanism whereby preferred orientations are usually developed in tectonites, since many minerals show similar patterns of preferred orientation without having a suitable shape.

Veins in a deformed granite or migmatite mass could be used to determine the deformation of the mass if they were originally randomly distributed. Such veins would probably fold ptygmatically and suffer boudinage or pinch-and-swell as described by Ramberg (1959) owing to competence differences between them and their host. Nevertheless, each vein would still rotate as a whole into its appropriate attitude to form the preferred orientation.

(ii) *Folding*. Straight lines and planes rotate towards the longest axis of the deformation ellipsoid but remain straight and planar. Parallel lines rotate but remain parallel. A curved surface, however, rotates bodily and either becomes more or less curved depending on its attitude. Thus, a folded layer rotates bodily so that the fold-axis and the axial plane approach the longest axis of the deformation ellipsoid and the layer thickens or thins and folds up further or unfolds about the same axis. If the folded layer had parallel boundaries before the deformation, the boundaries must be parallel afterwards and thus the folding is of similar type.

During finite homogeneous strain no folds can be generated, because straight lines must remain straight. If there is a competence difference between the layers then, as shown by Ramberg (1959), the layers may be expected to buckle. As soon as buckling is present it will be acted on by the homogeneous strain and further folding will be a mixture of similar folding due to homogeneous strain and concentric folding due to the buckling. Which type predominates depends on the competence difference. The axis on which buckling takes place can be predicted from the deformation ellipsoid since it will be one of the two principal directions for that plane. What cannot be predicted is the attitude of the axial plane in a newly generated fold. It is possible that the newly generated axial plane will be normal to the layer being folded. As soon as the axial plane is formed it rotates towards the longest axis of the deformation ellipsoid.

The combined effect of the rotation of the axial plane and of the layer being folded is to form overturned folds. Overturning in similar folds may thus be due to an original asymmetrical orientation of the layering to the strain-axes.

Folds can open or close, either actively or passively, during deformation. Active folds are those forming at principal directions, while passive folds open or close about an old fold-axis inherited from an earlier deformation. Ephemeral folds are active folds that form and then completely unfold in the course of a single deformation, or would completely unfold if the deformation continued long enough. De Sitter (1958) has described a particular type of ephemeral folding called *parasitic folding*. It occurs when layers originally parallel to X–Y are folded during deformation of $k = 1$ type.

(iii) *The deformation ellipsoid and the tectonic axial cross.* Neither fold-axes nor axial planes bear any special relation to the axes of the deformation ellipsoid. When they are observed to be parallel to one or more of these axes during or after deformation, it is because they were parallel to them before the deformation. Fold-axes and axial planes are therefore useless as indicators of movement-directions or directions of flow in rocks. To use them is to assume that at the start of the deformation the layering had a special orientation relative to the future pattern of flow, which is thereby fixed in direction quite arbitrarily.

No use has therefore been made in this paper of Sander's tectonic axial cross (*abc*) which is fixed in orientation by the symmetry and form of folds. In tectonic analysis the tectonic axial cross must be replaced by the deformation ellipsoid, that is by a cross whose axes have not only direction but also magnitude and whose orientation is fixed parallel to the three chief strain-axes. The determination of the shape and orientation of the deformation ellipsoid for a particular deformation must be made from the structures formed during that deformation. Such structures can be divided into three classes according to their relation to the deformation ellipsoid :

(1) Z' and X' structures ; for example folds and boudinage. The axes are parallel to symmetry axes of the deformation ellipsoid only under special circumstances.

(2) Structures parallel to symmetry planes or axes of the deformation ellipsoid, for example, Z-lineations formed by girdle or partial girdle preferred orientations of micaceous minerals, or X-schistosities formed by a preferred orientation of micaceous minerals parallel to the Z–Y plane.

(3) Structures giving a direct indication of the shape and orientation of the deformation ellipsoid, e.g. oolites, pebbles, spots, fossils, grain shapes, strain shadows, and so on.

The first class of structures has been the object of study in this paper, the other two classes having been considered elsewhere (Flinn 1956, 1958, 1961).

(iv) *Deformation by shearing and by homogeneous strain.* The formation of similar folds by finite homogeneous strain is an attractive alternative to current hypotheses involving shearing parallel to the axial plane, although it is necessary to appeal to non-planar layering, competence differences between the layers, or changes in homogeneous stain from place to place, in order to generate the folding. A somewhat similar objection applies, however, to the shearing hypotheses throughout the whole course of the folding, since there seems to be no explanation for the regular and continuous reversals in sense of shearing every half-wavelength. Moreover, there seems to be no cause for similar folding by shearing parallel to the axial plane, since no change in dimension normal to the axial plane is apparently necessary. Further, shearing hypotheses are only applicable to two-dimensional deformation, which probably explains the general lack of consideration of three-dimensional stain in structural geology. This paper provides a method for handling both two- and three-dimensional strain or flow without appeal to shear-planes. Hypotheses involving shearing thus become superfluous and even detrimental, leading as they often do to an uncritical search for structurally realized shear-planes in the rock and a limitation of strain to two dimensions. It is usual to suppose, on very little evidence, that axial-plane cleavage represents the shear-plane on which the shearing which caused the folding has taken place (de Sitter 1956, p. 182 ; Hills 1953, p. 84). This paper, by showing that similar folds can develop without shearing, supports a conclusion that folds and cleavage are two simultaneous but mutually independent responses of the same rock to the same stresses ; neither is the cause of the other. This conclusion seems to accord with the ideas long maintained by Foumarier (1956, p. 317).

(v) *Complexity of structure.* If fold-axes and axial planes have no special orientation relative to the chief strain-axes, it is to be expected that when fold-systems are subjected to a later deformation the resultant structures will be very complicated. Much of the paper is taken up with the analysis and synthesis of such structural complexes. It is predicted that they will indeed be very complicated—more complicated, in fact, than any described field example. There is, of course, a tendency to simplify data when collecting them. Lineations are expected to be parallel to one another and to fold-axes, and contemporaneously formed folds are expected to have parallel axial planes. Where these expectations are not realized, and especially where lineations are observed to intersect, the structures can be arranged in a time series in such a way that the expectations appear to be realized.

It is nevertheless clear that in many regions the structures are less complicated than is to be expected from the synthetic examples given above. This is probably because the layering and fold-axes occupied special positions in the deformation ellipsoid at the start of the deformation. If, for example, the layering is parallel to Y–X at the start of the deformation, and if the fold-axis that forms at Y remains parallel to Y during the deformation, then the structures produced are much simpler than those described above ; lineations and boudinage form and continue to develop parallel, and possibly also normal to, the fold-axis, which itself remains unfolded, that is, it does not rotate towards Z.

Equations (23) predict that if the fold-axis is oblique to Y by only a few degrees considerable rotation towards Z is to be expected as the result of quite moderate deformation. Axes must often deviate by a few degrees or more from Y, and yet in many regions little rotation (folding) of the fold-axes has occurred. This is probably an effect of inhomogeneity. The layers, once they have folded on an axis near Y, the strongest axis of folding, are sufficiently strengthened to resist folding on the same scale about an axis normal to Y ; hence no rotation takes place. This strengthening might not prevent cross-folding on a very much smaller scale or a very much larger scale. Inhomogeneity also tends to simplify structure by rotating the stress-axes, and hence the deformation ellipsoid, into a more symmetrical position relative to the layering and other structures already present.

Thus the observation that the structure of an area is simpler than might be expected from this paper probably signifies that the rocks were more inhomogeneous than was assumed above, and that the layering was more symmetrically oriented with respect to the deformation ellipsoid. To assume that this special orientation existed is, of course, to assume the orientation, and the deeper in the crust the structures originated, the more likely is the assumption to be wrong. Weiss (1955) showed how this assumption of special orientation could lead to false conclusions. The great complexity of structure often encountered in migmatite gneiss areas may be due to decreasing competence differences between the layers, rather than to increased mobility and turbulent flow.

(vi) *Lineations in repeated folding*. The angular relationship of lineations belonging to different deformation phases taking place in the same rock has recently attracted attention. In this paper it is shown that the angle between an early lineation folded by a later fold and the later fold-axis varies systematically over the fold and that, where the second deformation is not specially oriented relative to the older lineation and the layering, the angular variation is distributed asymmetrically over the fold. It is also shown that the same angular variation with the same distribution over the fold is to be expected when a later micro-fold lineation is superimposed on an earlier fold. It seems to be impossible to distinguish between the two cases on geometrical grounds alone, except in certain favourable conditions of deformation or orientation. The relationship predicted here between early lineations and later fold-axes agrees rather well with the relationship found from the examination of some field examples (Ramsay 1960). The field examples show the early lineation symmetrically deflected over the later fold. This seems to indicate that the deformation ellipsoid of the second folding either had a special shape ($k = 0$ or ∞) and/or a special orientation relative to the layering which has been folded.

Acknowledgements.—This work was carried out in Moscow at the Institute of Ore Deposit Geology, Petrology, Mineralogy, and Geochemistry during 1960, when I held a research fellowship under an exchange agreement between the Royal Society and the Academy of Sciences of the U.S.S.R. I am grateful to the Royal Society, the Academy of Sciences of the U S.S.R., and the Institute for this opportunity.

Author's note.—After this paper was submitted for publication, a paper by W. F. Brace entitled 'Mohr construction in the analysis of large geologic strain' appeared (1961, *Bull. Geol. Soc. Amer.* **72**, 1059–80). Brace offers an alternative scheme of calculating finite homogeneous strain based on totally different principles.

VI. List of References

Burri, C. 1950. *Das Polarisationsmikroskop.* Birkaüser, Basel.

Clifford, P. 1959. The geological structure of the Loch Luichart area, Ross-shire. *Q.J.G.S.*, **115**, 365–88.

Cloos, E. 1947. Oölite deformation in the South Mountain fold, Maryland. *Bull. Geol. Soc. Amer.*, **58**, 843–918.

Flinn, D. 1956. On the deformation of the Funzie conglomerate, Fetlar, Shetland. *Journ. Geol.*, **64**, 480–505.

——. 1958. On the nappe structure of north-east Shetland. *Q.J.G.S.*, **114**, 107–36.

——. 1961. On deformation at thrust planes in Shetland and the Jotunheim area of Norway. *Geol. Mag.*, **98**, 245–56.

Foumarier, P. 1956. Schistosite et forme des plis. *Ann. Soc. Géol. Belge*, **74**, 317–64.

Goguel, J. 1948. Introduction a l'étude mechanique des déformations de l'écorce terrestre. *Mem. Serv. Carte Géol. Fr.*, Paris.

Hills, E. S. 1953. *Outlines of structural geology.* Methuen, London.

Jaeger, J. C. 1956. *Elasticity, fracture and flow, with engineering and geological applications.* Methuen, London.

Joel, N. & I. Garaycochea, 1957. The 'extinction curves' in the investigation of the optical indicatrix. *Acta Cryst.*, **10**, 399–406.

Ramberg, H. 1955. Natural and experimental boudinage and pinch-and-swell structures. *Journ. Geol.*, **63**, 512–26.

——. 1959. Evolution of ptygmatic folding. *Norges Geol. Tidskr.*, **39**, 99–152.

——. 1960. Relationships between length of arc and thickness of ptygmatically folded veins. *Amer. Journ. Sci.*, **258**, 36–46.

Ramsay, J. G. 1960. The deformation of early linear structures in areas of repeated folding. *Journ. Geol.* **68**, 75–93.

Sander, B. 1948. Einführing in die Gefügekunde der geologischen Körper. I. Vienna.

——. 1950. Einführing in die Gefügekunde der geologischen Körper. II. Vienna.

Sitter, L. U. de. 1956. *Structural geology.* McGraw-Hill, London.

——. 1958. Boudins and parasitic folds in relation to cleavage and folding. *Geol. Mjnbouw*, N.S. **20e** (8), 277–86.

Voll, G. 1960. New work on petrofabrics. *Liverpool & Manchester Geol. Journ.*, **2**, 503–67.

Wahlstrom, E. E. 1951. *Optical Crystallography.* [2nd Edn.] Wiley, New York.

Weiss, L. E. 1955. Fabric analysis of a triclinic tectonite and its bearing on the geometry of flow in rocks. *Amer. Journ. Sci.*, **253**, 225–36.

——. 1959. Geometry of superposed folding. *Bull. Geol. Soc. Amer.*, **70**, 91–106.

13

Reprinted from pages 843, 853, 854, 856–865, 874–875, 877–879, 879–882, 883–886, 888–892, 894–897, 909, 910–911, 912, and 915–917 of *Geol. Soc. America Bull.* **58:**843–918 (1947)

OÖLITE DEFORMATION IN THE SOUTH MOUNTAIN FOLD, MARYLAND

BY ERNST CLOOS

[*Editor's Note:* In the original, material precedes these excerpts.]

DESIGNATION OF CO-ORDINATES AND SYMBOLS

The following designations have been used: A = maximum strain—ellipsoid axis; B = mean axis; C = smallest axis (Sander, 1930, p. 52). Co-ordinates *a, b, c,* according to Sander and many other workers (E. Cloos, 1946, p. 5) are arranged so that *b* parallels the fold axis, *a* is the direction of tectonic transport, and *c* is normal to the *ab* plane. If *b* is an axis of rotation it becomes B (Sander, 1930, p. 51; Knopf and Ingerson, 1938, p. 46). The maximum, mean, and minimum half axes of the deformed oöids are a, b, and c. Prior to deformation a = b = c, the oöids are spheres. *l* is lineation (E. Cloos, 1946, p. 4).

[*Editor's Note:* Material has been omitted at this point.]

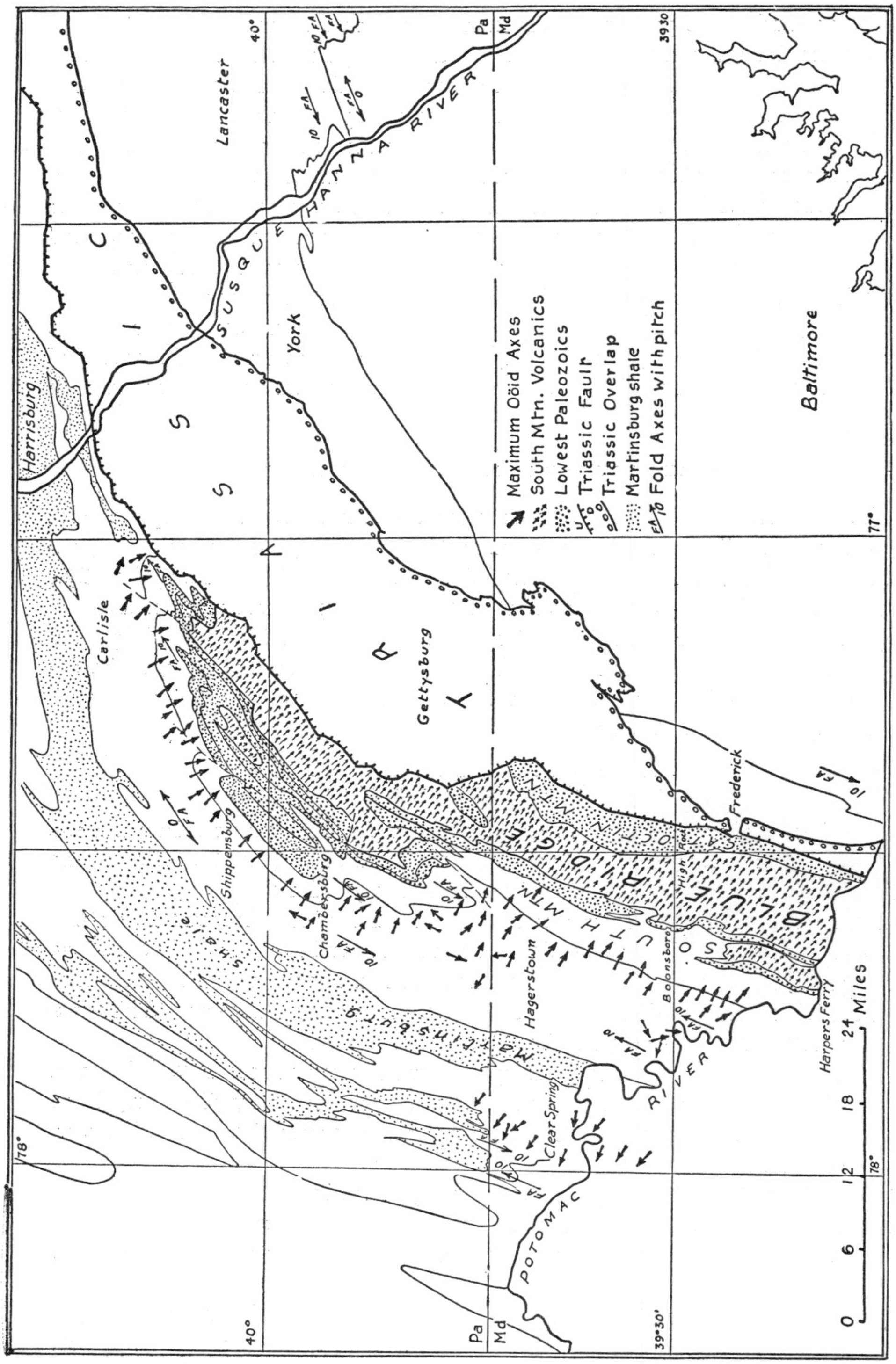

FIGURE 1.—*Index map of Appalachian Valley*

Localities, maximum oöid axes, and general trend of structural units.

METHOD OF INVESTIGATION

GENERAL REMARKS

The average diameter of the oöids is below one millimeter and ranges from 0.33 mm to 1.2 mm. Only in one layer southwest of Clearspring have oöids of over 1 cm diameter been observed. Wherever deformation is intense, oölites cannot be recognized but either by hand lens or binocular and many specimens were collected which are not oölites but turned out to be fine-grained fragmental limestones.

Recognition of directions of deformation is difficult if exposures are not three-dimensional and ratios small. Thin sections made in wrong directions may be misleading, and one of the most difficult steps is the determination of maximum and minimum directions in odd-shaped specimens or exposures.

Great care and thus a good deal of time has to be invested if accuracy is to be attained. A mistake in any one of the stages will never be cancelled out. Careful check and recheck by another investigator and plotting into three-dimensional diagrams prevents errors, and the author feels confident that the accuracy here attained is satisfactory.

H. R. Gault has recalculated 90 per cent of all figures used here and remeasured many of the slides, checked orientations repeatedly, and in general eliminated many errors.

Specimens have been collected at 241 localities. Two hundred and twenty-seven of these are oriented. Four hundred and four thin sections have been examined and measured. Fifty-five localities furnished nonoölitic limestone, deformation beyond measuring, or material which could not be used. Reliable measurements are available for 186 localities, and these have been plotted on the maps (Pls. 9–11).

All but two samples are limestones with varying amounts of dolomite, quartz, carbonaceous matter, and very fine muddy shale. Many are fossiliferous, and all contain quartz. Some hematite, fluorite, pyrite, and chert was found in small

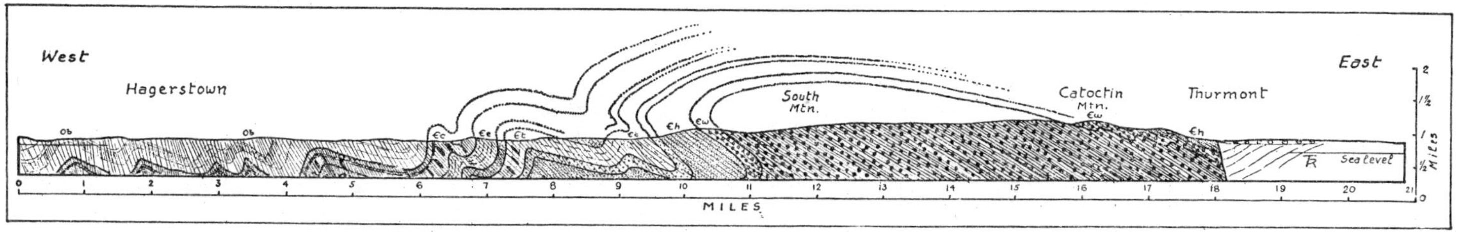

FIGURE 2.—*Generalized cross section*

Across South Mountain, Middletown Valley, and Catoction Mountain, showing overturned anticline with gently-dipping lower limb. Cleavage dips east in all rocks and transects formational boundaries far into Hagerstown Valley.

amounts. Two examples of siliceous oölite are interpreted as chert replacements The author has searched for good exposures and then searched the exposures for oölites. A good horizon may be followed and investigated for good measurements at even intervals.

Deformed oöids serve the same purpose whether they are in the Beekmantown limestone, the Conococheague, or the Tomstown. However, certain horizons are more oölitic than others and were thus more frequently encountered. The maps (Pls. 9–11) are, therefore, not geological maps even if the distribution of localities reflects the regional arrangement of formations and especially of the Conococheague limestone.

In view of the intricacy of the observations and the smallness of the average oöid, it was necessary to verify field observations by thin-section study. A microscope was, therefore, taken into the field, and thin sections made as soon as practicable. The sections were then measured and calculated, and the results applied in further collecting and checked in the field. In this manner, field and laboratory work alternated, revisions and corrections were made and much unnecessary mileage was eliminated, and the search directed to the most essential points.

First a rather loose net of scattered observations was spread over a larger area. Fixed points were established, and then observations filled in wherever necessary or desirable. Where orientations and ratios of deformed oöids are constant over many miles, gaps were left in favor of more intricate areas. In this manner the number of thin sections was reduced to a useful minimum, and they are now placed where they provide most useful information.

One hundred and twelve samples show spherulitic oöids with typical radial growth and extinction cross under crossed nicols. Eighty-four samples also contain ample muddy or shale pellets or lumps, or are without spherulites. Detrital dolomite pebbles occur in 28 samples some of which are not measurable because of lack of spherulites or pellets. Good bedding can be seen in 34 samples. Euhedral quartz is common in 26 specimens and zonal growth of quartz can be seen in 20 specimens.

FIELD WORK

Prior to the study of oölite deformation the author had completed geological mapping of Washington County (E. Cloos, 1942) and was therefore familiar with the rock types, oölite locations, and the terrane in general. The area was then extended northward across the Mercersburg-Chambersburg quadrangles (Stose, 1909) and northeastward to the east end of the South Mountain fold southeast of Carlisle.

A complete structural inventory at every locality is indispensable. Bedding, cleavages, fold axes, intersections of bedding and cleavage or cleavages, lineations in bedding and cleavage planes were measured and plotted in the field. Drawings, block diagrams, and descriptions are of great assistance in the evaluation of the measurements later. Errors in later stages may be eliminated if the locality is well known and the accuracy of orientation can easily be checked against field measurements.

Joints and fractures of all kinds were measured wherever possible.

SPECIMEN SELECTION AND ORIENTATION

Since oöids are hard to see in fresh exposures or freshly trimmed specimens, field observations are mostly from slightly weathered exposures which are whitish gray, bluish gray, or otherwise bleached even on almost black limestones or dolomites.

Selected specimens are best loosened from solid rock and then refitted because they frequently break in trimming across marked orientations, and a new specimen has to be taken. In refitting, all dust and small pieces must be blown away from the fracture to assure perfect relocation. The piece can easily be held in place by a strip of adhesive tape which permits all following operations without continuous falling off of the specimen.

The sample should show two surfaces normal to each other to permit determination of deformation under a binocular and thus enable the investigator to choose the proper directions for sawing. Strike and dip of the smoothest and largest even surface are then measured, and the directions scratched on that surface with a sharp-pointed tool, and also marked with adhesive tape. Directions are noted, and in order to avoid later confusion the surface is also plotted in its proper location with north at the top of the page. An arrow for the strike with attached dip sign will exclude all other possibilities like east or west dip or confusion of top and bottom of specimen.

White bases were painted on the specimen with quick-drying paint, and orientations marked with india ink. A coat of shellac protects this permanent label against damage during shipping, handling, and grinding. All pertinent data were entered in a card catalogue. The same data were later also entered in a book, and the records kept separate as insurance against loss.

A rough determination of the future saw cuts in the field is most serviceable, particularly if these are then checked against lineation, cleavage, or bedding which may be only faintly visible in the sample but excellent in the large exposure. It is very much easier and also more accurate to mark a line on a specimen in the field than to re- orient the specimen in the laboratory and to determine the cuts there.

The author used an average of one hour to an hour and a half for one location, including all measurements, orientation, labeling of specimens and notes. A field day with 10 specimens collected was a good one and unfortunately, there were days with none.

As a rule specimens are comparatively large and should not weigh less than half or three-quarters of a pound. In taking samples, one should bear in mind that sawing will be necessary, and odd-shaped pieces should be avoided since they cannot be clamped readily for sawing. Sawing is easiest and also cheapest across specimens which are bounded by two roughly parallel planes. Cuts parallel to thin slabs, for instance parallel cleavage, are difficult to obtain accurately.

DETERMINATION OF CUTS

Proper orientation of thin sections with respect to the axes of deformation is one of the critical steps in the investigation, particularly if distinct lineation is not visible in the field or if exposures are not three-dimensional.

Deformed oöids are triaxial ellipsoids, and two thin sections perpendicular to each other may contain the three axes if properly oriented. They may, however, also contain three differing axes which are not the principal ones, and nothing in the thin sections indicates at what angle to the principal axes the section might have been cut.

In the area here analyzed, orientation is rather easy where cleavage is obvious. There should be one section perpendicular to cleavage, to the intersection of cleavage and bedding, and one parallel to cleavage and the intersection. One section should always contain visible lineation, and one the fold axis. Wherever cleavage is not visible a section parallel to the axial plane of the fold and one perpendicular to it and the fold axis is likely to include the principal axes of the ellipsoid.

It may become necessary to try several sections, and in such cases three or more sections are essential.

If two sections are made, one for instance containing the longest and shortest axes and a second one the longest and mean axes, a third one perpendicular to the two others will provide an excellent control and should contain the shortest and mean axes. In calculating ratios and the radius or diameter of the original sphere, the combination of either two sections should furnish the same result. Even small deviations from the principal axes may introduce large errors if the ratios are high and the maximum axis is large.

In the present investigation, the author observed a remarkable regularity of orientation along the entire western slope of South Mountain. Maximum elongation of oöids parallels lineation and is in the steepest dip of cleavage. Since most fold axes are nearly horizontal, a vertical cut perpendicular to cleavage and fold axes furnished the principal axes of deformation. In the less deformed western area where structures are less distinct, cuts were oriented in the same manner unless elongations in other directions were visible in the field. Wherever calculations disagreed, additional cuts were made.

MICROSCOPIC STUDY

Microscopic investigation includes a study of the mineral content of each slide, and measurement of oöid dimensions.

The mineralogical and petrographical study was done largely by H. R. Gault, who listed all minerals present in each slide and noted their textural relations and peculiarities.

The dimensions of the oöids were measured with a one-tenth millimeter grid ocular which greatly facilitates measurements because two co-ordinates are always given, and readings in the horizontal and vertical are quickly made. The slide was placed on a mechanical stage, and the sections covered systematically.

It is advisable to draw the slide and its orientation arrow at the top of each sheet on which measurements are noted and next to the arrow to note the degrees read on the microscope stage referring to the zero position in reference to which degrees are read. The dimensions can then be read north-south and east-west whereby it is best to always read the longest dimension either one way or the other in order to avoid confusion of notes later. An average maximum orientation can be read with low magnification, thus providing a value which can then be checked against the measured average.

At first 100 readings were made per slide, but it was later found that 35 readings provide a value close to the average obtained from 100 measurements especially if orientation is distinct. Only few slides show no orientation.

Most oöids are sharply limited either by shells, coloration, or crystal boundaries, but frequently some new crystals have grown in the direction of maximum deformation thus increasing the ratio considerably. It seems reasonable to limit measurements to the actually deformed oöid and not to include secondary growth and enlargement. Calculations outlined below are based on deformation without change of volume. Secondary growth, however, increases the volume, and it would be difficult to include the increase in a determination of the size of the original sphere and its distortion.

In listing the measurements, the author first read the direction of maximum extension and then the minimum. The columns were then added mechanically and divided by the number of readings.

COMPUTATIONS

Oöids can be assumed to have been originally very nearly spherical bodies. Deviations can be seen in many slides as due to primary irregularities as, for instance, the shape of centers, irregular growth, or close packing. Such shapes are readily recognized and have not been used (Pls. 1, 2). In more than 400 slides there is no case in which primary eccentricity parallels a tectonic direction. Slight eccentricity which is not due to deformation remains unoriented. Elongation due to deformation can easily be detected because of uniform orientation (Pls. 4, 5, 6).

If the diameters of nonspherical undeformed oöids were plotted in a co-ordinate system, a sphere would result due to the random orientation of the eccentricity. This sphere can be used because random orientation has been changed to tectonic orientation in all grains of one specimen or location.

Tectonic orientations are obvious on the maps and illustrations, and deformation has more than doubled diameters. Primary eccentricity may introduce errors of 10 per cent, which carries little weight in deformations exceeding 100 per cent. The sensitivity of orientations even in the lowest deformations and the regional regularity seems sufficient guarantee that errors due to assumed sphericity of oöids would not alter the final result.

Fluctuation of orientation has been plotted against deformation ratio in Figure 3 for 144 slides. The range indicates the degree of parallelism of the longest dimension as measured by turning the microscope stage. The ratio is a/c. Perfectly spherical oöids should have the ratio 1, and it is clear that there are almost no spherical oöids represented. The range is reduced as ratio increases, but no slide shows less than 10° of fluctuation. At a ratio of 2:1 the long axis fluctuates still 50°, or 25° from the average.

Such fluctuations are well known in the orientation of mica grains in schists where only relatively few grains parallel the schistosity.

Figure 1 of Plate 4 shows intense and even alignment, a ratio of 1.4 and fluctuation of 25°. The ratio of Figure 1 of Plate 6 is 3 and fluctuation 20°, and in Figure 2 of Plate 6, the ratio is 2 and fluctuation also 20°. The slightly elliptical oöids in Figure 2 of Plate 1 fluctuate 180°, or 90° on either side of the average, and the ratio is 1.2. The four points are marked in Figure 2 by circles.

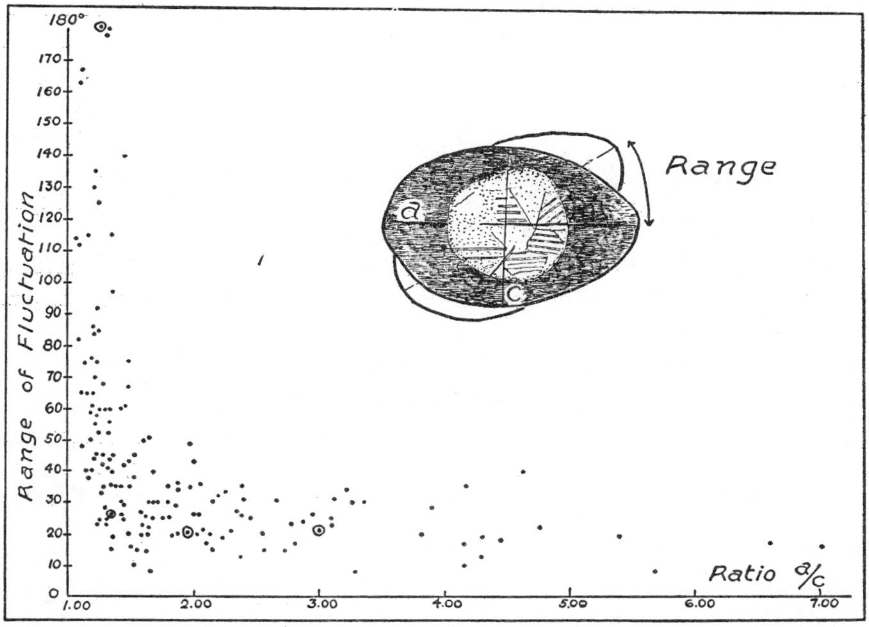

FIGURE 3.—*Range of fluctuation of orientation in relation to deformation ratio*

From the above it seems evident that eccentricity becomes significant only if the longest axes are oriented.

In the following procedure, a primary sphere was assumed as the original shape, and the deformation has been calculated with the diameter of the sphere as reference unit.

Two thin sections perpendicular to each other and containing the three semi-axes of the present ellipsoid a, b, and c permit calculation of the radius of the original sphere from which the ellipsoid has been derived provided that the volume has not changed. The volume of the sphere equals that of the ellipsoid:

$$\frac{4}{3}\pi r^3 = \frac{4}{3}\pi a.b.c.$$

$$r = \sqrt[3]{a.b.c.}$$

If, for example, the values measured at locality 300, northeast of Sharpsburg, Maryland, are used, the calculation is as follows:

2a = maximum diameter = 8.45, a = 4.23
2b = mean diameter = 6.74, b = 3.37
2c = minimum diameter = 5.06, c = 2.53
r = 4.23 × 3.37 × 2.53 = 3.3, d = 6.6

The components of deformation are, therefore:

$$2a = +1.85, \qquad 2b = +0.14, \qquad 2c = -1.54$$

In references to the original sphere the maximum extension in a is:

$$\frac{1.85 \times 100}{6.60} = 28\%$$

The extension in b is:

$$\frac{0.14 \times 100}{6.6} = 2.1\%$$

and the reduction of the c axis:

$$\frac{-1.54 \times 100}{6.6} = -23\%$$

One hundred twenty four specimens have been sawed in two directions, and all three components were determined. The average value of b is below 8 percent. At 22 localities it is zero, and at some localities where maximum elongation is in b and not a, deformation reaches 30 per cent. b varies with the accuracy of the measurements of a and c, and where deformation is strong, the ends of oöids are drawn out, torn off, and frequently damaged, thus making precise measurements impossible and at the same time preventing accurate measurements of the two other dimensions. In all such slides b will be positive, and if a is much longer than can be measured, the value for b will be large. Wherever recrystallization has taken place as for instance in Figure 1 of Plate 5, a becomes too large and b becomes negative.

The most difficult direction to measure accurately is a, and its computation decides the fate of the other two axes.

An abbreviated method where only one section in ac is cut can be used occasionally and the calculations are made with the assumption that b = 0. Thus the deformation ellipsoid becomes an ellipse (*see* Nadai, 1931, p. 294), and the unit sphere has the radius b. Calculation is much simplified, and the results are also reliable. Where b is large but not measured, values for a and c become impossible and the second sections must be cut later.

If extension in b equals 0, the semi-axis is 1 + 0. Thus:

$$r^2 = ac \qquad \text{or}$$
$$r = \sqrt{ac}$$

For instance in locality 204 only one section has been cut in the ac plane. 2a = 35.5, 2c = 6.7, thus:

$$r = \sqrt[2]{17.7 \times 3.35} = \sqrt[2]{159.295} = 7.69$$

The diameter of the original sphere is 15.38. The increase in the direction of a is thus 20.12. the decrease in c is 8.68.

The extension of a in per cent amounts to:

$$\frac{20.12 \times 100}{15.38} = 130.8 \text{ or } 131\% \text{ of the diameter of the unit sphere.}$$

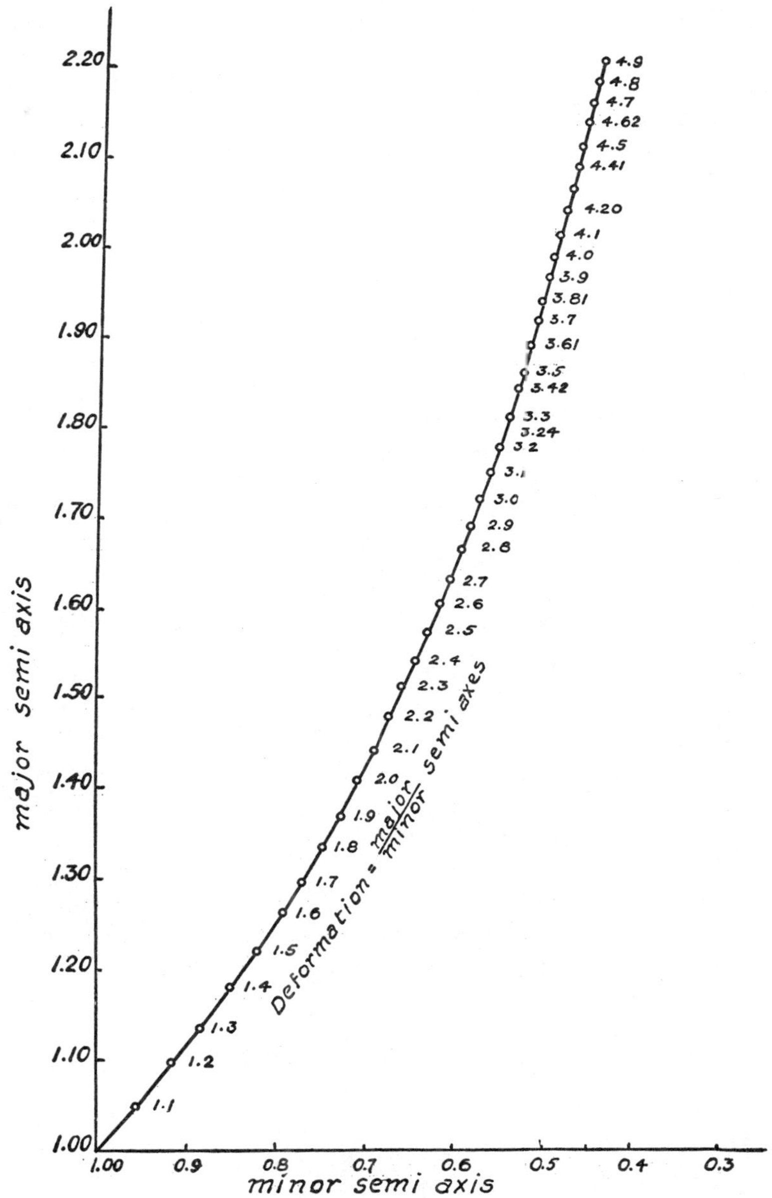

FIGURE 4.—*Plane strain with constant volume*

Curve shows major semi axes corresponding to minor semi axes, deformed from a circle with unit radius. Computed from Nadai 1931, p. 249 (2).

In order to reduce calculation, the ratio $\frac{a}{c}$ has been plotted into a curve from which the semi-axes and the extension can be read. The curve has been constructed by Edward Wisser and published here by his permission for the readers' convenience (Fig. 4).

One of the most important applications of the measurements of deformation is the increase or decrease of stratigraphic thicknesses in certain areas (E. Cloos, 1942). It is, therefore, necessary to calculate the components of deformation in relation to bedding. If a is perpendicular to bedding, the extension of a means increase of thickness by its full amount. If the angle becomes smaller, the component is reduced with the size of that angle.

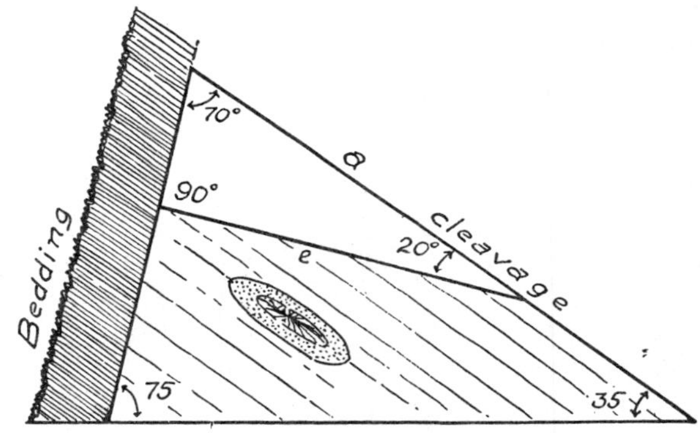

FIGURE 5.—*Relation of bedding to oöid extension*
a: cleavage, e: normal on bedding.

In Figure 5, bedding dips 75° west, cleavage and extension 35° east. The angle between the two planes is 180− (75+35) = 70. If the component of elongation normal to the bedding is e, then:

$$\sin 70° = \frac{e}{a} \text{ or } e = a.\sin 70°$$

$$\sin 70° = 0.9397 \text{ and } e = 0.9397.a$$

This value can very readily be plotted, and the extension read quickly for any angle between cleavage and bedding after the extension in a has been determined.

[*Editor's Note:* Material has been omitted at this point.]

BEDDING

Bedding is recognized in most of the larger exposures, particularly if deformation is moderate and cleavage lacking or only weak. With increasing distortion, however, stratification is obliterated, and cleavage becomes the only visible structure. If recrystallization or mineralization accompanies deformation, cleavage may readily be confused with bedding, and it becomes necessary to search for distinct lithologic differences. Where specimens are oriented in the field, the deformation directions can readily be related to the bedding, cleavage, lineations, and axes of folding.

In thin sections, bedding can also be recognized either in the orientation of primarily elongate forms like shells, fragments, and lenticular pebbles, or the planes themselves may have been cut by the section. Many specimens show good bedding in all degrees of deformation.

Stratification becomes visible as changes in lithology, fine shaly partings separating limestone beds, or as oölite beds in contact with nonoölithic mudstone, marble, or ordinary limestone. Bedding planes are always loci of solution phenomena like stylolites, or of deposition and replacement of silica or carbonate.

Figure 8 of Plate 7, shows very distinct bedding as thin shaly layers in otherwise clear limestone. The oöids are elongated at a large angle to the strata. In Figure 6 of Plate 7 a thin dark layer separates two layers of oölite of slightly differing color and grain size. Elongation is also almost at 90° to the plane. Intense crenulation is visible in Figures 12 and 13 of Plate 7 where the distortions of oöids are 40 and 30 per cent. Bedding is here visible as dark layers studded with quartz grains which have been left when carbonates were carried away. Shaly and carbonaceous materials are also remnants of pellets. As the lateral distance decreased the bedding plane was crenulated.

Solution along bedding planes is visible in Figure 5 of Plate 5, and Figure 7 of Plate 7, where oöids are cut and eaten into.

Extreme deformation of oöids and bedding is shown in Figure 14 of Plate 7. Lateral reduction forced the bedding plane into sharp folds, emphasized by flowage in the direction of maximum extension.

The close relation between oöid deformation and the degree of crenulation of bedding planes is a clue to quantitative analyses where oöids are absent. Systematic measurements of wrinkles in thin sections and counts of folds may provide rather valuable data on the distortion of a complex.

MATRIX

The matrix is probably the largest factor in the deformation of oöids and also the most difficult to evaluate. There are two principal types of material: crystalline carbonate and shale in varying proportions. Some oölites are dark and cryptocrystalline with only some light oöids of crystalline carbonate (Pl. 7, fig. 6); others consist of pellets in a clear and light crystalline matrix (Pl. 8, fig. 4). Between the two extremes are many gradations.

It has not been possible to determine the degree of deformation which is recorded in the matrix. Several possibilities have to be considered:

(1) The matrix may be much softer than the oöids. The latter would then be protected against deformation by the soft matrix and remain undistorted. This seems indicated in the relation between hard "pseudopebbles" and matrix.

(2) The matrix is crystalline, and the oöids are soft. This should be recognizable as mylonitization or granulation of matrix.

(3) Both oöids and matrix are crystalline and equally hard.

(4) Matrix and oöids are soft, pliable, and deformed at the same rate.

All other possibilities are probably represented, but they cannot always be clearly recognized because they grade into each other.

There seems to be little difference in the deformation intensity of matrix and oöids as shown in ratios of oöids and pellets within one slide. Figure 6 of Plate 7 for instance shows mud pellets and spherulites side by side with equal deformation ratio, but different sizes. In Figures 12 and 13 of Plate 7 oöids and pellets possess an equal ratio.

[*Editor's Note:* Material has been omitted at this point.]

DESCRIPTION OF DEFORMATION

The present appearance of the deformed oölite depends on its original composition, on the intensity of the distortion, and on postdeformational crystallization.

Crushed oölites have been found only in one slide.

Tectonic deformation ranges from 0 to several hundred per cent and is illustrated in the photographs and camera lucida drawings which are also intended to show the variability of the deformed oöids. Hardly two of over 400 thin sections are alike, and it is difficult to compare types. Orientation of deformation, however, is constant and easily recognized in large areas, in spite of the variability of the oöids. Bedding has become obsolete, and new alignments transsect it.

Two kinds of elements can be distinguished: (1) hard and resistant units composed of "pseudopebbles", and detrital sand grains and (2) deformed, pliable mud pellets, spherulites, and matrix. The former are undeformed or surrounded by pliable material but may also break up into angular fragments. Examples of this type have been described above and are shown in the following illustrations: Figure 3 of Plate 2; Figure 1 of Plate 3; Figures 1 and 10 of Plate 7; Figures 7, 8, 9, and 15 of Plate 8. Fragmentation of pebbles is indicated in Figure 7 of Plate 8.

All the elements of the second group are deformed by the same amount within one slide and there is little indication of differences of ratio between mud pellets or spherulites. Harder cores may, however, lag behind and show less distortion than the outer layers of an oöid. Irregularities can be overcome by measurement of many oöids and elimination of freaks. Only the best material should be used.

Undeformed oöids are recognized by absence of orientation of eccentricity and a wide range (Fig. 3). The matrix lacks flowage structures and orientation of components (Pls. 1, 2; Pl. 7, fig. 3, 4).

A deformation of 10 per cent can already be recognized easily in an alignment of the longest axes of oöids, but matrix and mud pellets remain unaffected (Pl. 7, fig. 9).

Twenty per cent deformation is very obvious and comprises spherulites, pellets, and matrix (Pl. 4, fig. 1; Pl. 8, figs. 1 and 5). If a bedding plane has been included in the thin section it is mildly undulating and due to lateral reduction of the maximum deformation axis at a large angle to the bedding plane as in Figure 6 of Plate 7.

Thirty per cent deformation includes all components, and cleavage begins to appear as in Figure 4 of Plate 4 or Figure 8 of Plate 7. Bedding planes may suffer considerable crenulation (Pl. 7, fig. 13), and solution plays an important part in the accumulation of residue in bedding planes. Cross fractures are filled with white carbonate in Figure 5 of Plate 7.

As intensity of deformation increases, distortion differs slightly in layered oöids and spherulites in that the former remains smooth and seems to conform more readily. The distortion of spherulites is accompanied by radial fractures. The layered oöid in Figure 6 of Plate 4 is deformed 50 per cent without difficulty, the ends are still smooth but a faint parallelism of particles in the cleavage direction appears (*see* also Pl. 4, fig. 5). The spherulite in Figure 2 of Plate 4 is deformed only 40 per cent but has begun to fracture radially. The same stage has been reached in Figure 9 of Plate 8 where the primary oöid is elongated and a secondary apron has grown beyond its limits. To the left of the distorted oöid is an undistorted "pebble".

The contrast between mud pellets and spherulites is visible in Figure 2 of Plate 6. Sperulites are here fractured and elliptical, but the mud layer has exceeded that deformation. The layer is thicker at the ends and has begun to fray.

Fraying becomes more obvious with increasing distortion (Pl. 6, fig. 1; Pl. 4, fig. 3). Measurements become more difficult at that state because it is probable that some of the material has been lost. In well-layered oöids the difficulty may be overcome by measuring an inner layer, but accuracy decreases with increasing deformation. This is partly compensated by increasing regularity of orientation.

Deformation beyond 100 per cent is rarely measurable in the ac plane, but a minimum value may be measured in the combination of the bc and ac planes. Figure 5 of Plate 4 pictures deformation exceeding 100 per cent in the ac plane but a well measurable distortion in bc. The oöids have become long and flattened spindles. A similar situation is shown in Figure 3 of Plate 8 where ac and bc cuts have been drawn for comparison.

Maximum deformation shown is in Figure 14 of Plate 7. The oöids are still recognizable, bedding intensely crenulated, a cleavage dominates, and measurements are barely possible and represent minima only, due to fraying in the direction of the longest axis.

Secondary replacement is localized and easily recognizable (Pl. 3, fig. 2; Pl. 8, fig. 12) Carbonate has filled available spaces and is generally restricted to the matrix, with little influence on the oöids. Much of this growth is in aprons around oöids and especially in the enlargement of spherulites. Postcrystalline deformation is visible in the matrix and spherulites in twinning of carbonate grains and herringbone pattern at the ends (Pl. 5, fig. 1). Since deformation of the oöids is many times that of the matrix the author thinks that the latter crystallized rather late and suffered deformation only during the late stage of deformation.

VARIATIONS OF DEFORMATION INTENSITIES

GENERAL REMARKS

Deformation intensity varies (1) with physical properties and thus with composition of oöids and matrix, (2) regionally with respect to location in the South Mountain fold, (3) in relation to other structure elements, like lineation and cleavage, and with position in a fold.

[*Editor's Note:* Material has been omitted at this point.]

VARIATION AND STRUCTURES

The differences due to position of a sample within a fold are well illustrated in the triangular area between Greencastle, Waynesboro, and Scotland. Axes in this region plunge southwestward and are offset in en echelon pattern. This is well shown on the Chambersburg quadrangle map (Stose, 1909). Between Scotland and Waynesboro at least three folds plunge to the southwest, one near Stonehenge, a second one at Elbrook, and a third at Waynesboro. In Plate 9 the Waynesboro

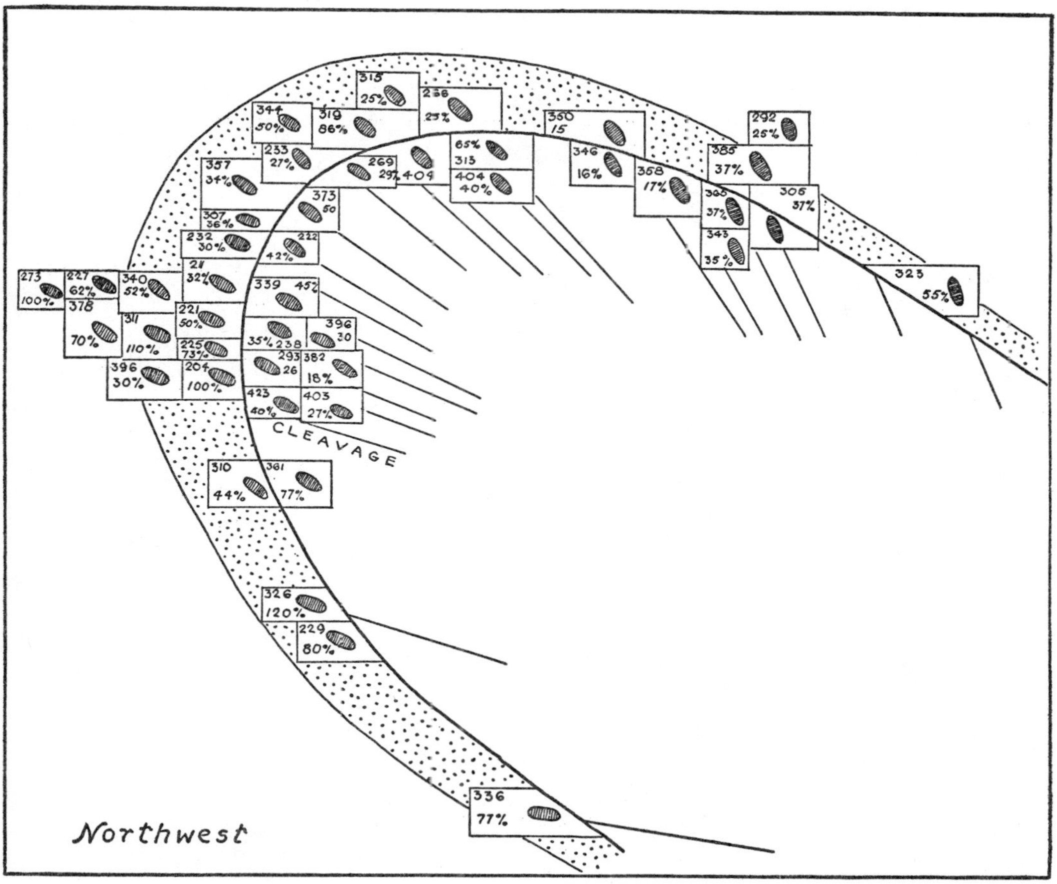

FIGURE 8.—*Reconstruction of fold*

Plat of *ac* sections. Pl. 12, figs. 1-5. Cleavage is arranged in a fan opening to the northwest. Most observations are in the steep limb of the asymmetric fold.

formation as copied from Stose's map reflects these folds. Oölite samples were consequently taken from various parts of those folds and not always from the lower limbs. Low and high values alternate therefore in a rather broad area and lines of equal intensity become confused and indented.

Positions of cöids are illustrated in Figures 1–5 of Plate 12 for the more important points. Intensities are in parentheses in per cent. The specimen are platted as cross sections (*ac*) normal to the fold axes, showing the maximum and minimum oöid axes, cleavage, and bedding in their position in space.

The same values have then been used in the construction of Figures 8 and 9. The

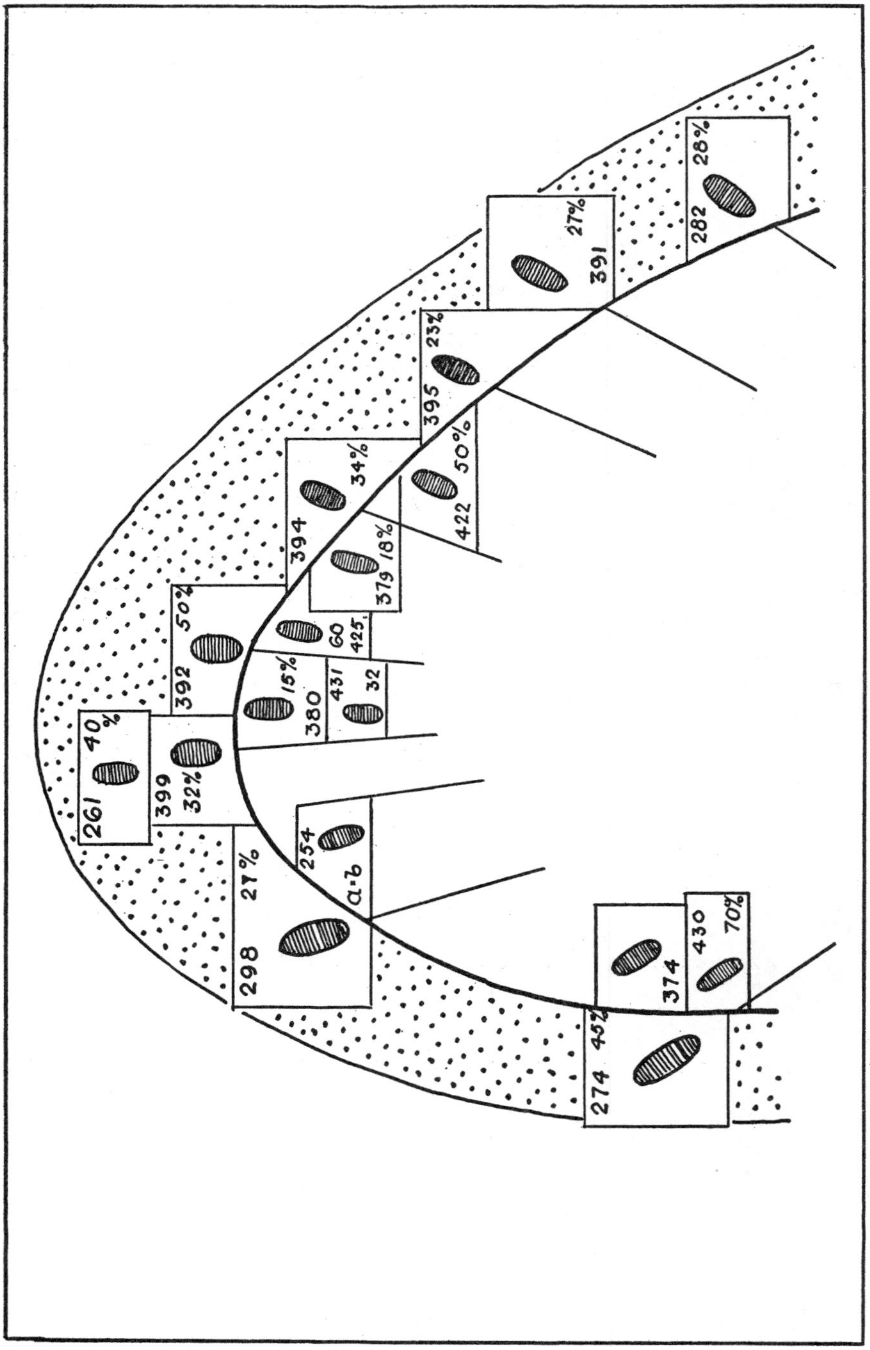

FIGURE 9.—*Reconstruction of upright fold*

Plat of *ac* sections Pl. 12, figs. 1-5 into an upright fold. Of the large number of observations only few are in an upright fold.

bedding has been pieced together to form an overturned fold and a slightly fan-shaped cleavage in Figure 8. In Figure 9 the same process produced an upright fold with steep cleavage. The western limb is somewhat steeper than the eastern one.

In spite of the rather schematical plats in Figures 8 and 9 the position of the oöids in relation to folds, the relation to cleavage, and the distribution of intensities appears very clearly.

Figure 8 shows a concentration of higher intensities in the lower limb. The average value between specimen 336 and the axial plane is 56 per cent. The upper limb average is 36 per cent. There are 4 values exceeding 100 per cent, 9 values over 70 per cent, and only 3 below 30 per cent. In the upper limb no value exceeds 90 per cent; only 2 exceed 65 per cent and there are 8 values below 30 per cent. In spite of the irregularities the upper limb shows generally lower percentages than the lower one. The average deformation intensity is obviously less: no value exceeds 70 per cent, the average is 36 per cent. The fact that this deformation is generally less is also shown in the map (Pl. 9) where all the points near South Mountain show greater deformation intensity and structural asymmetry, largely visible in an intense eastward-dipping cleavage and lineation.

Within the fold of Figure 8 are several groupings of values, which may not be significant in view of the small number of platted values. The crest of the fold shows low values to the east (right) and then higher ones, all in the thirties. Counter-clockwise from the axial plane, values are bunched together because that position of beds is the most frequently met in the field. From the distribution of values within this fold one may be tempted to conclude that intensity of deformation varies in different parts of a fold, in addition to the difference between the lower and upper limb.

Deformation is closely related to the fold and its position in space, and the symmetry of the fold is reflected in the distribution of deformation intensities.

Regional distribution has not been considered in Figures 8 and 9, but the distribution of intensities is closely related to the influence of the South Mountain fold.

ORIENTATION OF DEFORMATION AXES

[*Editor's Note:* Material has been omitted at this point.]

ORIENTATION IN FOLDS

Orientation of oöid axes a and c, bedding, and cleavage are shown in Figures 1–5 of Plate 12 for 74 localities.

Oöid orientation is without exception with a and b in the cleavage plane and with the shortest axis normal to it. Either axes may assume the direction of the fold axis, but as a rule a is normal to the fold axis. In Figures 1–5 of Plate 12 the longest axis parallel with cleavage dip is a maximum deformation axis; b is, as a rule, normal to the section shown.

Deformation axes show only accidental relations to bedding orientation. If cleavage parallels bedding, a and b are in the bedding plane as in Figure 3 of Plate 12, specimens 330 and 333, and in Figure 4 of Plate 12, specimen 258.

Where deformation is intense a lineation appears parallel to *a*, and in the South Mountain area a is down the dip of cleavage or nearly so. If cleavage is vertical as in symmetrical folds a is also vertical in the cleavage plane as for instance in number 392 and 399 of Figure 4 or 375 of Figure 5 of Plate 12. On Plate 10 several vertical axes can be seen near the Maryland-Pennsylvania border west of Waynesboro. Southeast of Chambersburg three arrows point northward and apparently parallel the strike of the fold axes. These arrows however are normal to southward-dipping fold axes. Steep axes at 70° or 80° are common in the Chambersburg-Greencastle-Waynesboro triangle and where upright folds and vertical or steep cleavages prevail.

Axial plane cleavage parallels the axial plane as a rule only in the vicinity of the axial plane. In the limbs of folds it deviates and forms a fan which opens toward the convex side of a fold,—that is upward in anticlines and downward in synclines. This is well shown in many Appalachian folds and has been observed frequently (Albert Heim, 1878; Leith, 1923, Fig. 47). Construction of Figure 8 would be impossible without a cleavage fan. Bedding and cleavage are in their mutual characteristic relative position in Figure 8. So characteristic is the relation between the two that knowledge of the shape permits the cleavage to be drawn; conversely, knowledge of the cleavage pattern would disclose the shape of the associated fold. Further, plotting of öolite elongation would disclose both the cleavage and the fold patterns.

Near South Mountain the overturned lower limb is exposed, hence the dip of cleavage and the pitch of elongation of oölites are gentler than in the valley (*see* also Figure 8).

Distribution of specimen numbers and oöids in Figure 8 shows that vertical or nearly vertical beds were sampled most frequently. Obviously the western slope of South Mountain exposes the vertical beds of an asymmetrical fold.

CHANGE OF VOLUME DURING DEFORMATION

In the calculations of deformation intensities the assumption was made that the volume remained constant because seemingly no material escaped or was added in the process. Since the proposition affects all conclusions so far as they are quantitative the author attempts to further analyze this point.

Several processes may result in volume changes: (1) crystallization, recrystallization, and replacement; (2) disruption of oöids and loss of parts during deformation; (3) compaction and loss of water.

In order to appraise the volume changes the average-sized unit sphere was plotted at an arbitrary scale (Fig. 10). Distortion ratios $\frac{a}{c}$ were then calculated and ellipses constructed with b = constant. The percentages in Figure 10 were chosen to represent some of the values observed frequently. Percentages have been calculated as extension of the unit sphere or $\frac{a}{r} \times 100$. This theoretical diagram shows the configuration to be expected with deformation without change of volume.

The measured values in Figure 11 differ from the theoretical diagram in many respects: The three highest percentages—85,100, and 125—are much smaller than the theoretical values. Unit sphere and ellipse are equal to the calculated values up to 58 per cent. From there on upward they are increasingly undersized.

Several possible reasons present themselves: Oöid substance may be lost with increase of deformation intensity; measurements may become less reliable particularly in the direction of maximum deformation when ends begin to fray as in Figure 2 of Plate 6, and Figure 5 of Plate 4. In Figure 11 the source material may have been of unequal size which explains the difference between the 10- and 23-per cent diagrams and their unit spheres. That difference would scarcely be as large as it is in the last

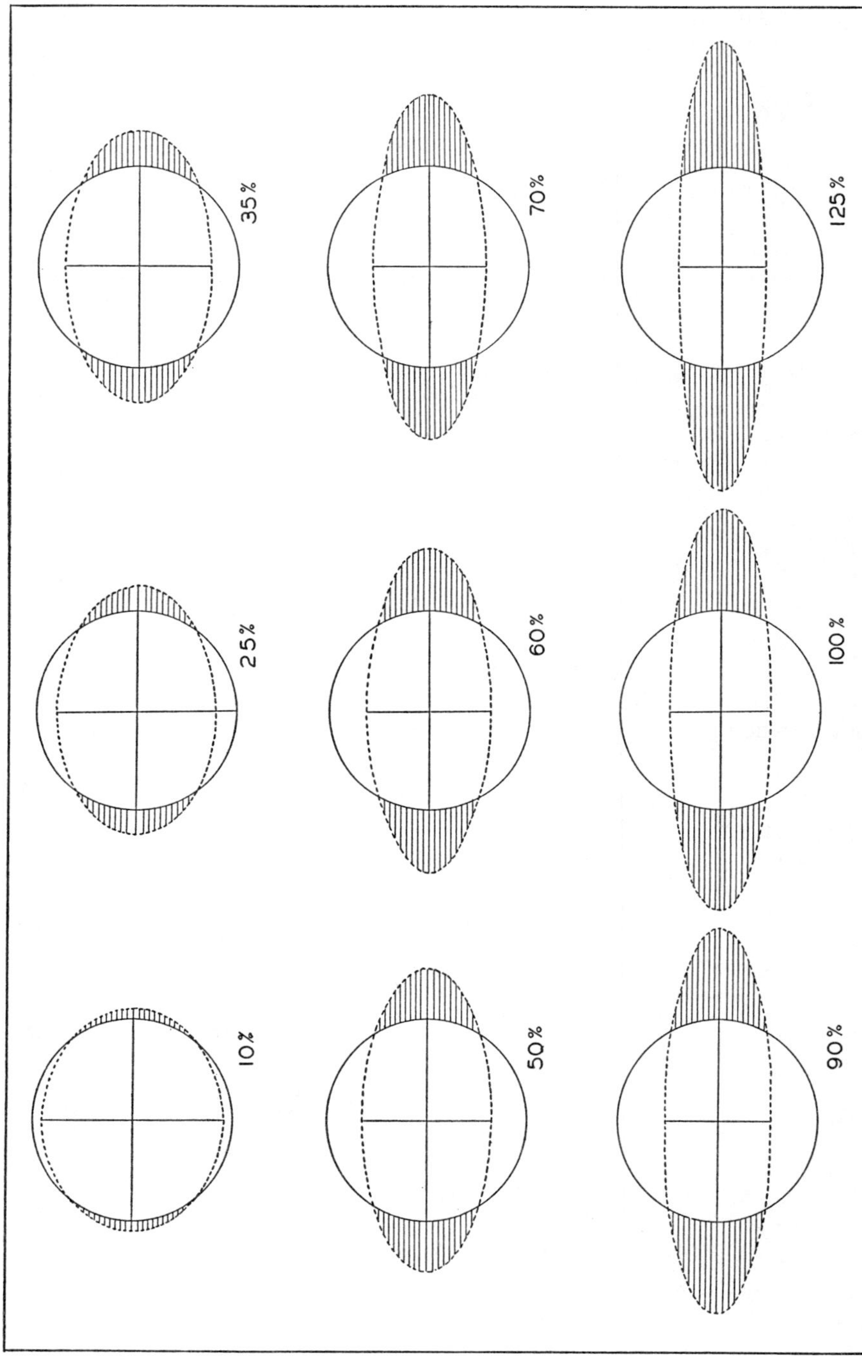

FIGURE 10.—*Geometrical solution of theoretical deformation*
In reference to one unit sphere for constant volume.

three examples. The author thinks that there is a loss of material and inaccuracy due to deformation of the ends in the a direction. It is advisable to use a section normal to the one shown, including the b and c axes, in order to verify the calculations of a and r. In spite of this inaccuracy, ratio and orientation remain the same for large or small oöids or unit spheres as long as the shapes remain similar. The large sphere and ellipse in Figure 10 show a 100 per cent deformation like the values in Figure 11 which are only half that size.

The author feels that the absolute size may not be decisive as long as values are computed as ratios or in percentages, and the third dimension b remains nearly constant. Distortion of b is fortunately small particularly in case of high $\frac{a}{c}$ ratios. Deformation of b (in the direction of the fold axes) is discussed below.

Loss due to compaction or loss of water tends to affect all oöids and matrix and would seem to be harmless because its value enters all calculations equally.

Crystallization and replacement are very common and can be seen in every slide. Due to ease of detection they are harmless as long as the original outlines of oöids remain visible. Almost all the long axes have grown as in Figure 1 of Plate 5 and in Figures 1 and 10 of Plate 8. Growth of aprons and rims does not always indicate increase of diameter because aprons grow as well in c as in a, even if the c axis has shrunk noticeably. Growth and crystallization are postdeformational and therefore only important if original textures have been destroyed or obscured. Even where the entire slide is crystallized, original outlines remain well visible, but thick slides may be necessary to detect the yellow-brown tint of the original oöid.

[*Editor's Note:* Material has been omitted at this point.]

DETERMINATION OF THE THIRD DIMENSION

Deformation in *b* can be determined in longitudinal sections or on maps and locally by observation of distortion of bodies of known shape. The author has here determined *b* at 119 localities where two mutually perpendicular thin sections were made. The values were calculated also in per cent of extension of the unit sphere. Figure 12 is the graphical evaluation. Deformation in *b* is shown horizontally, deformation in *a* vertically. Since both scales are equal, the line representing *a* equal *b* is at 45°. Below that line, *b* is larger than *a*, above it *a* is larger than *b*. At 15 localities the 2 axes are equal.

The amounts of deformation in *a* and *b* are mutually independent. The author had feared that errors may have entered and would appear as proportional increase of deformation in *a* or that in high values of a, b might decrease because of inaccurate measurements. Any proportional interdependence between the two axes would have to be investigated for errors. Eighty-seven per cent of all values for *b* are between 0 and 10 per cent and independent of distortion in *a*. Three *b* values are less than 10 per cent in spite of high values for *a* of more than 100 per cent. The highest deformation for *b* is 32 per cent and in the two cases *a* equals *b*. The oöids were merely flattened to discs.

The insert of Figure 12 shows percentages of deformation in *b* plotted horizontally and the number of observations is shown vertically as bars. The graph shows the summation of observations as intervals: 0, 1–5, 6–10, 11–15, 16–20, 21–25, and 26–30. The low numbers at 1, 2, and 3 are due to computation procedure. The author considered one per cent equal to zero because the accuracy of measurements is within 5, but scarcely one per cent. Twenty-six values are zero, 24 between 1 and 5, and 29 between 6 and 10 per cent. From there the values drop and remain low beyond 20 per cent. The peak between 6 and 10 per cent is significant because it points to a real maximum and probably means that deformation is three dimensional at many localities.

In an area of this size and with the rather intricate arrangement of axes it can be expected that deformation in the third dimension will vary considerably.

REGIONAL ARRANGEMENT

The (b = *b*) axes were platted on Plate 11, and an attempt was made to compare these with field measurements and to study possible relations between pitch and deformation intensities.

Plate 11 shows the intersections of cleavage and bedding, measured *b* axes, and fold axes. Double arrows are horizontal axes. Where *b* is horizontal, the *ac* plane is vertical and therefore normal to the axial trend.

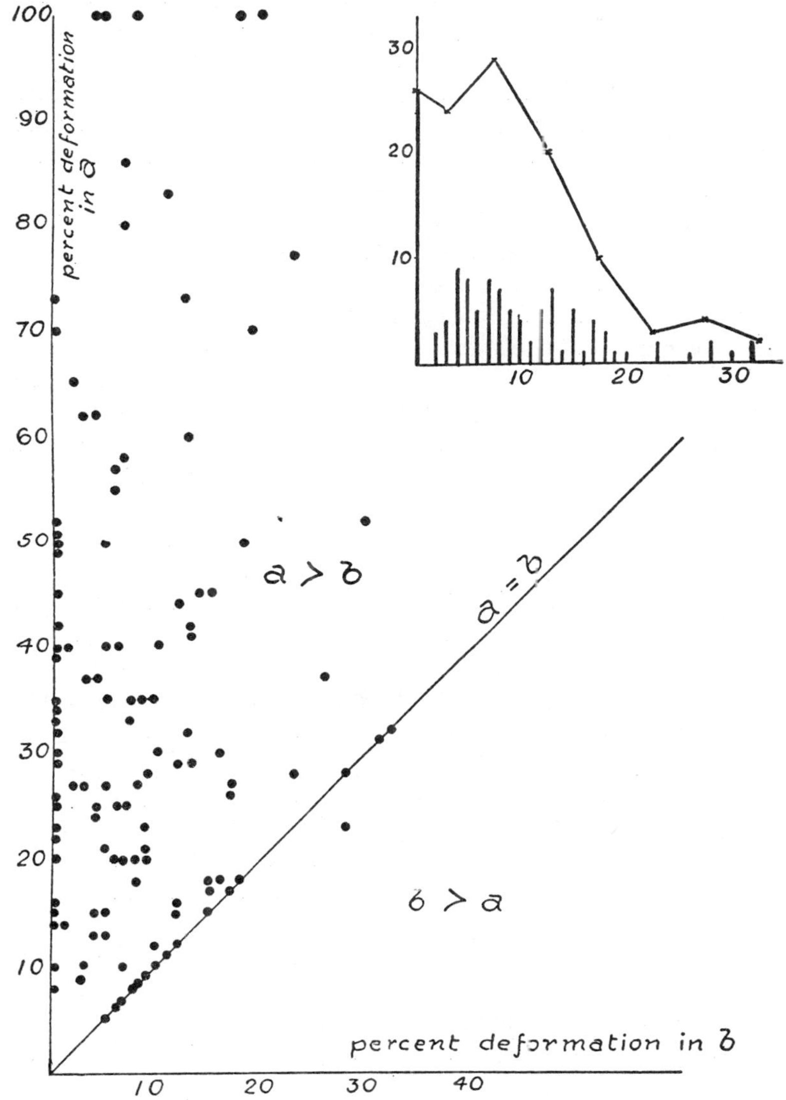

FIGURE 12.—*Deformation of* b *axes in relation to deformation in* a

Insert: number of measurements shown as bars.

This trend is constantly northnortheast and northeast in spite of arcuation of South Mountain. Between the Potomac River and Waynesboro the axes parallel the large fold and uplift to the east and remain essentially horizontal. In the vicinity of Waynesboro and to the north, axes pitch south-southwest or are horizontal, and at Stonehenge near Chambersburg the directions are confused and deviated at the southwest end of a plunging fold. From here to the northeast, the trend is to the northeast. South of Newville and Carlisle the northeast trend continues, and the axes dip off South Mountain.

Little of the curvature of the South Mountain fold is accomplished by curving axes, but mostly by en echelon arrangement of subparallel trends. Between Waynesboro and Chambersburg the western boundary of the uplift is north-south, but accomplished by southwestward pitching folds with constant strike. The northern portion of the uplift seems to have advanced further northwestward by addition of at least three large folds which are well shown in the trend of the Waynesboro formation (Pl. 11).

Intensities are shown on Plate 11 in large figures. Near Keedysville there is considerable but rather irregularly distributed three-dimensional deformation. Between Waynesboro and Greencastle the fold axes are undistorted; southeast of Chambersburg at the end of the plunging fold near Stonehenge deformation is considerable. Near Shippensburg there is little or no distortion, and from here on east there is always some elongation where the axes plunge off the northern slope of the uplift. Here the larger values are at greater distance from the uplift than the lower ones. South of Carlisle, the values decrease from 17 to 13, south of Newville from 6 to 2, and 12 to 4 toward the Mountain. Whether this distribution is significant or accidental is uncertain. More detailed information may provide a more complete picture. Near Smithville the values increase eastward, and in the southern portion they seem to increase east and westward from a central zone of low values.

The irregular distribution of intensities shows that the lineation parallel to *b* cannot very well be due to elongation of fold axes due to intense deformation in the folding process. The linear structures, presence or absence of cleavage, or intensity of folding show no relation to the distortion in *b*. In such a large, uniform, and asymmetrical fold the author had expected to find a much more intense and uniformly distributed distortion in the subordinate direction (E. Cloos, 1946, p. 12).

Lack of deformation is very significant and shows that lineation in *b* is due to intersecting shear planes rather than elongation especially in the areas of low values over a wide area with recognizedly intense deformation as shown by asymmetry of folds, cleavage, intense lineations, and thrusting. This may mean that elongation of *b* may generally be low also in very intensely deformed areas. The author thinks that this determination shows for instance that pebbles of extreme distortion values as described by Goldschmidt (1916), Strand (1945), Kvale (1945) and others, cannot be the result of stretching in the direction of fold axes due to movement normal to them. The author has recently discussed this problem at some length (E. Cloos, 1946, p. 30).

ELONGATION OF b IN ARCUATION

The existence of regions without elongation in the direction of the fold axes indicates that Appalachian arcuation is not entirely the result of bending of an originally straight line into a curve. It also illustrates that the larger forward surge of folds in the center of the arc may be accomplished without intense stretching. The outer portions of an arcuated orogen should show larger elongation of *b* than the

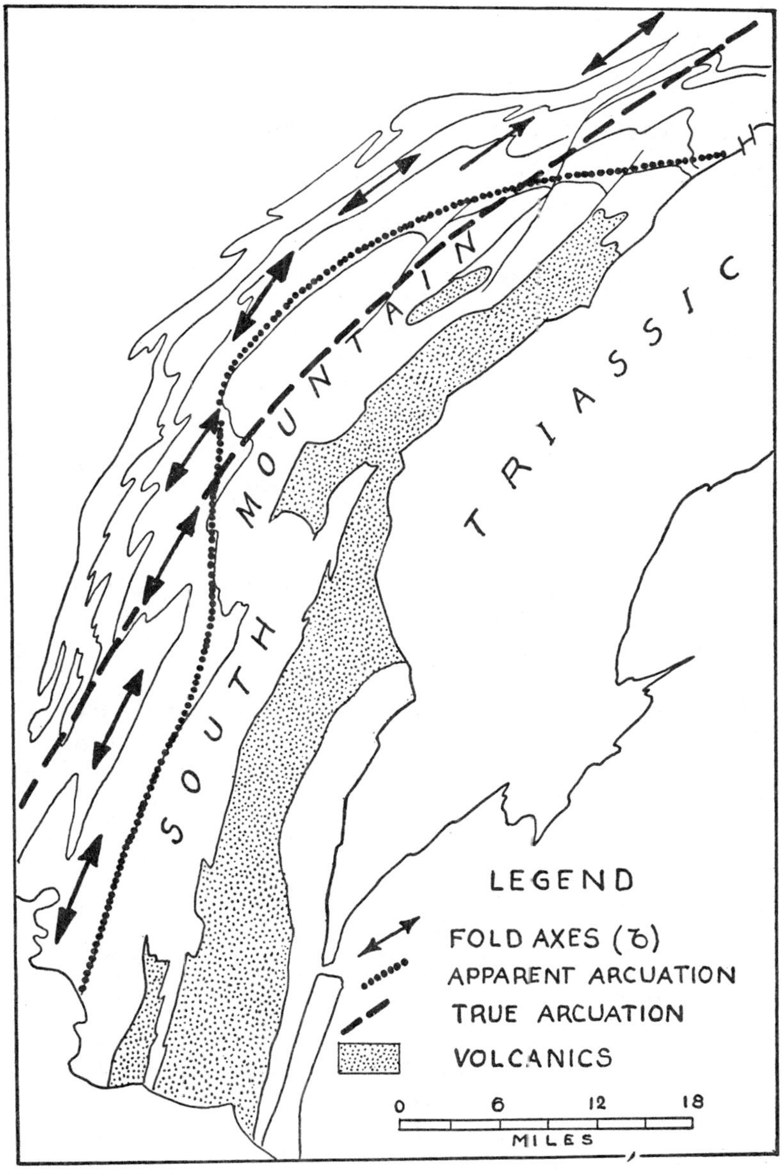

FIGURE 13.—*Apparent and true arcuation of South Mountain fold*

inner ones. The author knows of no systematic investigation of this question but feels that it would be most fruitful to measure deformation in *b* at many points in an arcuated orogen as the Appalachians, the Jura, or the Alps. Such a study may throw much light on the order of magnitude of lateral movement because folding would have to be radially outward with constantly increasing distances between points along the strike of the axes. The amount of increase will then be an indication of divergence and also arcuation.

Application of this principle to the present area is shown in Figure 13. A rough outline copied from the State maps of Pennsylvania and Maryland (1933) shows the apparent arcuation of the area as seen physiographically and in the trend of the formations. If the fold axes are plotted and folds traced, however, they show that the forward bulge is produced by a number of additional folds in the northern half of the map area and that the axes plunge off the uplift southwest and northeastward.

The remaining arcuation is probably due to varying folding intensity and the number of folds that account for the lateral shortening. The rather small elongation of fold axes seems to indicate absence of true arcuation of a larger order of magnitude. Such extension as has been described by Albert Heim (1921) in the Jura is lacking. There, diagonal faults increase the extension of the orogen.

SUMMARY

The third dimension should not be neglected since it can be a rather large factor. In the South Mountain area there is no large-scale and uniform elongation parallel to the trend of folds, but there is considerable local extension.

Extension in *b* is unrelated to extension in *a*, and neither value is related to the axial plunge.

The absence of uniform axial extension shows that the linear structures paralleling fold axes are not due to stretching and that the very intense lineation parallel *a* is due to movements in that direction.

Actual arcuation in the area is smaller than apparent arcuation. The limits of South Mountain fold are determined by en echelon folds.

[*Editor's Note:* Material has been omitted at this point.]

DEFORMATION ANALYSIS

THE DEFORMATION PLAN

Sander (1930, p. 57) described several deformation plans ("movement pictures", Knopf, 1938, p. 26) and studied their symmetries, for the deformation plan may reflect a setup of forces.

However, not all plans are directly indicative of the movement picture, and the author has recently tried to emphasize that the movement picture may be correlated only indirectly and through subordinate movements with principal directions (E. Cloos, 1946, p. 22). It is therefore most important to include an area large enough to comprise subordinate as well as principal movements. A thin section is mostly too small to show the whole plan, and detailed field work is always necessary in order to properly place the thin section within the master plan.

THE SOUTH MOUNTAIN PLAN

As seen in Plate 11 and Figure 15 the fold axes are largely horizontal or pitch very gently. Maximum deformation of oöids (a) is normal to the fold axes and pitches southeast where folds are overturned to the northwest (Pl. 9). a is a megascopical lineation, whose intensity and visibility also grows with intensity of deformation. The cleavage plane is determined by a and b without exception. Cleavage also grows more conspicuous with increasing deformation intensity. Bedding is transected by cleavage, and their intersection always parallels the fold axes. The deformed oöids are strictly correlated to this plan. The maximum axes a parallels co-ordinate *a*, the mean axis b is coordinate *b* (also *B*), and c is normal to the *ab* plane, which is at the same time the cleavage plane. Symmetry is monoclinic with the ac (and *ac*) plane as symmetry plane.

Figure 15 shows the axial part of an overturned fold as found along the western slope of the Mountain. The mutual positions of structure elements is constant in a large area, and the only variation is in the orientation of all elements in space. In upright folds cleavage is vertical or nearly so, and the bedding is horizontal in the crest.

The deformation is accompanied by rotation of bedding into a vertical position. This rotation may be flexure or slippage on subparallel planes and subsequent shearing. Rolling and rotation are absent as indicated by the regular subparallel arrangement of elongate oöids, whose positions remained unchanged since early deformation.

LINEATION AND CLEAVAGE

Lineation has been discussed recently at some length (E. Cloos, 1946) and is very prominent in the entire South Mountain fold.

Parallelism of oöid extension and lineation permits quick orientation in the field and facilitates choice of section for sawing and thin sections. Even the more massive

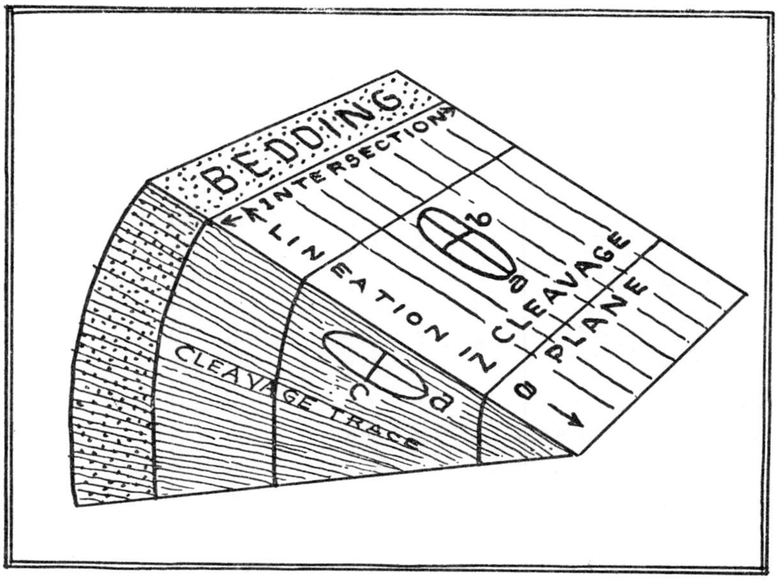

FIGURE 15.—*Deformation plan in South Mountain*

Axial portion of overturned fold showing mutual relations of bedding, cleavage, oöid extension, and lineation.

lower Cambrian quartzites show streaks, lines, clusters of magnetite, and stretched pyrite, and under the microscope quartz needles in *a* as described by Fellows (1943).

The volcanics in the core of the uplift are also strongly lineated. Cleavage planes are covered with chlorite blebs about 1 or 2 inches long, $\frac{1}{4}$ or $\frac{1}{2}$ inches wide, and paper thin. In the more massive greenstones amygdaloidal cavities or deep-green epidote hornblende or chlorite blebs are very common and also well oriented. The author has called attention to this structure repeatedly (E. Cloos, 1940, p. 866, fig. 8A; Cloos and Hietanen, 1941, Pl. 8; 1946, Pl. 2).

Maximum elongation of oöids is lineation in *a* and without exception normal to *b* (*B*) and the fold axes.

A second pronounced lineation parallels the intersection of bedding and cleavage and the fold axes *b* (*B*).

Cleavage is absent where deformation is feeble and grows stronger with increasing deformation. Approaching South Mountain, cleavage and lineation appear first as mere oöid orientation which intensifies southeastward. It is possible to identify the deformation plan in the orientation of oöids long before cleavage or lineation become megascopically visible.

In the oölites and associated Paleozoic rocks cleavage and lineation mark stages in a progressively more intense deformation of soft or maybe unconsolidated rocks. Limestones were formed under strain of varying intensity, and, as shown by the microscopic picture, oöids were deformed according to a uniform tectonic plan and pattern before crystallization of the matrix. Tectonites formed before rocks.

Oöid deformation may be compared with formation of flow structure in igneous rocks. Here constituents were arranged in patterns prior to consolidation during a viscous phase, and therefore a normal crystallization sequence exists without deformation, but with orientations in various intensities. Synkinematic intrusions furnish the most intensely textured rocks, and they are hard to distinguish from metamorphics. The oölites may be compared with a synkinematic intrusion of which only a portion was affected by deformation during intrusion. At one end there are primary flow structures and an undisturbed crystallization sequence. At the other end strong cleavages and lineations and intense deformation seem to indicate a crystalline schist. Between the two are gradations, but the whole intrusive shows one deformation plan. Intensely deformed oölites are schistose rocks, but carbonate growth and pebbles are undisturbed. The undeformed oölite shows the same crystallization phenomena, but there are fossils and no cleavage or lineation in *a*.

DISCUSSION OF CLEAVAGE

Definitions.—Cleavage as here used is the planar discontinuity which is defined by the position of the a and b axes of the ellipsoidal oöids, whether visible in slightly or strongly deformed oölites. It coincides with the *ab* plane of the co-ordinate system, and its dip fluctuates about the fold axis. Its strike parallels the axial trends where axes are horizontal. It also contains the lineation in a (*a*). Finally c = *c* is normal to this cleavage. Where oöids are only slightly deformed this cleavage remains invisible.

Near the axial plane of folds, cleavage is parallel to it and thus becomes axial plane cleavage. Since it is due to flowage it may also be called a flow cleavage.

The cleavage is therefore a secondary *s*-plane parallel to ab = *ab* and interesecting a primary bedding-*s* (depositional *s*) in *b* (fold axis). It grades from an incipient latent stage through faint visibility into a very prominent and dominating structure which obliterates bedding and leads geologists to believe it were bedding. In all its stages it maintains constant relationship to all other structure elements within the deformation plan (Fig. 15).

Cleavage interpretation.—A very useful summary and discussion of cleavage theories has been presented by Turner (1936; 1942). A more complete but more intricate discussion is by Sander (1930).

A review of cleavage theories shows that the discussion focuses on the question whether the cleavage, schistosity, or foliation formed as a shear plane at an angle to the major extension or strain, or in the plane including the major and mean strain axes.

One major difficulty in the interpretation of cleavage as shear plane is the onesidedness of most cleavages where there should be two shear planes in intersecting directions. This difficulty may be overcome in rotational deformation in which the two directions rotate at unequal rate, thus producing deformation of different kinds, possibly flow and fracture cleavage. This would require that the two cleavages, if correlated intersect in *b* (*B*).

The oöids were spherical at the beginning and have become ellipsoidal due to deformation which is clearly related to the folding of South Mountain and the Appalachian in general. Crystalization is later and adds to the visibility of the orientation. Cleavage forms in the AB plane of Becker's strain ellipsoid and not in a shearing direction.

This possibility has been discussed by Sander (1930, p. 79) for material of "very low rigidity" as for instance a dough which is deformed by its weight or for viscous liquids. In this case the strain ellispoids become so flat that their circular sections almost coincide with the AB plane equal *ab*. The cleavage plane then becomes *ab*. Becker's strain tectonics may be less applicable than the idea of tectonic flow (*tektonisches Strömen*) and that nonrigid bodies be preferably analyzed by application of the mechanics of viscous flow.

Cleavage in the South Mountain fold is under these premises due to laminar flow in flow planes (ab=*ab*), and the lineation is a cylindrical flow in flow lines.

Within the asymmetrical fold the cleavage (and ab planes) dip toward the center of the fold, and cleavage as well as lineation grow more intense in the same direction. If intensity of a deformation is proportional to the amount of (distance) movement that took place, maximum intensity is then located near the center of the large fold where movements must have been larger because the distance from the horizontal to the crest of an asymmetrical fold is larger than to the flanks.

Oöid extension by laminar flow at a soft stage would rule out the possibility of a flexure fold. The implication of this conslusion in thickening and thinning of stratigraphic sections is discussed below.

[*Editor's Note:* Material has been omitted at this point.]

STRATIGRAPHIC THICKNESSES

At an earlier occasion (E. Cloos, 1942) the author has called attention to the fact that stratigraphic thicknesses as generally measured in the field may include considerable distortion in folded areas and that such measurements cannot always be trusted to represent original thicknesses. It is necessary to analyze the deformation that a section may have suffered to a much larger extent than is customary, since stratigraphic thickness enters all discussions of depths of geosynclines and stratigraphic correlations.

As far as the author knows there are very few if any sections which include analyses of distortion, unless it is so intense that it has led the author to conclude that measurements are hopeless due to recrystallization.

The attitude of beds is rarely indicated which renders construction of cross sections impossible, and the reader must depend entirely on the interpretation of the author as shown in his sections.

Due to the deplorable habit of omitting structure symbols in maps many authors have also not indicated cleavage attitudes. This makes an appraisal of possible

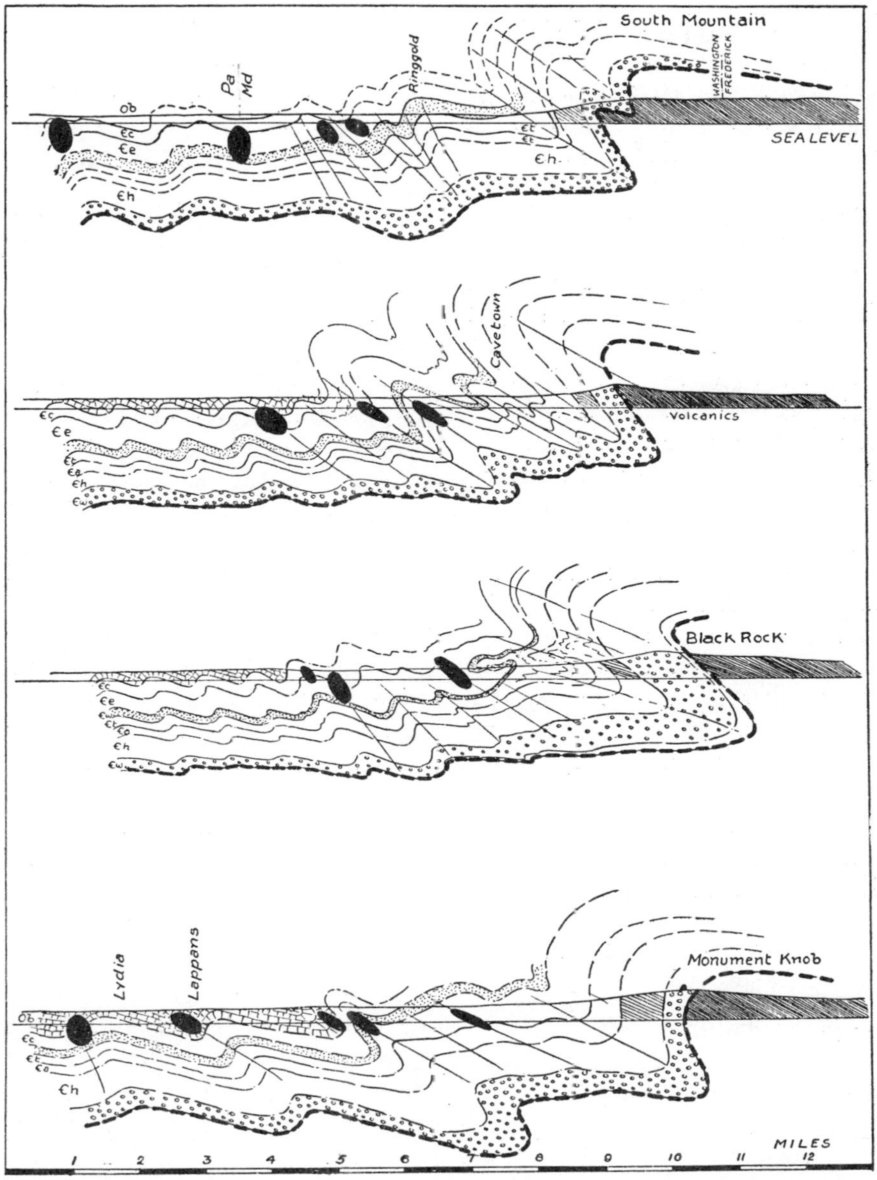

FIGURE 20.—*Four cross sections through the western slope and crestal portion of the South Mountain fold*

Black ellipses are constructed from oöid measurements.

distortion due to folding impossible. The practice to record geological evidence only selectively and incompletely is objectionable and has been partly abandoned by paleontologists and mineralogists-petrographers who measure dimensions of fossils or refractive indices respectively. There is little use, however, to be accurate to the third decimal in the fabrication of bricks if the architect omits the use of plumb line and rectangles in the construction of the house.

Analysis of the South Mountain fold shows that all the stratigraphic thicknesses in the area are distorted. In many sections this distortion can be shown directly and is demonstrated in Figure 21 A–C.

In Figure 21 a block of constant volume has been deformed up to 100 per cent, thus the bed becomes thicker as the block becomes lower. The third dimension remained unchanged.

Distortion of sections varies, however, with its position in the fold. Oölite extension normal to, or at large angles to, bedding is common because vertical beds are encountered along the western slope of the uplift and are near the crest of the fold as shown in Figure 20. Cleavage dips gently east.

All observations on oölites prove that in the South Mountain fold maximum extension is in the cleavage plane and also normal to the fold axes. In the limestone, cleavage appears where oöid deformation exceeds approximately 20 per cent. In the shales it becomes visible earlier. This may mean then that cleavage at a large angle or normal to bedding indicates increased sections. It would be of considerable interest and importance to determine relations between cleavage, bedding, and stratigraphic thicknesses systematically at many localities.

Few stratigraphers and geologists include statements on cleavage, an indication of its intensity, angular relation in space, or its relation to bedding.

The many obvious fluctuations are meaningless without such information. They may not be due to differences in depth of deposition but may be a function of distortion due to folding particularly if diastrophism introduced diagenesis.

This relationship is most important and rarely recognized as a problem. It means, however, that where soft sediments have been folded traces of deformation are obliterated later, and a stratigrapher may never suspect distortion.

Undistorted fossils may indicate lack of deformation, but they may also have been hard-shelled fossils in a soft matrix like the calcite pebbles and crystals in the oölites.

Thickening in anticlinal crests is well known, recognized easily, and pictured frequently. It proves that beds have thickened at a large angle to bedding and that it may have done so at many points without being detected. In the presence of cleavage great care is called for.

[*Editor's Note:* Material has been omitted at this point.]

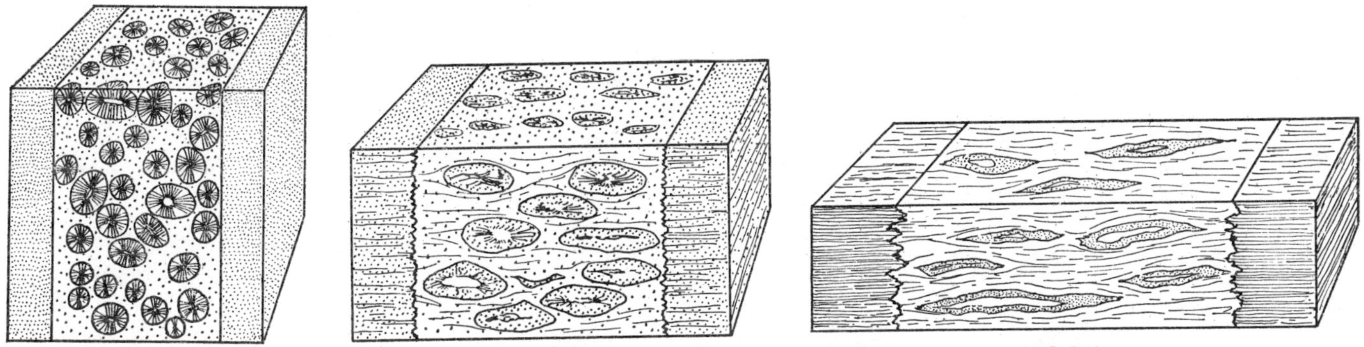

FIGURE 21.—*Elongation and bedding*

A: block diagram of undeformed oölite. B: deformation 50 per cent. C: deformation 100 per cent. The block is doubled in width and half as high. Bedding shows intense crenulation and has also doubled in thickness

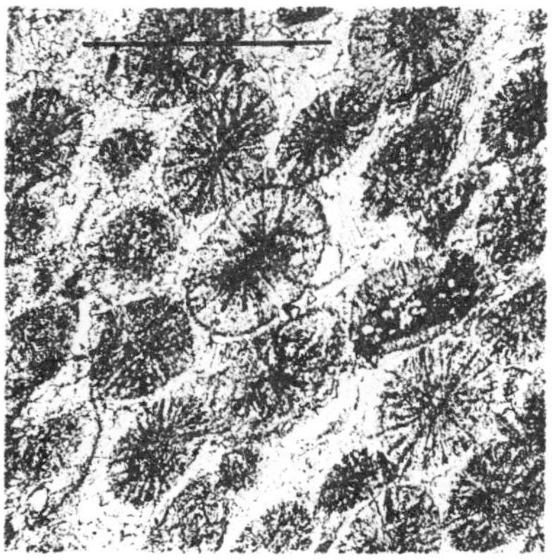

FIGURE 1. DEFORMED SPHERULITES IN POSTDEFORMATIONAL MATRIX
Loc. 397.

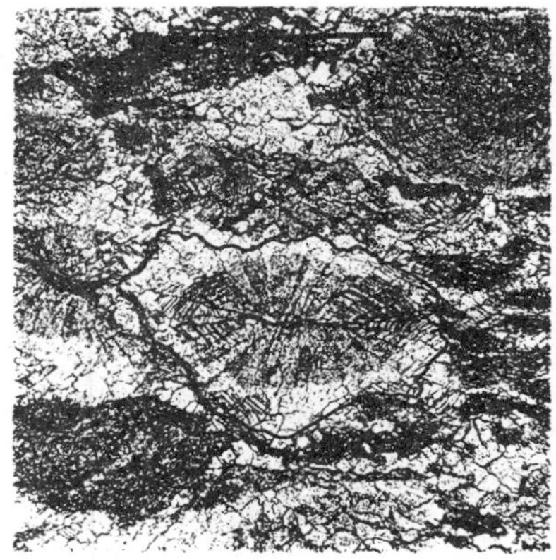

FIGURE 2. SPHERULITE WITH RADIAL FRACTURES
Probably due to compression. Growth aprons grow into matrix. Deformation 40 per cent. Loc. 344 A.

FIGURE 3. OÖLITE WITH STRONGLY DEFORMED OÖIDS
Partly with chert centers and solution in cleavage planes. Loc. 404.

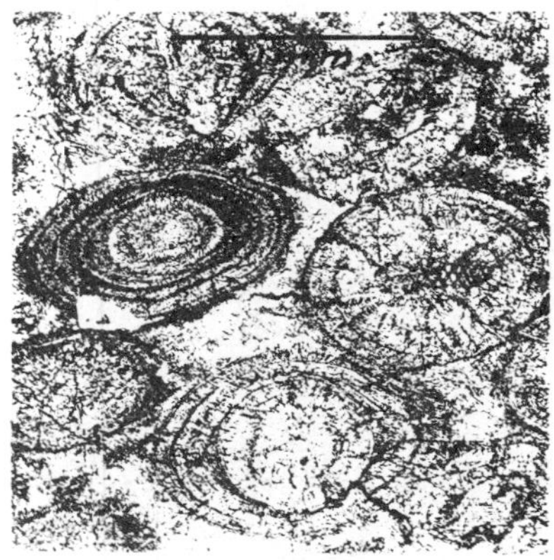

FIGURE 4. RINGED SPHERULITES WITH APRONS
Cleavage becomes visible from left to right. Deformation 30 per cent. Loc. 266.

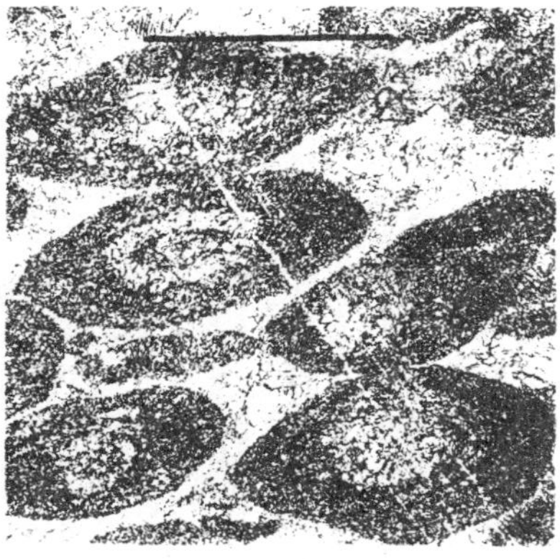

FIGURE 5. COMBINATION OF SPHERULITES WITH MUD RINGS
Deformation in *ac* exceeds 100 per cent. That oöids are flattened spindels is shown in *bc* section. Loc. 204.

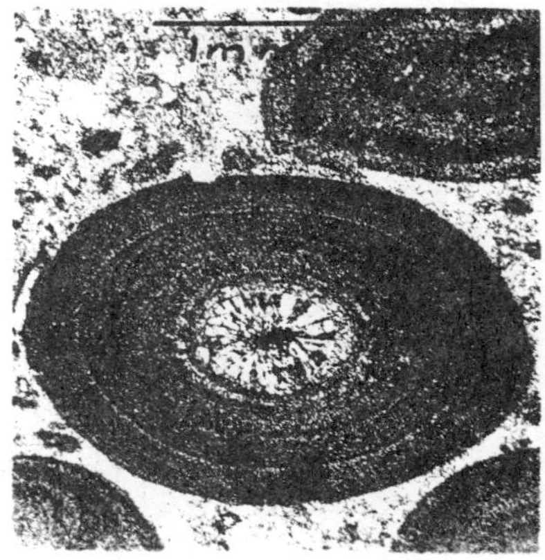

FIGURE 6. CONCENTRIC MUD-PELLET OÖLITE WITH SPHERULITE CENTER AND EVEN DEFORMATION
Parallel texture is faintly visible within oöid in the muddy portions. Loc. 287½.

DEFORMED OÖIDS

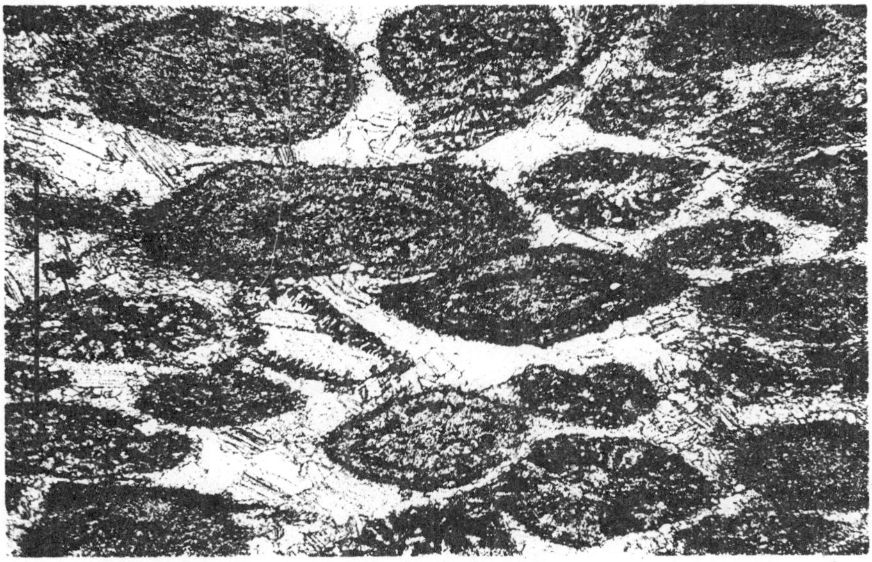

FIGURE 1. ELONGATION OF CONCENTRIC AND SPHERULITIC OÖIDS
Has reached 60 per cent, and the ends begin to fray. Matrix clear, little-deformed twinned calcite shows some strain. Loc. 313 N. of road intersection 2 miles west of Scotland, Chambersburg quadrangle.

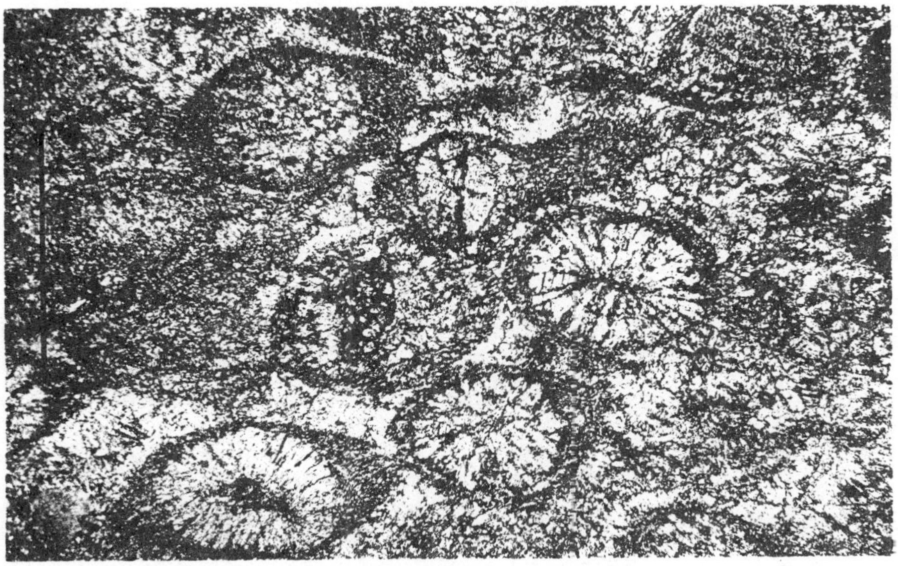

FIGURE 2. SPHERULITES WITH MUD RINGS WHICH PREVENT GROWTH OF APRONS
Deformation results in thickening of mud rings in the direction of elongation. Deformation 40 per cent. Loc. 328, Carlisle-Dillsburg road, north slope of Stony Ridge, Carlisle quadrangle.

DEFORMED AND FRAYED OÖIDS

REFERENCES CITED

Becker, G. F. (1893) *Finite homogeneous strain, flow, and rupture of rocks*, Geol. Soc. Am. Bull., vol. 4, p. 13–90.

Behre, Charles H., Jr. (1933) *Slate in Pennsylvania*, Pa. Geol. Survey, Bull. M 16, 400 pages.

Billings, M. P. (1942) *Structural geology*, Prentice Hall, New York, 473 pages.

Bornemann, J. G. (1886) *Beiträge zur Kenntnis des Muschelkalks, insbesondere der Schichtenfolge und der Gesteine des Unteren Muschelkalks in Thüringen*, Jahrb. Preuss. Geol. Landesanstalt 1885, p. 267–320.

Brown, T. C. (1914) *Origin of oölites and the oölitic texture in rocks*, Geol. Soc. Am. Bull., vol. 25, p. 745–780.

Bucher, Walter H. (1918) *On oölites and spherulites*, Jour. Geol., vol. 26, p. 593–609.

Busk, H. G. (1929) *Earth flexures*, Cambridge Univ., Press, 104 pages.

•Butts, Charles (1933) *Geologic map of the Appalachian Valley of Virginia with explanatory text*, Va. Geol. Survey, Bull. 42.

Chapman, Frederick (1893) *On oölitic and other limestones with sheared structure from Ilfracombe*, Geol. Mag., new ser., dec. 3, vol. 10, p. 100–104.

Cloos, Ernst (1932) *Feather joints as indicators of the direction of movements on faults, thrusts, joints, and magmatic contacts*, Nat. Acad. Sci., Pr., vol. 18, p. 387–395.

——— (1937) *The application of recent structural methods in the interpretation of the crystalline rocks of Maryland*, Md. Geol. Survey, vol. 13, p. 25–108.

——— (1940) *Crustal shortening and axial divergence in the Appalachians of Southeastern Pennsylvania and Maryland*, Geol. Soc. Am Bull. 51, pp. 845–872.

——— (1941) *Flowage and cleavage in Appalachian folding*, Trans. N. Y. Acad. Sci., vol. 3, p. 185–190.

——— (1942) *Distortion of stratigraphic thicknesses due to folding*, Nat. Acad. Sci., Pr., vol. 28, p. 401–407.

——— (1942) *Geologic map of Washington County, Maryland*, Geol. Survey, 1:62 500.

——— (1943) *Method of measuring changes of stratigraphic thicknesses due to flowage and folding*, Am. Geophys. Union, Tr., p. 273–280.

——— (1944) *Correlation of lineation with rock-movement*, Am. Geophys. Union, Tr., p. 660–662.

——— (1946) *Lineation, a critical review and annotated bibliography*, Geol. Soc. Am. Mem. 18, 122 pages.

Cloos, Ernst, and Hietanen, Anna (1941) *Geology of the "Martic overthrust" and the Glenarm series in Pennsylvania and Maryland*, Geol. Soc. Am., Spec. Papers, no. 35, 207 pages.

Cloos, Hans (1936) *Einführung in die Geologie*, Borntraeger, Berlin, 503 pages.

———, and Martin, Henno (1932) *Der Gang einer Falte*, Fortschr. Geol. Palaeont., H. 433, p. 74–88.

Corin, F. (1932) *A propos du boudinage en Ardenne*, Soc. Belge Geol., Bull., vol. 42, p. 101–117.

Dana, E. S. (1932) *Textbook of Mineralogy*, 4th Ed., John Wiley & Sons, New York.

Daubré, A. (1876) *Experiences sur la schistosité des roches et sur les déformations de fossiles corrélatives de ce phénomène; conséquences géologiques qu'ont peut déduire*, Geol. Soc. France, Bull., 3d ser vol. 4, p. 529–551.

Drescher, F. K. (1932) *Über Quartzgefügeregelung im Dattelquartzit von Krummendorf (Schlesien)*, Mineral Petrog. Mitt., vol. 42, p. 217–263.

Dufet, Henry (1875) *Note sur les déformations des fossiles contenus dans les roches schisteuses, et sur la détermination de quelques espèces due genre Ogygia*, Annales école Norm. Supérieure, Paris, 2d ser., vol. 4, p. 183–190.

Fairbairn, H. W. (1936) *Elongation in deformed rocks*, Jour. Geol., vol. 44, p. 670–680.

Fellows, Robert E. (1943) *Recrystallization and flowage in Appalachian quartzite*, Geol. Soc. Am., Bull. 54, p. 1399–1432.

Forrester, J. D. (1946) *Principles of field and mining geology*, John Wiley, New York, 647 pages.

Frantzen (1881) *Terebratula Eckinoi. sp. und das Lager dieser Versteinerung bei Meiningen*, Jahrb. Preuss. Geol. L. A., p. 157–174.

Gaub, Friederich (1910) *Die Jurassischen oölite der Schwäbishen Alb.* Geol. Paleont. Abh., N.F., Heft, 1, vol. 9, 79 pages.

Goldschmidt, V. M. (1916) *Konglomeraterne inden Høifjeldskvartsen*, Norges Geol. Unders., no. 77, 61 pages

Greenly, Edward (1919) *The geology of Anglesey*, Geol. Survey Great Britain, Mem., 980 pages.

Hadley, Jarvis B. (1942) *Stratigraphy, structure, and petrology of the Mt. Cube area, New Hampshire*, Geol. Soc. Am., Bull., vol. 53, p. 113–176.

Hatch, F. H., and Rastall, R. H. (1938) *The petrology of the sedimentary rocks*, George Allen and Unwin Ltd., London, 383 pages.

Haughton, Samuel (1856) *On slaty cleavage, and the distortion of fossils*, Philos. Mag., vol. 12, p. 409–421.

Heim, Alberg (1878) *Untersuchungen über den Mechanismus der Gebirgsbildung*, 2 vols., 1 atlas, B. Schwabe Basel.

——— (1905) *Das Säntisgebirge*, Beitr. Geol. Karte Schweiz K. F. XVI, Lieferung.

——— (1921) *Geologie der Schweiz*, Tauchnitz, Leipzig 2 vols.

Heim, Arnold (1917) *Monographie der Churfirsten-Mattstock-Gruppe*, Beitr. Geol. Karte Schweiz, N.F. 20, Lieferung, 272 pages.

Hills, E. S. (1941) *Outlines of structural geology*, Nordeman Pub. Co., New York.

Hitchcock, E., Hitchcock, C. H., and Hager, A. D. (1861) *Report on the geology of Vermont*, 2 vols., Claremont, N. H., 988 pages.

Ingerson, Earl (1937) *Accurate orientation of thin sections*, Am. Mineral., vol. 22, p. 760–772.

Jannettaz, Ed. (1884) *Mémoire sur les clivages des roches (schistosité, longrain)*, Soc. Géol. France, Bull., 3d ser., vol. 12, p. 211–236.

Jonas, Anna I., and Stose, G. W. (1938) *Ceologic map of Frederick County*, Md. Geol. Survey, 1: 62, 500.

Keith, Arthur (1893) *Geology of the Catoctin belt*, U. S. Geol. Survey, 14th Ann. Rept., pt. 2, p. 285–395.

Knopf, E. B., and Ingerson, Earl (1938) *Structural petrology*, Geol. Soc. Am., Mem. 6, 270 pages.

Krech, K. (1909) *Beitrag zur Kenntnis der oölithischen Gesteine des Muschelkalks um Jena*, Jahrb. Preuss. Geol. Landesanstalt 1909, p 59–133.

Kvale, Anders (1945) *Petrofabric analysis of a quartzite from the Bergsdalen quadrangle, western Norway*, Norsk. Geol. Tidskr., vol. 25, p. 193–215.

Leith, C. K. (1923) *Structural geology*, Henry Holt, New York, 390 pages.

Loretz, H. (1882) *Über Transversalschieferung und verwandte Erscheinungen im Thüringischen Schiefergebirge*, Jahrb. Preuss. Geol. Landesanstalt. 1881, p. 258–306.

Lotze, Franz (1932) *Über Beziehungen zwischen Faltung und Schieferung*, Nachr. Ges. Wiss. Göttingen IV, Nr. 19, p. 113–121.

Maryland Geological Survey (1933) *Geological map of Maryland*, 6 miles equal one inch.

Mehnert, Karl-Richard (1939) *Die Metakonglomerate des Wiesenthaler Gneiszuges im sächsischen Erzgebirge*, Mineral. Petrog. Mitt., vol. 50, p. 194–272.

Nadai, A. (1931) *Plasticity*, McGraw Hill, New York, 349 pages.

Nevin, Charles M. (1942) *Principles of structural geology*, John Wiley and Sons, New York, 320 pages.

Peach, B. N., Gunn, W., *et al.* (1912) *The geology of Ben Wyvis, Carn Chuinneag, Inchbae, and the surrounding country*, Geol. Survey Scotland, Mem. Expl. of sheet 93, 189 pages.

Pettijohn, F. J. (1934) *Conglomerate of Abram Lake, Ontario and its extensions*, Geol. Soc. Am., Bull., vol. 45, p. 479–506.

Phillips, J. (1843) *On certain movements in the parts of stratified rocks*, Brit. Assoc. Adv. Sci. Meeting at Cork Aug. 1843, p. 60–61.

Quirke, T. T., and Collins, W. H. (1930) *The disappearance of the Huronian*, Geol. Survey Canada, Mem. 160, 129 pages.

Reusch, Hans (1888) *Bømmeloen og Karmøen*, Stensballes, Kristiania, 422 pages.

Rüger, L. (1933) *Gefügekundliche Untersuchungen an den Geröllgneisen von Obermittweida (Erzgebirge)*, Neues Jahrb. Beil. Bd., Abt. A, vol. 66, p. 275–293.

Runner, J. J. (1934) *Pre-Cambrian geology of the Nemo District, Black Hills, South Dakota*, Am. Jour. Sci. 5 ser., vol. 28, p. 353–372.

Sander, Bruno (1912) *Über einige Gesteinsgruppen des Tauernwestendes*, Jahrb. Geol., Bundesanstalt 62, p. 219–287.

——— (1930) *Gefügekunde der Gesteine*, Julius Springer, Wien, 352 pages.

Scheumann, K. H. (1936) *Sudetische Studien III, Konglomerattektonite und ihre Begleitgesteine in der epizonalen Schieferscholle südlich von Strehlen in Schlesien*, Mineral. Petrog. Mitt., vol. 48, p. 325–372.

Scholtz, Hermann (1931) *Über das Alter der Schieferung und ihr Verhältnis zur Faltung*, Jahrb. Preuss. Geol. Landesanstalt 1931, vol. 52, p. 303–316.

Schmidt, Walter (1932) *Tektonik und Verformungslehre*, Borntraeger, Berlin, 208 pages.

Sharpe, Daniel (1846) *On slaty cleavage*, Quart. Jour., vol. 3, p. 74–104.

Sieburg, Richard (1909) *Über transversale Schieferung im thüringischen Schiefergebirge*, Zeitschr. Prakt. Geol., vol. 17, p. 233–262.

Sorby, H. C. (1853) *On the origin of slaty cleavage*, New Philos. Jour. (Edinburgh), vol. 55, p. 137–148.

Stose, George W. (1909) *Mercersburg-Chambersburg Folio*, U. S. Geol. Survey, Folio 170.

Strand, Trygve (1938) *Nordre Etnedal*, Norges Geol. Unders. Beskr. Geol. Gradteigskart, no. 152, 71 pages.

——— (1945) *Structural petrology of the Bygdin conglomerate*, Norsk Geol. Tidskr., vol. 24, p. 14–31.

Twenhofel, W. H. (1932) *Treatise on sedimentation*, 2d ed., Waverly Press, Baltimore, 926 pages.

Turner, F. J. (1936) *Interpretation of schistosity in the rocks of Otago*, Roy. Soc. N. Z., Tr., vol. 66, p. 201–224.

——— (1942) *Structural Petrology of quartzose veins in the schists of eastern Otago*, Roy. Soc. N. Z., Trans., vol. 71, p. 307-324.

Wegmann, C. E. (1932) *Note sur le boudinage*, Soc. Géol. France, Bull., 5th ser., vol. II, p. 477–489.

Wettstein, A. (1886) *Ueber die Fischfauna des tertiären Glarnerschiefers*, Schweiz. Paläont. Gesell., Abh. 13, 101 pages.

Willis, Bailey (1893) *Mechanics of Appalachian structure*, U. S. Geol. Survey, 13th Ann. Rept., pt. 2.

———, and Willis Robin (1934) *Geologic structures*, McGraw Hill, New York, 544 pages.

THE JOHNS HOPKINS UNIVERSITY, BALTIMORE 18, MD.
MANUSCRIPT RECEIVED BY THE SECRETARY OF THE SOCIETY, AUGUST 22, 1946
PROJECT GRANT 375/41

14

Copyright ©1969 by Elsevier Scientific Publishing Company
Reprinted from pages 117–135 and 136 of *Tectonophysics* **7**:117–136 (1969)

A TECHNIQUE OF FINITE STRAIN ANALYSIS USING ELLIPTICAL PARTICLES

D. DUNNET

Geology Department, Imperial College of Science and Technology, London (Great Britain)

(Received July 1, 1968)

SUMMARY

When initial elliptical objects are deformed homogeneously with their matrix they change both their ellipticity and axial orientation. Mathematical expressions are derived which relate the final elliptical ratio and orientation of particles to their initial form and the magnitude and orientation of the finite rotational or irrotational strain. Curves are constructed from these expressions and examples given of their use to determine finite strain and initial shape of deformed aggregates.

INTRODUCTION

In an aggregate of non-spherical, sub-ellipsoidal particles (oölites, pisolites, conglomerates) subject to a finite homogeneous strain, the final shape and orientation of each particle will depend on five independent variables. These are: (a) the initial shape of the particle; (b) the initial axial orientation; (c) the strain intensity; (d) the strain orientation; (e) the ductility contrast of the particles to the total particle/matrix system. From the resultant fabric (the shape and orientation of the group of particles) it is not possible to determine the values of all these variables.

Ramsay (1967, pp.202–209) has presented a mathematical treatment of the problem by means of a two dimensional transformation assuming no ductility contrast. This analysis is of practical application in the case of calcareous ooids, pisolites and conglomerate pebbles of similar composition to their matrix, where the ductility contrast will approach unity. Gay (1968b, pp.297–299) derives 'viscosity ratios' for various rock types from previously published conglomerate pebble shape data, assuming that the rocks behave as viscous fluids. These values for common rocktypes range up to 3.5:1 (with the exception of volcanic quartz: 4.2:1 to 12.0:1) relative to a shale or schist matrix. Gay (1968a, p.226) demonstrates that the viscosity ratio between a particle and the mean viscosity of a particle/matrix system is dependent on the volume concentration of the particles. The ratio is reduced for high volume concentrations. Thus in conglomerates where the proportion of pebbles to matrix is high, the ductility contrasts are probably low. Assuming that the ductility contrasts are negligible the remaining four variables which produce the final fabric can be determined.

The form of the equations presented by Ramsay (1967, eq.5.22,5.27) which define the transformation of elliptical particles during a finite strain have been modified here so that they are more readily applied to the practical problems of strain analysis. A mathematical treatment is presented below which defines the final shape and orientation of any initially random suite of elliptical particles after a finite strain. By comparing scatter diagrams of field data with the theoretically deduced curves the finite strain and range of shape of the initial particles are determined. Examples of data from oölites and grits are presented to demonstrate the applied technique. The limitations of the method and the variations to be expected from rotational strain, ductility contrast, initial preferred orientation and superimposed noncoaxial strains are also considered.

MATHEMATICAL TREATMENT

Ratio parameters

The finite transformation of elliptical particles does not depend on the absolute magnitudes of either the particles or the strain, thus only ratios of the particles are required and only ratios of the strain can be obtained. The strain ratio is the deviatoric component of the strain tensor; the area change (or volume change in three dimensions) is implied in any absolute strain parameter [$(1+e), \lambda$, etc.]. Particle axial ratios can be treated as deviatoric components of the absolute magnitude. Hence Ramsay (1967, eq.5.22,5.27) was able to derive two functions in ratios of the form:

$$\tan 2\phi = F\,(Ri,\ Rs,\ \theta) \tag{1}$$

and $$Rf = F\,(Ri,\ Rs,\ \theta,\ \phi) \tag{2}$$

where the symbols used here are defined as:

Ri = initial undeformed particle axial ratio.
Rs = finite strain axial ratio.
Rf = final deformed particle axial ratio.
θ = angle from the Ri long axis to the maximum principal strain direction.
ϕ = angle from the Rf long axis to the maximum principal strain direction.

During a deformation of an elliptical particle of ratio Ri (*A* of Fig.1) by a homogeneous strain of ratio Rs (*C* of Fig.1) the long axis of the particle changes through an angle $(\theta - \phi)$ degrees towards the maximum principal strain axis and the ratio of the particle changes to Rf (*B* of Fig.1).

From eq.1 and 2 a third relationship of the form:

$$Rf = F\,(Ri,\ Rs,\ \phi) \tag{3}$$

must exist (since θ and ϕ are not independent) relating the two final parameters, Rf and ϕ, to the two controlling variables Ri and Rs. In any two-dimensional section of a deformed aggregate normal to cleavage the angle ϕ is readily measured and is given by the angle from the long axis of a deformed particle to the trace of the cleavage, and Rf is the particle

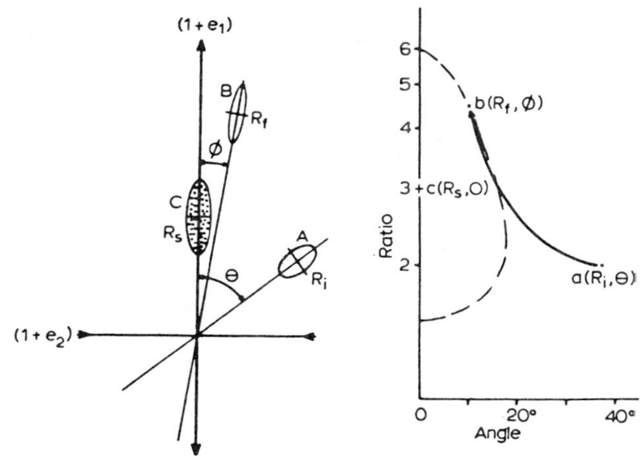

Fig.1. The change in shape and orientation of an initial elliptical particle (A) subject to a finite strain (C) to produce the final elliptical particle (B).

axial ratio. Over small areas the strain ratio (Rs) can be considered constant, the initial shape ratio (Ri) in most aggregates is restricted to a relatively small range of values, and major variation of the final shape (Rf) and orientation (ϕ) will result from variation in initial particle orientation (θ). Thus the form of eq.3 is suited to geological strain analysis in aggregates exhibiting a weak initial (sedimentary) orientation of particles.

An ellipse:

$$a_{11}X^2 + (a_{12} + a_{21})XY + a_{22}Y^2 = 1$$

can be represented by a Mohr diagram (Brace, 1961) as shown in Fig.2 and is the graphical representation of a two dimensional symmetric second order tensor (T) where:

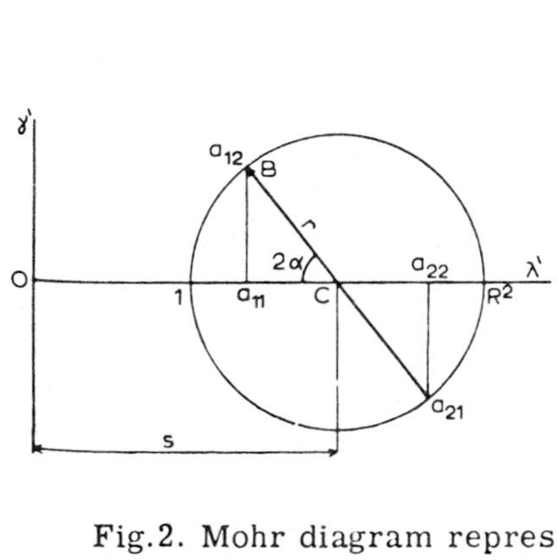

Fig.2. Mohr diagram representing a two dimensional symmetric second order tensor. Terms defined in text.

$$T = \begin{vmatrix} a_{11}X^2 & +a_{12}XY \\ +a_{21}XY & +a_{22}Y^2 \end{vmatrix} \tag{4a}$$

which can be written in abbreviated matrix form:

$$T = \begin{vmatrix} a_{11} & a_{12} \\ a_{21} & a_{22} \end{vmatrix} \tag{4b}$$

Written in matrix form tensor algebra is the same as matrix algebra (Brand, 1967, p.100). Then the following relationships may be derived from the Mohr diagram illustrated in Fig.2:

$$\left.\begin{array}{l} a_{11} = s - r\cos 2\alpha = \tfrac{1}{2}(1+R^2)-\tfrac{1}{2}(1-R^2)\cos 2\alpha \\ a_{22} = s + r\cos 2\alpha = \tfrac{1}{2}(1+R^2)+\tfrac{1}{2}(1-R^2)\cos 2\alpha \\ |a_{12}| = |a_{21}| = r\sin 2\alpha = \tfrac{1}{2}(1-R^2)\sin 2\alpha \end{array}\right| \tag{5}$$

When $\alpha = 0$ the ellipse is parallel to the reference coordinates and the tensor is 'diagonalised', that is the a_{12} and a_{21} (shear) terms vanish and the a_{11} and a_{22} terms are the reciprocal squares of the ellipse axes, the reciprocal quadratic elongation (λ') of finite strain. In deviatoric Mohr diagrams the principal value a_{11} is unity, whereas a_{22} becomes $R^2 = (\lambda' / \lambda_1')2$, which is always greater than or equal to 1.

The final ellipse (B of Fig.1) with reference to coordinate axes taken parallel to the ellipse axes may be written in abbreviated tensor notation as:

$$\begin{vmatrix} 1 & 0 \\ 0 & Rf^2 \end{vmatrix} \tag{6}$$

If the coordinate axes are positioned at angle ϕ to the principal axes of the final ellipse, parallel to the principal axes of strain, the tensor in eq.6 is given by:

$$\begin{vmatrix} \tfrac{1}{2}(1+Rf^2)-\tfrac{1}{2}(1-Rf^2)\cos 2\phi & \tfrac{1}{2}(1-Rf^2)\sin 2\phi \\ \tfrac{1}{2}(1-Rf^2)\sin 2\phi & \tfrac{1}{2}(1+Rf^2)+\tfrac{1}{2}(1-Rf^2)\cos 2\phi \end{vmatrix} \tag{7}$$

This final tensor is generated by the tensor product:

$$(C^{\frac{1}{2}})! \, A.(C^{\frac{1}{2}}) = B \tag{8a}$$

where A, B and C represent the initial shape tensor, the final shape tensor and the strain tensor respectively and all referred to the strain axis coordinates. Eq.8a can be rewritten:

$$A = (C^{-\frac{1}{2}})! \, B.(C^{-\frac{1}{2}}) \tag{8b}$$

where A, B and C are in terms of the same coordinate frame. Thus the initial shape tensor (A) referred to coordinates parallel to the strain axes is given by:

$$\begin{vmatrix} \tfrac{1}{2}(1+Ri^2)-\tfrac{1}{2}(1-Ri^2)\cos 2\theta & \tfrac{1}{2}(1-Ri^2)\sin 2\theta \\ \tfrac{1}{2}(1-Ri^2)\sin 2\theta & \tfrac{1}{2}(1+Ri^2)+\tfrac{1}{2}(1-Ri^2)\cos 2\theta \end{vmatrix} \tag{9}$$

The reciprocal of the displacement tensor ($C^{-\frac{1}{2}}$) referred to the same coordinates is:

$$\begin{vmatrix} 1 & 0 \\ 0 & 1/Rs \end{vmatrix} \qquad (10)$$

The tensor product $(C^{-\frac{1}{2}})!\, B.(C^{-\frac{1}{2}})$ (from eq.7,10) is:

$$\begin{vmatrix} \frac{1}{2}(1+Rf^2)-\frac{1}{2}(1-Rf^2)\cos 2\phi & \dfrac{\frac{1}{2}(1-Rf^2)\sin 2\phi}{Rs} \\ \dfrac{\frac{1}{2}(1-Rf^2)\sin 2\phi}{Rs} & \dfrac{\frac{1}{2}(1+Rf^2)+\frac{1}{2}(1-Rf^2)\cos 2\phi}{Rs^2} \end{vmatrix} \qquad (11)$$

Eq.9 and 11 are equivalent from eq.8b. Eq.11 expresses the initial undeformed elliptical particle in terms of the strain tensor and final elliptical particle ratio and with reference to the strain coordinate frame.

If J_1 and J_2 are the first and second invariants of the initial shape tensor in eq.9, where $J_1 = a_{11} + a_{22}$ and $J_2 = a_{11}\, a_{22} - a_{12}a_{21}$ (Ramsay, 1967, p.81), then:

$$Ri^2 = \frac{J_1 + \sqrt{J_1^2 - 4J_2}}{J_1 - \sqrt{J_1^2 - 4J_2}}$$

since:

$$\lambda'_1, \lambda'_2 = \frac{J_1 \pm \sqrt{J_1^2 - 4J_2}}{2} \qquad (12)$$

Rearranging eq.12:

$$Ri \times J_1 = \pm (Ri^2 + 1)\sqrt{J_2} \qquad (13)$$

The invariants of eq.11 can be equated with those of eq.9 (from 8b). The invariants from eq.11 are:

$$J_1 = [\tfrac{1}{2}(1+Rf^2)-\tfrac{1}{2}(1-Rf^2)\cos 2\phi] + \left[\frac{\frac{1}{2}(1+Rf^2)+\frac{1}{2}(1-Rf^2)\cos 2\phi}{Rs^2}\right] \qquad (14)$$

$$J_2 = [\tfrac{1}{2}(1+Rf^2)-\tfrac{1}{2}(1-Rf^2)\cos 2\phi]\left[\frac{\frac{1}{2}(1+Rf^2)+\frac{1}{2}(1-Rf^2)\cos 2\phi}{Rs^2}\right] - \left[\frac{\frac{1}{2}(1-Rf_2^2)\sin 2\phi}{Rs^2}\right]^2 \qquad (15)$$

Substituting eq.14 and 15 into eq.13 and simplifying:

$$\cos 2\phi = \frac{Ri(Rf^2+1)(Rs^2+1) \pm 2(Ri^2+1)Rs \times Rf}{Ri(Rf^2-1)(Rs^2-1)} \qquad (16)$$

Eq.16 relates the two measurable parameters Rf and ϕ to the initial shape ratio (Ri) and strain ratio (Rs). For any set values of Ri and Rs the locus of Rf/ϕ will reflect only variation in the initial orientation (θ) of the particle. Thus a suite of particles of constant initial shape, but variable orientation will have, after deformation, Rf/ϕ parameters which lie on a hyperbolic curve around the finite strain value (c of Fig.1 and Rs_1 of Fig.6).

Specific values of Ri and Rs substituted in eq.16 allow construction of theoretical Rf/ϕ curves which can be directly compared to Rf/ϕ diagrams derived from field data. The Rf/ϕ curves of Fig.3 have been drawn to cover the range of geological strain ratios Rs=1 to 8 and geological initial shape ratios Ri=1 to 4. They are plotted on log/linear paper to produce curves which are symmetric about the strain ratio Rs; a function of the transformation being a product, not addition, of the initial shape and strain tensors.

It will be noted that eq.11 is a symmetric tensor. This is the general tensor of the transformation and does not imply a component of external rotation in the pure shear transformation. The apparent rotation is due to the vertex of the ellipse occupying different material points before and after deformation. There is an angular difference between the Ri ellipse axes after deformation and the Rf ellipse axes (Ramsay, 1967, pp.57–64). The invariants are independent of this apparent rotation and thus can be equated.

Natural deviatoric parameters

Eq.16 can be expressed in a more simple form by the use of natural strain parameters. This form could be derived by direct substitution in eq.16 or by the complete solution given below.

By definition:

$$\epsilon = \log_e (1+e) \qquad (17)$$

where ϵ is the natural strain and e the engineers strain (Nadai, 1950). Then the strain ratio (R) is given by:

$$R = \frac{(1+e_1)}{(1+e_2)} = \frac{e^{\epsilon_1}}{e^{\epsilon_2}} \qquad (18)$$

or:

$$R = e^{(\epsilon_1 - \epsilon_2)}$$

To use natural deviatoric parameters we must define

$$\epsilon_1 = \epsilon m + \epsilon o$$

$$\epsilon_2 = \epsilon m - \epsilon o$$

where ϵm equals the mean natural strain $(\epsilon_1 + \epsilon_2)/2$ and half the natural area strain (ϵa), and ϵo equals the natural deviatoric strain (after Nadai, 1950, and D. Elliot, personal communication, 1966). Then the deviatoric ratio is related to the deviatoric natural strain by:

$$R = e^{2\epsilon o}$$

$$\log_e R = 2\epsilon o \qquad (19)$$

The terms of a symmetric second order tensor using reciprocal quadratic elongation parameters are (Ramsay, 1967, p.69):

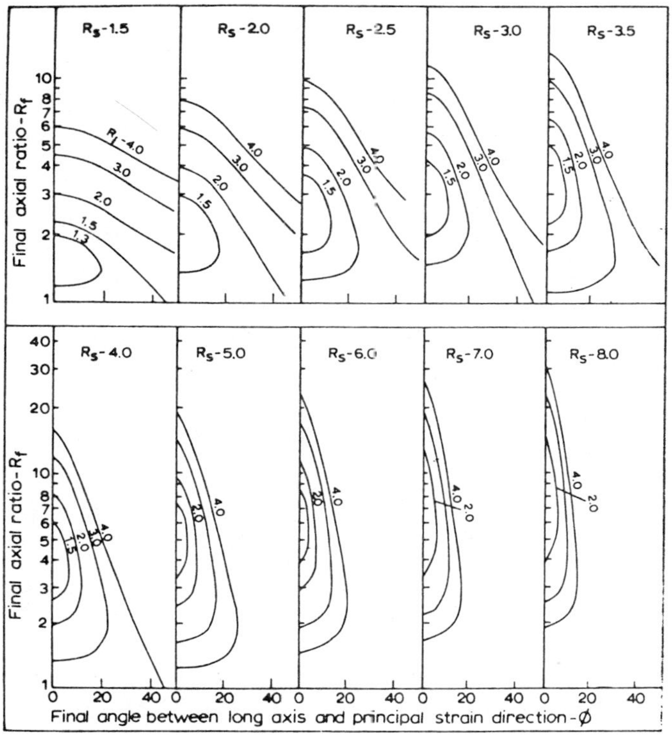

Fig.3. Curves of variation in Rf/ϕ for initial elliptical ratios (Ri) subject to various finite strain ratios (Rs). The curves are symmetric about the 0° ϕ axis.

$$\left.\begin{aligned} a_{11} &= \frac{(\lambda'_1 + \lambda'_2)}{2} - \frac{(\lambda'_2 - \lambda'_1)}{2}\cos 2\alpha \\ a_{22} &= \frac{(\lambda'_1 + \lambda'_2)}{2} + \frac{(\lambda'_2 - \lambda'_1)}{2}\cos 2\alpha \\ |a_{12}| &= |a_{21}| = (\lambda'_1 - \lambda'_2)\sin 2\alpha \end{aligned}\right| \quad (20)$$

Substituting natural strain parameters in eq.20:

$$\left.\begin{aligned} a_{11} &= (\cosh 2\epsilon o - \sinh 2\epsilon o \cos 2\alpha)\, e^{-\epsilon a} \\ a_{22} &= (\cosh 2\epsilon o + \sinh 2\epsilon o \cos 2\alpha)\, e^{-\epsilon a} \\ |a_{12}| &= |a_{21}| = (\sinh 2\epsilon o \sin 2\alpha)\, e^{-\epsilon a} \end{aligned}\right| \quad (21)$$

where:

$$\lambda'_1 = e^{-2\epsilon_1} = e^{(-2\epsilon m - 2\epsilon o)} = e^{(-\epsilon a - 2\epsilon o)}$$

$$\lambda'_2 = e^{-2\epsilon_2} = e^{(-2\epsilon m + 2\epsilon o)} = e^{(-\epsilon a + 2\epsilon o)}$$

or in Fig.2:

$$s = e^{-\epsilon a} \cosh 2\epsilon o \qquad r = e^{-\epsilon a} \sinh 2\epsilon o \qquad (22)$$

Defining from eq.19:

$$Ri = e^{2\epsilon i} \qquad Rs = e^{2\epsilon s} \qquad Rf = e^{2\epsilon f} \qquad (23)$$

Now as with the ratio parameters the natural deviatoric tensors can be derived from eq.21 and 23 so that the initial shape tensor (A) is:

$$\begin{vmatrix} \cosh 2\epsilon i - \sinh 2\epsilon i \ \cos 2\theta & \sinh 2\epsilon i \ \sin 2\theta \\ \sinh 2\epsilon i \ \sin 2\theta & \cosh 2\epsilon i + \sinh 2\epsilon i \ \sin 2\theta \end{vmatrix} e^{-\epsilon ai} \qquad (24)$$

or in terms of the final shape and strain tensors i.e. $(C^{-\frac{1}{2}})'.B.(C^{-\frac{1}{2}})$ (from 8b):

$$\begin{vmatrix} (\cosh 2\epsilon f - \sinh 2\epsilon f \cos 2\phi)\, e^{+2\epsilon s} & (\sinh 2\epsilon f \ \sin 2\phi) \\ (\sinh 2\epsilon f \ \sin 2\phi) & (\cosh 2\epsilon f + \sinh 2\epsilon f \ \cos 2\phi)\, e^{-2\epsilon s} \end{vmatrix} . e^{(\epsilon as \ - \ \epsilon af)} \qquad (25)$$

From eq.24:

$$J_1 = a_{11} + a_{22} = 2e^{\epsilon ai} \cosh 2\epsilon i \qquad (26)$$

and from eq.25, converting ϵs to hyperbolic functions:

$$J_1 = 2e^{+\epsilon as - \epsilon af} (\cosh 2\phi) \cosh 2\epsilon s - \sinh 2\epsilon f \sinh 2\epsilon s \cos 2\phi \qquad (27)$$

Equating J_1 in eq.26 and 27 and cancelling area terms since $\epsilon af = \epsilon ai + \epsilon as$, then:

$$\cos 2\phi = \frac{\cosh 2\epsilon f \ \cosh 2\epsilon s - \cosh 2\epsilon i}{\sinh 2\epsilon f \ \sinh 2\epsilon s} \qquad (28)$$

Noting that J_2 equals $1 \times e^{\epsilon a}$ in natural strain parameters.

Eq.16 and 28 are identical, and differ only in the parameters. The hyperbolic natural strain parameters of eq.28 are easier to manipulate for curve construction than those of eq.16, and the curves of Fig.3, although expressed in ratio parameters, were constructed by substitution of Ri and Rs values in eq.28. The resultant locus of Rf/ϕ reflects an unrestricted variation of initial orientation (θ) of the initial elliptical particles (Ri). Knowledge of the initial orientation of particles is not essential for the determination of finite strain. However, the initial orientation can be computed and is related to the initial shape, final shape and strain by the expression:

$$\cos 2\theta = \frac{\cosh 2\epsilon f - \cosh 2\epsilon i \ \cosh 2\epsilon s}{\sinh 2\epsilon i \ \sinh 2\epsilon s} \qquad (29)$$

Eq.29 is derived in a similar manner to eq.28, but with the three tensors referred to the final ellipse coordinate frame.

Cloos (1947) and Ramsay (1967, pp.202–208) have discussed fluctuation, the variation in ϕ, and Ramsay has derived the equation for maximum fluctuation (5.24):

$$\tan 2\phi_{max} = \frac{Rs(Ri^2-1)}{[(Ri^2 \times Rs^2-1)(Rs^2-Ri^2)]^{\frac{1}{2}}} \quad (30)$$

Eq.30 defines the value of the maximum fluctuation angle for any specific strain and initial shape ratios and defined the turning point on the ϕ ordinate of the Rf/ϕ curves. In construction of the curves of Fig.3 it is advantageous to determine the Rf value at the maximum fluctuation angle. By differentiation of eq.28, and equating to zero the expression:

$$\underset{(\text{at } 2\phi \text{ max})}{\cosh 2\epsilon f} = \frac{\cosh 2\epsilon s}{\cosh 2\epsilon i} \quad (31)$$

defines the Rf value at 2ϕ maximum. The expression:

$$\sin 2\phi \text{ max} = \frac{\sinh 2\epsilon i}{\sinh 2\epsilon s} \quad (32)$$

defines the maximum fluctuation angle, and is equivalent to eq.30. The graph of eq.31 is shown in Fig.4, and can be used in conjunction with the fluctuation graph of Ramsay (fig.5.26) to determine the particle shape at maximum fluctuation angle.

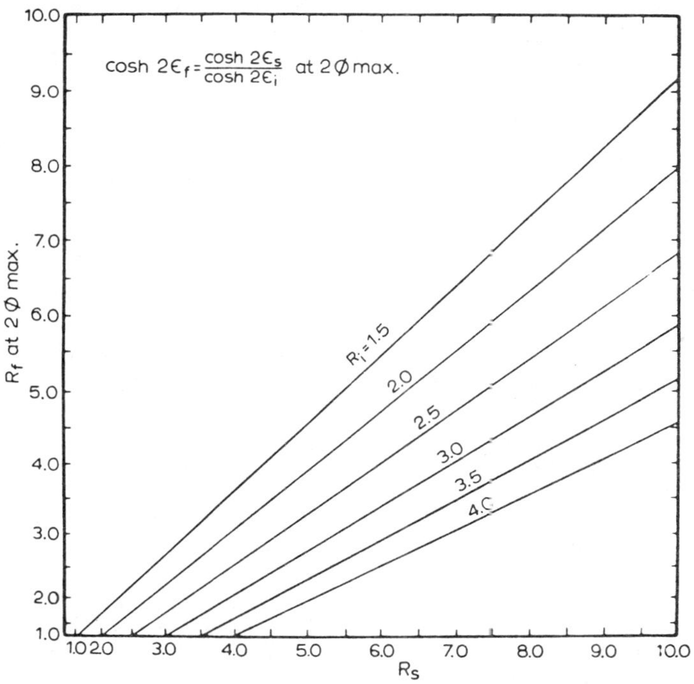

Fig.4. Graphical variation of Rf at maximum fluctuation for various initial elliptical shapes (Ri) and tectonic strains (Rs). Ordinates are a hyperbolic cosh scale.

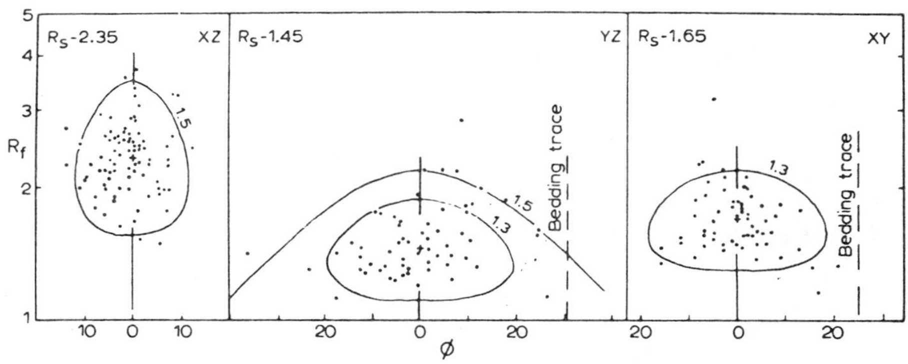

Fig. 5. Scatter diagrams of Rf/ϕ in oöids on the three principal planes of oölitic pebble No.16 from southwestern Sardinia. Three-dimensional finite strain is constrictional of ratio $X:Y:Z$ = 2.35:1.45:1.

EXAMPLES OF STRAIN EVALUATION BY Rf/ϕ DIAGRAMS

Introduction

A large number of Rf/ϕ diagrams have been plotted from deformed conglomerates, grits, oölites and pistolites, and the two-dimensional data has been combined to determine three-dimensional strain ratios. The geologic significance of these measurements will be presented in a later publication. Only the problems and limitations of the technique will be discussed here.

Diagrams can be prepared directly from field measurements on natural surfaces or in the laboratory from oriented polished specimens and thin sections. Thin sections placed directly in the enlarger have been used to produce large negative photographs for measurements. Measurement of the data on principal planes is not necessary (Ramsay, 1967, pp.142–148) but it greatly simplifies conversion to three-dimensional data. The position of the principal planes may be difficult to determine accurately in specimens of either very high or very low k values (Flinn, 1962), or of low deformation (Cloos, 1947). Faulty cuts during preparation of material can generally be assessed and corrected to ± 5° from the diagram symmetry after measurement. A ratio correction can then be made by a simple Mohr diagram construction. Rocks consisting of several particle lithologies should be avoided or measurement restricted to particular lithologies. Comparison of strain values (Rs) from particles of different lithologies in the same specimen may be used to determine the ductility contrast ratio by substitution in the equations presented by Gay (1968a,b). In an investigation of variation in strain, samples should be of similar rocktype and similar particle/matrix ratio. Specimens showing initial preferred planar orientation in bedding should be avoided or treated with caution (see p.134), and data collected from regions of rotational strain or superimposed deformation interpreted with care.

Deformed oölitic limestones

A typical set of Rf/ϕ diagrams from a deformed oölitic limestone pebble in the Ordovician rocks of southwestern Sardinia is shown in Fig. 5. The general geology of the region has been described by Poll (1966). The data has been measured on photographs prepared from three mutually perpendicular thin sections cut in the principal planes ($\pm$ 5°). Approximately 60 oöids were measured in each section by recording the two axis lengths and the angle of the long axis from an arbitrary line. Each ratio (Rf) was then calculated and plotted against the angle. The appropriate Rf/ϕ curves were then fitted to the scatter diagram of data, as described below, and the strain ratio and range of initial shape ratios derived. The majority of oöids show an initial shape ratio (Ri) of less than 1.3:1. The XZ section ($X > Y > Z$) indicates the greatest initial shape, which is consistent with its orientation at a high angle to bedding. All sections show a strong symmetry to the enveloping curves of best fit, but both the YZ and XY sections have a larger number of plots on the bedding side of the principal plane. The XY section shows the maximum and minimum values diagonally across the centre of the scatter which also suggests an initial fabric. The lack of points around the centre of the XY diagram is commonly encountered and must indicate an absence of initially circular particles.

To some extent the choice of Rf/ϕ curves for any natural data, and hence the choice of strain ratio, is subjective as it relies on a visual best fit. In Fig. 5 the XZ section strain ratio can be assessed within very close limits because of the symmetric distribution and high density of the plotted

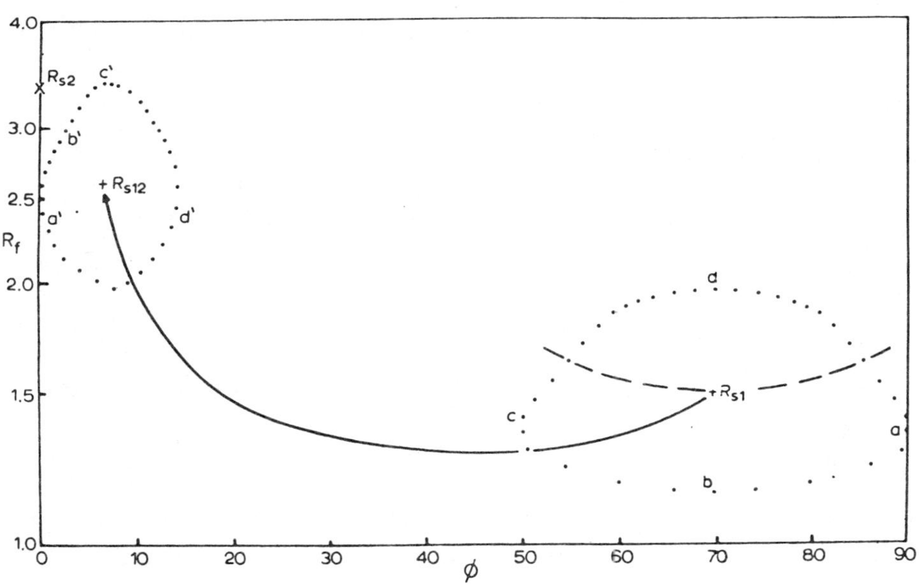

Fig. 6. Effects of a finite strain (Rs_2) on an initial preferred orientation group of particles ($abcd$). Curve through Rs_1 delimits 50% scatter of points. Diagram constructed from pure-shear curves by Gay (1968a, fig. 6).

points, but YZ and XY are less specific. A prime advantage of measurement on three principal planes in that only two are required to determine the three principal strain ratios, and the third gives an independent check on the accuracy ($Rs YZ \times RsXY = RsXZ$). The two-dimensional strain ratio (Rs) can generally be assessed using a combination of the following methods:

(a) The logarithmic mean of Rf maximum and Rf minimum should define the strain value (Ramsay, 1967, p.210). In practice this method may not give the most accurate results as the density of points on the periphery of the field of plotted points is generally low. Anomalous readings may occur which lie outside the true field because of particles being boudinaged, difficulty in determining margins of particles, or competent cores within particles.

(b) The scatter of plotted Rf/ϕ values can be halved numerically on both the logarithmic Rf and ϕ ordinates. The angular mean should coincide with the orientation of the principal plane trace and should also coincide with the maximum and minimum plotted Rf values. The logarithmic mean of the Rf values of all plotted points does not coincide with Rs, but takes a slightly higher value, since the 50% of data line is a hyperbolic function curve with its minimum value on Rs as shown in Fig.6, Rs_1.

(c) The Rf/ϕ curves of Fig.3 corresponding approximately to the strain value (Rs) can be placed over the plotted data and moved up the Rf axis to the position of visual best fit. This invariably permits a range of possible Rs values.

(d) Any low concentration of points around the centre of the scatter will generally enclose Rs and indicates a lack of initially circular sections. In some cases a restricted initial ratio range will clearly define all or part of a particular Rf/ϕ curve. This depends on the variance of the initial shape ratio frequency diagram. Even with a restricted initial shape ratio and equal angular distribution (θ), the final distribution of Rf and ϕ points is not of equal density. The points of Fig.6 around Rs_1 were initially all of ratio $Ri = 1.3$ and at $\theta = 5°$ intervals. It is seen that a greater concentration of Rf/ϕ values occurs on the upper part of the curve. Thus a restricted Ri range is generally more obvious in the upper part of the Rf/ϕ scatter diagram. Although it may be advantageous to contour such diagrams on an arbitrary density interval, the interpretation of such a contour diagram is difficult.

The normal procedure adopted to determine the strain value (Rs) is as follows. Firstly the mean of the scatter of plotted points is determined on both ordinates (method (b) above). The intersection of mean ϕ and logarithmic mean Rf is tentatively considered as the strain value ($Rs/0°$). The number of points in each of the four quadrants are counted and compared. They should be equal, and if not, define one type of asymmetry commonly encountered. Lack of coincidence of ϕ mean with the maximum and minimum Rf values defines a second type of asymmetry, and lack of coincidence of ϕ mean and the cleavage trace a third asymmetry. All these have been noted, particularly in the diagrams of clastic rocks discussed below. If the diagram is reasonably symmetric the appropriate Rf/ϕ curves are fitted to the data, remembering that the mean centre of the scatter should give a value higher than Rs, with an initially random fabric. The curves are adjusted if necessary to the visual best fit (method (c) above). In the case of asymmetric diagrams or where initial orientation in bedding

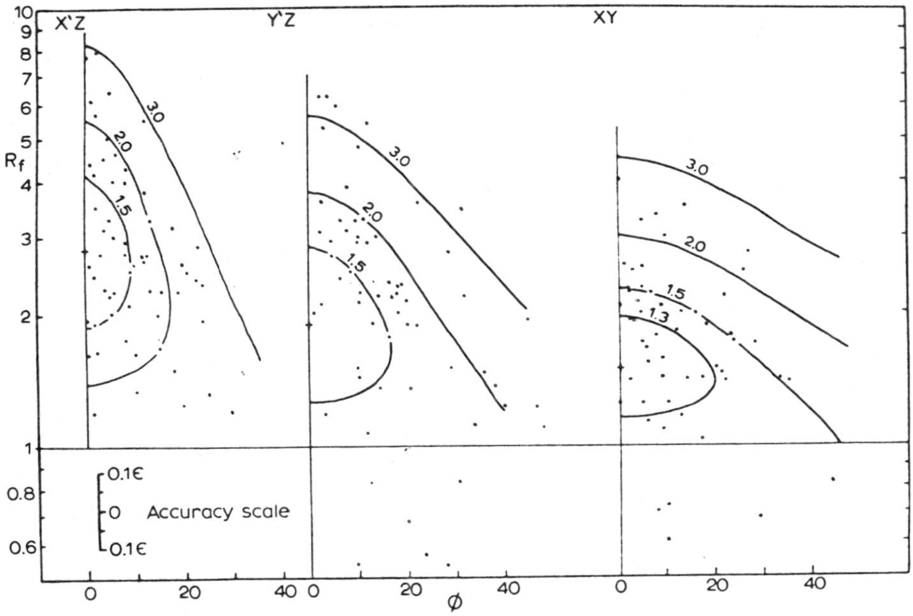

Fig.7. Scatter diagrams of Rf/ϕ on 3 mutually perpendicular planes of grit specimen 96,100. Planes $X'Z$ and $Y'Z$ are 12° from the principal planes. Accuracy scale of strain ratio assessment is shown in inset. Data has been rotated about the ϕ symmetry plane.

is suspected, the position of the bedding trace should be considered. Particles with their initial long axes in the bedding at a high angle to the maximum strain axis will produce Rf/ϕ values less than the strain ratio. Conversely particles in bedding at a low angle to the maximum strain axis will produce Rf/ϕ values greater than the strain ratio. Strain ratios from the three principal planes can be compared and if not consistent a reassessment can be made. Strain values determined by this method are reproducible, especially using oöids and pisolites because of their low initial shape ratio and lack of initial preferred orientation. The precision of assessment of the strain ratio using oöids appears to be within the limits of $\epsilon = \pm 0.03$ units.

Clastic rocks

The scatter diagrams of Fig.7 and 8 have been prepared from polymictic grits from the Ordovician of southwestern Sardinia. Both specimens are of similar lithology, consisting of approximately 40% quartzite particles, 30% ferruginous slate and siltstone particles, and about 30% matrix of similar composition to the ferruginous slate fragments. The diagrams have been prepared from data from the deformed slate fragments. The psammitic particles have been measured but are plotted separately because their different ductility as discussed below.

The fragments of these clastic rocks, unlike oöids and pisolites are not all of sub-elliptical shape. The other common shapes encountered are rectangular, rhombic, pear-shaped, pinch-and-swell, folded and irregular forms. The pinch-and-swell, folded, and irregular fragments have not been measured as they reflect inhomogeneous deformation on the scale of an individual particle. Rectangular, rhombic and pear-shaped particles are measured by estimation of the axes of the ellipse of equivalent area. Ductile particles adjacent to large competent fragments are not measured as the strain around these large fragments is obviously inhomogeneous.

Both the specimens in Fig.7 and 8 were collected from the overturned limbs of asymmetric folds, bedding was inclined at about 25° to the cleavage. The maximum pebble ratio (Rf max.) and the mean angular spread (ϕ mean) of sections normal to the cleavage coincide ($\pm$ 3°) with the cleavage trace, so that the cleavage is the XY principal plane. In the specimen of Fig.7 the angle between the bedding/cleavage intersection and the X direction on the cleavage surface is 20°, and in the specimen of Fig.8 the bedding/cleavage intersection makes an angle of 80° to X. The strain ellipsoids in both specimens have low finite k values, so that the initial shape variation, and the partial initial orientation of particles in bedding, tend to make the position of X difficult to determine accurately. Because of this difficulty thin sections were inaccurately prepared for the

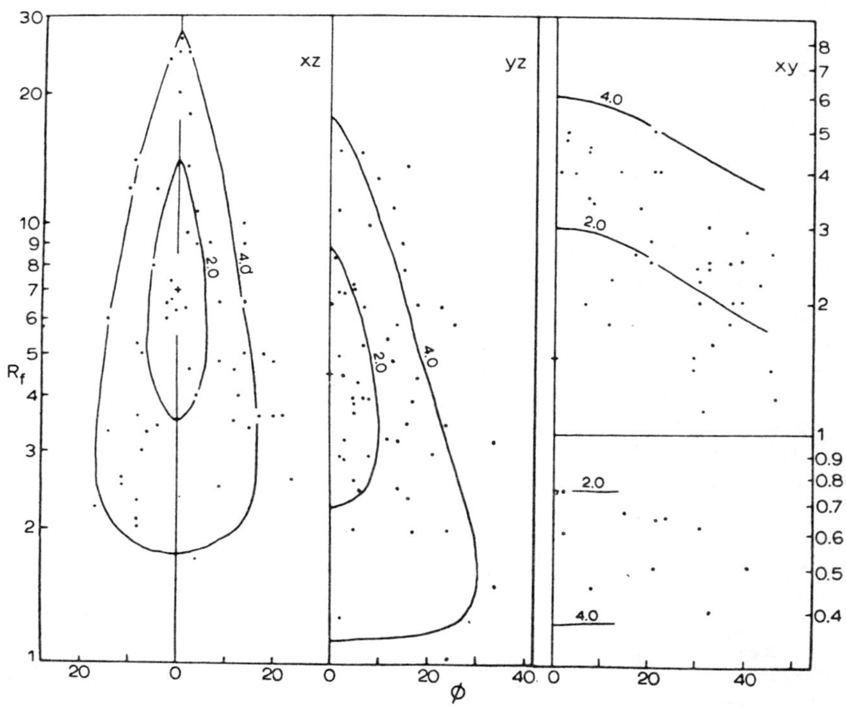

Fig.8. Scatter diagram of Rf/ϕ on 3 principal planes of grit specimen 480: a large finite flattering deformation (6.8:4.5:1) of an initial shape fabric ranging in excess of 4:1.

specimen of Fig.7 at an angle of 12° to the *XZ* and *YZ* planes. If the mistaken axes have coordinates *X'Y'Z* and the strain axes have coordinates *XYZ* then to derive the principal strain ratios a Mohr diagram construction can be used to rotate the *X'Z* and *Y'Z* strain values about *Z* to the true principal planes. Measurements on the *X'Y'* face (=cleavage face, *XY*) produce a symmetric diagram at 12° to the *X'* direction thus the orientation of *X* and *Y* and the strain ratio *XY* are defined. By constructing a deviatoric Mohr diagram for this face the actual ratio of *X'* to *Y'* can be determined at $2\alpha = 24°$ to *X* and the *X'Z* and *Y'Z* ratios scaled up to determine the *Z* value relative to *X* and *Y*.

The three sections of Fig.7 when originally plotted were symmetrical about the 0° axis shown on the diagrams. The data displayed in Fig.7 has been rotated onto one side of the 0° axis (*X* or *Y*) to accentuate the scatter pattern. This technique is useful when large initial ratios are involved, but is only valid when the total diagram is symmetric. The initial shape range of particles in both the *YZ* and *XY* sections exceeds the strain value for that section ($Ri > Rs$) and because of this feature some ϕ values are in excess of 45° and no maximum fluctuation angle exists. To retain the symmetry of the scatter diagram, and permit the methods (a) and (b) of strain assessment to be employed, all particles with fluctuation angle (ϕ) greater than 45° are plotted as reciprocals ($1/Rf, \phi$ -90°). The ratio of these points will then take a value less than 1, so that the total scatter of data on the logarithmic *Rf* axis is still symmetric about the strain value (Ramsay, 1967, p.210).

The specimen of Fig.7 has been analysed for magnetic susceptibility anisotropy at the University of Birmingham (R.F. King, personal communication, 1967). The magnetic susceptibility tensor approximates the strain ellipsoid in both orientation and symmetry of *k* value. Orientations of *X* differ by 10°, the magnetic *k* value is between 0.52 and 0.55 and the strain *k* is 0.57. The striking similarity of the *k* values of the two independent analyses indicates a correlation between magnetic and strain fabrics which warrants further investigation. As the two *k* values are not strictly comparable an exact agreement between them is not necessarily expected.

The specimen shown in Fig.8 exhibits the largest strain measured by the author in Sardinia, a strain ratio of 6.8:4.5:1. This is equivalent, assuming no volume change, to 68% shortening normal to the cleavage and because of the large strain, presents certain difficulties in measurement of the deformed fragments in the *XZ* section. Particles of *Rf* ratio in excess of 15:1 frequently have 'frayed ends' (Cloos, 1947) and their extremities are difficult to locate with precision. This difficulty in measurement sets an upper limit of approximately 10:1 to the determination of strain by Rf/ϕ diagrams. Although rocks of low strain do not present problems of this type it is found that ϕ may be difficult to define accurately when the deformed particle shape approaches that of a circle. The *XY* section of Fig.8 indicates the type of broad scatter of points derived from low strains in grits and the difficulty in defining the *X* direction in deformation of low *k* value. In this section the strain is approximately 1.5 and the range in initial shape ratio between 2.0 and 4.0. More easily interpreted diagrams have been obtained from slightly deformed oölitic limestones ($Rs = 1.2$) because of their much lower initial shape factor.

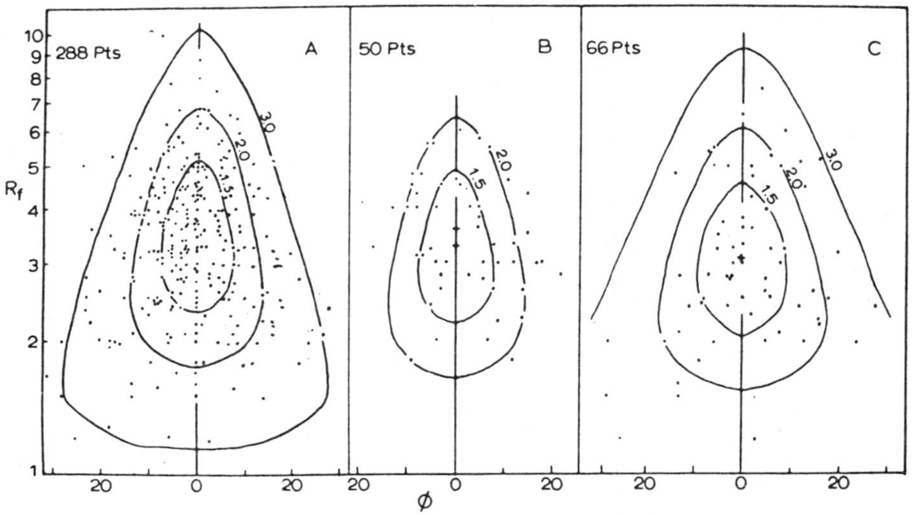

Fig.9. Scatter diagram of Rf/ϕ on the XZ plane of lithic greywacke specimen 1,132. A. Total lithic slate particles; B. most symmetric; C. least symmetric subgroup distribution.

Quantity of data

The number of measured particles required to define the strain value on an Rf/ϕ diagram depends on the range of the initial shape ratio, the degree of initial preferred orientation and the accuracy with which readings are taken. Previous strain values (Cloos, 1947; Flinn, 1956; Hossack, 1968) have been assessed from about 30 measurements. Evaluation from the arithmic mean of Rf is invalid and the geometric mean of 30 particles is considered to give crude and approximate results. From the results presented here it would appear that strains derived from 30 to 40 oöids give reproducible strain values, but a minimum of 50 or 60 particles seems necessary to define the tectonic strains in conglomerates and grits where the fragments initially had an irregular shape and an initial ratio up to 4.0:1. The author considers 40 to be the optimum number of oöids or pisolites required, and between 60 and 100 fragments in grits or conglomerates depending on the nature of the material.

The specimen whose data is shown in Fig.9 is a lithic grey greywacke consisting of 20% ferruginous slate fragments in a quartz-rich, quartz-chlorite-mica matrix. All slate particles in the three mutually perpendicular sections total over 900, and all were measured. The Rf/ϕ values of the 288 particles in the XZ section are shown in Fig.9A. The total data from this section was divided into six groups, each of about 50 particles, as it accumulated. The most and least symmetric sub-groups are shown in Fig.9B and C respectively. The total diagram defines an Rs value of 3.5 ± 0.3 using methods (a) and (c) and 3.4 using method (b) (p.128). The six sub-group scatter diagrams give values of Rs ranging from 3.0 to 3.8, with a maximum range of $\pm$ 0.45 by visual curve fitting (method (c)) as

TABLE I

Range of Rs determinations by methods (b) and (c) in sub-groups of specimen 1132 section XZ (Fig.9)

Group	Number of particles	Mean value of Rf	Rs value of curves by visual fit	Mean Rs value
1(B)	50	3.3	3.3–3.6	3.45
2(C)	66	2.85	2.8–3.2	3.0
3	40	3.7	3.2–3.8	3.6
4	41	3.8	3.2–3.8	3.5
5	50	3.7	3.5–4.1	3.8
6	41	3.4	3.2–4.1	3.65
Total (A)	288	3.4	3.2–3.8	3.5

shown in Table I. Fig.9B is symmetric and defines an *Rs* value within close limits (3.3–3.6) and of similar value to Fig.9A. Fig.9C is asymmetric, and the *Rs* value is lower than in all other groups. The suite of particles in Fig.9C probably includes a larger number than other sub-groups of particles of initial shape ratio (*Ri*) greater than 2.0:1, and initially oriented with their long axis in bedding. Internal consistency with the other two sections implies that the actual strain value in this section is greater than the 3.5 indicated in Fig.9A, and is in the range 3.7–4.0. This discrepancy despite the large number of particles measured, and the symmetric nature of the total diagram, is also probably a result of preferred orientation of particles in the bedding. Bedding makes a high angle to *XZ* (78°) and the trace of bedding on the section is at -40° to *X*, so that a high proportion of particles give low *Rf* values, and consequently a low apparent strain value. Consistency between *Rs* values on three principal planes, and appreciation of the effects of initial preferred orientation allow assessment of the *Rs* value as a valid finite strain ratio in each section. Thus a minimum of 50 conglomerate or grit particles can define *Rs* values reproducible within limits of about $\epsilon = \pm 0.05$ units. However, this *Rs* value is related to the total finite fabric, a product of the initial sedimentary fabric and the finite strain fabric, and in some cases it may be impossible to distinguish between them. Fortunately it would appear that grit particles of low initial shape ratio exhibit little tendency for an initial preferred orientation, and if these can be distinguished the strain can be determined.

Limitations

Some of the limitations imposed on the determination of finite strain ratios from Rf/ϕ diagrams have been discussed above. The main factors to influence the interpretation of the *Rs* value as a finite strain value are the effects of: (1) initial preferred orientation of fabrics; (2) superimposed deformation; (3) rotational strain; (4) ductility contrasts; and (5) inhomogeneous strain over the field of the analysed specimen.

Initial preferred orientation and superimposed strain produce superficially similar results. Ramsay (1967, p.217) has shown that a two dimensional section oblique to an initial planar oriented fabric produces an initial shape and angular distribution (Ri/θ) on the section of somewhat similar form to the Rf/ϕ curves presented here. Thus modification of the fabric by a finite strain is somewhat similar to the modification of the resultant fabric of an earlier deformation by a later finite strain. In the case of an initial planar sedimentary fabric modified by deformation the initial fabric is not defined by the mathematical expression of eq.16 and 28 and thus the final fabric is generally asymmetric. However, if the fabric is an initial perfect planar one the Rs value derived from the resultant fabric is, as with superimposed deformation, the finite product of the initial and induced fabric.

An example of the effects of two superimposed deformations is shown in Fig.6. A group of elliptical particles of axial ratio 1.3 and oriented at 5° intervals through the fabric are deformed by a finite strain (Rs_1) of magnitude 1.5. The deformed particles then occupy the Rf/ϕ locus *abcd*. This approximately corresponds to the range of particle shapes and orientations due to an oblique planar orientation. If the fabric produced by Rs_1 is subject to a finite strain, Rs_2, at 70° to Rs_1, then the Rs_1 fabric will move along strain paths to *a'b'c'd'*. This new locus of particle ratios and angles (*a'b'c'd'*) defines a total finite strain, Rs_{12}, which is the finite strain product of Rs_1 and Rs_2. The strain product Rs_{12} is oblique too, and has a strain ratio less than Rs_2. There is no direct evidence that two strains have produced the fabric. In practise additional evidence such as cleavage oblique to the Rs_{12} orientation may indicate an initial fabric. At the present time the author has been unable to determine if, in superimposed deformations, the observed cleavage is parallel to the total finite XY direction (Rs_{12}), or whether it is parallel to the second deformation plane XY (Rs_2). The Rf/ϕ technique applied to suitable material should clarify this point. Note that in both Rs_1 and Rs_{12}, greater than 50% of the particles have final ratios in excess of the strain value.

A perfect linear initial fabric will restrict the range of *abcd* to a single valued Rs_1 which on deformation will produce a single Rs_{12}. Fortunately initial fabrics are not perfect and in general an oblique asymmetric Rf/ϕ diagram across Rs_{12} and Rs_2 will result. If sufficient initially unoriented particles are present Rs_2 can be determined.

Rotational strain will similarly produce a finite strain value, oblique to the incremental axes and less than the simple product of the increments. Thus this technique is suitable for both irrotational and rotational strains and the *finite* strain can be derived.

The effects of ductility contrasts have been discussed by Gay (1968a,b). An increase in the viscosity of an initial elliptical particle greatly reduces both the change in shape and the rate of change of the long axis towards the maximum principal direction for a given finite strain. The final ratio and angular relations (Rf/ϕ) will define curves not unlike the Rf/ϕ curves presented here, but with the maximum fluctuation greatly reduced, and the Rs value obtained much less than the deformation undergone by the matrix. The effects of a high particle to matrix ratio is to lower strongly the absolute ductility contrast. The mathematics of the combined effects of particle interaction and ductility contrast is complex and little understood. It appears that analysis by the method described above can produce values for Rs in aggregates of high particle density which are approximately correct.

Gay (1968b) suggests a method to determine finite strain in deformed conglomerates which necessitates an estimation of both the 'viscosity ratio' of pebbles to matrix and the initial shape of the pebbles. The above technique when used to determine *Rs* values for particles of different composition, allows substitution of relative strain ratios into the equations presented by Gay to determine the viscosity ratio without estimation of initial shape. Preliminary measurement and comparison of quartzite and slate particles in a slate matrix give apparent 'viscosity ratios' in the same order as those presented by Gay (*Rv* quartzite/slate=2:1).

Inhomogeneous strain on the scale of a single thin section in the specimens analysed is not common. Local strain inhomogeneities adjacent to large competent particles have been avoided as discussed above. Any major strain discontinuities can generally be ascertained by inspection, since a gradual variation in strain intensity necessitates, by continuity of the material, a variation in the strain orientation. Thus cleavage divergence implies inhomogeneous strain.

[*Editor's Note:* Material has been omitted at this point.]

REFERENCES

Brace, W.F., 1961. Mohr construction in the analysis of large geologic strain. Bull. Geol. Soc. Am., 72: 1059–1060.

Brand, L., 1957. Vector Analysis. Wiley, New York, N.Y., 282 pp.

Cloos, E., 1947. Oolite deformation in the South Mountain fold, Maryland. Bull. Geol. Soc. Am., 58: 843–918.

Flinn, D., 1956. Deformation of the Funzie Conglomerate, Fetlar, Shetland. J. Geol., 64: 480–505.

Flinn, D., 1962. On folding during three-dimensional progressive deformation. Quart. J. Geol. Soc. London. 118: 385–428.

Gay, N.C., 1968a. Pure shear and simple shear deformation of inhomogeneous viscous fluids, 1. Theory. Tectonophysics, 5(3): 211–234.

Gay, N.C., 1968b. Pure shear and simple shear deformation of inhomogeneous viscous fluids, 2. The determination of the total finite strain in a rock from objects such as deformed pebbles. Tectonophysics, 5(4): 295–302.

Hossack, J.R., 1968. Pebble deformation and thrusting in the Bygdin area (southern Norway). Tectonophysics, 5(4): 315–339.

King, R.F., 1966. The magnetic fabric of some Irish granites. Geol. J., 5: 43–66.

Moore, F., 1965. Rheology of Ceramic Systems. Maclaren, London, 78 pp.

Nadai, A., 1950. Theory of Flow and Fracture of Solids. Eng. Soc. Monographs, McGraw-Hill, New York, N.Y., 572 pp.

Poll, J.J.K., 1966. The geology of the Rosas-Terresio area (Sulcis, South Sardinia). Leidse Geol. Mededel., 35: 117–208.

Ramsay, J.G., 1967. Folding and Fracturing of Rocks. McGraw-Hill, New York, N.Y., 568 pp.

15

Copyright ©1977 by Elsevier Scientific Publishing Company
Reprinted from pages T7-T13 and T15-T16 of *Tectonophysics* **42:**T7-T16 (1977)

Determination of geologic strain using randomly oriented strain markers of any shape

PIERRE-YVES F. ROBIN

Department of Geology and Erindale College, University of Toronto, Mississauga, Ont. L5L 1C6 (Canada)

(Submitted February 10, 1977; revised version accepted May 29, 1977)

ABSTRACT

Robin, P-Y.F., 1977. Determination of geologic strain using randomly oriented strain markers of any shape. Tectonophysics, 42: T7—T16.

Strain markers are geologic bodies within a rock which, during the deformation of that rock, have retained their identity but did not differ from their surrounding material in their mechanical behaviour. In sections they appear as closed contours. If it can be assumed that there was no preferred orientation of the markers, the axial-strain ratio in a planar section can be determined. Two lines are drawn through the centre of each marker, parallel respectively to the two axes of the sectional strain ellipse. If a_j and c_j are the segments defined by the intersection of these two lines with the contour of a marker j, then the axial-strain ratio is the logarithmic average of all a_j/c_j. As shown in an example, this method can also be used if the markers are ellipsoids. A statistical test of initial isotropy is proposed for this special case.

INTRODUCTION

A major concern of the geologist trying to unravel geologic deformations is the discovery, assessment and use of natural strain gauges. The ideal strain gauge is of course an originally spherical body having the same competence as the host, because its final shape is then that of the strain ellipsoid itself. This determines the state of strain except for a possible volume change. The deformed oöids analyzed in Cloos' (1947) classical study came very close to this ideal. Departures from this ideal case are of two kinds.

(1) There may be a competency contrast between gauges and host rock. Other methods have been developed for this case (eg. see Ramsay, 1967, Section 5—13) which require different assumptions.

(2) There is no competency contrast but the deformed markers are not spherical. Two situations may be distinguished:

(a) The initial shape of the markers is known, as in the case of deformed fossils; bilaterally symmetric fossils are particularly useful in this regard.

Here *finite-strain analysis* will yield the strain ellipsoid (except again for a possible volume change) almost as readily as if the fossil were a sphere.

(b) A more complicated case arises if the original shapes of the markers are not known a priori. Several methods have been developed to obtain strain when the markers are ellipsoids of unknown initial axial ratios and orientations. The ellipsoids are either assumed randomly oriented (Ramsay, 1967; Dunnet, 1969; Elliot, 1970; Shimamoto and Ikeda, 1976), or assumed to have initial distributions symmetric with respect to a recognizable bedding plane (Matthews et al., 1974).

The method presented here belongs to this latter group of strain methods, but the markers can have any arbitrary shape rather than be constrained to be ellipsoids. The assumptions are thus as follows:

(1) A sufficiently large number of inclusions in the rock, having varied shapes, acted as passive *markers* in a deforming rock; that is, there was no significant competency contrast between inclusions and host.

(2) The rock had an initially isotropic fabric:

$$\infty\,\infty\,\infty\text{ , or }\frac{\infty}{m}\frac{\infty}{m}\frac{\infty}{m}$$

at least in so far as the original orientation of these markers is concerned.* This means that their shapes and their 'directions' (if such directions can be unequivocally defined) are randomly oriented. A more operational definition, useful in the demonstration of the Appendix, is as follows. If a proportion of the markers have certain shape characteristics with respect to a given direction, a statistically equal proportion of the markers have the same characteristics with respect to any other direction.

(3) The rock underwent no volume change. This assumption is common to most gauges of paleostrain but does not restrict their usefulness (eg. see Ramsay, 1967, p. 161). It is made here for ease of exposition and writing. If a volume change were to be known or assumed independently the present method could still be used, mutatis mutandis.

The assumptions are less restrictive than those of Ramsay and Dunnet cited earlier, in that initial (and also final) shapes are not necessarily ellipsoidal. The method presented is of course applicable also if the markers are ellipsoidal, and that case will be discussed.

PRINCIPLE OF THE METHOD

The present treatment is restricted to the two-dimensional problem. The results obtained from several planar sections can be combined into a single three-dimensional strain tensor (see Ramsay, 1967, Section 4—7). Chiefly for expository convenience again, the major and minor axes of the sectional strain ellipse are designated as maximum and minimum principal strain

*In other respects the rock may have a strongly anisotropic fabric, with for example the markers concentrated in discrete layers, as shown schematically in Turner and Weiss' (1963) figure 2—8(b).

directions, and the corresponding quadratic elongations are designated as λ_1 and λ_3, respectively. Such designation would be strictly justified only if the plane contained the true maximum and minimum principal directions of the three-dimensional strain.

Initial unstrained state

Consider a number of markers, having various outlines within a planar section through an unstrained and isotropic rock. A center C can be determined for each marker. C can be the center of gravity, although the only requirement at this stage is that C can be determined independently of the orientation of the body. This means that if two markers are congruent but differ in orientation, their centers, determined according to the same rule, should come into coincidence if the shapes are themselves rotated into coincidence. For example, the midpoint between the two points that are the farthest apart on the contour would satisfy this requirement.

Consider now two orthogonal directions within the section, designated *1* and *3*. Through the center C of each marker we can draw two diameters across the body, parallel to the *1*- and *3*-axes, respectively (Fig. 1). In each marker, j, the diameters have lengths a_j, parallel to *1*, and c_j, parallel to *3*.

A direct consequence of isotropy can be stated in the form of a theorem, demonstrated in the Appendix:

If the fabric of the rock is isotropic in the sense stated above, and if there is a sufficiently large number of markers, then:

$$\prod_{j=1}^{n} \frac{a_j}{c_j} = \frac{a_1}{c_1} \times \frac{a_2}{c_2} \times \ldots \frac{a_j}{c_j} \times \ldots \frac{a_n}{c_n} \simeq 1 \tag{1}$$

Strained state

Now consider the rock after it has undergone a homogeneous strain. Directions *1* and *3* in the unstrained state were arbitrary; they can therefore be chosen to coincide with the principal directions of the strain, the corresponding quadratic elongations being λ_1 and λ_3. The line segments a_j and c_j defined for marker j have become a_j' and c_j' given by:

$$a_j' = \sqrt{\lambda_1}\; a_j \tag{2a}$$

$$c_j' = \sqrt{\lambda_3}\; c_j \tag{2b}$$

Therefore:

$$\prod_{j=1}^{n} \frac{a_j'}{c_j'} = \left(\sqrt{\frac{\lambda_1}{\lambda_3}}\right)^n \prod_{j=1}^{n} \frac{a_j}{c_j}$$

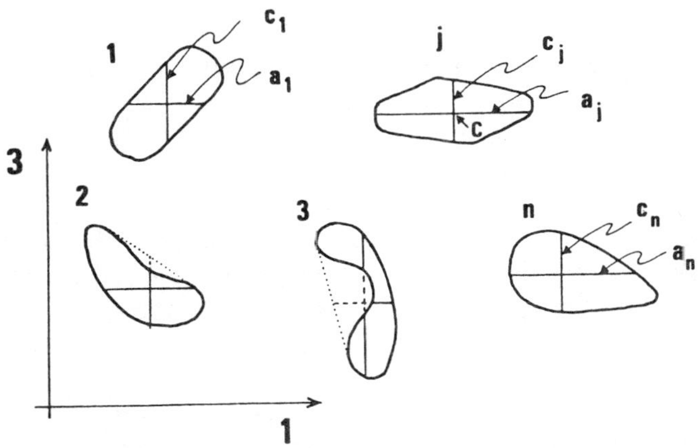

Fig. 1. Isotropically oriented markers in the unstrained rock. Lines drawn parallel to the two directions *1* and *3* through the center c_j of each marker define segments a_j and c_j respectively. If markers have such shapes that their contours are intercepted more than twice by the lines (e.g. marker No. 3), a_j or c_j can be defined as the sum of the two or more subsegments. Alternatively, all markers which are not uniformly convex outward can be redefined by replacing the concavity by a straight line, as shown on Nos. 2 and 3.

or, using (1):

$$\prod_{j=1}^{n} \frac{a_j'}{c_j'} \simeq \left(\sqrt{\frac{\lambda_1}{\lambda_3}}\right)^n \tag{3a}$$

In logarithmic form, (3a) becomes:

$$\sum_{j=1}^{n} \log \frac{a_j'}{c_j'} = n \log \sqrt{\frac{\lambda_1}{\lambda_3}} \tag{3b}$$

Determination of the axial strain ratio, $\sqrt{\lambda^1/\lambda^3}$ is therefore straightforward in principle. The segments a_j' and c_j' are measured on n markers in the planar section, the ratios a_j'/c_j' calculated, and the axial strain ratio is given by their logarithmic average (eq. 3b).

The method requires that principal directions of strain *1* and *3* be known. In principle, they could always be found by evaluating eq. 3 for many directions and choosing the direction for which its value is maximum. In practice, it is probably sufficient to estimate the direction of maximum principal strain from a histogram of the directions of the longest diameter of each deformed marker, or from the trace on the planar section of a foliation associated with the deformation.

A difficulty arises from the fact that the center C for each marker must be determined in the strained state rather than in the undeformed one. Equation 1 theoretically requires that the choice of C for each marker *in the unstrained state* be independent of the orientation of the marker, as explained above. The center of gravity in the unstrained state satisfies this requirement; furthermore, it remains the center of gravity of the strained marker, a property of homogeneous strain. Therefore, if the center of gravity in the strained state is taken as center of the marker, we know that the point chosen was also the center of gravity in the unstrained state; this is a theoretically proper choice. However, the center of gravity is not easy to determine, except for centrosymmetric markers such as ellipsoids. Other rules to choose a center, such as the midpoint between two points that are the farthest apart on the contour of the marker, are much easier to use. However, because this determination must be made in the strained state rather than in the unstrained state, and because two markers which were initially identical except for a rotation, are no longer identical in the strained state, the centers determined by such rule would not coincide in the two 'destrained' markers. The demonstration of eq. 1 would therefore not be exactly valid. For most markers it is unlikely that choosing centers with such a simple rule would lead to significant errors in practice, as a/c is generally not very sensitive to the exact location of C.

CASE OF INITIALLY RANDOMLY ORIENTED ELLIPSOIDS

Ramsay (1967) and Dunnet (1969) have proposed a method to calculate the strain from the final axial ratio, R_j, and axis orientations, ϕ_j, of initially randomly oriented ellipsoids. The principle of the method is rather complex, essentially because the relationships between initial and final axes of each elliptical marker are complex. Furthermore, its practice requires a subjective fitting of curves around a cluster of points, which is done by successive approximations (Dunnet, 1969). Shimamoto and Ikeda (1976) showed how the strain could be calculated, uniquely and objectively, from the same data. The mathematical justification of their method is somewhat more involved than that of the present one but its actual use is as simple. Application of the present, more general method to ellipsoids is therefore given here mostly as an example which lends itself to easy verification. The ratio a_j'/c_j' required to evaluate eq. 3 can be calculated from R_j and ϕ_j by:

$$\frac{a_j'}{c_j'} = \sqrt{\frac{\tan^2 \phi_j + R_j^2}{R_j^2 \tan^2 \phi_j + 1}} \tag{4}$$

Once these ratios for n ellipses are calculated with eq. 4 from R_j and ϕ_j, eq. 3 can again be used to estimate the strain. As in the general case, the direction of maximum principal strain, if not known or assumed, must be determined by trying several directions. For ellipses, however, trials to find the direction which maximizes eq. 3 can be done numerically, without new measurements on the rock face. A new origin for the orientation angles

ϕ_j, i.e. a new *1*-direction, can be tried by simply adding a uniform value (eg. 1°, 2°, . . .) to all ϕ_j, and repeating the above calculations. In principle, for ellipses, this method is thus somewhat less elegant than Shimamoto and Ikeda's (1976) which finds the principal directions of strain algebraically by diagonalization of a 2 × 2 matrix.

A CASE EXAMPLE

Bau (1977) has used strained cordierite crystals, feldspar crystals and volcanic amygdules from several sites in the Superior Province of Ontario as strain gauges. Bau treated these bodies as ellipsoidal markers and collected his data accordingly, i.e. in the form R_j, ϕ_j. Data from one site are plotted in Fig. 2. In this case, the origin for the angles ϕ_j has been taken as the average of the original measurements. The present method yields a value

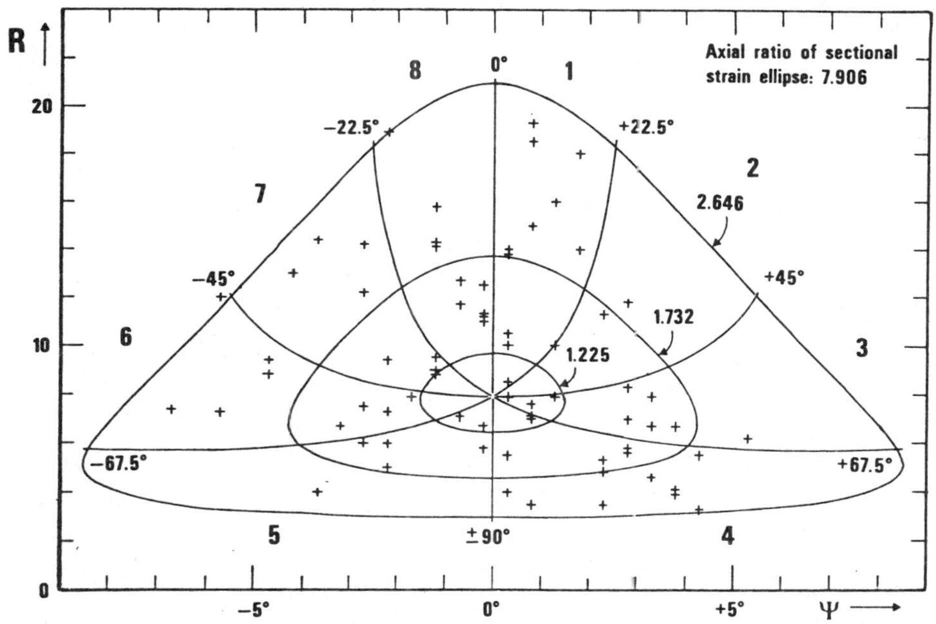

Fig. 2. Plot of final axial ratio vs. direction of major axis for 72 deformed corderiete crystals treated as ellipsoidal markers. Location: Hwy. 11, 10 miles west of Kashabowie, N.W. Ontario, Quetico sub-province, Superior Province (Bau, 1977). Measurements by A. Bau. R is the *final* axial ratio of the ellipses (Dunnet's 1969 notation) rather than the square of it as in Ramsay (1967). Orientation angles, ψ, have been recalculated taking their average as new origin and as the *1*-direction.

Numbers shown next to each of the three 'concentric' curves indicate the corresponding *initial* axial ratios. Numbers shown next to each 'radial' curve show the corresponding initial angle of major axes with the *1*-direction. The axial ratio of the strain ellipse, 7.906, used to compute these curves was calculated from the data by using eqs. 4 and 3.

Numbers *1—8* merely designate angular sectors for the purpose of the statistical analysis discussed in the text (see also Table I).

of 7.91 for the axial ratio of the sectional strain ellipse. Attempts to maximize this ratio by small changes of origin of ϕ were felt unnecessary. Applying Shimamoto and Ikeda's method to the same data yields a ratio of 7.94, with the direction of maximum principal strain making an angle of only 0.26° with the direction chosen here. It can therefore be concluded: (1) that these two methods, which should give identical results for infinitely large samples, do not differ significantly in the present case where $n = 72$; and (2) that choosing the average value of ϕ to approximate the direction of maximum principal strain was a satisfactory procedure.

After strain is calculated algebraically, a grid can be superimposed on the plot of Fig. 2. This grid is made up of two families of curves. One family shows the position on the plot of ellipsoids of same initial ellipticity but variable orientations; this family of curves is similar to that shown by Dunnet (1969, fig. 2). The other family of curves shows the position of ellipses of same initial orientation but variable ellipticity. The superposition of this grid to the data can serve several purposes:

(1) It shows that the strain determined objectively by the method proposed here, compares well with the strain that would be determined by fitting curves of the first family (equal initial ellipticity) around the data as proposed by Dunnet (1969).

(2) In principle it allows one to determine the distribution of initial ellipticities of the markers; this distribution can then be judged likely or unlikely by comparing it with that in unstrained rocks.

(3) More importantly, it suggests a test of whether the initial population was random or not.

TEST OF INITIAL RANDOMNESS OF ELLIPSOIDAL MARKERS

Let us consider the following situation. Our *hypothesis*, based on geological considerations, is that markers are randomly oriented, and we wish to see whether data are compatible with this hypothesis. Each angular sector in Fig. 2 should be equally populated, the departures from this equality being due to sampling chance alone. Choosing eight sectors as in Fig. 2 the expected number of points in each sector is $n/8$. We can calculate:

$$P = \sum_{k=1}^{8} \frac{(N_k - n/8)^2}{n/8}$$

where N_k is the actual number of points in sector k (Table I). This quantity should be distributed as χ^2 with five degrees of freedom (eg. Miller and Kahn, 1962, p. 114—117).* In the case shown in Fig. 2, $P = 9.11$; sampling from an isotropic population would give a value of P greater than 9.11 in 11% of cases. Using the usual confidence levels of 5 or 10%, we would have no reason to reject the isotropy hypothesis.

*If the strain were known a priori, that is, in effect, if the center of the family of curves in Fig. 2 were known independently of the data, the number of degrees of freedom would be 8—1 = 7. Here, the *two* coordinates of this center are determined from the data themselves and the number of degrees of freedom is reduced to 7—2 = 5.

[*Editor's Note:* Material has been omitted at this point.]

REFERENCES

Bau, A., 1977. History of Regional Deformation of Archean Rocks in the Kashabowie Lake—Lac des Mille Lacs Area, Northwest Ontario. Doctoral dissertation, University of Toronto.

Cloos, E., 1947. Oölite deformation in South Mountain Fold, Maryland. Geol. Soc. Am. Bull., 58: 843—918.

Dunnet, D., 1969. A technique of finite-strain analysis using elliptical particles. Tectonophysics, 7 (2): 117—136.

Dunnet, D. and Siddans, A.W.B., 1971. Non-random sedimentary fabrics and their modification by strain. Tectonophysics, 12 (8): 307—325.

Elliot, D., 1970. Determination of finite strain and initial shape from deformed elliptical objects. Geol. Soc. Am. Bull., 81: 2221—2236.

Matthews, P.E., Bond, R.A.B. and Van den Berg, J.J., 1974. An algebraic method of strain analysis using elliptical markers. Tectonophysics 24: 31—67.

Miller, R.L. and Kahn, J.S., 1962. Statistical Analysis in the Geological Sciences. Wiley, New York, 483 pp.

Ramsay, J.G., 1967. Folding and Fracturing of Rocks. McGraw-Hill, New York, 568 pp.

Shimamoto, T. and Ikeda, Y., 1976. A simple algebraic method for strain estimation from deformed ellipsoidal objects. I. Basic theory. Tectonophysics, 36: 315—337.

Turner, F.J. and Weiss, L.E., 1963. Structural Analysis of Metamorphic Tectonites. McGraw-Hill, New York, 545 pp.

16

Copyright ©1979 by Pergamon Press Ltd

Reprinted from pages 155–156 and 159–162 of *Jour. Structural Geol.* **1**:155–162 (1979)

A comparison of methods of strain determination in rocks from southwest Dyfed (Pembrokeshire) and adjacent areas

S. S. HANNA and N. FRY

Department of Geology, University College, Swansea SA2 8PP, Wales, U.K.

INTRODUCTION

FRY (in press, a) describes a method of measuring natural strain in rocks from the relative locations of marker objects. It involves the creation of an 'autocovariance' or 'all-object-separations' plot, on which the distance and direction from each marker to each other marker in the rock are recorded by a point at that distance and direction from the origin. From an isotropic distribution such a plot has a concentric circular pattern. From a deformed distribution which was initially isotropic, the pattern is concentric elliptical, with the ellipse axial ratio and orientation corresponding to the strain ellipse of the deformation.

Fry (in press, a) is largely concerned with theoretically justifying his method but he also suggests that the best procedure is numerical handling of marker coordinates. Hanna, as part of a study of regional rock deformation, has used a manual graphical procedure and has shown it to have practical advantages not foreseen by Fry (in press, a). This is the subject of the present paper.

TECHNIQUES USED IN THE COMPARISON

Many techniques for measuring the finite strain of rocks have been documented in the last twelve years. These generally involve the determination of two-dimensional strain on planar sample sections through a rock, with the intention that three such planes may be combined to yield the three-dimensional strain. This paper briefly discusses several methods for determining two-dimensional strain (see Fig. 2 and Table 1). These methods are later referred to informally or by our letters A, B, C, etc. References are only given fully in this section.

Method A—Method 1 *of Ramsay* (1967, *p.* 193) *and Cloos* (1947)

This method uses as strain markers objects which were approximately circular in their initial state. Direct measurement is made of the lengths of the principal axes of the strained object ellipses. Either the arithmetic mean of the individual observations is calculated or the data are plotted as a graph of the short axis length vs the long axis length. The slope of a line through these points is taken to be the ratio of the principal strains within the particular section.

Method B—Method 3 *of Ramsay* (1967, *pp.* 195–197), *'centre-to-centre' method*

This method assumes that the length of a line from the centre of an object to the centre of one of its nearest neighbours is initially statistically independent of the line's direction. In no other respect is the initial distribution necessarily assumed to be isotropic. The method considers only the relative displacement of the centres of adjacent objects by the deformation, and does not rely on elliptical shapes of strained objects to indicate the strain ellipse. It is useful for rocks composed of packed objects (ooids in Ramsay 1967) which have had the regularity of their external shapes destroyed by pressure solution. The strain ratio is estimated from a graph of length (d) against orientation (α), relative to a common azimuth, of every measured line between centres of adjacent objects.

Method C—Dunnet (1969) *and Dunnet & Siddans* (1971), Rf/φ *method*

This method assumes a homogeneous deformation of initially elliptical marker objects with their matrix. Initially randomly variable in both ellipticity and orientation, both these features are changed by the deformation. A plot is made of log Rf, where Rf is the final ellipse ratio, against φ, the final long axis orientation, of each ellipse. These parameters depend on five variables (Dunnet 1969, p. 117); initial particle shape, initial particle orientation, strain intensity, strain orientation and the degree of ductility contrast between particles and the entire particle/matrix system. The method is extended by Dunnet & Siddans (1971) to include planar, linear and imbricate sedimentary fabrics. They outline two computer programs, one of which constructs standard Rf/φ curves and '50% of data' curves for different Rf values. The other programme (our method F) is a procedure for determining the strain ratio and orientation of sections through certain homogeneously deformed planar and semiplanar fabrics by re-establishing initial symmetry.

Method D—Fry (in press, a)

This method assumes that before deformation the locations of the centres of intersections of marker objects by a sample plane would on any sample plane have an isotropic, but non-Poisson, distribution.

If the position-marking objects behave passively during deformation, the ellipticity and orientation of an 'autocovariance' or 'all-object-separations' plot correspond to those of the strain ellipse. Fry (in press, a) suggests a numerical approach, but the following purely manual plotting procedure has been adopted in this study. Place a sheet of tracing paper, on which a series of parallel reference lines have been drawn, over the sample and mark the centre of every marked object within the chosen sample field. Take a second sheet of tracing paper with a centre point marked on it and a set of marked lines which can be kept aligned parallel to those on the first sheet. Place the central point of the second sheet on one marked object position on the first, and mark on the second sheet all the positions of points on the first. Now place the central point of the second sheet on a different point on the first, and again mark on the second sheet all positions of points on the first. Continue accumulating points by this procedure, maintaining the same orientation, until all points on the first sheet have been used as centres on the second. The resulting plot on the second sheet has an ellipticity of the same ratio and orientation as the strain ellipse. The long and short axial lengths and orientations are measured directly with ruler and protractor. Theoretical limitations, with a discussion of edge effects and other complications, are given by Fry (in press, a).

Method E—Shimamoto & Ikeda (1976)

Shimamoto & Ikeda (1976) assume all objects to be initially ellipsoidal and to comprise an initially isotropic population. Using as data the dimensions and orientations Rf and φ (Dunnet 1969) of a large number of object intersections, the descriptive parameters of an overall ellipse for the sample plane can be numerically determined. This ellipse should converge in the limit precisely to the strain ellipse, provided a correct numerical procedure is adopted. In use by a number of workers in the past as a method for measuring strain, Shimamoto & Ikeda appear to be the first to have published the procedure fully, adopting a description of it in terms of shape matrices. The existing University College of Swansea computer programme for this procedure was used for our study.

Method F—Dunnet & Siddans (1971), numerical version of Rf/φ

This computation, using a programme similar to their "STRANE" (p. 322) checks symmetry about "$\varphi = 0$" and "50% of data" curves for tested strain values, assigning the optimum tested value of strain.

Method G—Mimran (1976) Density distribution technique

This technique is based on the principle that the longer the dimension of any strain marker object in a given orientation, the higher the probability of its being intersected by sample planes perpendicular to this orientation. Provided that a population of marker objects is initially isotropic, the density of intersections of objects by a plane in the strained state is a measure of the perpendicular dimension of the strain ellipsoid. Mimran's development of the technique involves counting objects on three perpendicular sample planes, but his method of constructing three two-dimensional strain ellipses from such data is theoretically unsound (Fry in press, b). A correct use of intersection density data does not permit direct specification of two-dimensional strains for direct comparison with the results of the other methods used in this paper.

COMPARISON OF RESULTS

The methods outlined above have been applied to three mutually perpendicular plane sections of an oolitic limestone sample from West Angle Bay, southwest Dyfed (formerly Pembrokeshire). The results are summarized in Table 1. Although judged subjectively to be principal planes of the strain before sectioning, there is no way to corroborate or refute such a relationship independently of the determined two-dimensional strain ratios of the sample sections. Therefore comparison of the values for the different sections on the assumption of principal planes is not a valid test of the methods, and we do not here refer these orientations to know principal directions. That the product of the three strain ratios is not unity indicates that these are not principal planes, and a forthcoming paper (Fry in press, b),

Table 1. Measured strain ratios on three studied sections through an oolitic limestone from West Angle Bay, Dyfed for the methods discussed. The sections 1, 2 and 3 were chosen to lie approximately in XY, YZ and XZ, respectively

Sample plane:	1	2	3
Method: A: Ramsay (1967, p. 193) long/short	1.74	2.52	3.31
B: (1967, p. 195) centre-to-centre	1.78	2.44	3.50
C: Dunnet & Siddans (1971) Rf/φ	1.65	2.40	3.25
D: Fry (1979) our version	1.66	2.42	3.27
E: Shimamoto & Ikeda (1976) (numerical)	1.68	2.41	3.26
F: Dunnet & Siddans (1971) (numerical)	1.69	2.42	3.24
G: Mimran (1976) density distribution	1.23	1.42	1.76

[*Editor's Note:* Figure 1 has been omitted, and, in the original, page 158 is blank.]

which discusses errors due to misjudging principal planes, would here indicate misorientations of at least 10°. This matter is not considered further because it is methods of two-dimensional strain ratio measurement which are the subject of this paper, the forthcoming paper discusses three-dimensional strain determination and errors. Internal consistency of the data for each plane is our only judge of the accuracy of the particular two-dimensional strain ratios.

General comments on the methods

It should be noted that methods A, C, E and F rely on the measurement of elliptical sections of objects assumed also to have been elliptical in their initial state. If this assumption applies to a particular rock, valuable information would be lost by not using one of these methods. On the other hand, these methods determine strain of objects, which may not correspond to the overall rock strain. Methods B, D and G do not rely on individual object shapes and are therefore more generally applicable. The similarity of results for our sample indicates the homogeneity of the deformation of this particular rock, which is illustrated in Fig. 1.

Even in an oolitic limestone, object shape may be difficult to use. Difficulties noted by Cloos (1947) and cited by Ramsay (1967, p. 193) are: (a) frayed ooid ends or pressure dissolved sides; (b) unidentifiable ooid ends at high strains and (c) difficulty in locating principal ellipse axes when strain is slight.

Note also that it is only methods C, D and F (our method and the Rf/φ method) which provide information about the validity of the assumptions on which they are based.

The data plots for the graphical methods A–D for each of the sample planes are shown in Fig. 2.

Remarks on each method

Method A. The long-to-short method is often quick and easy to plot, but problems of definition of a best-line offset these advantages. Whatever definition is used, this line is more reliable if it is calculated. Its slope can be shown on theoretical grounds to be an overestimate of the strain ratio by an amount which is unknown and is dependent upon dispersion of original shapes from truly circular, and upon any correlation of either this dispersion or of competence contrast with grain size. These latter variations receive different weightings in the definitions given below, with values shown for this sample:

	Plane 1	Plane 2	Plane 3
$\Sigma X/\Sigma Y$	1.739	2.55	3.40
$\dfrac{\Sigma(X/Y)^{\frac{1}{2}}}{\Sigma(Y/X)^{\frac{1}{2}}}$	1.754	2.57	3.41
$\left(\dfrac{\Sigma X/Y}{\Sigma Y/X}\right)^{\frac{1}{2}}$	1.741	2.55	3.40
'Eye-balled'	1.740	2.52	3.31

Method B. The 'centre-to-centre' method is slow and laborious, and it is not always clear how best to define the maximum and minimum distances from the plot (see discussion of method D).

Method C. The Rf/φ methods are slow but can provide interpretable information on initial fabrics, and on the relative orientation of strain to bedding and cleavage directions, in addition to giving strain values. These relationships give these methods considerable advantage (shared by our method D) in the unravelling of large-scale folds or other structures.

The number of objects required varies. Estimates include: 40 ooids or 60–100 pebbles in conglomerates (Dunnet & Siddans 1971); 30 ooids (Tan 1976, p. 167) and 20 ooids for strain ratios of 4 : 1 or greater in this study. This study was made using the D-mac digitizer on-line to a Digico Micro 16V minicomputer at Leeds University (Siddans 1976).

Method D. This version of the method described by Fry (in press, a) is much quicker than any of the other methods because the data are accumulated graphically as the final plot without being first rendered in numerical form. Definition of the ellipse ratio appears to lack precision, but in practice the reproducibility is found to be good and values obtained compare well with other methods. This applies particularly here because of the very sharp cut-off to neighbour distances caused by touching of well-sorted spherical objects (ooids) in the initial state. These same maximum and minimum distances to the same cut-off are also the best definition of the ellipse ratio when data is handled using Ramsay's 'centre-to-centre' method (our method B). On Ramsay's plot the cut-off should be sinusoidal, while by this method it is elliptical. In practice it is easier to visualise the ellipse form. Thus our method is of advantage in both the preparation and interpretation of the graphical form. Note that the methods are only directly comparable when it is, as with this oolitic sample, the initially nearest-neighbour cut-off which is being measured (see later discussion).

Method E. The time taken by the numerical methods (E and F) is only that required to convert the dimensions and orientations Rf and φ of object intersections into digital form. An important limitation to most such calculations is that although the results are correct only for an initially isotropic population, this assumption is not tested by the method. Such a limitation does not apply to the numerical method of Matthews *et al.* (1974) which we have not tried. In as much as their method tests for symmetry of anisotropic configurations it is equivalent to method F with a simpler mathematical base.

Method F. Siddans' (1976) version of the Dunnet & Siddans (1971) calculation is complicated but does test for initial isotropy. However, if isotropy is established the methods of Matthews *et al.* (1974) and method E are superior.

Method G. In view of the incorrect development by Mimran (1976) of his density distribution technique, and its inability to give true two-dimensional strain ellipses other than by working back from three-dimensional strain (Fry in press, b), we include values here only to show that they are out-of-line with other methods.

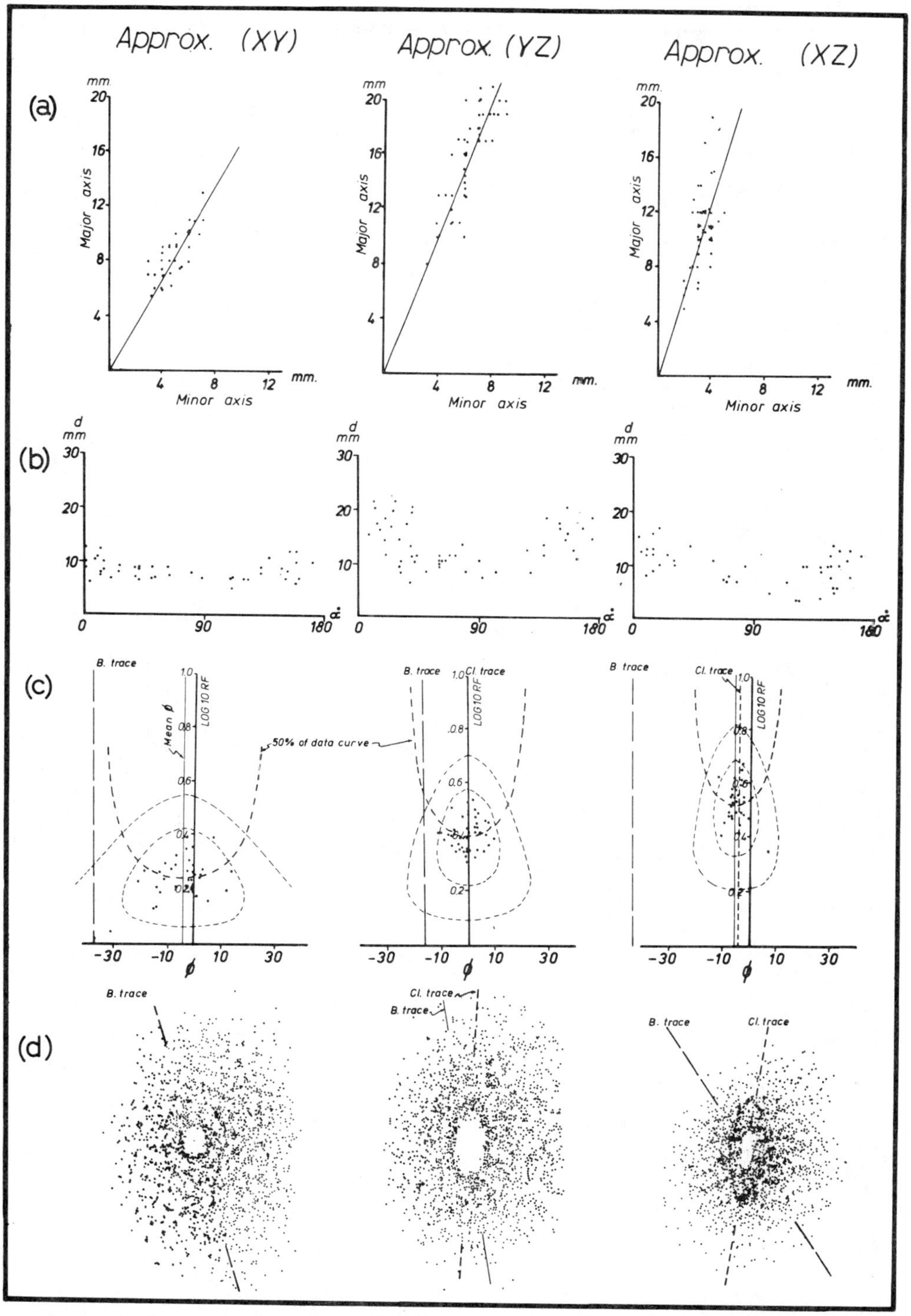

Fig. 2. Plots for each of the three sample planes of the oolitic limestone specimen from West Angle Bay, Dyfed, for the four graphical methods discussed. Letters (a)–(d) refer to methods A–D in the text.

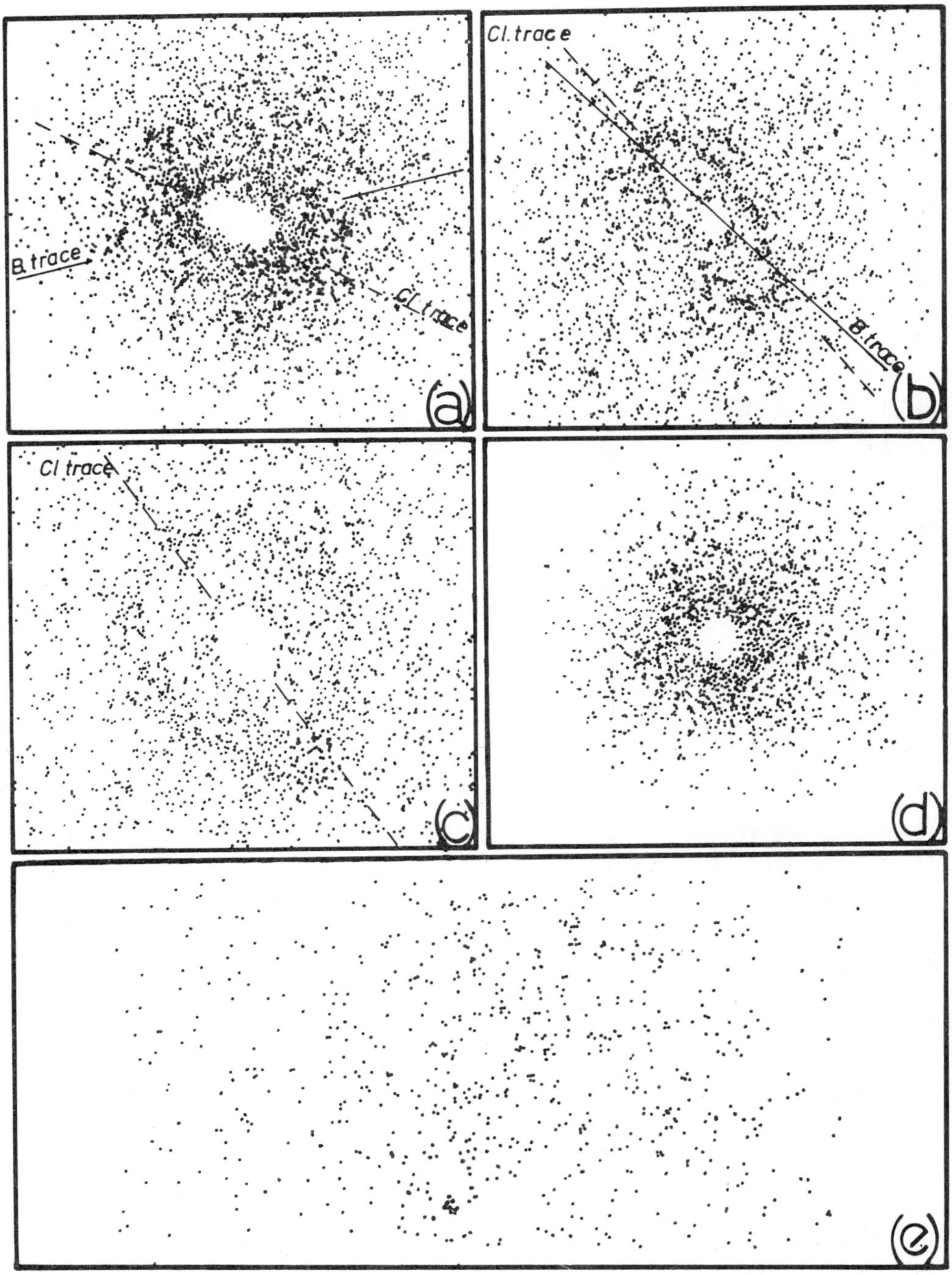

Fig. 3. Plots for the rock samples mentioned in the further discussion of our method. (a) Oolitic Lower Carboniferous limestone from Caswell Bay, Gower. (b) Reduction spots in Old Red Sandstone mudstones, Ballycotton, Co. Cork. (c) Ferruginous rhizoid-fill nodules in Coal Measures sandstone, Amroth, Dyfed. (d) Calcareous concretions in Old Red Sandstone mudstones, Llanstephan, Dyfed. (e) Pyrite aggregates in Carboniferous slates, Bantry, Co. Cork.

ADVANTAGES OF THE MANUAL GRAPHICAL METHOD

The manual plotting version of the method of Fry (in press, a) has been applied to many types of marker objects. Those reported here from South Wales are ooids in oolitic Lower Carboniferous limestones from West Angle Bay, Dyfed (Fig. 2) and Caswell Bay, Gower (Fig. 3a), rhizoid-fills in Westphalian sand-stone from Amroth, Dyfed (Fig. 3c) and calcareous concretions in Old Red Sandstone mudstones from Llanstephan, Dyfed (Fig. 3d). Also illustrated are reduction spots in mudstones (Fig. 3b) and pyrite aggregates in slates (Fig. 3e) from localities in Ireland.

Acetate peels were made from etched cut surfaces of the oolitic limestones and then projected onto a ground glass screen. The other rocks were photographed in the field and a transparency projected in the same manner. Thin sections may be similarly treated, as the only thickness effect is a modification of the object distribution's density but not its anisotropy. We see no reason why, given a flat rock surface with a suitable density of marker objects, the method should not be used directly in the field.

The concentric similar elliptical pattern may have some very clear initially circular feature, such as a cut-off due to ooid size. This is found to be clear and easily measurable with a sample size of about 60 ooids, taking about half an hour to complete. Fry (in press, a) mentioned that for poorly-defined patterns a sample of several hundred objects may be necessary, and computer handling of digitized coordinates would then become worthwhile. In practice, this study has shown that the sample size need only be raised to about 140 for concretions, and the time taken never exceeded 1 h. Fry (in press, a) mentioned that Ramsay's 'centre-to-centre' method (method B) gives greater accuracy for small sample size. We find that this is not true for ooids, and that it is quicker to handle a large sample by our method than a selected small part of it by Ramsay's method.

Some distributions of markers give a plot which has a gradationally edged girdle of higher density of points instead of a sharp cut-off. Fry (in press, a) states as the main advantage of his method that it produces true strain ratios from such distributions, whereas nearest-neighbour methods give false results. This study confirms this advantage, Fig. 3(b) being such a plot.

Figures 3(c) and 2(d) are of particular interest in having both an inner-neighbour cut-off ellipse and a denser outer ellipse. In general, these may be different in ellipticity and orientation, the cut-off showing the average strain of marker object shapes that method B would determine, while the outer dense girdle shows the overall strain of the rock. Their similarity in Fig. 2(d) is confirmation of the homogeneity of the strain of this sample.

Figure 3(c) deserves particular attention, although the denser ellipse is not accurately located by such a small number of plotted points. The strain of this rock is by pressure solution striping (Beach & King 1978). The preferred spacing of markers is greater than that of the stripes, so that the ellipse of Fig. 3(c) portrays the overall rock strain resulting from the pressure solution. We are not aware of any other method that can determine the strain of this rock.

Figure 3 illustrates that this method is able to show the relationship of the traces of bedding and cleavage to ellipse girdle. Figure 3 also shows two samples for which the method fails to give a value for the strain. Figure 3(d) shows an isotropic pattern of centres of packed competent calcareous nodules in a deformed rock, in which we deduce that sliding at nodule boundaries has accommodated the strain. Figure 3(e) shows a pattern of centres of pyrite aggregates sufficiently close to Poisson (see Fry in press, a) such that no information on strain is forthcoming, whatever the strain. This puts our method at no disadvantage, as all methods fail on these two samples.

CONCLUSIONS

We conclude that our method and the Rf/φ methods are the best of those tested, as information is directly available about the relationship of strain to other features, and about the validity of assumptions in the methods. Our method is the quicker. It does not employ all usable information, however it is applicable to many rock types for which Rf/φ methods cannot be used, and in some cases for which there is no other method. We would recommend trial of our method even if Rf/φ is being used, as there is a possibility of showing discrepancies between marker object strains and overall rock strain, whether as a result of normal ductility contrast or of differences in proportion of volume loss between objects and matrix.

A critical comparison of manual and numerical versions of Fry's method will be the subject of a forthcoming paper.

Acknowledgements—We are indebted to Dr R. H. Graham for encouraging us to try our method on anything and everything in rocks from South Wales and Ireland.

REFERENCES

Beach, A. & King, M. 1978. Discussion on pressure solution. *J. geol. Soc. Lond.* **135**, 649–651.

Cloos, E. 1947. Oolite deformation in South Mountain Fold, Maryland *Bull. geol. Soc. Am.* **58**, 843–918.

Dunnet, D. 1969. A technique of finite strain analysis using elliptical particles. *Tectonophysics* **7**, 117–136.

Dunnet, D. & Siddans, A. W. B. 1971. Non-random sedimentary fabrics and their modification by strain. *Tectonophysics* **12**, 307–325.

Fry, N. in press, a. Random point distributions and strain measurement in rock. *Tectonophysics* 60, 1979, 89-105.

Fry, N. in press, b. On density distribution techniques and strained length methods for determination of finite strains. *J. Struct. Geol.*

Matthews, P. E., Bond, R. A. B. & Van den Berg, J. J. 1974. An algebraic method of strain analysis using elliptical markers. *Tectonophysics* **24**, 31–67.

Mimran, Y. 1976. Strain determination using a density-distribution technique and its application to deformed Upper Cretaceous Dorset Chalk. *Tectonophysics* **31**, 175–192.

Ramsay, J. G. 1967. *Folding and Fracturing of Rocks*. McGraw-Hill, New York.

Shimamoto, T. & Ikeda, Y. 1976. A simple algebraic method for strain determination from deformed elliptical objects—I. Basic theory. *Tectonophysics* **36**, 315–337.

Siddans, A. W. B. 1976. Deformed rocks and their textures. *Phil. Trans. R. Soc.* **A283**, 43–54.

Tan. B. K. 1976. Oolite deformation in Windgallen, Canton Uri, Switzerland. *Tectonophysics* **31**, 157–174.

17

Copyright ©1972 by Geological Society of America
Reprinted from *Geol. Soc. America Bull.* **83:**2621-2638 (1972)

Deformation Paths in Structural Geology

DAVID ELLIOTT *Department of Earth and Planetary Sciences, The Johns Hopkins University, Baltimore, Maryland 21218*

ABSTRACT

The evolution of folds and the origin of mineral orientations and schistosity are analytically related to the deformation paths or histories in deforming rocks. The several different finite and incremental methods of representing deformation paths are connected by matrix algebra, and graphical presentation in natural strain space is found to be particularly clear. New methods permit calculation of various components of deformation paths from boudinage, pressure shadows, and inclusion trails in naturally deformed metamorphic rocks. In coaxially accumulating deformation paths the principal strains remain parallel to the same material lines in the rock and have been referred to inappropriately as irrotational. Straight inclusion trails in metamorphic minerals may be produced by synkinematic grain growth during a coaxially accumulating path and do not necessarily indicate postkinematic grain growth. Synkinematic inclusion trails can be produced by growth of the grain through its pressure shadow rather than matrix schistosity, and the quartz protected from recrystallization by being included in snowball garnets may sometimes form this way.

INTRODUCTION

A given finite deformation can be reached by an infinite number of different paths only one of which actually occurred. A deformation history or path indicates the states through which a deformation progresses en route to its final value.

Familiar fluids such as water, air, tar, and plasticine preserve none of the history of the deformations they have been through. Metamorphic rocks are different, as their texture may be utterly dominated by the effects of deformation history.

Understanding the history of deformation plays a crucial role in structural geology, yet the concept was not discussed in any definite way until Flinn (1962). To work out the development of a particular fold, for example, one has to know either the deformation history for a sufficient number of points, or the history of velocity boundary conditions and there is simply no way to get velocity boundary conditions *a priori.*

This paper outlines new ways of defining deformation history in naturally deformed rocks and critically reviews methods in current use. In order to accomplish this in a unified way, it is necessary first to extend and review the mathematical formulation of deformation paths and then to relate this in turn to folding, boudinage, pressure shadows, inclusion trails in minerals, and the formation of schistosity or cleavage. This paper, then, is an attempt to interpret deformation paths directly from observations in the field and in thin section; two other different approaches can also be used to determine deformation history. One of these approaches is to assume all the material properties and boundary conditions as given, and then produce the folds by performing computer experiments (for example, Chapple, 1969; Dieterich, 1969). The results may then be compared to the natural structures. Another possible way is to seek special situations, such as ductile shear zones where the deformation is restricted to simple shear paths by the undeformed rock on both sides (Ramsay and Graham, 1970). All of these different methods have important common ground and are complementary.

DEFORMATION PATHS AND COORDINATE FRAMES

The next six sections of this paper deal with the mathematical framework behind the concept of deformation paths. Unfortunately, not much of this material is available in the geological literature, and it is only partly available and widely scattered in the mechanics literature. Analytical relations are established between the various ways of defining and describing deformation history and several new terms are introduced. Much of the treatment is two-dimensional for ease of presentation, but extension into three dimensions is straightforward. First we must examine the matrix algebra in three different types of coordinate frames.

Fixed External Frame

Externally fixed coordinate axes can be visualized as east-west and north-south grid lines scratched on a sheet of glass. We are interested in what is happening in a particular mass of rock, so as the rock deforms beneath the glass the coordinate origin is kept fixed over the same material point but the coordinate frame maintains the same east-west and north-south grid lines and scale. Let the mass of rock be small enough that the strains remain homogeneous. A homogeneous but otherwise perfectly general rotational strain will be called deformation, and can always be represented by a matrix **D**.

The deformation path is the history of **D** and can be analyzed as two separate components: the pure strain history and the rotation history. This is because any deformation matrix can be factored by left polar decomposition (Elliott, 1970a):

$$\mathbf{D} = \mathbf{T}\mathbf{R}\,. \tag{1}$$

The simplest description of the symmetric left stretch matrix **T** is when it is in diagonal form **Td** (all diagonal forms are distinguished by d).

$$\mathbf{T} = \boldsymbol{\theta}\,\mathbf{Td}\,\boldsymbol{\theta}^{-1}$$
$$\mathbf{Td} = \begin{bmatrix} T_I & 0 \\ 0 & T_{II} \end{bmatrix} \tag{2}$$

T_I and T_{II} are the long and short semiaxes of the strain ellipse.

The orthogonal matrix θ in equation (2) is an orientation matrix defining the angle θ of the long axis of the strain ellipse to the externally fixed x coordinate (Fig. 1).

From (1) and (2):

$$\mathbf{D} = \boldsymbol{\theta}\,\mathbf{Td}\,\boldsymbol{\theta}^{-1}\mathbf{R}\,. \tag{3}$$

When a rock deforms, an original embedded sphere changes progressively through a sequence of continuously varying finite strain ellipses, and each instant is characterized by a unique ellipse. We can see that the left stretch

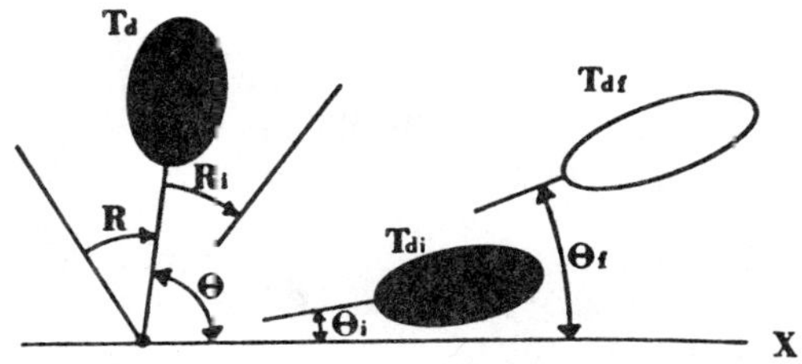

Figure 1. Original unit circle rotates R and then strains Td at an angle θ to the X axis, then rotates an increment Ri and is strained an increment Tdi at an angle θi. The total final strain ellipse Tdf is oriented θf. This is a geometric interpretation of deformation in a fixed external frame of equation (6b).

matrix **T** is concerned with this finite strain ellipse. The pure strain history is the history of the magnitudes of the axes of the finite strain ellipses T_I, T_{II}, and also the history of the orientation θ of the finite strain ellipse.

The orthogonal rotation matrix **R** is concerned with the rotation through an angle R°of a material line in the undeformed rock into parallelism with the long axis of the strain ellipse in the deformed rock. The history of the angle R is the rotation history.

Let us specify at the outset that the deformation path must be smooth and continuous during one period of deformation. No sudden reversals in principal direction or jumps in amount of finite strain occur. Different deformation periods are probably separated by changes in physical conditions, so each period should be described by its own unique deformation path.

Deformation history is frequently composed of information on successive portions, or increments, of the total deformation. After a general initial deformation **D** allow a subsequent deformation increment **Di** (incremental matrices are distinguished by i). In general, this increment **Di** can be any size.

The total final deformation **Df** (final or total deformation matrices indicated by f) from the original geometry into the current shape is

$$\mathbf{Df} = \mathbf{Di}\,\mathbf{D}\,. \tag{4}$$

Note that the deformation **Di** premultiplies the earlier deformation **D**, following the standard convention that matrices representing later events accumulate toward the left (Truesdell and Toupin, 1960, p. 246; Elliott, 1970a).

Any given finite deformation can be regarded as a product of incremental deformations successively compounded according to equation (4). Each new increment modifies the total deformation. Assuming that the rock was undeformed at the beginning, then a sequence of superimposed incremental deformations (indicated by subscripts) is

$$\mathbf{Df} = \mathbf{Di_n} \cdots \mathbf{Di_3\, Di_2\, Di_1}\,, \qquad (5)$$

and **Df** is the total finite deformation at time *t*.

Equation (5) is the finite deformation accumulation equation for a general path and represents the entire history of deformation increments. One must multiply all the increments and there is no shortcut except for special deformation paths. In summary, deformation history is completely known if one has either the history of the total finite deformation **Df** or the sequence of incremental deformations **Di**.

Equation (4) can be expanded by left polar decomposition, from (1) and (2):

$$\mathbf{Df = Tf\,Rf = Ti\,Ri\,T\,R}\,; \qquad (6a)$$

$$\mathbf{\theta f\, Tdf\, \theta f^{-1}\, Rf = \theta i\, Tdi\, \theta i^{-1}\, Ri\, \theta\; Td\, \theta^{-1}\, R}\,. \qquad (6b)$$

Equation (5) can be expanded the same way:

$$\mathbf{Df = \theta i_n\, Tdi_n\, \theta i_n^{-1}\, Ri_n \cdots \theta i_2\, Tdi_2\, \theta i_2^{-1}\; Ri_2\, \theta i_1\, Tdi_1\, \theta i_1^{-1}\, Ri_1}\,. \qquad (7)$$

With time the incremental strain ellipse is continuously changing its axial lengths, orientation, and rotation. There is a geometric interpretation of this incremental deformation history (Fig. 1).

Clearly, the incremental deformation history can be determined if the finite deformation history is available, and vice versa.

Now

$$\mathbf{Di = Ti\,Ri = Df\,D^{-1}}\,,$$

Multiplying each side of the equation by its transpose:

$$\mathbf{(Ti)^2 = Df\,D^{-1}\,(D^{-1})^T\,Df^T}\,.$$

Having **Ti** from this equation we can now calculate **Ri**:

$$\mathbf{Ri = Ti^{-1}\,Df\,D^{-1}}\,.$$

Rather surprisingly, to calculate the pure strain **Ti** (or the rotation **Ri**) we must not only have the finite pure strains **T** and **Tf**, but also the rotations **R** and **Rf**.

Frame of Incremental Principal Directions

There is another way of interpreting incremental deformation history. Adjacent orthogonal matrices can always be compounded into one orthogonal matrix, so defining

$$\mathbf{R_1' = \theta_1^{-1}\,Ri_1;\; R_2' = \theta_2^{-1}\,Ri_2\,\theta_2;\; R_3' = \theta_3^{-1}\,Ri_3\,\theta_2},\ \text{etc.}$$

Substituting into (7), the finite deformation accumulation equation (5) now becomes

$$\mathbf{Df = Tdi\,R_n' \cdots Tdi_3\,R_3'\,Tdi_2\,R_2'\; Tdi_1\,R_1'}\,. \qquad (8)$$

A perfectly general finite deformation path may be interpreted as a sequence of incremental rotations interspersed with incremental strains whose strain ellipses are all oriented parallel to the coordinate frame (Fig. 2).

Rigidly Moving Frame

There is often no particularly useful way of choosing a coordinate frame which maintains constant spatial orientation throughout the deformation, such as the east-west x coordinate used up to now. In many cases the equations show more of the physics with a rigidly moving coordinate frame undergoing all the solid body motions of the rock. Choose the rigidly moving x coordinate as the one which is eventually oriented in the east-west direction in the final deformation state, and defining

$$\mathbf{\phi f = \theta f;\; \phi i = \theta i;\; \phi = Ri\,\theta;\; Rm = Rf\,R^{-1}\,Ri^{-1}};$$

substitute these equations into equation (6b) and after rearranging

$$\mathbf{\phi f\,Tdf\,\phi f^{-1}\,Rm = \phi i\,Tdi\,\phi i^{-1}\,\phi\,Td\,\phi^{-1}}\,,$$

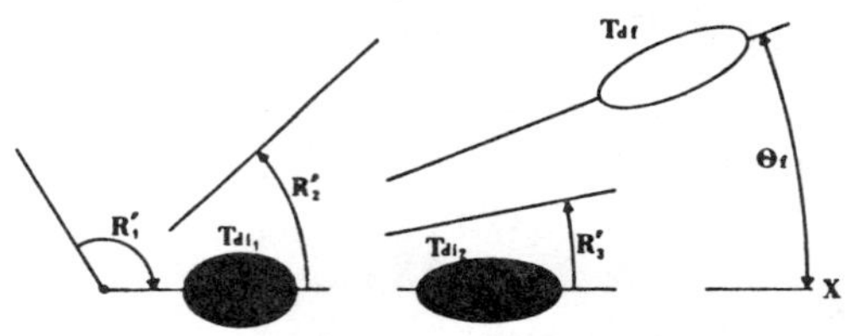

Figure 2. Deformation in a frame of incremental principal directions. The progress of the deformation is geometrically interpreted by reading the sequence of matrices toward the left in Df = Tdf R'f = R₃' Tdi₂ R₂' Tdi₁ R₁' (Equation 8).

the left and right sides of this equation define

$$\mathrm{Dfm} = \mathrm{Tim\ Tm}\ ;$$

the deformation accumulation equation now becomes

$$\mathrm{Dfm} = \mathrm{Tim}_n \cdots \mathrm{Tim}_3\ \mathrm{Tim}_2\ \mathrm{Tim}_1\ .$$

The horizontal sheet of glass is now keeping the grid origin over the same material point and turning from one direction to another with the solid body rotations of the rock (Fig. 3). Deformation matrices calculated in such a frame clearly cannot "see" incremental rotations. In a rigidly moving coordinate frame all the solid body motions are removed and greater attention is focused on changes of shape and size.

The calculation of the pure strain increment is a little simpler than before

$$\mathrm{Tim} = \mathrm{Tfm\ Rfm\ Tm}^{-1}\ .$$

Yet another type of coordinate frame is one in which the x axis is kept always parallel to the same material line throughout the deformation history. A special case where the chosen material line is also a simple shear direction has received considerable attention in structural geology and is not considered further here (*see* Ramsay and Graham, 1970).

There is an infinite choice of coordinate frames which could be used to describe deformation history and from a mathematical point of view it does not matter which is used, but some coordinate frames lend themselves to a simpler physical interpretation of the processes.

NATURAL TIME

Even if the increments making up a deformation path are known, it is still not possible to plot the path unless the order in which successive increments occur is known. History implies time ordering and some sort of time unit must be considered.

The time unit could be true time but it is difficult to imagine geological information which could provide a record of true time during deformation under metamorphic conditions. Order-of-magnitude estimates of real-time deformation rates, such as the often quoted $\dot{e} \simeq 10^{-12}$ to 10^{-16} per sec, are interesting but cannot be defined around individual folds. In natural folds deformation rates must be based on suitable continuously increasing parameters other than true time.

Suppose some scalar parameter t ("natural" time) increases continuously during deformation. The total deformation at a particular instant has a specific value of t, and increasing values of t unambiguously order the progress of the deformation. The continuously increasing arc length of growing quartz fibers is used later in this article as an appropriate time-like parameter t, and if

$$t = \frac{\text{current arc length}}{\text{typical arc length}}, \qquad (9)$$

then the natural time t is used in a dimensionless sense. The growing radius of a metamorphic mineral, a steadily increasing strain axial ratio near a boundary of a region being studied, or the increasing dip of a bed could also be used as dimensionless time-like parameters. Indeed, any monotonically increasing quantity which can be correlated with the progress of the deformation provides a built-in clock. The parameter t takes on a broader meaning than is usual in geology, but this broader interpretation is widespread in mechanics.

DIFFERENTIAL DEFORMATION AND RATES

The initiation or development of a structure depends upon what is happening at an instant. The concern is with the rate of change of finite deformation. The derivative with respect to natural time t of each of the elements of the finite deformation matrix in an externally fixed oriented frame defines the deformation rate matrix $\dot{\mathrm{D}}$ at that instant. For example:

$$\dot{\mathrm{D}}_{11} = \lim_{\Delta t \to 0} \frac{\Delta \mathrm{D}_{11}}{\Delta t} = \frac{d}{dt}\mathrm{D}_{11}\ ,$$

or as a matrix

$$\dot{\mathrm{D}} = \frac{d}{dt}\mathrm{D}\ .$$

Consequently, the finite deformation up to a natural time t can be determined by integrating in a common externally fixed frame each of the elements of the deformation rate matrix $\dot{\mathrm{D}}$,

$$\mathrm{Df} = \int_0^t \dot{\mathrm{D}}\, dt\ .$$

This integral only applies to a region of homogeneous deformation.

The deformation rate matrix $\dot{\mathrm{D}}$ is most suitably factored by Euler-Cauchy-Stokes decomposition (Truesdell and Toupin, 1960, p. 362).

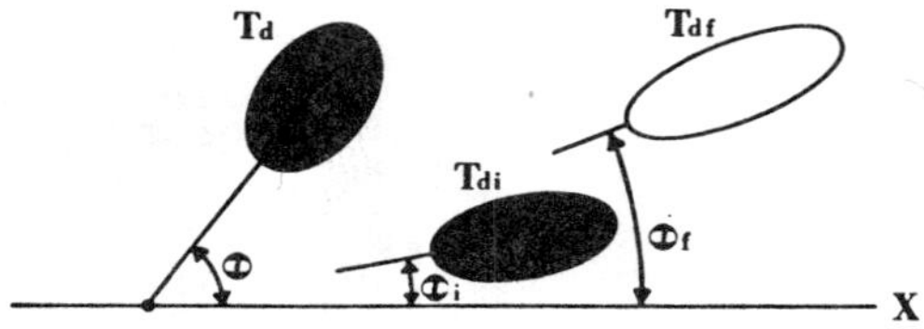

Figure 3. Deformation in a rigidly moving frame. Progress of the deformation is geometrically interpreted by reading the sequence of matrices toward the left in Dfe = $\phi f\ Tdf\ \phi^{-1} = \phi i\ Tdi\ \phi i^{-1}\ \phi\ Td\ \phi^{-1}$.

$$\dot{D} = \dot{e} + \dot{W} \text{ where } 2\dot{e} = \dot{D} + \dot{D}^T,$$

and

$$2\dot{W} = \dot{D} - \dot{D}^T.$$

The strain rate matrix $\dot{e}$ is symmetric, and any principal value, say $\dot{e}_I$, is defined as

$$\dot{e}_I = \frac{1}{\ell}\frac{d\ell}{dt} = \frac{d}{dt}\ln\left(\frac{\ell}{\ell o}\right),$$

when ln is the natural logarithm, ℓ and ℓo are original and final lengths.

The strain-rate matrix is represented by a strain-rate ellipse whose longest semi-axis ($1 + \dot{e}_I$) has an orientation angle θr to the externally fixed coordinate frame.

The spin $\dot{W}$ is the rate of rotation of material lines parallel to the axes of the strain-rate ellipse at a particular instant **t**. The spin matrix is always skew-symmetric.

If a rigidly moving coordinate frame is used, then similarly

$$\dot{D}m = \dot{e}m = \frac{d}{dt} Dfm.$$

The strain rate ellipse is oriented at an angle ϕr to this coordinate frame.

Paths of constant deformation rate $\dot{D}$ when the parameter **t** is true time are particularly important for experimental determination of material properties because stress strain-rate relations are then largely free of the effects of previous history or memory. Such paths are known in mechanics as constant stretch history, viscometric, or substantially stagnant flows. The constant strain-rate test widely used in experimental structural geology can only qualify insofar as one accepts that

$$\frac{d\ell}{\ell o} \simeq \frac{d\ell}{\ell}$$

is sufficiently accurate for the strain considered.

A deformation increment can have any size, as it is simply the total deformation which has occurred over some interval. If the increment **Di** is relatively small compared with the total final deformation **Df**, then a good approximation to the true deformation increment **Di** is the differential deformation increment **dD**:

$$\mathbf{Df} = \mathbf{dD} + \mathbf{D}$$

where **D** and **Df** are now the finite deformations at the beginning and end of a short interval. This equation for a differential increment is simpler than for a finite increment and is widely used. We can now analyze the deformation **Df** into what has happened (**D**) and what is now happening (**dD**). The matrix algebra of **dD** is similar to $\dot{D}$, and

$$\mathbf{dD} = \dot{e} + \mathbf{W}$$

where the Cauchy strain matrix $\dot{e}$ has as principal values the ratio of change in length $\Delta\ell$ over the current length ℓ during the differential deformation; for example

$$e_I = \frac{\Delta\ell}{\ell}.$$

"Infinitesimal" deformation is often used in the sense of a differential deformation (for example, Flinn, 1962, 1965; Ramsay, 1967).

The small masses of material making up a structure each have deformation rates or differential deformations related to the velocity field or small displacement field at that instant; an aspect thoroughly reviewed by Hsu (1966) in mechanics and Howard (1968) and Ramsay and Graham (1970) for structural geology.

NATURAL STRAIN SPACE

A particularly useful measure for longitudinal strain of a material line is natural strain

$$\mathbf{H} = \ell n\left(\frac{\ell}{\ell o}\right).$$

The Hencky strain matrix **H** with respect to the deformed state has the same principal directions as the left-stretch matrix (equation 3), and its principal values, $\mathbf{H}_I$, $\mathbf{H}_{II}$, and $\mathbf{H}_{III}$ in three dimensions, are defined by the last equation. This matrix will only be used in a diagonal form in this paper:

$$\mathbf{Td} = e^{\mathbf{H}} \text{ or } \ln \mathbf{Td} = \mathbf{H}.$$

The rate of change of the natural strain H is precisely the longitudinal strain rate $\dot{e}$:

$$\frac{dH}{dt} = \frac{d}{dt}\ln\left(\frac{\ell}{\ell o}\right) = \dot{e}.$$

For small strains the Cauchy strain is approximately the same as natural strain:

$$H = \ln\frac{l}{lo} \simeq \frac{l - lo}{lo} = e.$$

The history of the magnitude of strain is most clearly interpreted on a suitable graphical plot. In three dimensions plot the finite strain at an instant by the three principal natural strains H_I, H_{II}, H_{III} as cartesian components of a strain vector **h** (Fig. 4). This representation will be called natural strain space.

Volume change is measured by mean natural strain **hm** which is the projection of the strain vector onto the space diagonal. The shape of the strain ellipsoid measures the deviatoric component of the strain matrix and is the projection **O′P** of the strain vector onto a plane normal to the space diagonal and called the deviatoric plane (Fig. 4). All finite strain ellipsoids with given axial ratios have the same shape and lie on the same straight line in natural strain space.

The strain vector **h** always has its tail at the origin of the natural strain space and its tip at the current value of the finite strain. As t advances, the deformation progresses, and the tip of the strain vector traces out a curved line in three dimensions as a graph of the strain history (Fig. 5).

Principal stresses and strain rates can be plotted in the same way as finite strain, and these types of representation are all examples of Haigh-Westergaad Space (for stress references *see* Hill, 1950, p. 17). Haigh-Westergaad space is an outstanding graphic representation of second order tensors and is as useful as the Mohr Diagram. Algebraic manipulations involving invariants lose their fussiness and reduce to simple geometry. Observational and experimental results are clear and can easily be transformed into various invariant formulations.

One of the attractive features of the Hencky strain matrix and natural strain space is its close analogy with stress. Mean natural strain measures volume changes in the same way that mean stress measures hydrostatic pressure. Deviatoric natural strain and stress both measure shear and distortion. This correspondence does not hold with any other common strain measure.

In structural geology only Hossack (1968) has used natural strain space to date, and most workers use a more complex graph due to Flinn (1962) and originally proposed by Zing (1935) for use in sedimentology.

Information on the strain path in structural geology is often restricted to a plane section containing two principal directions, so only the plane in natural strain space containing the appropriate principal values is used; often this is the $H_I H_{III}$ plane. The two-dimensional deformation path is the projection of the three-dimensional path (Fig. 5). Nothing is implied about the third strain in this type of plot. Information on variation of the third strain

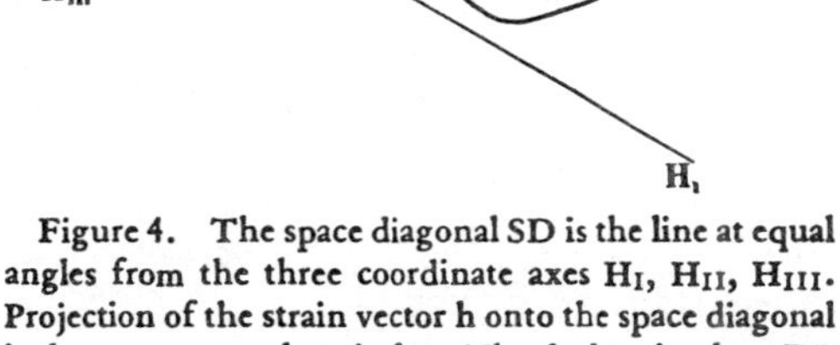

Figure 4. The space diagonal SD is the line at equal angles from the three coordinate axes H_I, H_{II}, H_{III}. Projection of the strain vector h onto the space diagonal is the mean natural strain hm. The deviatoric plane DP is normal to the space diagonal and projection of h onto the deviatoric plane gives the deviatoric strain vector O′P.

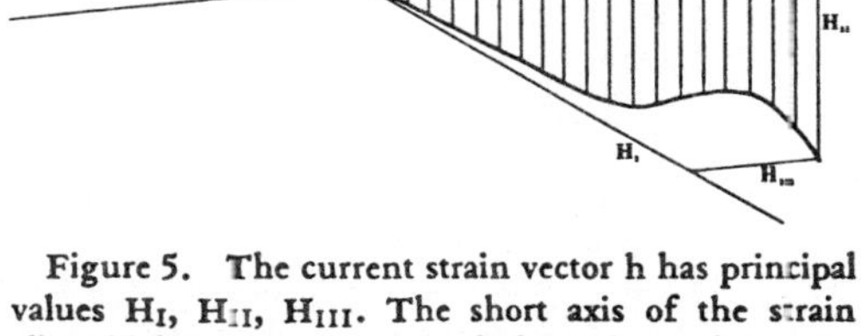

Figure 5. The current strain vector h has principal values H_I, H_{II}, H_{III}. The short axis of the strain ellipsoid has been compressed throughout the strain path so that H_{III} is negative. Note that the strain path observed in the H_I H_{III} plane can be derived by projection from the three-dimensional path.

can be determined either from differently oriented plane sections or from the volumetric strain history if known.

The two-dimensional deformation path of a particular material plane which has not remained a principal plane can be shown in a similar way.

PRINCIPAL STRESS DIRECTIONS

A widely used technique attempts to identify principal stress directions at a particular instant from features formed inside a crystal at this instant, such as twins, kinks, and deformation lamellae (*see* Carter and Raleigh, 1969). More deformation **Di** generally follows and unless the deformation **Di** is known it is *impossible* to determine what the orientation of the principal stress direction was at the time the particular stress state was active, so any attempt to deduce principal stress directions with this method must always make some assumption about the deformation path.

Further, individual crystals can start and stop deforming throughout the history, and there is no way of distinguishing when a particular lamella or twin was formed. The fabric diagrams composed from a large number of deformed crystals record some complex cumulative effects with no necessary or unique relation to any stress or strain principal directions (Dieterich and Carter, 1969).

However, if the deforming rock is essentially isotropic, the orientation ϕr of a principal strain rate is the same as the principal stress direction at that instant. Consequently, if the curve of ϕr versus t can be found, the history of principal stress directions is known. Some methods will be discussed later in this paper.

LINEARITY AND COAXIALITY OF PATHS

If the same material lines remain axes of all the finite and strain rate ellipsoids during an interval, then this portion of the deformation path is defined as *coaxial* (Hsu, 1966). The material line parallel to a strain ellipse long axis might at a later time become a short axis, but the change would be smooth, and one of the strain ellipses in between would have to be a circle.

Coaxially accumulating deformation paths can sometimes be recognized by inspection; several striking examples in blueschist grade rocks from the Cordillera Coast Ranges are illustrated by Misch (1969, Figs. 1, 4, 5, 10; 1970). These crystals show several successive pull-aparts, each separated by optically distinctive grain growth and fracture filling, and if all the differently oriented crystals in the thin section show similar features, then throughout the increasing deformation the same material lines always remained parallel to the principal strains.

The most common type of coaxial path is one in which the finite strain ellipsoid short axis has always been a shortening material line. In such a path it would not be possible to distinguish whether cleavage formed normal to the maximum compressive stress or to the short axis of the finite strain ellipsoid, since they would remain parallel. A noncoaxial path is a better place to investigate the origins of cleavage, both experimentally and in the field.

In an externally fixed coordinate frame the orientation of the accumulating finite strains can vary even though successive strain increments are coaxial, as a result of rotation **Ri**:

$$\mathbf{Ri} = \theta f \theta^{-1} .$$

The rotation accumulation equation for several increments is simply the product of the incremented rotation

$$\mathbf{Rf} = \mathbf{Ri_3\, Ri_2\, Ri_1} .$$

Coaxially accumulating paths have usually been referred to as "irrotational." This is an inappropriate description because (as many have realized) material lines acting as axes for all the finite strain and strain-rate ellipses can still be rotating in an externally fixed coordinate frame by the last two equations.

The calculation of a final pure strain after an increment is

$$\mathbf{Tdf} = \mathbf{Tdi\, Td} ,$$

using the Hencky strain matrix gives an even simpler version

$$\mathbf{Hf} = \mathbf{Hi} + \mathbf{H} .$$

We shall now see that the special nature of coaxial paths gives rise to an important graphical visualization, some aspects of which have also been discussed from the point of view of mechanics by Hsu (1966). The equation above can be re-expressed using the principal strains as cartesian components of a final strain vector **hf** in natural strain space (Fig. 6):

$$\mathbf{hi} = \mathbf{hf} - \mathbf{h} .$$

Therefore, finite pure strain increments have a

geometric interpretation as chords on the strain path (Fig. 6).

The chords begin and end at true points on the deformation path, but as the increments get larger there is increasing uncertainty about the curve between the end points.

The strain accumulation equation for a coaxial path is

$$\mathbf{Tdf} = \mathbf{Tdi_n} \cdots \mathbf{Tdi_3}\,\mathbf{Tdi_2}\,\mathbf{Tdi_1}\,,$$

or, in natural strain

$$\mathbf{hf} = \mathbf{hi_n} + \cdots \mathbf{hi_3} + \mathbf{hi_2} + \mathbf{hi_1}\,,$$

so that coaxially accumulating strain increments may be computed by vector addition in natural strain space. The simplicity of coaxially accumulating deformation paths leads to particularly clear representation.

Since $\mathbf{h} = f(t)$, one can take the derivative of the vector function with respect to the scalar variable t.

$$\frac{d\mathbf{h}}{dt} = \dot{\mathbf{e}} = \lim_{\Delta t \to 0} \frac{\Delta \mathbf{h}}{\Delta t}\,.$$

The strain-rate vector $\dot{\mathbf{e}}$ in natural strain space is a vector tangential to the path, but to determine the magnitude of the strain rate vector requires the values of the principal strain rates.

$$/\dot{\mathbf{e}}/ = (\dot{e}_I^2 + \dot{e}_{III}^2)^{\frac{1}{2}}\,.$$

In a coaxial path the maximum principal strain rate $\dot{e}_I$ is the rate of change of the long axis of the finite strain ellipse.

$$\dot{e}_I = \frac{d\,H_I}{dt}:$$

similarly,

$$\dot{e}_{III} = \frac{d\,H_{III}}{dt}\,.$$

The angle of slope γ of the strain-rate vector $\dot{\mathbf{e}}$ at a particular point on the path is the ratio of the principal strain rates (Fig. 6).

$$\tan\gamma = \frac{d\,H_{III}}{d\,H_I} = \frac{d\,H_{III}}{dt}\cdot\frac{dt}{d\,H_I} = \frac{\dot{e}_{III}}{\dot{e}_I}\,.$$

To determine the magnitude of the principal strain rates one must know the values of t along the path, but the ratio of the principal strain rates can be determined knowing only the succession of finite strains.

Several important measures of the degree of coaxiality are available, and in some cases it is possible to determine them in naturally deformed rocks. We will now define three measures of path coaxiality.

1. If a coaxial path has the finite, incremental, and strain rate ellipses all parallel at any one instant in a rigidly moving frame, $\phi i = \phi r = \phi f$ = constant at any t. So one measure of coaxiality is the angle between current finite and strain ellipses (ϕf-ϕr) and its history with t.

2. The angle between two successive finite strain ellipses is $\Delta\phi f$, and the rate at which new material lines became axes of successive finite strain ellipses is $\frac{d\,\phi f}{dt}$. In other words, as the finite strain ellipse develops, the ends (or vertices) of the ellipse move around occupying new material points on the perimeter at a rate $\frac{d\,\phi f}{dt}$.

3. New strain-rate ellipses are forming at each instant, and the rate with which new material lines become axes of the strain-rate ellipses $\frac{d\,\phi r}{dt}$ is a third measure of coaxiality.

The concept of coaxiality has important implications for the material properties of rock. For example, Lifshitz (1963), Green (1970), and Weertman (1970) argue that purely diffusive flow laws of the Nabarro-Herring or grain-boundary diffusion type can never lead to steady-state flow, because the shape of the grains changes with time. Green (1970) and Weertman (1970) considered the special case of a coaxial deformation path. Now an incremental strain on the grain is reflected by an incremental change in the grain's shape and

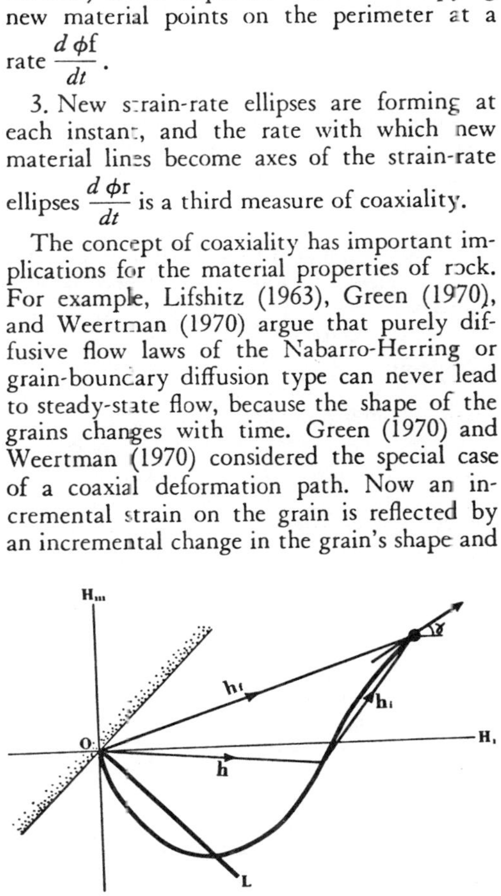

Figure 6. The linear plane strain path OL separates the H_I H_{III} plane into two sectors, and all strain states found above or below OL have an area increase or decrease. Since $H_I \geq H_{III}$, strain states in the stippled region are impossible.

size, and the largest possible alteration to a grain's dimensions with a given number of increments is accomplished in a coaxial deformation path. Consequently, the approximation to steady-state flow gets better and better as the path becomes more and more noncoaxial.

Linearity is the approach of the strain path to a straight line in natural strain space, and ideally is specified by the radius of curvature of the strain path. *Linear coaxial deformation paths* have constant axial ratios of the strain-rate ellipsoids as well as being coaxial.

$$\frac{H_{II}}{H_I} = \frac{\dot{e}_{II}}{\dot{e}_I} = \text{constant for all t.}$$

Such paths plot as straight lines through the origin in natural strain space (Fig. 6).

Linear coaxial deformation paths have been aptly called the "simplest possible strain paths" by Flinn (1962, 1965) and have been more precisely defined by Ramsay (1964, 1967). In the mechanics literature such paths are described as "proportional," "radial," or "linear."

It is apparent that a coaxial path is very much simpler than a noncoaxial one. For noncoaxial paths the finite strain path can still be plotted in natural strain space and the linearity described, but strain increments are no longer chords to the strain path, and strain rates cannot be shown as vectors tangential to the strain path. The matrix equations must be used.

FOLDING AND PLANE STRAIN

Let us apply the preceeding analysis of deformation paths to the important and frequently assumed case of folds growing in plane strain. In the widest sense of plane-strain flow, the normal to a flow plane is a principal strain which is of constant magnitude everywhere at any one instant. Rotation is only possible about an axis normal to the flow plane. All lines normal and all planes parallel to the flow plane remain straight. Folds developing in such a general plane-strain field are cylindrical with hinge lines parallel to the rotation axis. The hinge-line direction remains a principal direction throughout and is a coaxially accumulating principal strain. Consider two-dimensional deformation paths on the profile plane of symmetrical folds developing in a general plane-strain field. From symmetry, material points lying on a fold-hinge surface must be undergoing coaxially accumulating deformation paths, whereas the limbs have new material lines becoming principal strains at each instant, and the paths are noncoaxial.

Plane strain is frequently used in a stricter sense, with no volume strain and the principal strain normal to the flow plane specified as a direction of unchanged length throughout the deformation. If the H_{II} principal direction is parallel to the hinge line of a fold, then in the stricter sense of plane strain $H_{II} = 0$, and $H_I = -H_{III}$. So although the strain paths are everywhere linear it is only at special spots that the deformation paths are coaxial.

Theoretical and computer studies of folding have always assumed cylindrical folds whose hinge lines parallel the axis of rotation in plane strain. But the cylindrical nature of a fold is by itself insufficient to prove plane strain, and one must use the additional evidence provided by grooves and striations on bedding planes (Wegmann and Schaer, 1957) or fibrous syntectonic mineral growths (Wickham and Elliott, 1970).

Why are deformation paths so important for the study of natural folds? Consider a bed with a nonlinear flow law in a more ductile medium. The folds developed in the bed will have a different geometry every time a different history of velocities at the boundaries is used. Consequently the history of velocity boundary conditions is amongst the most important features to be determined on a natural fold (Chapple, 1969).

Examine the profile of a symmetrical fold growing in a plane-strain field. The local velocity $\mathbf{v}'$ (in natural time) of one point Q relative to another point P a distance away $\mathbf{ds}$ is (Fig. 7)

$$\mathbf{v}' = [\dot{\mathbf{D}} - \mathbf{1}]\,\mathbf{ds}\,.$$

Any convenient length $\mathbf{s}$ will serve; for example, the arc length measured from the hinge along any particular bed. The velocity varies smoothly from point to point along the bed, so considering the hinge $\mathbf{h}$ as stationary, the velocity $\mathbf{vi}$ of the inflection point on the limb is

$$\mathbf{vi} = \int_0^{s} [\dot{\mathbf{D}} - \mathbf{1}]\,\mathbf{ds}\,.$$

Clearly, if the deformation paths at a sufficient number of points in a fold are known, the history of the velocity boundary conditions may be determined. This result holds regardless of the type of material properties of the rock.

Velocity boundary conditions cannot be derived *a priori* from folding theory; they must be measured from the deformation paths. It is not an intractable problem, but something which can be measured with careful field work and thin-section study. Even a very few points with known deformation paths have enormous significance for the theoretical interpretation of a fold.

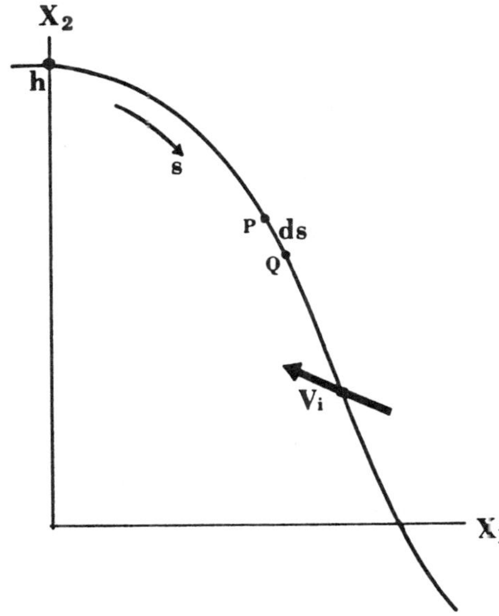

Figure 7. Relation between velocity boundary conditions and deformation paths on a symmetrically folded bed.

DEFORMATION PATHS FROM MICROBOUDINAGE

The remainder of this paper shows how many features of deformation history can be calculated in a practical way from boudinage, pressure shadows, and inclusion trails, and that the concept is important in any discussion of the origin of folded boudins and of cleavage and schistosity. An interesting field example described by Badoux (1963) illustrates some of the possibilities of measuring deformation paths. Because his paper might not be readily accessible, some of his data is reviewed along with the further conclusions derived here.

Badoux measured 300 deformed belemnites from a small area in the Morcles Nappe. The belemnites all lie in bedding surfaces approximately parallel to a strong cleavage (Fig. 8). Let us assume that this cleavage is developed parallel to the intermediate and long axes of the total finite strain ellipsoid. A mineral lineation on the cleavage is parallel to both the long axis of the finite strain ellipsoid and the fold hinge line. The belemnites have all been fractured, and because belemnites break after very small extension, the original and final lengths ℓo, ℓ are measured. From the orientation α and ratio $\frac{\ell}{\ell o}$ for each belemnite the arithmetic mean $\frac{\ell}{\ell o}$ for each 5° of orientation α is calculated, then the total strain ellipse constructed (Fig. 9). The ellipse is well defined, demonstrating that the area over which the belemnites were measured is small enough for the strain to be homogeneous.

Badoux deduced that the fractures were initially filled by quartz segregating into the neck, then metamorphic conditions changed and calcite started crystallizing into the growing crack between quartz and belemnite. It is possible to construct the finite strain ellipse for two separate events: the quartz infilling, and the calcite infilling (Fig. 9).

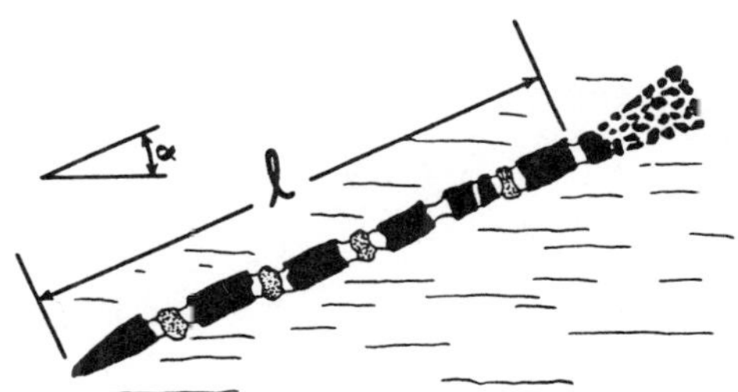

Figure 8. Boudiné belemnite (black) of current length ℓ lying in a cleavage plane at an angle α to a mineral lineation quartz (stippled) and calcite (white) infilling boudin necks. (Redrawn after Badoux, 1963, Fig. 1.)

The axes of the two strain ellipses are parallel; therefore, the two-dimensional deformation path was *coaxial*. When the results are plotted, we see that the strain path is *linear* as well as coaxial (Fig. 10).

The belemnite fragments are only slightly squashed normal to the cleavage; however an independent estimate of the third principal strain is provided, as Badoux considers that the density of the rock remained constant during the deformation. Consequently, if the quartz and calcite are locally derived from adjacent host rock there is no volume change and

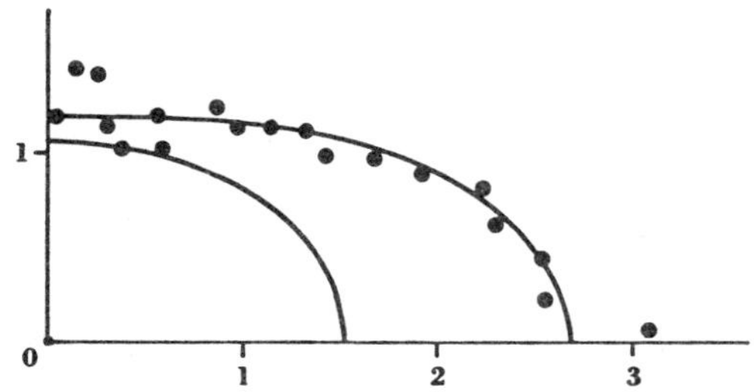

Figure 9. On a polar graph the radius vector is indicated by dots representing the arithmetic mean $\left(\frac{l}{l_0}\right)$ of all belemnites after the calcite finished infilling for each 5° sector of α. The outer strain ellipse for the calcite infilling fits the data well. The inner strain ellipse for the quartz infilling is shown without data points (redrawn *after* Badoux, 1963, Fig. 3).

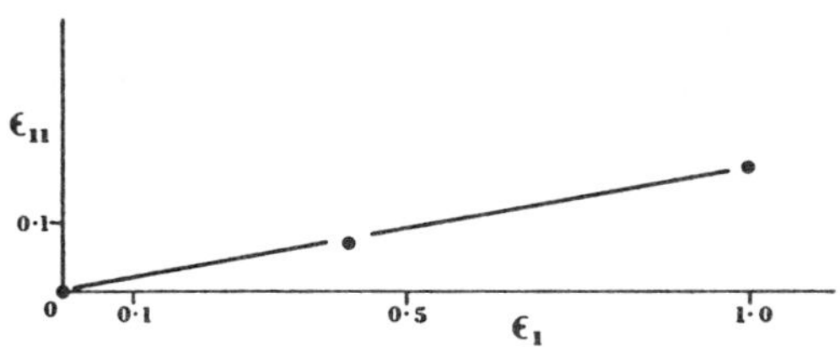

Figure 10. Linear strain path in ϵ_I ϵ_{II} plane defined by the unstrained state at the origin and the strain states after the quartz and after the calcite finished infilling. Without further information it is not possible to define the correspondence between ϵ_I ϵ_{II} and the H_I H_{III} planes.

$$H_{11} + H_{22} + H_{33} = 0 \text{ for all t.}$$

When the strain perpendicular to cleavage is calculated, it appears that very large compressive strains are not recorded by belemnites, probably due to their greater resistance to compression than to extension.

In summary, we know the three finite principal directions and strains after the infilling of calcite if we assume cleavage is a principal plane. For the earlier deformation recorded by the quartz infilling we have the principal direction and magnitude of the strain parallel to the fold hinge line, but the angle of cleavage to bedding at this time is unknown, so the two other principal directions and values are unknown. The deformation path in the profile section might be evaluated by the pressure shadow method discussed below. Suitable boudinage and microboudinage is available from a wide variety of areas which can be analyzed in similar fashion (Misch, 1969, 1970).

DISTRIBUTION OF FOLDS AND BOUDINS

It has been suggested that noncoaxial deformation paths might be indicated by the angular distribution of folded boudins. This is because in an area undergoing homogeneous deformation, layers in some directions may be extending and necking down while in other directions they will be shortening and folding (Flinn, 1962; *reviewed in* Ramsay, 1967). With coaxial deformation paths, any layer undergoing shortening always moves toward that sector of the strain ellipsoid where it will be extended. However during noncoaxial deformation it is possible that layers undergoing extension can later undergo shortening, but even then only in that sector of the finite strain ellipsoid which the long axis has been moving away from. This is where folded boudins should be found (Ramsay, 1967, p. 120, 175).

Talbot (1970) devised an ingenious way of determining the orientation and shape of the finite strain ellipsoid using only the orientation of folded and boudiné veins. Folded boudins produced during one period of deformation were never found, and he concluded that the deformation paths are coaxial in all the areas he investigated.

A layer must undergo appreciable extension for a boudin to be finite and observable, and the path must be sufficiently coaxial for the layer to remain in the extension field during this interval. Similarly, for a fold to become visible, the layer must be shortened for an appreciable time along a path sufficiently coaxial not to remove the layer from the shortening field. As a result, the path required for folded boudins is an initial finite strain on a nearly coaxial path followed by an increment of finite strain on another nearly coaxial portion of the path, with a large angle between the two finite-strain ellipsoids. The overall path necessary to produce folded boudins is a noncoaxial path of a very special type, and the absence of folded boudins only means that this special type of noncoaxial path is uncommon during one period of deformation. During *two* periods of deformation the two finite strain ellipsoids are frequently at a large angle, and the combined path may approach the one required. Folded boudins produced during two periods of deformation have been observed by Talbot (1970).

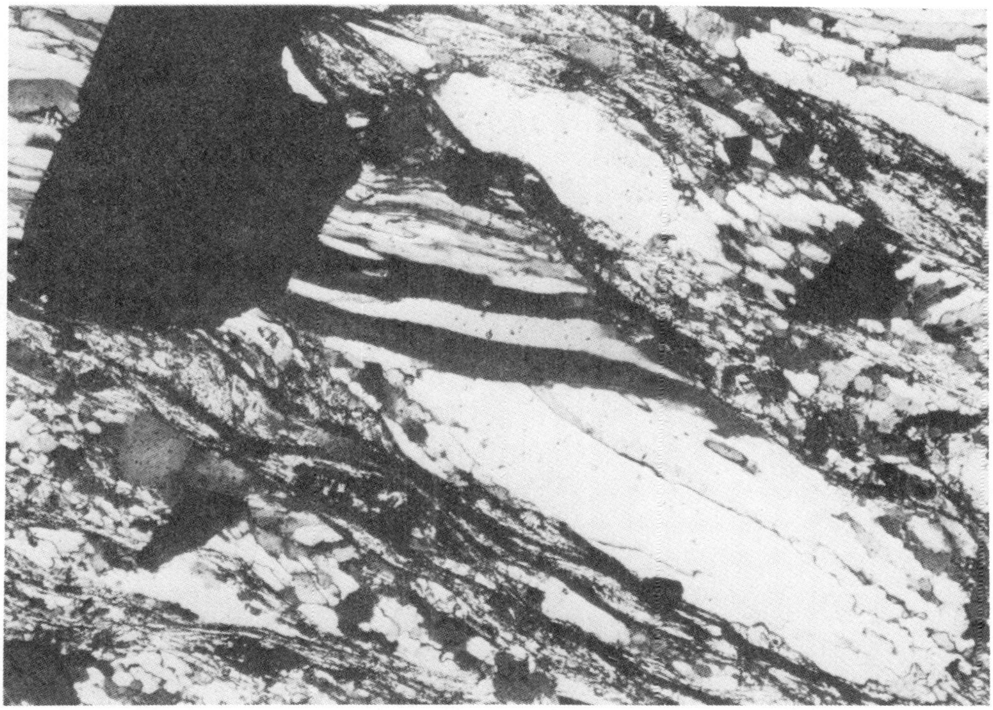

Figure 11. Pressure shadow on a magnetite crystal, crossed nicols. Thin section cut perpendicular to the hinge line of an isoclinal fold in the Weaverton Formation, South Mountain, Maryland.

It appears that the distribution of folds and boudins is not sensitive enough to be of help in distinguishing coaxial and noncoaxial paths.

PRESSURE SHADOWS

Relatively rigid particles such as small pebbles, phenocrysts, porphyroblasts, magnetite or pyrite crystals are often present in a rock at the outset of a deformation. The rigid particle is deflecting the streamlines of the flowing matrix, introducing inhomogeneity into the flow pattern, and giving rise to dissipative forces and a boundary layer about the rigid particle. The deformation path being recorded by growing pressure shadows is that of the boundary layer adjacent to the rigid particle but will show the same trends as in the matrix outside the boundary layer.

The interface between rigid particle and ductile matrix is a sharp discontinuity in material properties. During a deformation increment, any part of the rigid particle's surface whose normal is a direction of incremental extension tends to unstick from the matrix, and this is where crystal growth occurs (Mügge, 1930).

Segment boundaries separate the pressure shadow into portions within which the minerals have similar angular relations to a face of the rigid grain (Spry, 1969, p. 240). Quartz, for example, always tries to grow normal to the face of magnetite crystal, and as the crystal moves during deformation the quartz fiber grows curved to remain normal to the face (Figs. 11 and 12). A sharp curve or junction of two faces on the rigid grain boundary traces out a segment boundary as this point slowly drifts away from its previous positions (small black circles on Fig. 12). Pressure shadows, then, are swept out by the boundary of a rigid grain, and certain points on the grain boundary leave a recognizable path line (Fig. 13). The arc length of the path line is a convenient time-like quantity (*see* equation 9).

Successive orientations of the rigid grain can be traced by (1) maintaining a uniform angle between the rigid grain and the adjacent minerals within the segment while (2) keeping

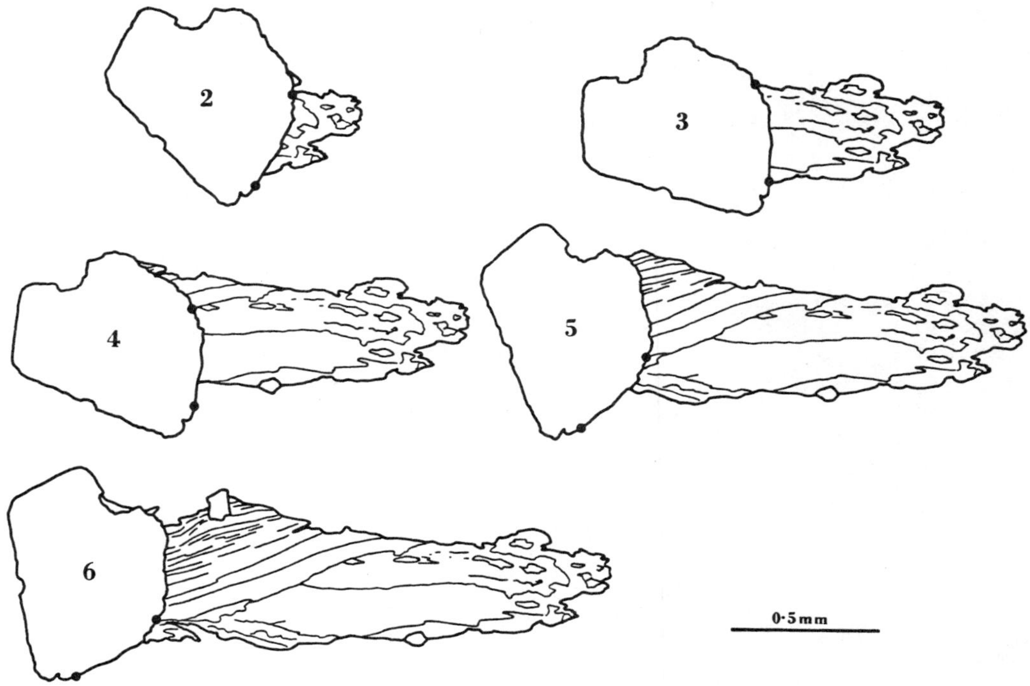

Figure 12. Growth from the initial state without pressure shadow (step 1 not shown) through to final result (6) shown in photograph (Fig. 11).

points of the grain boundary on their path lines (Figs. 12, 13). At any instant the pressure shadow is limited by two points on the rigid grain boundary outside of which minerals do not grow. The normal to the grain boundary at these two points are directions of zero longitudinal strain rate. The variation of the angle between these directions (2V′) in a plane strain situation is ideally the 2V of the strain rate ellipsoid (Fig. 14). Because axes of the strain rate ellipsoid are of very similar length the 2V should in principal be close to 90°. The measured 2V′ is about 15° smaller than the ideal value. This discrepancy could be either a result of seeing a finite increment rather than a strain rate or to complex boundary effects, but the trend of 2V′ may be indicative of how the true 2V is varying.

The principal directions of strain rate bisect the 2V′. In the example illustrated the deformation path is clearly noncoaxial. Accuracy may be established by comparing pressure shadows from both sides of the same grain and from different grains in the same slide.

Published illustrations of pressure shadows suggest that coaxial paths are not rare; good examples are in Burger and others (1970), and Ramsay (1967, p. 181). Pressure shadows suggesting noncoaxial paths around folds are illustrated by Zwart and Oele (1966) and Langheinrich (1964), but unfortunately these

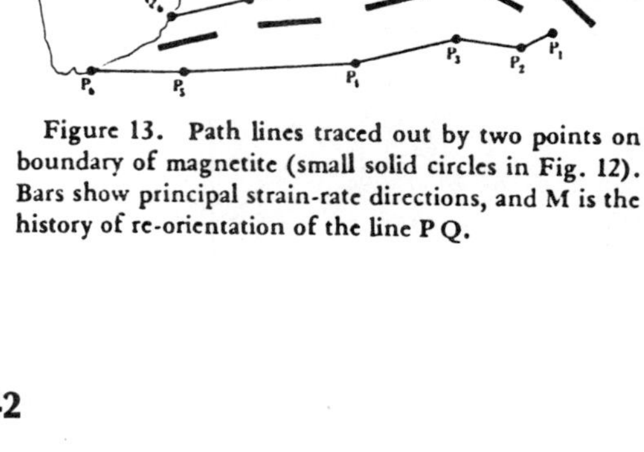

Figure 13. Path lines traced out by two points on boundary of magnetite (small solid circles in Fig. 12). Bars show principal strain-rate directions, and M is the history of re-orientation of the line P Q.

two examples are not really detailed enough to allow analysis in depth.

INCLUSION TRAILS AND FORMATION OF SCHISTOSITY

Garnets and many other porphyroblastic minerals may show inclusion trails running through them. Inclusions in garnet are commonly made up of rounded or elongate tapered blebs of quartz in discontinuous stringers, flat bands, or lines of flattened groups of crystals. Several inclusion trails are usually present in one garnet and all trails curve sympathetically with one another.

The garnet grows by extending fingers among the trails of quartz in the surrounding matrix, and these quartz trails are then incorporated into the garnet (Spry, 1963; 1969, p. 255). The orientation of the trail at the time it was swallowed by the growing garnet is thus preserved.

The continuously increasing ratio $\frac{\text{current radius}}{\text{final radius}}$ of a growing garnet serves as a dimensionless time-like parameter t with which to mark the progress of the deformation.

Changes in color of the garnet or breaks in inclusion trails show up any discontinuities in grain growth and are aptly termed "metamorphic unconformities" by Rosenfeld (1969; 1970). They represent discontinuities in the deformation path as well.

The orientation of the inclusion trail at a particular radius is measured with respect to a coordinate x_1 embedded on fixed material particles of the garnet. However, this is not the straightforward angular measurement it would appear to be; an aspect treated in some detail by Powell and Tregus (1970), and Rosenfeld (1970).

The maximum angular difference $\Delta\psi$ between the innermost and outermost part of inclusion trails reaches values for individual garnets as high as 540° in the Scottish Highlands (Flett, 1912; *see* Spry, 1969, p. 256) and 625° in the Northern Appalachians (Rosenfeld, 1968).

An inclusion trail clearly tells something about the deformation path, but what it tells depends upon interpretation of how the quartz trail orientation formed in the matrix before it was enveloped by the growing garnet. This is summarized in Table 1. Many quartz trails seem to form as part of a schistosity which develops in the matrix as the garnet grows, so the inclusion trails preserved in the garnet record the orientation history of the current schistosity. The problem now is to determine the cause of the schistosity.

Most strain determinations measured from snowball garnets involve the assumption that schistosity is a set of finite simple shear planes. An equation derived originally by Schmidt in 1918 was somewhat modified by Spry (1963; 1969, p. 256), but Rosenfeld (1970) shows that all these formulas are incorrect and that the amount of finite simple shear γs is

$$\gamma s = 2\Delta\psi .$$

If the schistosity planes have always been the simple shear planes, then a simple shear deformation path is indicated. But can schistosity be regarded as forming parallel to simple shear surfaces, and what is the origin of schistosity?

There is very strong evidence for rocks deformed once that schistosity develops normal to the short axis of the finite strain ellipsoid. In particular whenever it is possible to calculate the strain from deformed oölites, fossils, and conglomerates, schistosity is always found normal to the short axis of the total finite strain ellipsoid (evidence reviewed by Ramsay, 1967).

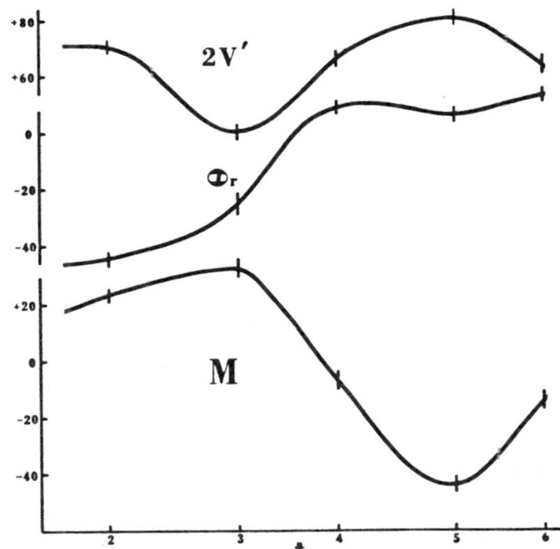

Figure 14. Variation with t in 2V′, orientation of principal strain-rates ϕr, and orientation of the magnetite crystal. Angular datum is arbitrary (ϕ_r = 0, M = 0).

TABLE 1. INCLUSION TRAILS IN METAMORPHIC MINERALS

Origin of included fabric	Assumptions	Results
		Attitude of schistosity versus t
Older Schistosity or	Principal directions of finite strain	Axial ratio
Inherited Lithological	Principal directions, coaxial path	Axial ratio versus t
Layering	Finite simple shear on known shear planes	γs
	Simple shear deformation path on known shear planes	γs versus t
	Finite simple shear on schistosity planes	γs
Current	Simple shear deformation path on schistosity planes	γs versus t
Schistosity	Schistosity parallels current finite principal strains	ϕf versus t
	Schistosity parallel current incremental principal strains	ϕi versus t
Growth over Pressure Shadow	Analyse as Pressure Shadow	ϕr and 2V' versus t

Further evidence is provided from the strain distribution across ductile shear zones (Ramsay and Graham, 1970). *If* schistosity is always normal to the short axis of the strain ellipsoid then the garnet inclusion trails give a history of lines which have been in principal planes of the finite strain ellipsoids, and a graph of the coaxiality of the deformation path can be constructed, ϕf versus t.

In rocks that have been deformed by two separate periods the total finite strain ellipsoid in general has a different orientation from either individual strain ellipsoid. From the theory above *only one schistosity* should be expected, normal to the total finite strain ellipsoid. I only know of one instance when this effect is claimed to occur (Ramsay, 1963, p. 375), and in all other areas each individual deformation phase develops its own schistosity and cross-cuts any earlier ones. It appears that schistosity cannot always be normal to the shortening direction of the total finite strain ellipsoid.

There are three other independent possible causes of schistosity: (1) After a certain definite amount of finite strain, deformable grains and clastic particles suddenly and discontinuously acquire a preferred orientation and cleavage as a result of the "shape factor" (Elliott, 1970a). (2) Schistosity is sometimes due to rigid body reorientation of platy minerals through a ductile matrix, and this has no necessary simple relation to any stress or strain ellipsoids. (3) Subparallel pressure solution surfaces contribute toward a visible cleavage and are the dominant process in some rocks.

Another possibility is that schistosity may appear normal to the short axis of the last incremental strain ellipsoid which was just sufficiently large to produce the three above effects to a noticeable degree. In this case the mineral inclusion trails record the material lines which have been axes of the incremental strain ellipse, so ϕi versus t could be measured. This incremental strain ellipsoid may even be large enough to approximate the orientation of the total finite strain ellipsoid.

The quartz trails in the matrix could be part of a lithological layering such as bedding, or an old foliation inherited from an earlier period of metamorphism. The inclusion trails give us the history of orientation of a lithological layer but without further information nothing more can be said. If the finite principal directions at any instant can be identified, the axial ratio of the total finite strain ellipse can be calculated from Harker's formulae (Ramsay, 1962).

It is not clear whether cleavage and schistosity are tracking stress, strain rate, finite strain, or an incremental strain of some finite magnitude. Distinction between the various hypotheses is only possible in a region of noncoaxial deformation paths.

Several authors have noted that internal inclusion trails do not look like remnants of the schistosity presently seen in the matrix as the shape and size of the quartz in the trails is different (Spry, 1963). It is well established that garnet and other porphyroblasts have remained essentially rigid during deformation, and quartz-filled pressure shadows are found on garnets (Harker, 1939, p. 219; Wilson, 1971, Figs. 1 and 3). The garnet growing over the pressure shadow would include the quartz fibers and a new pressure shadow would form at the end of the newly enlarged garnet (Fig. 15). Inclusion trails which were "frozen" pressure shadows could be interpreted in the same way as the shadows discussed earlier, but there is a practical difficulty as they are much less complete. Approximations have to be made, such as assuming that the fibers grew roughly in the direction of current incremental principal extension.

If inclusion trails in snowball garnets are remnants of fibrous growth in pressure shadows this is one of the few bits of direct evidence that diffusion-flow laws might be operative under fairly high-grade metamorphic conditions—evidence that usually would be destroyed by the annealing recrystallization so widespread at these temperatures (Elliott, 1970b, 1972).

An important case arises when the deformation path is coaxial during the garnet growth. The inclusion trail within the garnet would then be straight and parallel to the external schistosity—regardless of whether the quartz inclusion trails formed as a schistosity parallel to the strain ellipse or as quartz fibers in a pressure shadow. But such textures have traditionally been interpreted as postkinematic (Fig. 16a). Posttectonic recrystallization episodes postulated on this evidence alone can be reinterpreted as synkinematic grain growth during a coaxially accumulating deformation path.

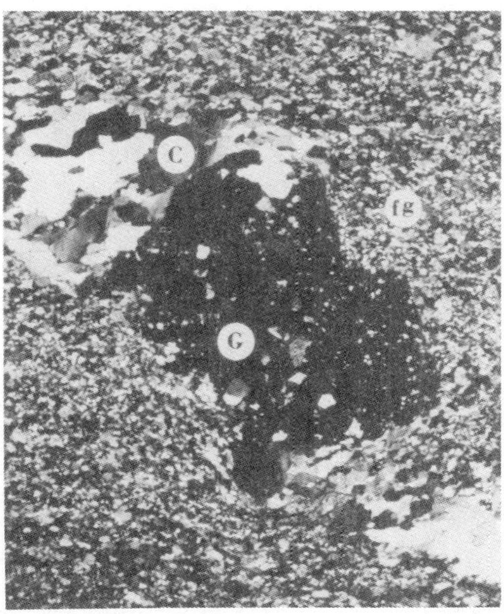

Figure 15. Photo of coarse quartz (c) in pressure shadow at upper left and bottom right of garnet, surrounded by fine-grained quartz (fg). Garnet (G) is engulfing the pressure shadow quartz. Crossed nicols. Garnet-mica gneiss collected by R. Balk in the Benardston quadrangle, Massachusetts.

A particularly strong case for synkinematic grain growth arises when the inclusion trails and the porphyroblast composition are either smoothly varying or constant (Fig. 16a, b).

Inclusion trails which change from curved to straight are conventionally interpreted as indicating a change from synkinematic to static postkinematic grain growth. Alternatively, the grain growth could be entirely synkinematic and reflect the change from a noncoaxial to a coaxial deformation path (Fig. 16c).

The initiation and development of structures in higher-grade metamorphic terrains may be followed by an extended interval of approximately homogeneous flattening (Ramsay, 1962; Mukhopadhyay, 1965). Homogeneous flattening implies a coaxially accumulating deformation path. Is it possible that the straight portions of inclusion trails recorded from such terrain as "static postkinematic" are in fact evidence of the coaxial portions of such paths?

In the spectacular snowball garnets of southeastern Vermont—some of the finest in the world—Rosenfeld (1969) reports that inclusion trails in garnets measured around the major Ascutney fold have different senses of curvature on each limb, except at one place where the inclusion trails are straight. At this locality he concludes that there was static postkinematic grain growth, although synkinematic grain growth holds everywhere else. Unlike the symmetric folds discussed earlier, material points which have coaxial deformation paths in an asymmetric fold will not lie on the hinge surface; if they exist they are somewhere on the long limb side of the hinge. Is it possible that the straight inclusion trails found at one spot

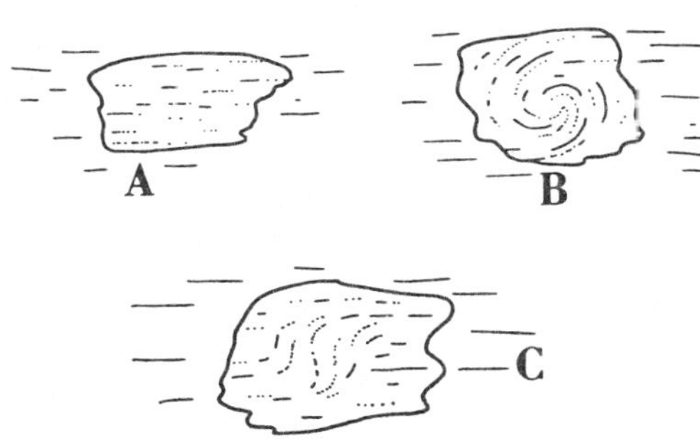

Figure 16. Typical patterns of inclusion trails in metamorphic minerals when thin section is cut normal to the rotation axis.

on the major Ascutiny fold are at a point of coaxially accumulating deformation path, and that the garnet growth here is synkinematic as it is everywhere else?

There are still valid criteria for distinguishing synkinematic from postkinematic grain growth. Straight inclusion trails produced during synkinematic grain growth along a coaxially accumulating path should only occur in restricted zones bordered by regions with oppositely curved synkinematic inclusion trails. The habit remains useful, for true postkinematic crystals may grow randomly across the schistosity, while synkinematic grains often are elongated in the schistosity.

DISCUSSION

Analysis by means of matrices and natural strain space gives a clear picture of the various aspects of deformation paths. Linearity and coaxiality emerge as two particularly important properties which are significant not only for the evolution of cleavage, boudinage, and folds but also have ramifications on the flow laws or stress-strain rate relations. There is a clear need for many more field observations from a variety of rocks to see which of the various possible systematic patterns of deformation history are important in naturally deformed rocks.

One of the most powerful ways of comparing computer experiments on folding with natural folds is by means of their deformation paths, something which has not as yet been attempted. Several new methods for determing deformation paths from boudinage, pressure shadows, and inclusion trails are proposed. Unfortunately some of the methods in current use, such as interpreting principal stresses within crystals or the distribution of folded boudins, do not appear to have much promise for describing deformation paths. Suitable material might be available in metamorphic rocks of most grades; examples noted in this paper range from greenschist and blueschist up to garnet-grade rocks. Inclusion trails in metamorphic minerals may be more valuable for measuring path coaxiality than they are for defining the time relations between metamorphic and deformation periods.

ACKNOWLEDGMENTS

I would particularly like to thank Ernst Cloos, Peter Hudleston, John Wickham, C. B. Raleigh, Neville Carter, and James Talbot for constructive and critical reviews of the manuscript. This research was supported by the National Science Foundation.

REFERENCES CITED

Badoux, H., 1963, Les bélemnites tronconnées de Leytron (Valais): Bull. Laboratoires de Géologie, Minéralogie, Géophysique et du Musée Géologique, Université de Lausanne, Bull. 138, p. 1–7. *Also in* Bull. de la Sociéte vaudoise des Sciences Naturelles, no. 311, v. 68, 1963.

Burger, H. R., Perry, B. E., and White, B., 1970, Relationship between pressure shadows and shear fracture orientation: Am. Mineralogist, v. 55, p. 1791–1793.

Carter, N. L., and Raleigh, C. B., 1969, Principal stress directions from plastic flow in crystals: Geol. Soc. America Bull., v. 80, p. 1231–1264.

Chapple, W. M., 1969, Fold shape and rheology: The folding of an isolated viscous-plastic layer: Tectonophysics, v. 7, p. 97–116.

Dieterich, J. H., 1969, Origin of cleavage in folded rocks: Am. Jour. Sci., v. 267, p. 155–165.

Dieterich, J. H., and Carter, N. L., 1969, Stress History of Folding: Am. Jour. Sci., v. 267, p. 129–154.

Elliott, D., 1970a, Determination of finite strain and initial shape from deformed elliptical objects: Geol. Soc. America Bull., v. 81, p. 2221–2236.

——1970b, The determination of flow laws in naturally deformed quartzites: Geol. Soc. America, Abs. with programs (Ann. Mtg.), v. 2, p. 546.

——1972, Diffusion flow laws in metamorphic rocks: Geol. Soc. America Bull., (in press).

Flinn. D., 1962, On folding during three-dimensional progressive deformation, Geol. Soc. London Quart. Jour., v. 118, p. 385–433.

——1965, Deformation in metamorphism, *in* Pitcher, W. S., and Flinn, G. W., eds., Controls of metamorphism: Edinburgh, Oliver and Boyd, 368 p.

Green. H. W., 1970, Diffusional flow in polycrystalline materials: Jour. Appl. Physics, v. 41, p. 3899–3902.

Harker, A., 1939, Metamorphism: London, Methuen and Co., 362 p.

Hill, R., 1950, The mathematical theory of plasticity: London, Oxford University Press, 355 p.

Hossack, J. R., 1968, Pebble deformation and thrusting in the Bygdin area (southern Norway): Tectonophysics, v. 5, p. 315–339.

Howard. J. H., 1968, The role of displacements in analytical structural geology: Geol. Soc. America Bull., v. 79, p. 1846–1852.

Hsu, T. C., 1966, The characteristics of coaxial and

non-coaxial strain paths: Jour. Strain Analysis, v. 1, p. 216-222.

Langheinrich, G., 1964, Vergleichende Untersuchungen über das Verhältris der Schiefening zur Faltung unter Berucksichtigung des Stockwest-Problems: Neues Jahrb. Geologie u. Paläontologie Abh. 120, p. 41-80.

Lifshitz, I. M., 1963, On the theory of diffusion-viscous flow of polycrystalline bodies: Soviet Physics, JETP, v. 17, p. 909-920.

Misch, P., 1969, Paracrystalline microboudinage of zoned grains and other criteria for synkinematic growth of metamorphic minerals: Am. Jour. Sci., v. 267, p. 43-63.

—— 1970, Paracrystalline microboudinage in metamorphic reaction sequence: Geol. Soc. America Bull., v. 81, p. 2483-2486.

Mügge, O., 1930, Bewegungen von Porphyroblasten in Phylliten und ihre Messung: Neues Jahrb. Geologie u. Paläontologie Abh., v. 61, p. 469-510.

Mukhopadhyay, D., 1965, Effects of compression on concentric folds and the mechanism of similar folding: Geol. Soc. India Jour., v. 6, p. 27-41.

Powell, D., and Treagus, J. E., 1970, Rotational fabrics in metamorphic minerals: Mineralog. Mag., v. 37, p. 801-814.

Ramsay, J. G., 1962, The geometry of mechanics of formation of similar type folds: Jour. Geology, v. 70, p. 309-327.

—— 1963, Structural investigations in the Barberton Mountain land, eastern Transvaal: Geol. Soc. South Africa Trans. and Proc., v. 66, p. 353-401.

—— 1964, Progressive deformation in tectonic processes [abs.]: Am. Geophys. Union Trans., v. 45, p. 106.

—— 1967, Folding and fracturing of rocks: New York, McGraw-Hill, p. 568.

Ramsay, J. G., and Graham, R. H., 1970. Strain variation in shear belts: Canadian Jour. Earth Sci., v. 7, p. 786-813.

Rosenfeld, J. L., 1969, Garnet rotations due to major Paleozoic deformations in southeast Vermont, *in* White, W. S., Zen, E., Hadley, J. B., and Thompson, J. B., eds., Studies of Appalachian geology: Northern and maritime: New York, Wiley-Interscience Publishers.

—— 1970, Rotated garnets in metamorphic rocks: Geol. Soc. America Spec. Paper 129, p. 1-105.

Spry, A., 1963, The origin and significance of snowball structure in garnet: Jour. Petrology, v. 4, p. 211-222.

—— 1969, Metamorphic textures: Oxford, Pergamon Press, 350 p.

Talbot, C. J., 1970, The minimum strain ellipsoid using deformed quartz veins: Tectonophysics, v. 9, p. 47-76.

Truesdell, C., and Toupin, R. A., 1960, Classical field theories, Encyclopedia of Physics: Berlin, Springer-Verlag, v. 3/1.

Wegmann, E., and Schaer, J. P., 1957, Lunules tectoniques et traces de Mouvements dans les plis du Jura: Eclogae Geol. Helvetiae, v. 50, p. 491-496.

Weertman, J., 1970, The creep strength of the earth's mantle: Rev. Geophysics, v. 8, p. 145-168.

Wickham, J. S., and Elliott, D., 1970, Rotation and strain history in folded carbonates, Front Royal Area, northern Virginia [abs.]: Am. Geophys. Union Trans., v. 51, p. 422.

Wilson, M. E., 1971, On syntectonic porphyroblast growth: Tectonophysics, v. 11, p. 239-260.

Zingg, T., 1935, Bietrage zur Schotteranalyse: Schweizer. Mineralog. u. Petrog. Mitt., v. 15, p. 39-140.

Zwart, H. J., and Oele, J. A., 1966, Rotated magnetite crystals from the Rocroi Massif (Ardennes): Geologie en Mijnbouw, v. 45e, p. 70-74.

MANUSCRIPT RECEIVED BY THE SOCIETY SEPTEMBER 1, 1971

REVISED MANUSCRIPT RECEIVED DECEMBER 13, 1971

18

Copyright ©1976 by Springer-Verlag
Reprinted from *Contr. Mineralogy and Petrology* **59:**203–226 (1976)

A Quantitative Study of Deformation Mechanisms and Finite Strain in Quartzites

Shankar Mitra*

Department of Earth and Planetary Sciences, Johns Hopkins University, Baltimore, Maryland, USA

Abstract. The Weverton quartzites in the Maryland Blue Ridge are deformed by one major period of greenschist-grade deformation. The components of finite strain due to different independent mechanisms have been measured for these rocks. The total strain is split up into two major components:

$$\varepsilon^t = \varepsilon^p + \varepsilon^d.$$

The finite natural strain caused by dislocation creep (ε^d) is measured by a new technique using folded and stretched rutile needles which are good strain markers within the quartz crystals. Pressure solution strain (ε^p) is measured from the ratio of the area of new crystals and fibers to the whole rock area in principal sections. Grain boundary sliding is a dependent process which accompanies both mechanisms. Pressure solution obeys a linear Newtonian flow law, $|\dot{\gamma}_0^p| = A_p|\tau_0|$, while dislocation creep obeys a power law of the form $|\dot{\gamma}_0^d| = A_d|\tau_0|^n$ where $\dot{\gamma}_0^p, \dot{\gamma}_0^d$ are octahedral shear strain rates, τ_0 is the octahedral shear stress and A_p, A_d and n are constants. A direct correlation between finite strain measurements and the operating flow laws can be made. Application of these methods and principles to a few field examples indicates that the rocks obey a flow law partly governed by each mechanism. Any set of physical conditions defines a unique flow law and there is a transition in creep behavior from dominantly Newtonian to a power law with increasing strain rate.

Introduction

In recent years, finite strain analysis has become an important part of Structural Geology and numerous methods of measuring strain have been described. In this paper, a new approach to this subject is attempted: the finite strain is determined from measurements of the strain components due to different mechanisms. This enables a direct evaluation of the relative importance of the deformation mechanisms and the flow laws operating during deformation. This method is

* *Present address:* Department of Geological Sciences, Cleveland State University, Cleveland, Ohio 44115, USA

applied to the Eocambrian Weverton quartzites exposed in the Harpers Ferry area of the Central Appalachians. The quartzites have been deformed by one major period of late Palaeozoic greenschist-grade deformation at a temperature of 625 °K and under an overburden of about 10 km, corresponding to a hydrostatic pressure of 3.5 kbar. The two important independent deformation mechanisms in these rocks are pressure solution and dislocation creep. Methods of isolating and measuring the strain due to the two mechanisms are derived using the grains which constitute the rocks themselves and differs from most methods which use polycrystalline objects embedded in the rocks as strain markers. The flow laws associated with each of the mechanisms are discussed and the complex flow law for deformation governed by both the mechanisms is derived. From finite strain measurements a fairly accurate picture of the flow law and the relative importance of the two mechanisms is obtained for a few illustrative examples.

Partition of Strain into Components

Although the study of deformation mechanisms in rocks has recently become quite common, the emphasis is either on a qualitative statement about the operating mechanisms, or on a descriptive account of a single dominant mechanism. For geological deformations in which several mechanisms contribute significantly to the finite strain, it is essential to quantitatively isolate and measure the effects of each mechanism. In this section, the scheme of partition of finite strain into components due to different mechanisms will be developed.

Consider a material line of length l_0 which is deformed successively to lengths l^A and l^B by two deformations A and B (Fig. 1a). Then, the total 'stretch' (T) is given by the equation:

$$T = \frac{l^B}{l_0} = \frac{l^B}{l^A}\frac{l^A}{l_0} = T^B\,T^A. \tag{1}$$

This theory can be extended for the case of two simultaneously operating independent mechanisms a and b. Let the line l_0 be deformed alternately by the two

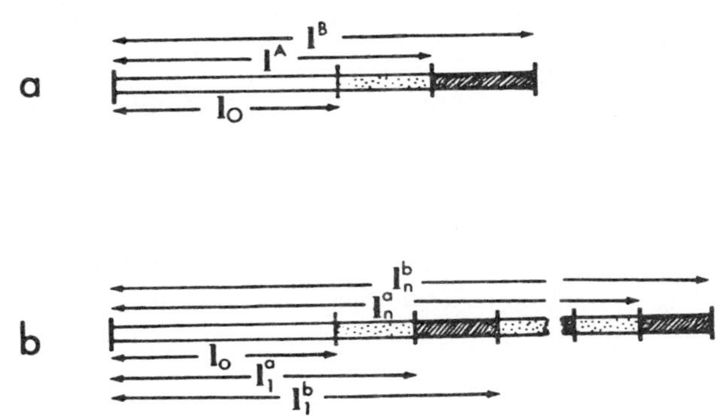

Fig. 1a and b. Deformation of a line of length l_0 **a** by two superimposed deformations A and B to final lengths l^A and l^B. **b** by two mechanisms a and b through small increments to final lengths l^a_n and l^b_n

mechanisms through a large number of infinitely small increments (Fig. 1 b). If $l_1^a, l_1^b, l_2^a, l_2^b \ldots l_n^a, l_n^b$ are the final lengths after each increment, then

$$\begin{aligned} T &= \frac{l_n^b}{l_n^a} \cdot \frac{l_n^a}{l_{n-1}^b} \cdot \frac{l_{n-1}^b}{l_{n-1}^a} \cdots \frac{l_1^b}{l_1^a} \cdot \frac{l_1^a}{l_0} \\ &= \left(\frac{l_n^b}{l_n^a} \cdot \frac{l_{n-1}^b}{l_{n-1}^a} \cdots \frac{l_2^b}{l_2^a} \cdot \frac{l_1^b}{l_1^a} \right) \left(\frac{l_n^a}{l_{n-1}^b} \cdot \frac{l_{n-1}^a}{l_{n-2}^b} \cdots \frac{l_2^a}{l_1^b} \cdot \frac{l_1^a}{l_0} \right) \\ &= T^b T^a. \end{aligned} \tag{2}$$

Equation (2) can also be written as the sum of natural strain components and generalized for m independent mechanisms:

$$\varepsilon^t = \varepsilon^a + \varepsilon^b + \cdots + \varepsilon^m = \sum_{k=1}^{m} \varepsilon^k. \tag{3}$$

Differentiating with respect to time, the summation of components of the natural strain rate ($\dot{\varepsilon}$) also holds, i.e.

$$\dot{\varepsilon}^t = \sum_{k=1}^{m} \dot{\varepsilon}^k. \tag{4}$$

Equations (1) to (4) hold for a material line and cannot be adapted directly for the principal directions of the respective strain ellipsoids. This is because although the principal incremental directions for the two mechanisms will always be parallel, the integrated strain ellipsoids will have different orientations for a non-coaxial deformation path. If the strain ellipsoids for a and b have identical principal directions, the two diagonal stretch matrices T_d^a and T_d^b can be multiplied to directly obtain T_d^t:

$$T_d^t = T_d^b T_d^a. \tag{5}$$

Since the matrices are diagonal, their multiplication is commutative, and the sequence of strains does not affect the results. This case holds only if the deformation is coaxial or steady-state. If a and b have different principal directions, the total strain contains a component of rotation and cannot be determined uniquely from the two matrices, since it is dependent on the deformation path.

In the case of the Weverton quartzites, two mechanisms have been identified to be responsible for most of the deformation:

(1) Pressure solution, involving diffusion of material from the point of maximum compressive stress to the point of minimum stress where it precipitates as new crystals or fibers (Fig. 2a).

(2) Dislocation creep, which includes any kind of creep associated with the movement of dislocations by glide or climb through the crystal (Fig. 2b). Equation (3) now simplifies to a form involving only two components:

$$\varepsilon^t = \varepsilon^p + \varepsilon^d. \tag{6}$$

The non-coaxiality of deformation is restricted to the $a-c$ principal plane, with the b axis, which is parallel to the regional fold axis, being constantly oriented. In the $a-c$ plane, although the principal axes for the two mechanisms do not

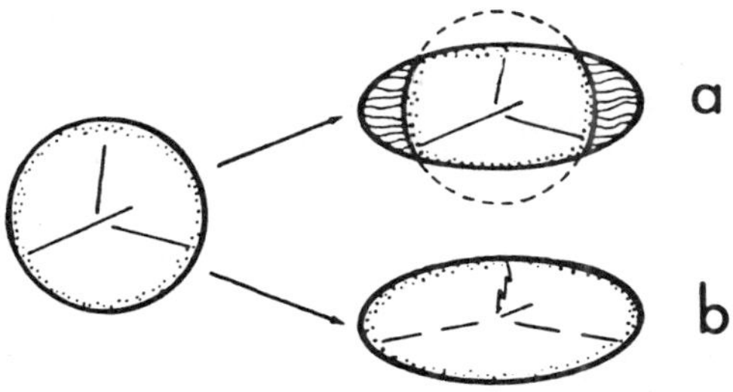

Fig. 2. a Deformation of a circular grain by pressure solution. Fibers in the pressure shadows are separated from the detrital clast by a dust ring. **b** Same deformation by dislocation creep. Dust rings and deformed rutile needles reflect the internal strain

coincide, the angle between the two (θ) seldom exceeds 10^0 (see examples in Fig. 14), so that taking into account the errors of measurement, the application of Equation (5) is a good approximation for the determination of the total strain. The slight misorientation of the principal axes rules out strictly coaxial or steady state deformation, but nonetheless imposes constraints on the history of deformation.

The strain due to the two mechanisms and the calculated total strain are represented in terms of two parameters: the natural octahedral shear strain (γ_0) and Lode's number (ν) which are defined as follows:

$$\gamma_0 = \tfrac{2}{3}[(\varepsilon_1-\varepsilon_2)^2+(\varepsilon_2-\varepsilon_3)^2+(\varepsilon_3-\varepsilon_1)^2]^{\frac{1}{2}}, \quad (7)$$

$$\nu = \frac{2\varepsilon_2-\varepsilon_1-\varepsilon_3}{\varepsilon_1-\varepsilon_3}. \quad (8)$$

γ_0 defines the magnitude of the three-dimensional strain while ν determines the shape of the strain ellipsoid. The relative importance of the two mechanisms is compared more directly by computing the terms Γ^p and Γ^d which are defined as follows:

$$\Gamma^p = \frac{\gamma_0^p}{\gamma_0^p+\gamma_0^d} \qquad \Gamma^d = \frac{\gamma_0^d}{\gamma_0^p+\gamma_0^d}. \quad (9)$$

Pressure Solution as a Deformation Mechanism

The importance of pressure solution in low grade deformation is now widely recognized (Durney, 1972; Groshong, 1972; Elliott, 1972, 1973). Fibers growing in pressure shadows adjacent to detrital clasts and in tension veins have been used to determine deformation paths (Wickham and Elliott, 1970; Elliott, 1972; Wickham, 1973; Durney and Ramsay, 1973) and their mode of formation is a subject of much speculation (Stromgaard, 1973). Elliott (1973) suggested that the stress-strain-rate relationship for pressure solution is identical to that described for diffusion in the metallurgical literature. There are two possible diffusion mechanisms: (1) Lattice diffusion, or Nabarro-Herring creep which involves the movement of vacancies through the crystal lattice (Nabarro, 1948; Herring, 1950); and (2) Grain boundary diffusion or Coble creep, in which case the grain boundaries act as the diffusion path (Coble, 1963; Lifshitz, 1963). Both processes obey linear flow laws. Lattice diffusion requires considerably higher tempratures than grain boundary diffusion (Ashby, 1972) and under our estimated conditions of $T<0.5\,T_m$ (where T_m is the melting temperature) is probably insignificant in comparison to the latter. The flow law for Coble creep is of the form (Raj and Ashby, 1971):

$$\dot{\varepsilon}=\frac{K\pi D_{gb}h\Omega}{X^3 r^{3/2} kT}\sigma_d \tag{10}$$

where D_{gb} is the grain boundary diffusion coefficient, h is the width of the grain boundary, Ω is the atomic volume, X is the grain size, r is the grain shape, k is Boltzman's constant, σ_d is the deviatoric stress, and K is a constant.

Diagnostic features of pressure solution are stylolitic zones resulting from the dissolution of material in the direction of compression and clear new grains and fibers in the direction of tension (see Elliott, 1973, Figs. 4–7). The new grains and fibers can be distinguished from the detrital grains by the fact that they contain very few inclusions. Some fibers also accumulate in tension veins and adjacent to rigid inclusions such as pyrite, haematite and magnetite crystals. Pressure solution only affects the surfaces of grains, so that material points within the grain remain unaffected after deformation.

Measurement of Pressure Solution Strain

The isolation and measurement of pressure solution strain is more complicated than might seem at first sight. For example, a direct measure of the shapes of deformed grains plus the adjacent fibers is not practicable because (1) Some of the pressure-solved material may accumulate in tension veins, (2) At high strains the pressure shadows lose their elliptical outlines, (3) A non-uniform grain size produces uneven effects and (4) The method does not eliminate the flattening effects of dislocation creep. Instead of this approach, we will consider the case of a spherical volume of rock of unit radius being deformed by pressure solution to an ellipsoid of constant volume. With progressive deformation, the volume of fibers increases, so that the ratio of fibers to detrital grains is directly related to the axial ratio (T^p) of the pressure solution strain ellipsoid.

For this method, two principal planes ($a-c$ and $b-c$) are considered. Let Ao and At be the original and final whole rock areas and let Af and Ac be the areas of fibers produced and currently observed in the $a-c$ plane. Similarly, we will define Bo, Bt, Bf and Bc for the $b-c$ plane. Let $a>b>c$ be the semi-axes of the triaxial ellipsoid. For a constant volume, $abc=1$. In the $a-c$ plane, the c direction will undergo pressure solution while fibers will grow along a in different parts of the specimen with an area equal to $2MPNP'$ (Fig. 3a). The coordinates of $P(x, y)$ which is the point of intersection of the original circle and the final ellipse are

$$\left(a\sqrt{\frac{c^2-1}{c^2-a^2}},\ c\sqrt{\frac{c^2-1}{c^2-a^2}}\right).$$

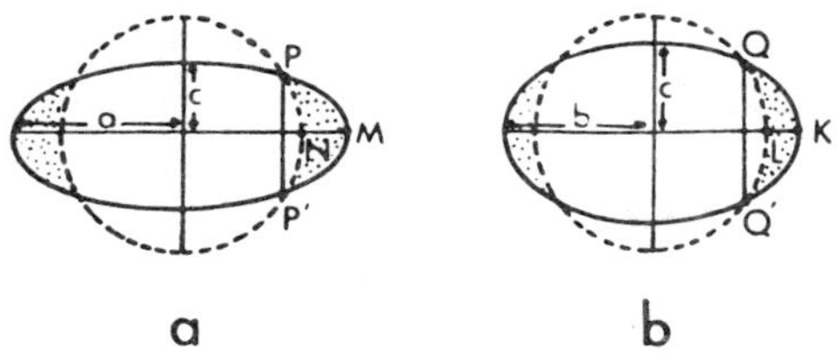

Fig. 3. a Unit circle deformed by pressure solution to an ellipse in the $a-c$ plane generating an area of fibers equal to $2MPNP'$. **b** Same in the $b-c$ plane, with an area of fibers equal to $2KQLQ'$

The area of segment

$$MPP' = ac \arccos \frac{x}{a} - xy = ac \arccos \sqrt{\frac{c^2-1}{c^2-a^2}} - xy.$$

The area of the segment

$$NPP' = \arccos x - xy = \arccos a \sqrt{\frac{c^2-1}{c^2-a^2}} - xy.$$

Therefore, the area of the crescentic segment

$$MPNP' = MPP' - NPP' = ac \arccos \sqrt{\frac{c^2-1}{c^2-a^2}} - \arccos a \sqrt{\frac{c^2-1}{c^2-a^2}}.$$

The total area of fibers produced,

$$\mathrm{Af} = 2MPNP' = 2\left(ac \arccos \sqrt{\frac{c^2-1}{c^2-a^2}} - \arccos a \sqrt{\frac{c^2-1}{c^2-a^2}}\right).$$

Since $abc=1$, substituting $1/ab$ for c and $1/b$ for ac,

$$\mathrm{Af} = 2\left(\frac{1}{b} \arccos \sqrt{\frac{1-b^2a^2}{1-b^2a^4}} - \arccos a \sqrt{\frac{1-b^2a^2}{1-b^2a^4}}\right). \tag{11}$$

This gives us the area of fibers produced assuming that only detrital grains are pressure solved. However, with progressive deformation, older fibers also undergo pressure solution and the area of recycled fibers increases with strain, so that the area of fibers currently observed, $\mathrm{Ac} \neq \mathrm{Af}$. We will correct for this by estimating the amount of fibers pressure-solved during each of a large number of increments of strain. Let Af_x and Ac_x $(x=1, \ldots, n)$ be the area of fibers produced and observed after each increment of strain respectively, and let $d\mathrm{Af}_x$ be the incremental change in Af_x at each instant.

Then,

$$\mathrm{Ac}_1 = d\mathrm{Af}_0 = \mathrm{Af}_1$$

$$\mathrm{Ac}_2 = d\mathrm{Af}_0 + d\mathrm{Af}_1 - d\mathrm{Af}_1\left(\frac{d\mathrm{Af}_0}{\mathrm{Ao}}\right) = \mathrm{Af}_2 - d\mathrm{Af}_1\left(\frac{\mathrm{Af}_1}{\mathrm{Ao}}\right),$$

$$\mathrm{Ac}_3 = \mathrm{Af}_3 - d\mathrm{Af}_1\left(\frac{\mathrm{Af}_1}{\mathrm{Ao}}\right) - d\mathrm{Af}_2\left(\frac{\mathrm{Af}_2}{\mathrm{Ao}}\right)$$

and

$$\begin{aligned}\mathrm{Ac}_n &= \mathrm{Af}_n - \frac{1}{\mathrm{Ao}}(\mathrm{Af}_1 d\mathrm{Af}_1 + \mathrm{Af}_2 d\mathrm{Af}_2 + \cdots + \mathrm{Af}_{n-1} d\mathrm{Af}_{n-1}) \\ &= \mathrm{Af}_n - \frac{1}{\mathrm{Ao}} \int_0^{n-1} \mathrm{Af}_x d\mathrm{Af}_x \\ &= \mathrm{Af}_n - \frac{(\mathrm{Af}_n)^2}{2\mathrm{Ao}},\end{aligned} \tag{12}$$

since integration over $n-1$ infinitely small increments is equal to integration over n increments. The ratio of the current area of fibers observable (Ac) and the current total area (At) =

$$\mathrm{Ac/At} = \frac{\mathrm{Af}}{\mathrm{At}} - \frac{\mathrm{Af}^2}{2\mathrm{Ao\,At}}. \tag{13}$$

Since $\mathrm{At} = \mathrm{Ao}/b$, and $\mathrm{Ao} = \pi$

$$\frac{\mathrm{Ac}}{\mathrm{At}} = b\left(\frac{\mathrm{Af}}{\pi} - \frac{\mathrm{Af}^2}{2\pi^2}\right). \tag{14}$$

In this derivation, we have assumed that Ao is constant, whereas it is actually decreasing. However, since this involves only a second order correction, it does not influence our results.

In the $b-c$ plane (Fig. 3b), the coordinates of Q are

$$\left(b\sqrt{\frac{c^2-1}{c^2-b^2}},\ c\sqrt{\frac{c^2-1}{c^2-b^2}}\right).$$

Proceeding in the same way as above, total area of fibers produced,

$$\mathrm{Bf} = 2KQLQ' = 2\left(\frac{1}{a}\arccos\sqrt{\frac{1-b^2a^2}{1-a^2b^4}} - \arccos b\sqrt{\frac{1-b^2a^2}{1-a^2b^4}}\right) \tag{15}$$

and

$$\frac{\mathrm{Bc}}{\mathrm{Bt}} = a\left(\frac{\mathrm{Bf}}{\pi} - \frac{\mathrm{Bf}^2}{2\pi^2}\right). \tag{16}$$

Equations (14) and (16) contain four variables: a, b, $\frac{\mathrm{Bc}}{\mathrm{Bt}}$ and $\frac{\mathrm{Ac}}{\mathrm{At}}$. On a plot of a against b, a series of lines representing different values for $\frac{\mathrm{Ac}}{\mathrm{At}}$, and $\frac{\mathrm{Bc}}{\mathrm{Bt}}$ are constructed (Fig. 4). Using this plot, any two values of $\frac{\mathrm{Ac}}{\mathrm{At}}$ and $\frac{\mathrm{Bc}}{\mathrm{Bt}}$ uniquely define a and b. Since $abc=1$, the value of (T^p) can be determined. An assumption in this derivation is that b is a direction of elongation, which is the only case we are presently concerned with. However, the derivation can be extended to fit a case where $b<1$. In order to determine the principal directions for pressure solution strain, the directions of maximum and minimum growth of fibers are measured for a number of grains in the $a-c$ section and a mean value is calculated.

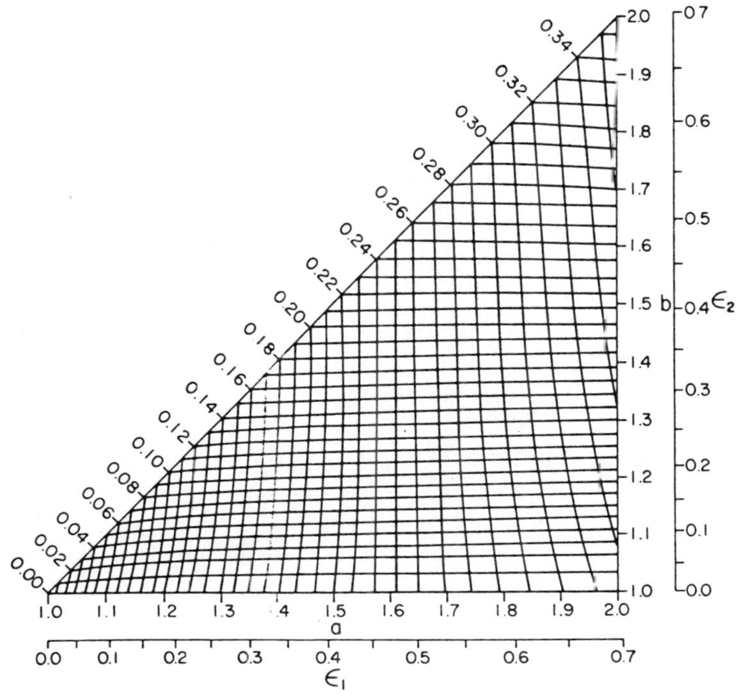

Fig. 4. Graph for determining T^p in a three-dimensional case. The horizontal set of curves represents values of (Bc/Bt) for the $b-c$ principal section; the vertical set represents values of (Ac At) for the $a-c$ section. Intersection of the curves for any set of values defines a, b and the axial ratio

In the case of plane strain, $b=1$ and $a=1/c$. Therefore, $T^p=a^2$. and our equations can be derived using only the $a-c$ principal section.

$$\mathrm{Af}=2\,\mathrm{arc\,cos}\left(\frac{1}{\sqrt{T^p+1}}-\mathrm{arc\,cos}\sqrt{\frac{T^p}{T^p+1}}\right). \tag{17}$$

Since Ao = At,

$$\frac{\mathrm{Ac}}{\mathrm{At}}=\frac{\mathrm{Af}}{\pi}-\frac{\mathrm{Af}^2}{2\pi^2}. \tag{18}$$

A plot of T^p against Ac At (Fig. 5) defines the axial ratio for any value of the area ratio.

Dislocation Creep Mechanisms

The most generalized flow law for all dislocation creep mechanisms is of the form (Weertman, 1970):

$$\dot{\varepsilon}=\frac{K' D_l \Omega}{kT}\sigma_d^n \tag{19}$$

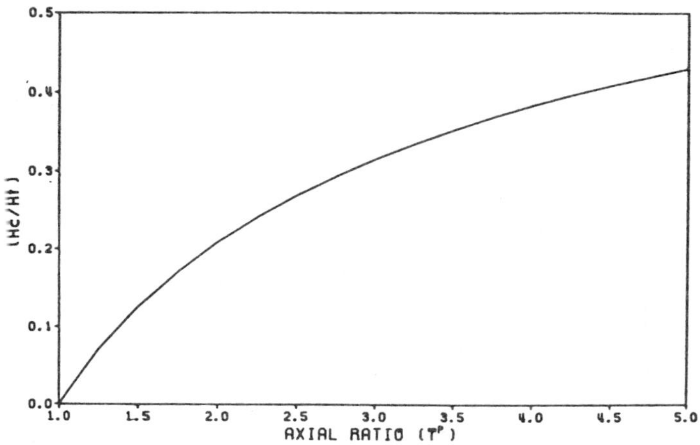

Fig. 5. Graph relating (Ac/At) to the pressure solution axial ratio (T^P) for plane strain

where K', n are constants for a particular mechanism and D_e is the lattice diffusion coefficient. When a crystal is stressed above its yield point and dislocations move along suitable glide planes, they multiply and increase the dislocation density. Within a short period the dislocations encounter obstacles and other mobile dislocations, and work hardening sets in. For deformation to continue either the stress has to be raised for continuing dislocation glide or "recovery" must take place so that the dislocations can climb out of their slip planes and allow the process to achieve a steady state. Climb is slower and energetically more difficult than glide. Since the two are dependent processes, continuous glide cannot occur until climb takes place so that the latter is rate determining. The maximum creep rate occurs when the rate of recovery is equal to the rate of work hardening (McLaren and Retchford, 1969). The only other alternative is glide controlled creep where the dislocations must carry with them a Cottrell cloud of impurity atoms. Flow laws for all these mechanisms have been discussed by Weertman (1970) and Stocker and Ashby (1973).

Griggs and Blacic (1964) first documented that introduction of trace amounts of OH into the silica lattice decreases its strength by an order of magnitude. Later Griggs (1967, 1974) and Blacic (1975) showed that above a critical temperature (T_c), quartz exhibits hydrolytic weakening and suggested a mechanism of dislocation glide by breakage of hydrolysed Si–O–Si bonds. McLaren and Retchford (1969) found that below T_c, deformed crystals showed considerable work-hardening while above T_c there was a low dislocation density and signs of recovery. They concluded that hydrolytic weakening assists climb rather than glide and enables steady-state deformation by continuing recovery of tangles. T_c depends principally on the hydrogen content and also on the strain rate. For a sufficiently high H content, even at high strain-rates ($\dot{\varepsilon} \simeq 10^{-5}$–$10^{-7}$ per s) values of T_c close to 350°C have been found (Griggs, 1967), so that at a much lower strain rate ($\dot{\varepsilon} \simeq 10^{-14}$ per s), T_c will be well within the range of geological deformation conditions that we are presently concerned with and hydrolytic weakening should be considered as rate determining for dislocation creep mechanisms.

Evidence of Dislocation Creep in Thin Sections

Optical features previously explained in terms of dislocation substructures such as deformation lamellae and deformation bands are widespread in the Weverton quartzites (Fig. 6a). Deformation lamellae are narrow subplanar zones having slightly different refractive indices from the host quartz while deformation bands are usually wider subplanar zones which have a slightly different crystallographic orientation from the host grain. Boehm lamellae which are defined here as trails of minute bubble following slip planes are also present (Fig. 6b). Both types of lamellae occasionally offset rutile needles in the quartz grains giving evidence of dislocation motion. The relations of optical features to dislocation structures have been described by several authors (Carter and others, 1964; Christie and others, 1964; Green and Radcliffe, 1972; White, 1973; Christie and Ardell, 1974). Preston (1958) suggested that deformation lamellae are actual planes of slip while bubble trails are formed in adjacent areas due to drag exerted by the slip. However, it seems much more plausible to explain these features using the concept of hydrolytic weakening.

The theory states that hydrogen can be introduced into the quartz lattice by hydrolysis of Si – O bonds to form

$$Si-OH \cdot HO-Si.$$

The hydrogen bond between the two groups is much weaker than the Si – O or Si – OH bonds and above a critical temperature (T_c) will break with the condensation of a H_2O molecule. Whe the H_2O molecule is trapped at a dislocation, the latter can move by glide (Griggs, 1967) or climb (McLaren and Retchford, 1969). The zones of minute bubbles could have been formed by the coalescence of a

Fig. 6. a Deformation lamellae showing as thin lines of contrasting colors with deformation bands at right angles to them. Upper Weverton. Crossed nicols. **b** Same as (**a**) using a dark field condenser. Trails of bubbles and inclusions occur along the lamellae.

large number of H_2O molecules as the dislocations glided along the slip plane or climbed out of it. Such recovery probably takes place at each of an infinite number of slip planes, but the trails are only visible on planes which have the maximum concentration of dislocations causing the spacing between the trails. Deformation lamellae seldom occur at exactly the same sites as the bubble trails (Fig. 6), but the two features can often be traced into one another. All these observations suggest that the two are different manifestations of the same feature; with partial recovery the optical features associated with active glide planes disappear and bubble trails are formed. Stated in another way, Boehm lamellae are partially or totally annealed deformation lamellae.

Measurement of Dislocation Creep Strain

Although transmission electron microscopy studies have added immensely to our knowledge of dislocation substructures in crystals, the role of dislocations during deformation and the mechanisms of creep, they have not enabled a direct measure of the cumulative effects of dislocation creep during large natural deformations. Also T.E.M. studies are restricted to the observation of very small areas, typically a few microns wide. McLean (1951, 1952) and Green and others (1970) suggested a method for measuring dislocation strains for metals using structures such as slip bands and polygonized crystals, by estimating the dislocation concentrations in such zones. However, both this method and T.E.M. studies take into account only those dislocations that are present in the final deformed state and miss those that have passed through the crystal. Elliott (1973) suggested a method for measuring dislocation creep strains using the reconstructed shapes of deformed dust rings which were present at the boundary of originally circular detrital clasts (Fig. 2b). However, this method loses its practicability when good dust rings are not clearly discernible. This is the case whenever extensive pressure solution occurs on a non-coaxial path.

In the present study a new method of measuring dislocation creep strain has been found using deformed rutile needles in the quartz grains. Rutile needles are widespread in most quartzites and Preston (1958) found that slip planes often offset the needles. In this study, it was found that the needles are actually stretched in the direction of tension (Fig. 7a, b) and folded or kinked in the direction of compression (Fig. 7c, d) and are excellent markers of finite strain within the crystal structure. The needles are extremely fine and brittle and therefore record the deformation within the host crystals. The concentration of needles in any grain is high so that there is no lack of data. By measuring only those needles that lie in the principal planes the strain measurements can be simplified.

The needles may have been introduced into the quartz lattice in several ways. The most prevalent hypotheses (Frondel, 1962) are that they may have been exsolved from igneous quartz or included into precipitating vein quartz. A third hypothesis is that they may have been exsolved from the quartz lattice during a charnockite grade of metamorphism affecting the basement rocks from which the present sediments were derived. Regardless of which of these processes is most common, the important point is that the needles were present in quartz

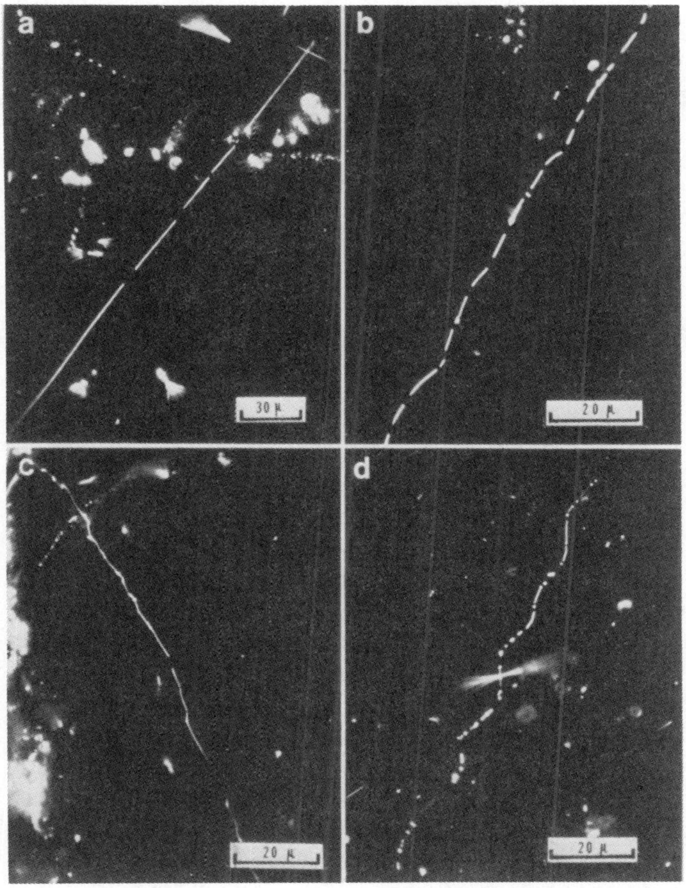

Fig. 7a–d. Dark field photographs of deformed rutile needles. **a** Stretched needle cut by a trail of bubbles at an angle of about 30°. **b** Rutile needle first boudinaged by tension and later folded by compression. **c** Kinked and folded thin needle with zones of overlap at the points of maximum strain. **d** Strongly folded and boudinaged needle

grains derived from underlying basement rocks and were *undeformed* prior to the Palaeozoic South Mountain deformation.

The needles are best observed using a dark field condenser under high power (600x), and show up as fine and brightly illuminated lines against a dark background. Using a graduated eyepiece the original (l_0) and final lengths (l') as well as the angle (α) with respect to a reference direction are measured for each needle. For each grain a number of measurements, usually 20 to 30 are made; also for each thin section. measurements on a number of grains are made since different grains have different principal values and directions. By plotting $T(=l'/l_0)$ against α on a polar graph and constructing the best fit ellipse through the points, the strain in the individual grains and that of the section is determined (Fig. 8). The inhomogeneity of deformation within a single grain together with the variation in principal directions of different grains produces a slight scatter in the data. In fact, if we plot each grain separately, we get a number of different ellipses

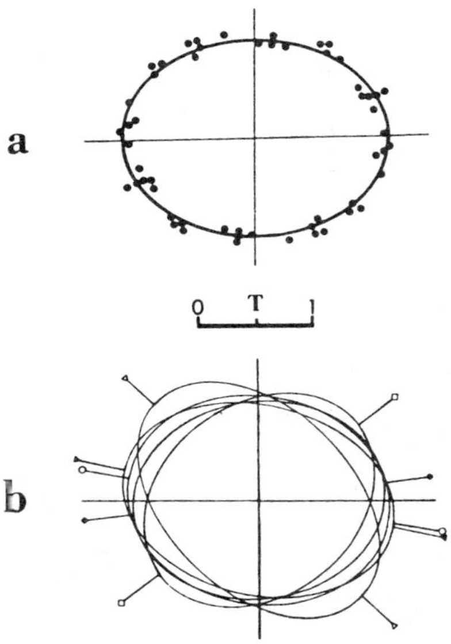

Fig. 8. a Plot of deformed rutile needles and the best fit ellipse for the *a–c* section of a sample of Weverton quartzite. **b** Same as (**a**) with different ellipses drawn for each grain measured

converging to the strain ellipse (Fig. 8b). The three dimensional value of (T^d) can be determined using the equation:

$$T_1^d \, T_2^d \, T_3^d = 1. \tag{20}$$

By measuring T_1^d and T_3^d in the $a-c$ plane, the value of T_2^d and therefore of the axial ratio can be found. The measurements can be checked by using two principal sections (Fig. 9) and comparing the values of the axial ratio determined from each section. Rutile needles can be used as strain markers even in rocks which have been strongly polygonized or recrystallized. The needles cut across sub-grain boundaries and are also displaced by them (Fig. 10) so that in addition the original dislocation strain they record the dislocation movements during the formation of sub-grains.

An interesting point is the manner in which the rutile needles are deformed. Figs. 7b and 7d show needles which are both boudinaged and folded. In the case of Fig. 7b this effect seems due to a non-coaxial deformation path. The needle was first boudinaged by tension but later entered a field of compression and low amplitude folds were formed. But this feature can also be attributed to a different reason. In order to fold a thin brittle rod elastic strains are set up at certain points which will tend to fracture it and restore the lattice to the unstrained state (Fig. 11). Supporting evidence for this mechanism is that for most grains, the percentage of folded boudins is higher in the direction of compression than in the direction of tension. The strain required to break a needle will depend on its thickness. Very thin needles remain unbroken after being folded to low amplitudes. With sufficient strain weak zones are formed (Fig. 7c) along which the needle breaks. Thick needles are broken more easily; in fact some of them are not folded and instead respond to the compression by kinking and overlapping.

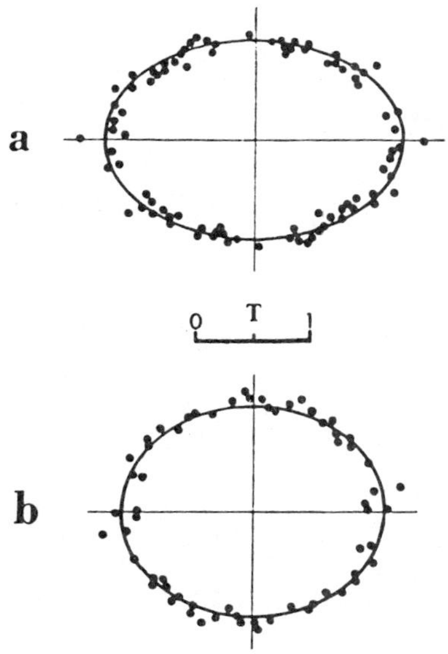

Fig. 9a and b. Plot of rutile needles for the $a-c$ section (**a**) and the $b-c$ section (**b**) of a sample of Lower Weverton quartzite

Another important question is the homogeneity of strain within a single crystal. Is the strain concentrated in certain planes along which shearing takes place and breaks the needles, or is it distributed more evenly through the crystal? Occasionally the deformation lamellae and Boehm lamellae behave as strong shear zones, breaking the needles into small fragments (Fig. 12). However, most of the deformation of the needles seems to be the result of the overall stress field due to dislocations in the crystal, rather than the effect of single shear zones. In thin sections, many of the breaks in the needles are unrelated to lamellae, and lamellae often intersect the needles without displacing them. Dislocations pile up at the needle on different slip planes and when their concentration exceeds a certain value, the net stress field causes the needle to deform. The homogeneity of deformation is reflected in the remarkably low scatter of the data. A logical conclusion from the strain homogeneity is that the dislocations are more or less evenly distributed throughout the crystal lattice and not concentrated in narrow zones. This would probably be the optimum configuration for minimum energy expenditure with maximum strain.

The Role of Grain – Boundary Sliding

Voluminous metallurgical literature exists on the relative movement of grains past each other, a mechanism referred to as grain boundary sliding. What is the exact meaning of this term and what role does it play during deformation? At low temperatures grain boundaries are stronger than the crystals while at high temperatures they become weaker zones; at an intermediate "equicohesive

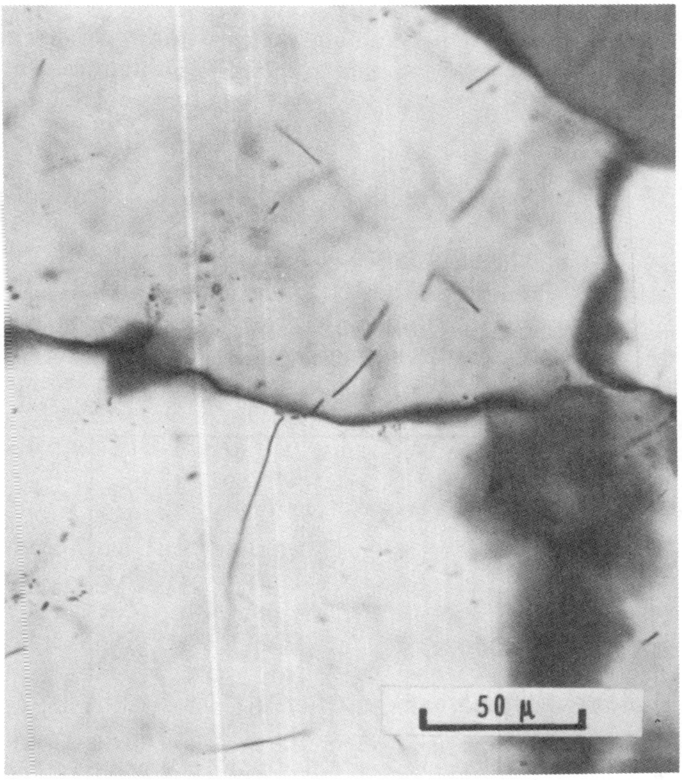

Fig. 10. Rutile needle deformed at a subgrain boundary by the movement of dislocations during polygonization. Crossed Nicols

temperature", the two are equal in strength (Jeffries, 1917). Crussard and Tamhankar (1958) have demonstrated that for steel this temperature is well above $0.5\,T_m$. So at low temperatures, grain boundary sliding will be a dependent process and only occur to the extent dictated by other mechanisms, while at high temperatures, it will act as an important independent mechanism. It is appropriate to describe two separate grain boundary mechanisms:

(1) *Grain Boundary Accomodation.* At low temperatures when an aggregate is deforming by diffusion, the grains slide past each other to maintain continuity. Again, if the grains are flattening by lattice deformation, sliding must occur along grain boundaries. This follows from Von Mises' theory (1928) that unless a crystal has five independently operating slip systems, grain boundary movement must take place to prevent the development of voids. The characteristic feature of this kind of sliding is that the grains have the same shape change as the specimen. In other words, it is a dependent process which must accompany diffusion and dislocation glide, but does not iself add to the total strain.

(2) *Superplastic Deformation.* At high temperatures, the grain boundaries become weak zones and most of the strain is taken up by grains sliding past each

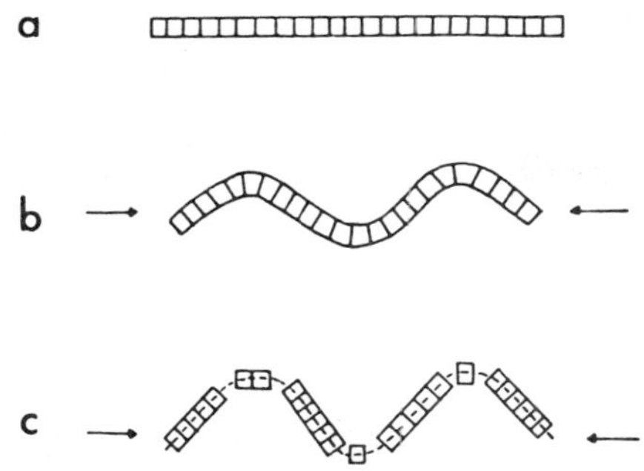

Fig. 11. a Undeformed needle. **b** On folding the needle the lattice is bent considerably especially in the hinge areas. **c** With further compression the crystal breaks into fragments to release the elastic energy. (Adapted from Nadai. vol. I., p. 200)

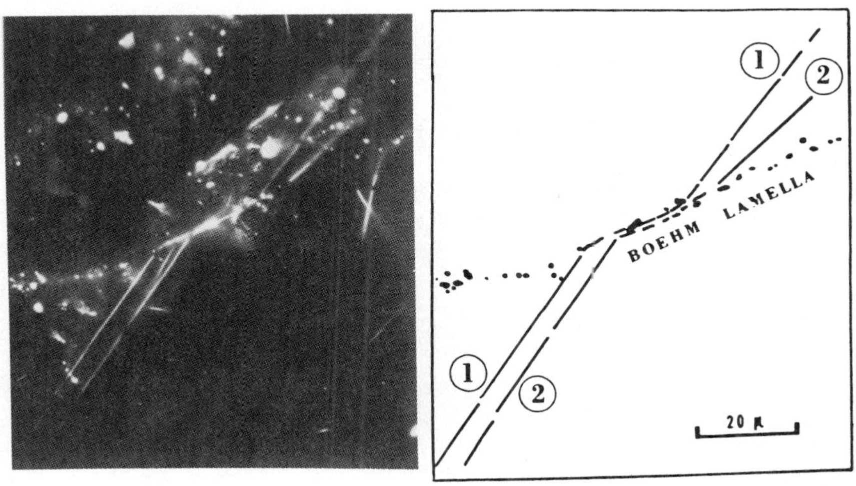

Fig. 12. Bubble trail acting as a "shear zone" and displacing two rutile needles considerably. Parts of the photograph are out of focus because the needles do not lie in the plane of the thin section. Sketch shows the important features observed under the microscope

other without changing their shapes appreciably (Langdon, 1974). This independent creep mechanism is made possible by a zone of high concentration of dislocations adjacent to the grain boundaries forming "lubricated" surfaces. This mechanism has been documented for the high temperature creep of metals and has also been suggested as a mechanism for the development of mylonites (Boullier and Gueguen, 1975). Since the effectiveness of this process is enhanced by a high grain boundary surface area, it is usually confined to very fine grained aggregates. Several authors (Rachinger, 1952; McLean, 1970; Langdon, 1970; Stevens, 1971) by different approaches based purely on geometrical considerations came to the conclusion that this process can account for as much as two-thirds of the total strain.

For most of the regime in which pressure solution is dominant, the effects of superplastic deformation are probably insignificant, but it may be more important for fine grained aggregates in the dislocation creep regime. In our present analysis, we are probably concerned more with grain boundary accomodation which plays an important but passive role during deformation – it enables grains to attain the necessary position to undergo pressure solution; during dislocation creep it enables rotation of grains so that operating slip planes can attain the optimum position for continuing glide.

Flow Laws

It has been shown previously (Gifkins, 1970; Elliott, 1973; Stocker and Ashby, 1973) that the components of the strain rate tensor $[\dot{\varepsilon}_{ij}]$ due to different independent mechanisms can be summed. In this paper we will use this to derive a three-dimensional flow law for deformations governed by more than one mechanism in terms of octahedral components.

The generalized uniaxial flow law derived from the experimental deformation of ice and metals is of the form:

$$\dot{\varepsilon} = B\sigma_d^n \tag{21}$$

where B and n are constants. The flow law is unaffected by the hydrostatic pressure so that the deviatoric stress (σ_d) is used. The three-dimensional flow law which has been derived theoretically (Odqvist, 1936, 1966) as well as from studies of glaciers and ice-sheets (Nye, 1957; Glen, 1958) is of the form:

$$|\dot{\gamma}_0| = A|\tau_0|^n \tag{22}$$

where A and n are constants, and $|\dot{\gamma}_0|$ and $|\tau_0|$ are absolute values of the octahedral shear components of the natural strain rate and non-dimensionalized deviatoric stress tensors respetively:

$$\begin{aligned} \dot{\gamma}_0 &= \tfrac{2}{3}[(\dot{\varepsilon}_1 - \dot{\varepsilon}_2)^2 + (\dot{\varepsilon}_2 - \dot{\varepsilon}_3)^2 + (\dot{\varepsilon}_3 - \dot{\varepsilon}_1)^2]^{\frac{1}{2}} \\ \tau_0 &= \tfrac{1}{3}[(\sigma_{d_1} - \sigma_{d_2})^2 + (\sigma_{d_2} - \sigma_{d_3})^2 + (\sigma_{d_3} - \sigma_{d_1})^2]^{\frac{1}{2}}. \end{aligned} \tag{23}$$

For a Newtonian material $n = 1$, for a perfectly plastic material $n = \infty$, and for a power law material $1 < n < \infty$ (Fig. 13).

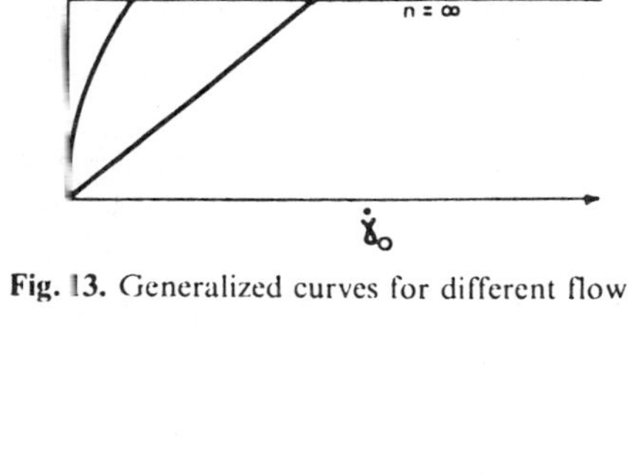

Fig. 13. Generalized curves for different flow laws on a plot of $\dot{\gamma}_0$ against τ_0

We will now derive the flow law for two simultaneously operating mechanisms – pressure solution which follows a linear flow law of the form $|\dot{\gamma}_0^p| = A_p|\tau_0|$ and dislocation creep which follows a power law of the form $|\dot{\gamma}_0^d| = A_d|\tau_0|^n$. We will assume that the quartzites approximate a Levy-Mises material i.e. at every point, the strain rate components are proportional to the deviatoric stresses:

$$\dot{\varepsilon}_{ij} = \lambda \sigma_{d_{ij}} . \tag{24}$$

This is a good approximation for an isotropic material with no Bauschinger effect (for example, a polycrystalline aggregate with random orientation and grain size distribution). The flow laws for the two mechanisms are:

$$\dot{\varepsilon}_{ij}^p = \lambda_p \sigma_{d_{ij}} , \tag{25}$$

$$\dot{\varepsilon}_{ij}^d = \lambda_d \sigma_{d_{ij}} . \tag{26}$$

Substituting Equations (25) and (26) into (4), the combined flow law is:

$$\dot{\varepsilon}_{ij}^t = \dot{\varepsilon}_{ij}^p + \dot{\varepsilon}_{ij}^d = (\lambda_p + \lambda_d)\sigma_{d_{ij}} . \tag{27}$$

Substituting Equation (24) into Equation (23) we get:

$$\dot{\gamma}_0 = 2\lambda\tau_0 . \tag{28}$$

Proceeding in the same way as for Equations (25) to (27) we can write:

$$\dot{\gamma}_0^p + \dot{\gamma}_0^d = 2(\lambda_p + \lambda_d)\tau_0 = \dot{\gamma}_0^t , \tag{29}$$

$$\therefore |\dot{\gamma}_0^t| = A_p|\tau_0| + A_d|\tau_0|^n . \tag{30}$$

In general for m mechanisms,

$$|\dot{\gamma}_0^t| = \sum_{k=1}^{m} A_k|\tau_0|^n . \tag{31}$$

Application to Field Examples

Having developed the theory for partition of strain and outlined practical methods of measurement, I will now show how it can be applied to the study of deformation mechanisms and flow laws for a few natural examples from the Weverton quartzites. Three samples of widely varying strains from the Lower Weverton are chosen as representative examples. Figure 14 and Table 1 show the finite strain measurements for the samples, which are illustrated by low power photomicrographs. For each sample, the values of T^p, T^d and T^t together with the octahedral shear components (γ_0^p, γ_0^d, γ_0^t) and Lode's numbers (ν^p, ν^d, ν^t) are determined. The relative importance of pressure solution and dislocation creep are shown by the values of Γ^p and Γ^d.

Photomicrographs in Figure 14 of the $a-c$ sections of the samples clearly exhibit the increasing strain from A 82 to SM 10–18. In A 82, the grains are rounded and almost unstrained. Pressure solution is the only recognizable creep mechanism with fibers growing in pressure shadows adjacent to detrital grains. This is also

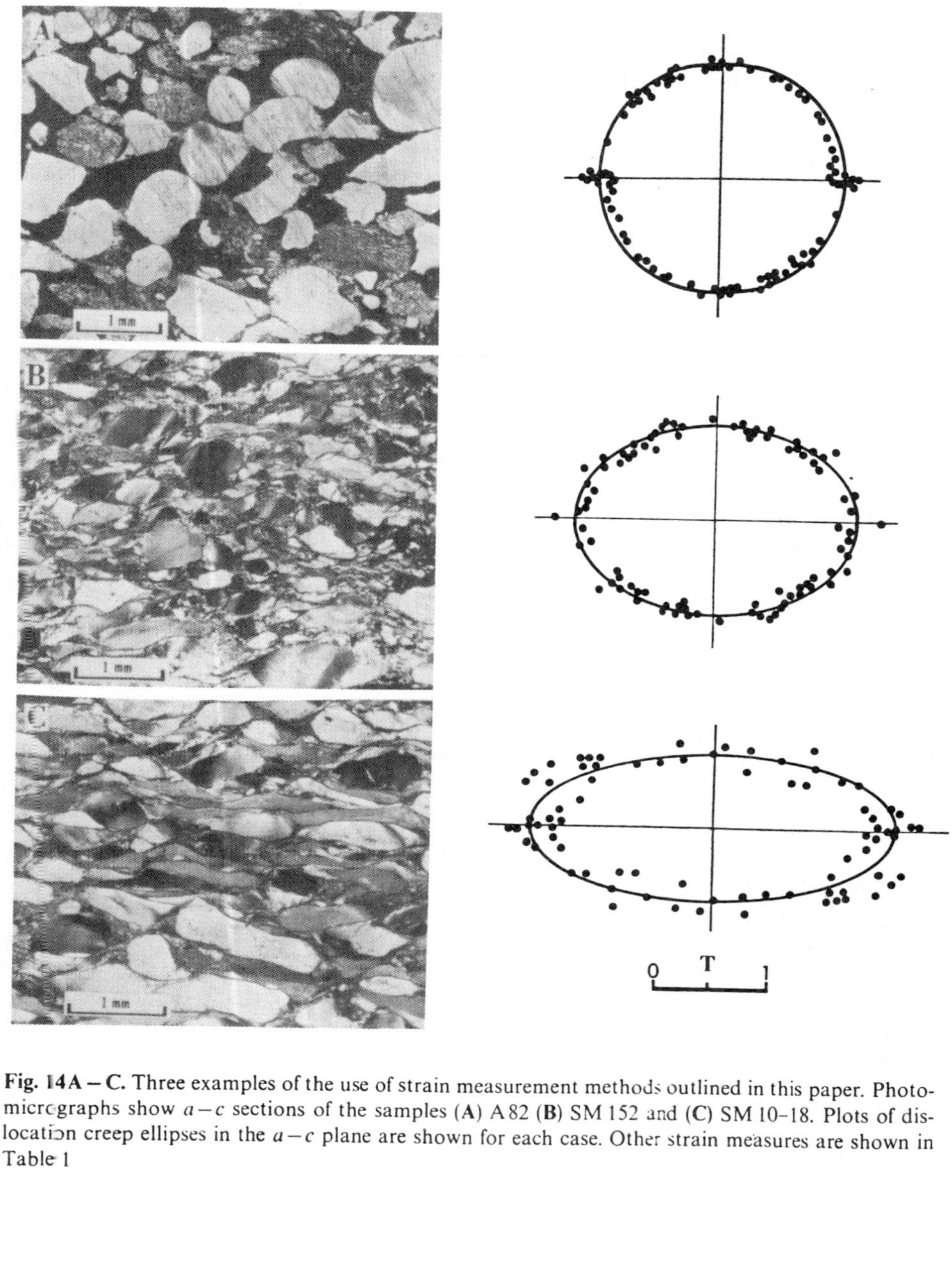

Fig. 14A – C. Three examples of the use of strain measurement methods outlined in this paper. Photomicrographs show $a-c$ sections of the samples (**A**) A 82 (**B**) SM 152 and (**C**) SM 10–18. Plots of dislocation creep ellipses in the $a-c$ plane are shown for each case. Other strain measures are shown in Table 1

Table 1. Strain measures for three samples of Weverton quartzites

	Sp.	T^p	γ_0^p	ν^p	T^d	γ_0^d	ν^d	T^t	γ_0^t	ν^t	θ	Γ^p	Γ^d
A	A 82	2.0:1.83:1	0.5	0.73	1.11:1.06:1	0.08	0.07	2.21:1.93:1	0.69	0.65	11°	0.86	0.14
B	SM 152	2.56:2.38:1	0.85	0.85	1.52:1.24:1	0.34	0.03	3.88:2.96:1	1.17	0.6	5°	0.71	0.29
C	SM 10–18	2.32:2.0:1	0.73	0.65	2.67:1.74:1	0.8	0.12	6.19:3.48:1	1.52	0.37	9°	0.48	0.52

shown by the almost circular plot for the dislocation strain ellipse in the $a-c$ plane. In sample SM 152, the strains are higher, and here dislocation creep is a more important mechanism as shown by the marked flattening of the grains; however, pressure solution is still the dominant mechanism. Specimen SM 10–18 is a mylonite showing elongated ribbon-shaped grains which have undergone extensive dislocation creep. In fact, in this specimen, it becomes the dominant creep mechanism as shown by the large axial ratio of the dislocation creep strain ellipsoid. The three samples therefore cover a wide range of strains and each shows a unique combination of components of the two creep mechanisms; Γ^p decreases with increasing strain (γ_0^t) while γ_0^d and Γ^d increase. Interestingly, the samples represent a transition from the normal to the overturned limb of a fold, therefore providing valuable information regarding the mechanism of folding. This will be the subject of a detailed analysis in a later paper.

Discussion – The Relation between Finite Strain and Creep Mechanisms

How do we relate the finite strain measurements obtained above to the stress-strain rate relationships outlined earlier in order to determine which creep mechanism is more important? If we assume a steady state creep the strain rate ($\dot{\varepsilon}$) is constant with time and can be determined directly by dividing the finite strain by the time; the flow law in such a case can be directly derived from the finite strain. For a varying strain rate, on the other hand, different mechanisms may dominate over different periods of time. The best approach to this problem is to define an average strain rate ($\bar{\dot{\varepsilon}}$), which determines the dominant mechanism over the entire deformation history. For steady state creep $\dot{\varepsilon}=\bar{\dot{\varepsilon}}$. This seems to be a reasonable approach. since we are actually more interested in determining the flow law *averaged over time*, rather than with respect to a single instant. Whether there has actually been a transition in creep behavior over time is an indeterminant problem. In using strain rates ($\dot{\varepsilon}$ or $\bar{\dot{\varepsilon}}$), we are in effect getting the relative importance of different mechanisms in *producing the finite strain*. For non-coaxial deformation, this does not necessarily imply which has been *more active*; this is determined by a parameter which I will define as the *strain activity* ($\hat{\gamma}_0$) which is obtained by addition of increments of strain along the deformation path. The octahedral shear strain and the strain activity may be distinguished as follows:

$$\gamma_0 = \int_0^t \dot{\gamma}_0 \, dt$$

while (32)

$$\hat{\gamma}_0 = \int_0^t \hat{\dot{\gamma}}_0 \, dt$$

where $\dot{\gamma}_0$ is measured with respect to the components in the principal directions of the finite strain ellipsoid, while $\hat{\dot{\gamma}}_0$ is computed with respect to the instantaneous principal values. The distinction between the two can be better explained by considering Figure 15. The circle is deformed by a sequence of strains with a non-coaxial path. The finite strain produced is small, but the integrated strain

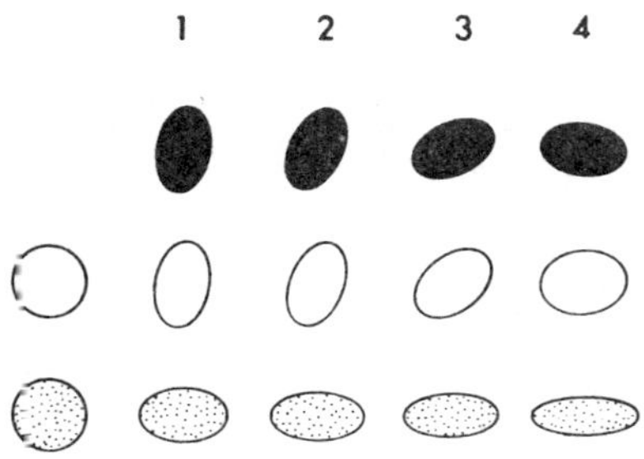

Fig. 15. Progressive deformation of a circle by a series of non co-axial strain ellipses (black). The finite strain at the end of the deformation (white ellipses) is very low, while the strain activity (stippled ellipses) assumes high values

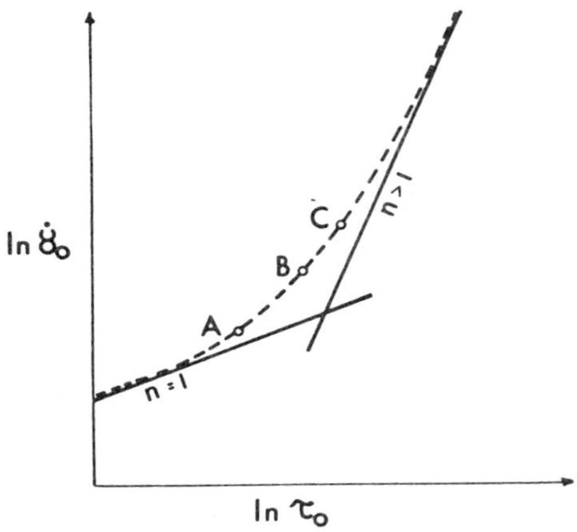

Fig. 16. Schematic logarithmic plot of $\dot{\gamma}_0$ against τ_0 showing two straight lines representing a linear flow law and a power law. The combined flow law followed by the rocks is shown by the dashed line. Relative positions of (*A*) A 82, (*B*) SM 152 and (*C*) SM 10–18 show a transition from Newtonian to power law behavior with increasing stress

along the deformation path is very high, showing that the process has been highly active. For an almost co-axial deformation path, the strain activity and finite strain would be equal. In thin sections, the lack of curvature of quartz fibers suggests that the path was close to co-axial for most of the deformation. Besides, for an approximately steady state deformation, both mechanisms should be equally affected by a non-coaxial path so that although we cannot determine the activity of the two mechanisms, a relative measure is still possible. For most measurements, the principal directions of the two ellipsoids do not vary appreciably ($< 10°$), so that the non-coaxiality presumably had the same effect on the two mechanisms. The extent of departure from coaxial and steady state deformation can be approximately determined by a large number of strain measurements on a regional scale.

The complex interaction between the two mechanisms can be best understood with the help of Figure 16. In this schematic diagram, the stress-strain rate relationships for the two mechanisms are represented by two straight lines on a logarithmic plot of $\dot{\gamma}_0$ against τ_0. The flow law followed by the deformation is a combination of the two and therefore follows the curved line. Natural examples do not necessarily approach either extremity where only one creep mechanism operates. Plots of the three measured samples show that with progressive deformation we migrate upwards along the curved line towards the right. Although for A82 and SM152 pressure solution is the more important mechanism, for SM 10–18 dislocation creep is more important. This transition is also reflected by the increase in Γ^d and the decrease in Γ^p with higher values of γ_0^t. An interesting point which emerges from this study is that the flow law is not purely Newtonian or a pure power law; instead, it is a complex combination of both. For each sample, there is a unique two term combination flow law which is entirely governed by physical constraints. Therefore, no simplified assumptions regarding a single term flow law for the deformation are applicable for these rocks.

Conclusions

The total strain in a rock which has undergone one period of deformation can be partitioned into components due to different mechanisms. Using this method of analysis for the greenschist-grade Weverton quartzites in the Harpers Ferry area, two independent mechanisms have been identified to be important: pressure solution and dislocation creep. Methods have been worked out for measuring both these components: dislocation creep is measured from deformed rutile needles which are good strain markers in the quartz grains; pressure solution is measured by a method involving the ratio of the area of fibers to the whole rock are a in principal sections in the specimen. These measurements can be used to determine the flow law operating in the rocks during deformation. It is found that the flow law is a complex combination of Newtonian and power law behaviour. A unique flow law defined by the physical conditions exists for each specimen. With increasing strain-rate, there is a transition to a progressively larger power law component. The study of deformed rutile needles suggests that dislocation glide and climb cause fairly homogenous deformation within the quartz crystals. The deformation is the result of the overall effect of dislocation creep rather than that of slip on individual glide planes. Deformation lamellae and Boehm lamellae are found to be manifestations of dislocation glide within the crystal; the bubble trails are hypothesized as forming due to annealing of lamellae by the climb of dislocation out of glide planes.

Acknowledgements. I would like to thank Dr. D. Elliott under whose supervision the research was carried out. Thanks are also due to Drs. H. Green and J. Tullis for their critical reviews of the manuscript and my colleagues at Johns Hopkins University for their suggestions. Lastly, I would like to thank Aparna Mitra for typing all versions of the manuscript.

This research was supported by NSF Grant. No. DES 74-17467 of the Earth Sciences section to D. Elliott.

List of Symbols

T	Stretch $(= l'/l_0)$
ε^t	Total natural strain $(= \ln T)$
$\varepsilon^p, \varepsilon^d$	Natural strain due to pressure solution and dislocation creep
$\dot{\varepsilon}, \bar{\dot{\varepsilon}}$	Natural strain rate and average strain rate
T	Axial ratio of strain ellipsoid

$a>b>c$	Principal semi-axes of strain ellipsoid
$\gamma_0, \dot{\gamma}_0$	Natural octahedral shear strain and strain rate
v	Lode's number
θ	Angle between principal axes of p.s. and d.c. strain ellipsoids
$\hat{\gamma}_0$	Octahedral strain activity
σ_d	Deviatoric stress.
τ_0	Octahedral shear component of deviatoric stress
Γ^p, Γ^d	Fractions of total strain contributed by pressure solution and dislocation creep
Ac/At, Bc/Bt	Area ratios of fibers to whole rock for $a-c$ and $b-c$ principal sections
D_{gb}, D_l	Grain boundary and lattice diffusion coefficients
X	Grain size
r	Axial ratio of grain
Ω	Atomic volume of diffusing species
h	Width of grain boundary
k	Boltzman's constant
T_c	Critical temperature for hydrolytic weakening
T_m	Temperature of melting

References

Ashby, M.F.: A first report on deformation-mechanism maps. Acta Met. **20**, 887–895 (1972)

Blacic, J.D.: Plastic deformation mechanisms in quartz: the effect of water. Tectonophysics **27**, 271–294 (1975)

Boullier, A.M., Gueguen, Y.: SP-Mylonites: Origin of some mylonites by superplastic flow. Contrib. Mineral. Petrol. **50**,93–104 (1975)

Carter, N.L., Christie, J.M., Griggs, D.T.: Experimental deformation and recrystallization of quartz: J. Geol. **72**, 687–733 (1964)

Christie, J.M., Ardell, A.J.: Substructures of deformation lamellae in quartz. Geology Vol. 405–408 (1974)

Christie, J.M., Griggs, D.T., Carter, N.L.: Experimental evidence of basal slip in quartz: J. Geol. **72**, 734–756 (1964)

Coble, R.L.: A model for boundary diffusion controlled creep in polycrystalline materials: J. Appl. Phys. **34**, 1679–1682 (1963)

Crussard, C., Tamhankar, R.: High temperature diffusion of steels: A study of equicohesion, activation energies, and structural modifications: Trans. AIME **212**, 718–730 (1958)

Durney, D.W., Ramsay, J.G.: Incremental strains measured by syntectonic crystal growths. In: De Jong, K.A., Scholten, R. (eds.), Gravity and tectonics, p. 502. New York: Wiley 1973

Elliott, D.: Determination of finite strain and initial shape from deformed elliptical objects: Geol. Soc. Am. Bull. **81**, 2221–2236 (1970)

Elliott, D.: Deformation paths in structural geology: Geol. Soc. Am. Bull. **83**, 2621–2638 (1972)

Elliott, D.: Diffusion flow laws in metamorphic rocks: Geol. Soc. Am. Bull. **84**, 2645–2664 (1973)

Frondel, C.: The system of mineralogy of Dana, J.D., and Dana, E.S. V.3, p. 334. New York: Wiley 1962

Gifkins, R.C.: Transitions in creep behavior. J. Materials **5**, 156–165 (1970)

Glen, J.W.: The flow law of ice: Symp. Chamonix, p. 171–183 (1958)

Green, H.W., Radcliffe, S.V.: The nature of deformation lamellae in silicates. Geol. Soc. Am. Bull. **83**, 847–852 (1972)

Green, W.V., Gillis, P.P., Zukas, E.G.: Subgrain boundary changes caused by creep deformation. Acta Met. **18**, 43–52 (1970)

Griggs, D.T.: Hydrolytic weakening of quartz and other silicates. Royal Astron. Soc. Geophys. J. **14**, 19–31 (1967)

Griggs, D.T.: A model of hydrolytic weakening in quartz. J. Geophys. Res. **79**, 1653–1661 (1974)
Griggs, D.T., Blacic, J.D.: The strength of quartz in the ductile regime (abstract). EOS Trans. AGU **45**, 102–103 (1964)
Groshong, R.H. Jr.: Deformation by pressure solution. Geol. Soc. Am. Abs. with programs **4**, 17–18 (1972)
Herring, C.: Diffusional viscosity of a polycrystalline solid. J. Appl. Phys. **21**, 437–445 (1950)
Jeffries, Z.: The amorphous metal hypothesis and equicohesive temperatures. J. Am. Inst. Metals **11**, 300–324 (1917)
Langdon, T.G.: Grain boundary sliding as a deformation mechanism during creep. Phil. Mag. **20**, 689–700 (1970)
Langdon, T.G.: The significance of grain boundaries in high-temperature creep. Can. Met. Quart. **13**, 223–228 (1974)
Lifshitz, I.M.: On the theory of diffusion-Viscous flow of polycrystalline bodies. Soviet Physics. JETP **17**, 909–920 (1963)
McLaren, A.C., Retchford, J.A.: Transmission electron microscope study of the dislocations in plastically deformed synthetic quartz. Phys. Stat. Sol. **33**, 657–668 (1969)
McLean, D.: Creep processes in coarse grained aluminium. J. Inst. Metals **80**, 507–519 (1951)
McLean, D.: Crystal fragmentation in aluminium during creep. J. Inst. Metals **81**, 287–292 (1952)
McLean, D.: Grain boundary sliding: A finite but unbounded limit. Met. Sci. J. **41**, 144–146 (1970)
Mises, R. von: Mechanik der Plastischen Formänderung von Kristallen. Z. Angew. Math. Mech. **8**, 161 (1928)
Mitra, S.: Deformation mechanisms and folding in the Weverton quartzites (abstract). EOS, Trans. A.G.U. **56**, 445 (1975)
Nabarro, F.R.N.: Deformation of crystals by the motion of single ions. *In:* Report of a conference on the strength of solids. Phys. Soc. (London) Proc. 75 (1948)
Nadai, A.: Theory of flow and fracture of solids. Vol. I, p. 572. New York: McGraw-Hill 1950
Nye, J.F.: The distribution of stress and velocity in glaciers and ice sheets. Proc. Roy. Soc. (London) Ser. A, **239**, No. 1216, 113–133 (1957)
Odqvist, F.K.G.: Theory of creep under combined stresses with application to Machinery. Roy. Swed. Acad. Eng. Res. Proc. **141**, 31 (1936)
Odqvist, F.K.G.: Mathematical theory of creep and creep rupture. Oxford University Press, p. 168, 1966
Preston, J.: Quartz lamellae in some Finnish quartzites. Bulletin de la Commision geologique de Finlande No. 180, p. 65–77 (1958)
Rachinger, W.A.: Relative grain translations in the plastic flow of aluminium. J. Inst. Metals **81**, 33–41 (1952)
Raj, R., Ashby, M.F.: On grain boundary sliding and diffusional creep. Metallurg. Trans. **2**, 1113–1127 (1971)
Stevens, R.N.: Grain-boundary sliding and diffusion creep in polycrystalline solids. Phil. Mag. **23**, 265–283 (1971)
Stocker, R.L., Ashby, M.F.: On the rheology of the upper mantle. Rev. Geophys. and Space Physics **11**, 391–426 (1973)
Stromgaard, K.E.: Stress distribution during formation of boudinage and pressure shadows. Tectonophysics **16**, 215–248 (1973)
Voll, G.: New work on petrofabrics. Geol. J. **2**, 503–567 (1960)
Weertman, J.: The creep strength of the earth's mantle. Rev. Geophysics **8**, 145–168 (1970)
White, S.: The dislocation structures responsible for the optical effects in some naturally deformed quartzites. J. Material Sci. **8**. 490–499 (1973)
Wickham, J.S.: An estimate of strain increments in a naturally deformed carbonate rock. Am. J. Sci. **273**, 23–47 (1973)
Wickham, J.S., Elliott, D.: Rotation and strain history in folded carbonates, Front Royal area, Northern Virginia. Trans. A.G.U. **51**, 422 (1970)

Received November 19, 1975/Accepted May 20, 1976

19

Copyright ©1980 by Pergamon Press Ltd
Reprinted from *Jour. Structural Geol.* **2:**83–99 (1980)

Shear zone geometry: a review

J. G. RAMSAY

Abstract—Shear zones may be classified into brittle, brittle–ductile, and ductile shear zones. The geometry and displacement boundary conditions of these zones are established. The geometric characteristics of ductile shear zones relevant to geological studies are described: orientations and values of principal finite strains, rotation, and deformation features of pre-existing planar and linear structures. Ductile shear zones show a fabric (schistosity and lineation) related to the finite strain state. The methods for determining strains and displacements from field studies are described.

Shear zones commonly occur in conjugate sets, but the two differently oriented sets do not seem to be able to operate synchronously. The angular relationships of conjugate ductile shear zones are different from those of brittle shear zones.

The termination of all types of shear zones poses complex compatibility problems, some solutions are suggested.

A synthesis of shear zone geometry in regions of crustal contraction and crustal extensions is made, and ideas on how deep level ductile shear zones relate to high level brittle shears and gliding nappes are put forward.

TYPES OF SHEAR ZONES

GENERATIONS of geologists studying the natural deformations of the Earth's crust have noticed that high deformations are often localised in narrow, sub-parallel sided zones, and those have been loosely termed *shear zones.*

Faults or *brittle shear zones* are special varieties of shear zones, where a clear discontinuity exists between the sides of the zone, and where the shear zone walls are almost unstrained or perhaps brecciated (Fig. 1a). Such fault zones are generally attributed to brittle failure controlled by the limiting elastic properties of the rock under orogenic stress. Other fault-like features show some ductile deformation in the walls, and are perhaps best termed *brittle–ductile shear zones.* The walls may show permanent strain for a distance of up to 10 metres on either side of the fault break (Fig. 1b). In the past these distortions have often been attributed to localised drag effects as the adjacent walls of the fault moved past each other, and the distortion of marker layers of rock entering this zone of flow have been used to determine the approximate movement sense of the displacement vector. This latter deduction is not always possible, because the geometry of the deflected marker surfaces in this zone is controlled by the line of intersection of the initial surface and the shear zone, and not by the movement vector. In brittle–ductile shear zones it is quite possible that the ductile part of the deformation history formed at a different time from that of the fault discontinuity. Another type of brittle–ductile shear zone is the extension failure of extension openings, usually filled with fibrous crystalline material, the openings generally making an angle of 30° or more with the shear zone and sometimes showing a sigmoidal form. The rock material between the extension fissures generally shows a certain amount of coherent permanent deformation, and it appears that the extension fissures developed at some stage of the deformation when a certain limit to coherent flow in the zone was reached. The third type of shear deformation is that of the *ductile shear zone.* Here the deformation and differential displacement of the walls is accomplished entirely by ductile flow, and on the scale of the rock outcrop no discontinuites can be seen (Fig. 1d). Marker layers in the country rock can be traced through the shear zone, they are deflected and may change their thickness, but they remain unbroken. Ductile shear zones are extremely common in deformed crystalline basement rocks (granites, gabbros, gneisses) which have been deformed under metamorphic conditions of greenschist, blueschist, amphibolite and granulite facies. They seem to be the dominant deformation mode whereby large masses of physically rather homogeneous rock can change shape under medium to high grades of metamorphism. The mineralogy of the rock in the ductile shear zone usually shows the characteristics of the metamorphic facies under which it developed. It is historically interesting as well as pertinent to this last point to note that the first understanding of the geological significance of hornblende schist was made from a study of shear zones in the Lewisian rocks of NW Scotland by Teall (1885). He was able to trace an unmetamorphosed basaltic dyke with normal igneous mineralogy though an amphibolite grade shear zone,

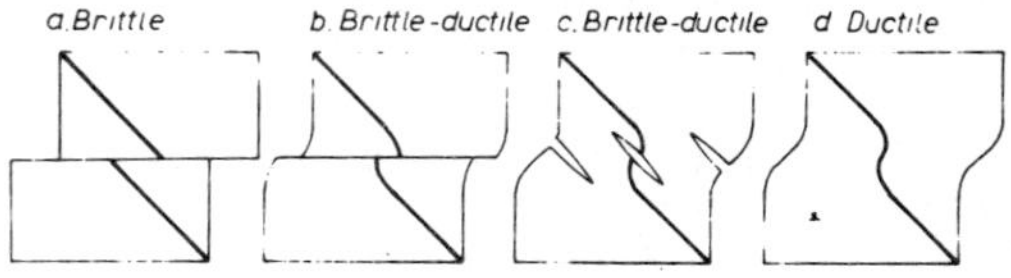

Fig. 1. Types of shear zones.

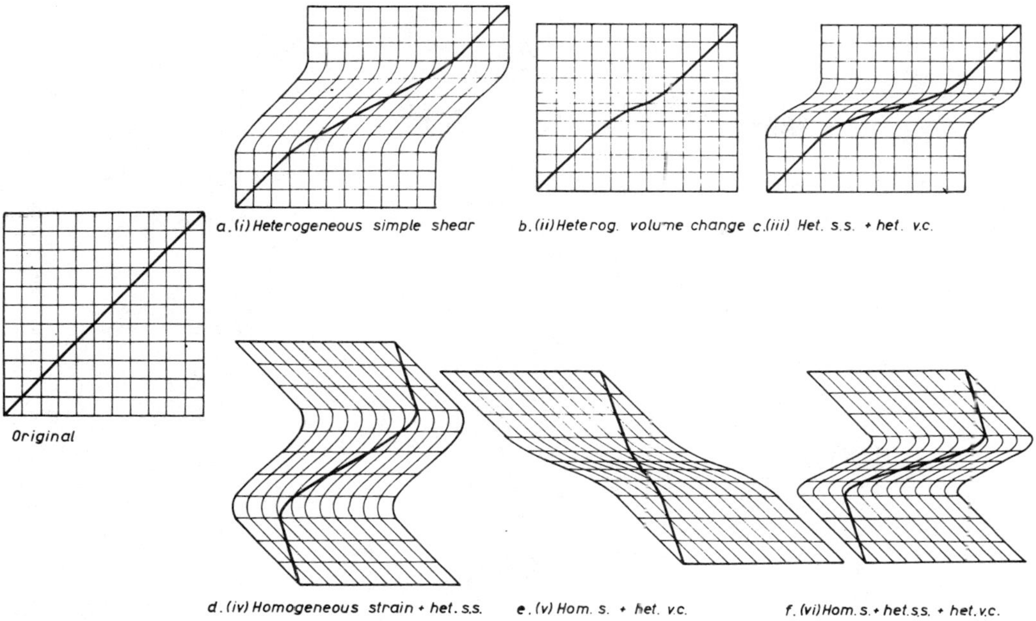

Fig. 3. Displacement fields of ductile shear zones.

where it was transformed, into a hornblende–biotite–oligoclase rock with a strongly schistose fabric.

The main aim of this contribution is to establish the geometrical features of shear zones, especially those of the ductile and brittle–ductile varieties, because in many regions these seem to be the deep level counterparts of brittle shear zones and faults seen at higher levels in the crust. Examples of these various types of shear zones are illustrated in Fig. 2.

SHEAR ZONE GEOMETRY

The boundary conditions for the geometrically simplest shear zones are first, that the shear zone is parallel sided, and second, that the displacement profiles along any cross sections of the zone are identical. This second condition implies that the finite strain profiles and the orientations and characteristic geometric features of small scale structural features across profiles are also identical — these being more easily recognised boundary conditions in practical geological terms. Although these conditions can never be completely realistic because all shear zones have to come to an end (their sides must eventually come together or splay apart, and their displacement profiles must change near their termination), it is my experience that very many shear zones often approximate closely to these over quite large zone lengths.

From the nature of the general displacement equations it can be shown that these boundary conditions constrain the types of displacement field (Ramsay & Graham 1970, pp. 796–798). If the walls of the shear zone are unstrained, then the following displacement fields are possible:

(i) heterogeneous simple shear (Fig. 3a);
(ii) heterogeneous volume change (Fig. 3b);
(iii) combinations of (ii) and (i) (Fig. 3c).

For the mathematical description of the third of these possibilities one should note that the order of the simple shear and volume change components should be specified. In this paper the simple shear component is followed by a volume reduction even though the two effects may have been geologically synchronous. This choice leads to mathematically simpler equations than when the two effects are reversed. In all three types the intermediate or Y_f axis of the finite strain ellipsoid (axes

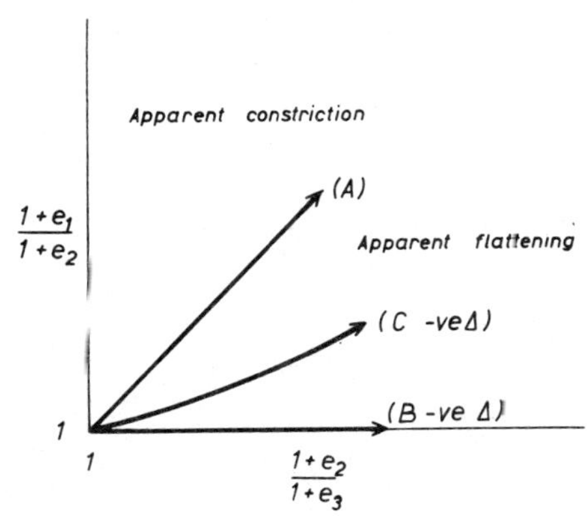

Fig. 4. Strain fields of the ductile shear zones of types (i), (ii), and (iii) shown in Fig. 3.

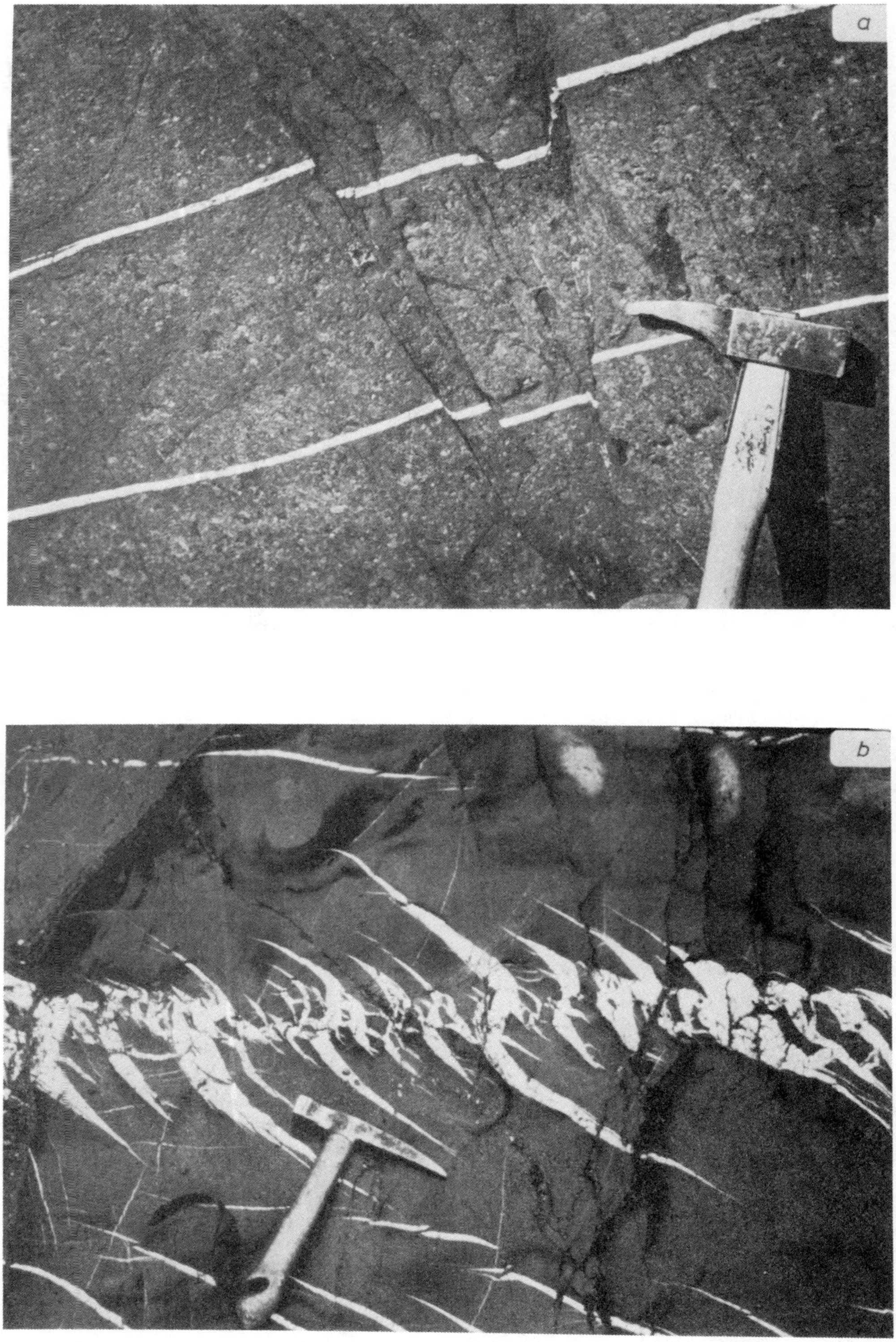

Fig. 2 (a) and (b). For caption see p. 87

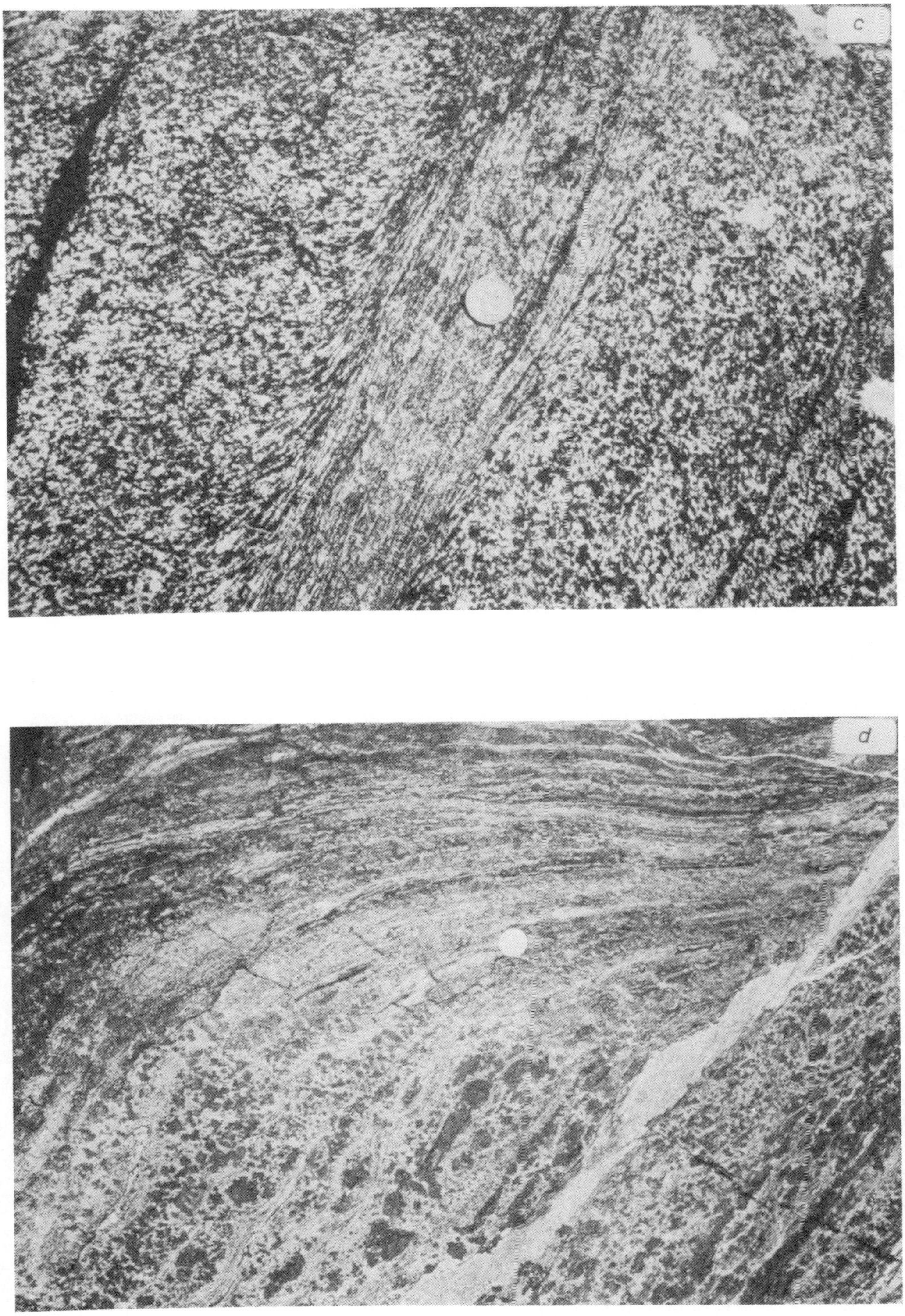

Fig. 2 (c) and (d). For caption see p. 87

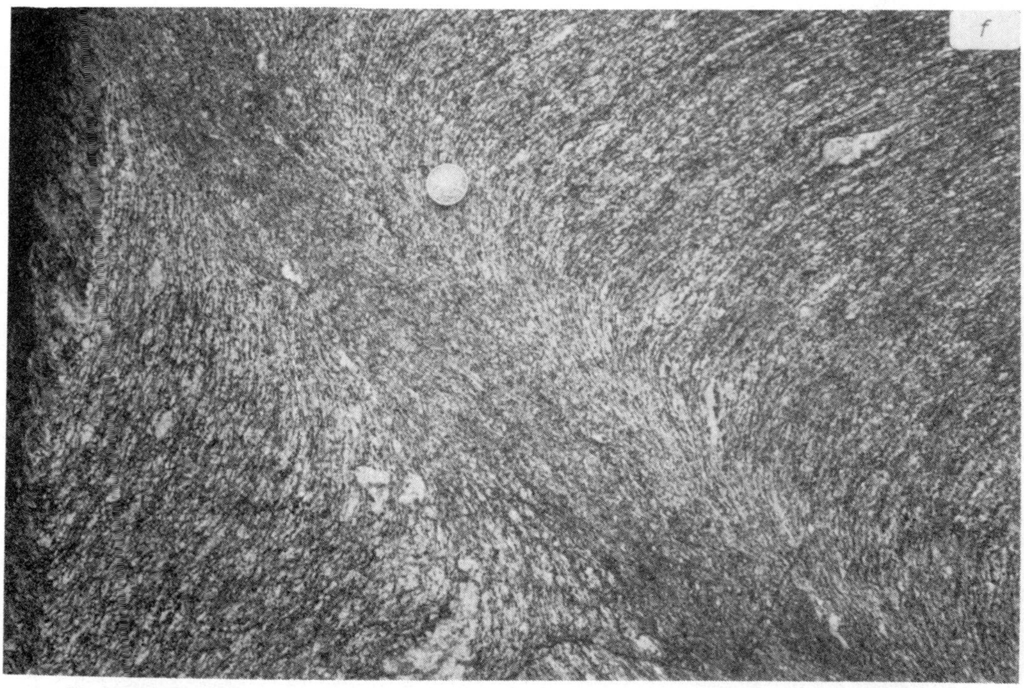

Fig. 2. Examples of shear zones. (a) Conjugate brittle shear zones, Hartland Quay, N Cornwall. (b) En echelon brittle–ductile shear zone; quartz filled vein fissures in a sandstone, Millook Haven, N Cornwall. (c) Ductile shear zone in Lewisian metagabbro, Castell Odair, N. Uist, Scotland. (d) Ductile shear zone in granulitic gneiss, showing modifications of mineralogy, N. Uist, Scotland. The wall of the shear zone shows dark garnet with orthopyroxenes, and this is transformed into amphibolite facies mineralogy (hornblende, biotite, clinopyroxene) inside the shear zone (e) Shear zones with volume reduction zones in Jurassic limestone, Gasterntal, Central Switzerland. (f) Ductile shear zones with deformed matrix gneisses, Cristallina, Ticino, central Switzerland.

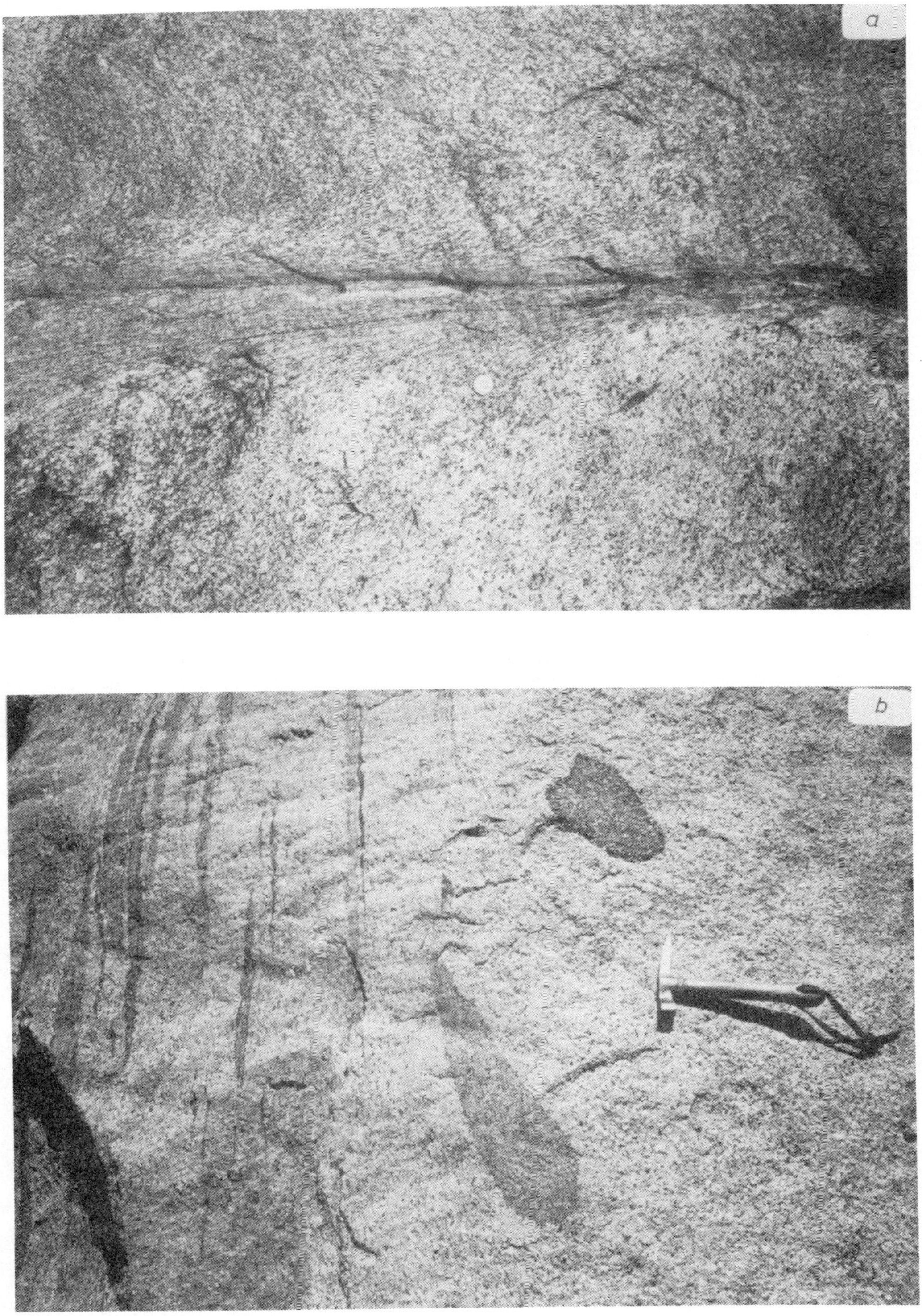

Fig. 10. (a) Sigmoidal schistosity fabric produced in a ductile shear zone. (b) Strained xenoliths in a shear zone cutting granitic rocks. Laghetti, Ticino, Central Switzerland.

$X_f \geq Y_f \geq Z_f$) is always contained in a plane parallel to the shear zone walls. The types of strain ellipsoid produced from these displacement fields are illustrated in the graphical plot of Fig. 4. Simple shear (Fig. 4, A) is clearly a plane strain ($e_2 = 0$). Heterogeneous volume reduction leads to progressive flattening ellipsoids (Fig. 4, B) and the combination leads to a deformation field in the apparent flattening sector of the diagram.

If the walls of the shear zone are themselves heterogeneously strained, then the displacement plans can be expressed mathematically as the same as (i) to (iii) above but with the addition (mathematically a matrix multiplication) of a homogeneous strain component inside and outside the shear zone, i.e.

(iv) homogeneous strain combined with simple shear (Fig. 3d);
(v) homogeneous strain combined with volume change (Fig. 3e);
(vi) homogeneous strain combined with simple shear and volume change (Fig. 3f)

With these three types the Y_f axis of the strain ellipsoids will have no specific relationship to the shear zone plane and all principal strain axes will have a variable orientation across a profile of the zone. The types of strain (i.e. apparent flattening or apparent constriction) have no general constraints, and all types are possible depending upon the orientations of the various 'componental' strains.

Although all six basic types are geometrically separable, it should be noted that their overall visual effect on marker surfaces crossing the shear zone is remarkably similar (see Fig. 3). This means that the field geologist has to be especially careful in avoiding jumping to premature conclusions as to the displacement field if he only observes deformed planar markers. In my experience natural shear zones accord very closely to these six models. Some geologists might object to the inclusion of models (ii) and (v) (Fig. 3b & e) which involve differential volume loss under the category of shear zones, but I think it is right to include them for two reasons. First their basic geometric boundary conditions accord with those of more 'orthodox' shear zones where shear is dominant (*S-bands*, Cobbold 1977a); second, their effects on oblique planar markers are geometrically very close to those of shear zones with differential shear displacement parallel to the zone walls. It will be shown later that the amount of volume change across a shear zone can be calculated using the displacement of initially plane markers. Deformation zones involving only volume loss are sometimes called *pressure solution zones* or *solution-seams* or *stripes* (Fig. 2e) or *P-bands* (Cobbold 1977a). These zones are usually deficient in certain mobile mineral species compared with their walls. In low to medium grade metamorphic environments, where such zones are common, quartz or calcite are usually removed from the rock leading to a proportional increase in relatively immobile components, such as clays, chlorite and micas. These changes in relative proportions of mineral species leads to a characteristic colour striping.

The deformation zones involving heterogeneous simple shear without differential volume change (types (i) and (iv), Fig. 3a & d) are commonly isochemical and the proportions of different mineral species across a zone profile is nearly constant. However, the mineral grain sizes and the rock fabric and texture do usually change with the strain variations across the zone.

The zones involving combinations of simple shear and volume change are usually variable in mineralogy, in mineral proportions and in texture and fabric (Fig. 2d).

SIMPLE SHEAR ZONES

The basic component in practically all shear zones is that of heterogeneous simple shear. It is therefore appropriate to investigate the geometric properties of this type of displacement, to establish how strains are related to displacement and how the structural features we see in deformed rocks within a shear zone can be related to displacement and strain.

A zone with heterogeneous simple shear can be considered to be made up of a number of infinitesimally small elements showing homogeneous simple shear. In such a small homogeneously strained element it is mathematically convenient to relate the x-coordinate direction parallel to the shear direction, and the z-axis normal to the plane of shear (xy) — see Fig. 5. If the top face of an initially cube element is displaced by a distance s, the shear strain γ (more exactly γ_{zx}) is related to the angular shear strain ψ produced by the deflection of the lines initially parallel to z such that

$$\gamma = \tan \psi \tag{1}$$

$$\text{and so } s = z \tan \psi = z\gamma. \tag{2}$$

The deformation matrix is then given by:

$$\begin{bmatrix} 1 & 0 & \gamma \\ 0 & 1 & 0 \\ 0 & 0 & 1 \end{bmatrix} \tag{3}$$

A circle of unit radius drawn on the xz face is deformed into a strain ellipse with principal semi-axis lengths along X_f of $1+e_1$ and along Z_f of $1+e_3$ (Fig. 6). The values of these strains are given by:

$$(1+e_1)^2 = \tfrac{1}{2}[2 + \gamma^2 + \gamma(\gamma^2+4)^{\frac{1}{2}}] \tag{4}$$

$$e_2 = 0 \tag{5}$$

$$(1+e_3)^2 = \tfrac{1}{2}[2 + \gamma^2 - \gamma(\gamma^2+4)^{\frac{1}{2}}]. \tag{6}$$

The orientations θ' of these principal strains measured from the x coordinate direction is given by:

$$\tan 2\theta' = \frac{2}{\gamma}. \tag{7}$$

These two directions originated in two other orthogonal directions θ in the unstrained state given by:

$$\tan 2\theta = \frac{-2}{\gamma}. \tag{8}$$

The strain is clearly a rotational strain, the finite rotation ω can be derived from:

$$\tan \omega = \tan (\theta' - \theta) = \frac{2}{\gamma}. \quad (9)$$

Any passive plane marker is displaced by the shear such that, if its trace on the xz plane makes an initial angle α with the x-direction before deformation and an angle α' after deformation (Fig. 6), then

$$\cot \alpha' = \cot \alpha + \gamma. \quad (10)$$

The mathematical relationships refer to a special choice of coordinate reference frame such that x is parallel to the shear direction. It is sometimes more useful to refer to the displacements with the shear direction at some angle φ to the x-direction. For example one may be interested in computing the effects of subjecting an element to superposed shear zones at differing angles (we will discuss this problem later).

If the shear zone is oriented at an angle φ to the x-direction, then from the geometry shown in Fig. 7 an initial point (x,z) is displaced to a new position $(x_1 z_1)$ according to the displacement matrix:

$$\begin{bmatrix} 1 - \gamma \sin \varphi \cos \varphi & 0 & \gamma \cos^2\varphi \\ 0 & 1 & 0 \\ -\gamma \sin^2\varphi & 0 & 1 + \gamma \sin\varphi \cos \varphi \end{bmatrix}. \quad (11)$$

The values of the principal strains and rotations are identical to those of (4), (5), (6) and (9), but the orientation of the principal strains is given by

$$\tan 2\theta' = \frac{2 + \gamma \tan 2\varphi}{\gamma + 2 \tan 2\varphi}. \quad (12)$$

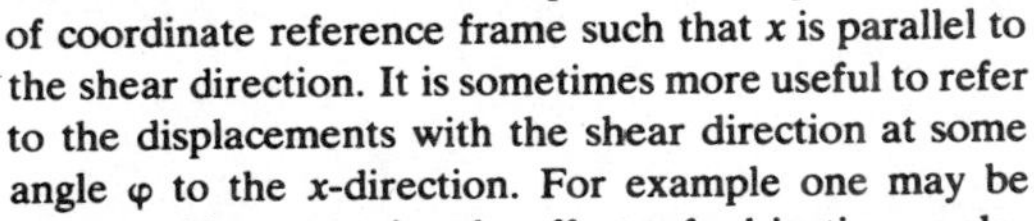

Fig. 5. The geometric features of simple shear, ψ, angular shear strain; s, displacement parallel to the x-axis.

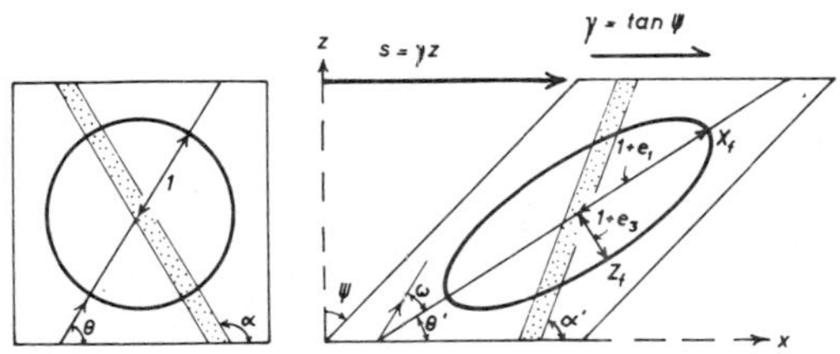

Fig. 6. The relation of the strain ellipse to shear in a simple shear system.

The positions of the principal strain axes for the first small increment of simple shear are from (7) oriented at 45° and 135° to the x-coordinate axis. The incremental strain axes X_i and Z_i are identically oriented and coincide with the principal strain rate axes. The principal

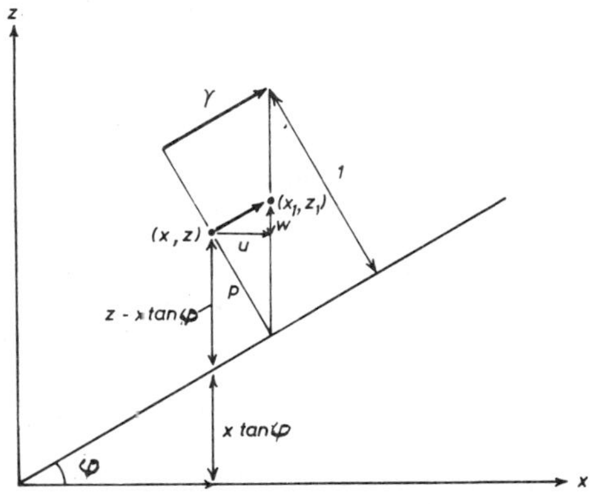

Fig. 7. Geometry of simple shear with shear oriented at an angle φ to the x coordinate axis.

strain rates $\dot{e}_1$ and $\dot{e}_3$ are related to the shearing strain rate $\dot{\gamma}_{zx}$ according to

$$\dot{e}_1 = -\dot{e}_3 = \frac{\dot{\gamma}_{zx}}{2}, \quad (13)$$

$$\dot{e}_2 = 0,$$

and the rate of rotation, or vorticity $\dot{\omega}$ is

$$\dot{\omega} = \dot{\gamma}. \quad (14)$$

During the first stages of deformation of an initially homogeneous isotropic body the principal axes of stress σ_1, σ_2 and σ_3 will coincide with the incremental strain axes X_i, Y_i and Z_i respectively, and the magnitudes of the strain rates will be some function of the magnitudes of the stresses. As deformation proceeds, however, it is very common for the deforming mass to become anisotropic as a result of the finite strains which build up with increasing values of shearing strain γ. During these mature stages of deformation, although the principal strain rate and incremental strain axes X_i and Z_i must remain at 45° and 135° to the x-direction because of geometric constraints of the process, the principal stress axes σ_1 and σ_3 will almost certainly be influenced by the increasing obliquity of the anisotropic fabric to these strain axes, and are therefore unlikely to coincide with the strain increment axes.

Fabrics produced by simple shear strain

As simple shear displacement gets larger, the angle θ' between the principal finite elongation X_f and the shear zone walls becomes progressively smaller (Fig. 8). In many naturally deformed rocks found in ductile shear zones, a statistically preferred orientation of the minerals is developed in this finite flattening plane, giving rise to schistosity or cleavage. Within this planar, tectonically induced fabric there is commonly a strain related linear orientation of the mineral components parallel to the finite greatest extension direction X_f on the schistosity plane. Progressive shearing leads to a progressive intensification of these linear fabrics and a change in

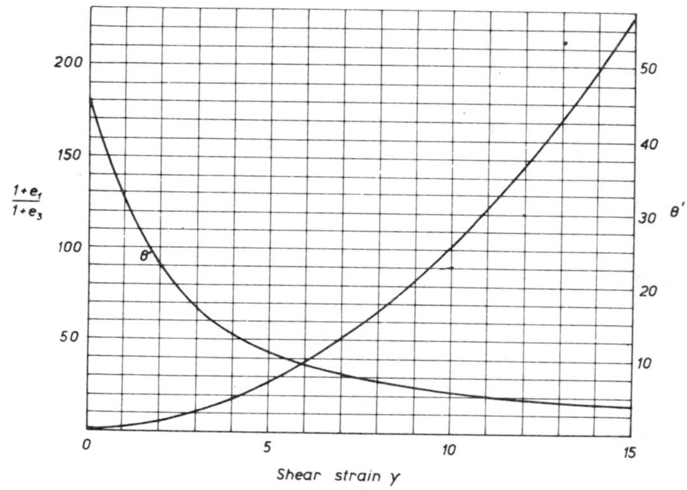

Fig. 8. Simple shear, values and orientations of the principal finite strains.

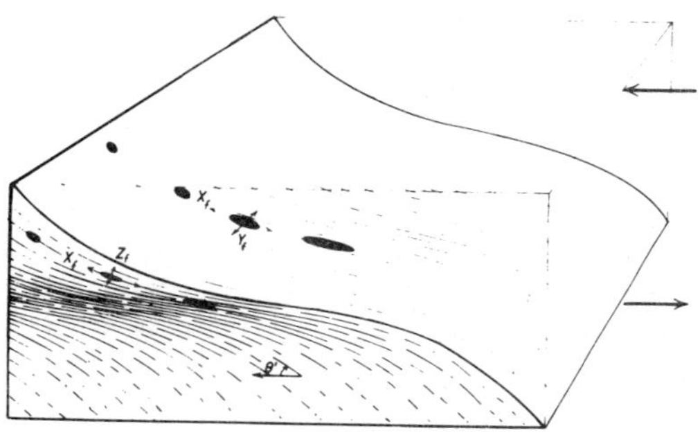

Fig. 9. Fabric in a ductile simple shear zone.

their orientation. Because ductile shear zones generally show a maximum displacement gradient in the zone centre (the gradient decreasing towards the margin), it follows that the tectonically induced planar fabrics of shear zones generally show a characteristic sigmoidal form as shown in Figs. 9 and 10.

Because the orientation of the X_fY_f surface with its coincident schistosity is a function of shear strain, it follows that the angle between schistosity and shear zone walls can be used to measure the shear zone parallel shear strain, and shape of the strain ellipsoid at that point. This technique, originally developed by Ramsay & Graham (1970), can be extended so as to integrate successive finite shear strains across a shear zone profile and, in so doing, to calculate the total differential displacement across a zone with heterogeneously developed shear strain (Fig. 11). The results of the application of this technique on a regional scale are often very striking. For example, Beach (1974) showed that the group of shear zones making up a large movement zone in the frontal part of the Precambrian Laxfordian orogenic belt of NW Scotland has a minimum displacement of 25 km, which can be resolved into a horizontal component of 18 km and a vertical component of 16 km.

The main problem of using this technique occurs where the shear displacement gradients are high. For example, angles between schistosity and shear zone wall of 5.7° and 2.9° imply a difference of displacement of 10γ and 20γ respectively. With such high displacement gradients (and high finite strains) very small observation errors can lead to gross errors in computing the total displacement across the shear zone.

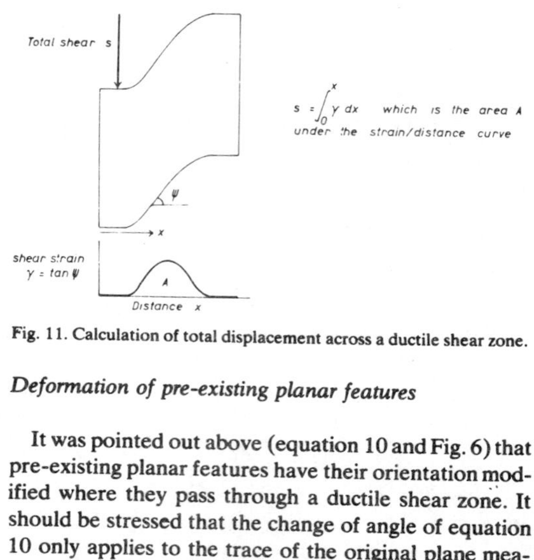

Fig. 11. Calculation of total displacement across a ductile shear zone.

Deformation of pre-existing planar features

It was pointed out above (equation 10 and Fig. 6) that pre-existing planar features have their orientation modified where they pass through a ductile shear zone. It should be stressed that the change of angle of equation 10 only applies to the trace of the original plane measured in the xz profile of the shear zone. Although it does not refer directly to the dihedral angle between two planes, the modification of such dihedral angles can be easily computed (Ramsay 1967, pp. 504–508). Where a set of sub-parallel planes crosses a shear zone with a varying displacement gradient profile, variations in the amount of local shear imposed on the layers sets up folding with a characteristic similar style (Fig. 12). The axes of these folds are controlled by the line of intersection of the initial plane structure and the xz plane of the zone. Their axial planes are parallel to the zone, and are located at positions dependent upon the local changes in shear gradient and the orientation of the plane before shearing (Ramsay 1967, pp. 508–509). Variations in orientation of the layers can be used to compute finite strain values at different positions in the zone (according to the relationships of equation 10), and can be used to determine the total shear across a zone. The practical application of this technique for calculating displacements has been shown in a beautiful study of basic dykes passing through shear zones along the Nagssugtoqidian orogenic front of west Greenland (Escher *et al.* 1975). Here it has been shown that shear zones have led to an average 66% crustal shortening of the deep level basement rocks of this orogen over a distance of about 100 km.

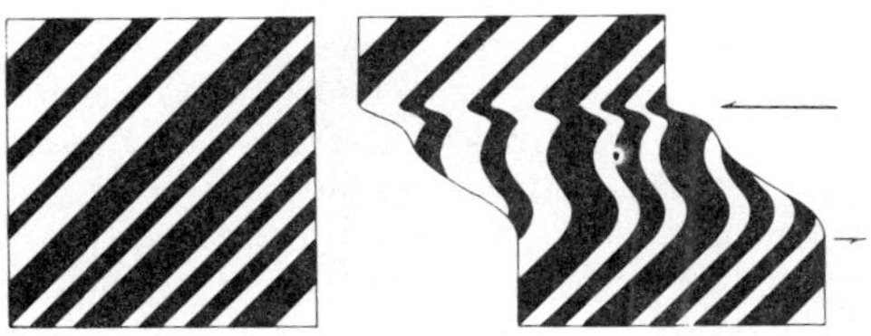

Fig. 12. Folds of 'similar' style produced by the shearing of passive layers.

The change in orientation of pre-existing planar features is accompanied by other geometric modifications. If there is no ductility contrast between the layer and its matrix it may either increase (Fig. 13a, i, ii) or decrease (Fig. 13b, i, ii) in thickness as a result of passive shearing. The conditions for increases or decreases in thickness are specified in Fig. 13. The change of thickness (t to t') is given by:

$$t' = \frac{\sin\alpha'}{\sin\alpha}t. \qquad (15)$$

If the angular conditions are such that $\alpha > 90°$ and $\alpha' > 180° - \alpha$ the progressively sheared layer first thickens, then thins to its original width (where $\alpha' = 180° - \alpha$) and then subsequently becomes thinner than its original value. The reverse sequence — thinning followed by thickening — can never occur in progressive simple shear. If the layers undergoing shear are not passive, but show a competence contrast with their matrix, then mechanical instabilities are set up in the competent layer which lead to the formation of new structures. If the layer is shortened during deformation (equivalent to the passive layer undergoing thickening), buckle folds develop with wavelengths characteristic of the thickness of the competent layer, and the extent of the competence contrast (Fig. 13a, iii). The orientation of the axes of these folds depends upon the obliquity of the layer to the shear zone. It depends upon the direction of maximum shortening within the surface of the contracting competent layer and is not simply related to the principal displacement features of the shear system (as with the similar folds produced by passive layer deflection). For example, the fold axes will not generally form parallel to y or Y_f and they need not lie in the xy shear plane or the X_fY_f plane of the finite strain ellipsoid (Fig. 14).

If the competent layer is oriented in a position such that it becomes extended by the shearing, then pinch and swell or boudinage structure may be developed (Fig. 13b, iii). The necking directions of these boudins will be controlled by the maximum extension directions in the layer being sheared and will not in general, be related to the xy plane of the shear zone. In fact the two-dimensional strains within the deflected competent layer are likely to involve both extensions and contractions. Both buckle folding and boudinage then develop together, probably one of these being dominant over the other (Fig. 14).

One further possible complication that can develop in competent layers occurs where the initial orientation and value of shear lead to layer shortening followed by extension. Here we have the possibility of layers developing buckle folds, which subsequently become unfolded or boudinaged (Fig. 13c, iii).

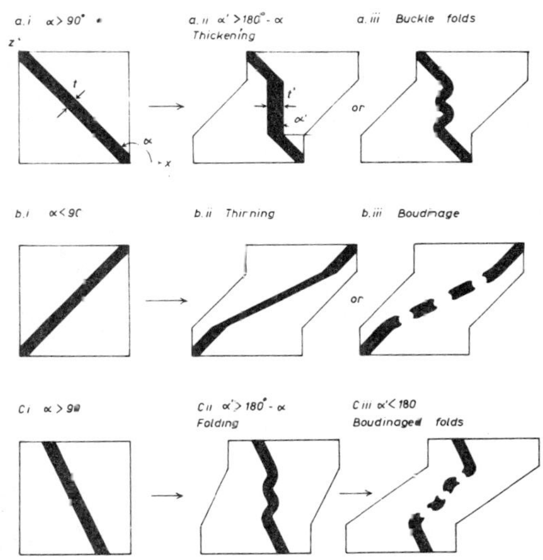

Fig. 13. Effects of shear zone deformation on competent layers.

Deformation of pre-existing linear features

Most pre-existing line elements (e.g. pre-shear zone fold axes) become deflected and take up new orientations in the shear zone. These lines come to lie closer to the direction of the shear (x) and move on a plane locus which connects their initial orientation with the x-direction (Fig. 15) (Weiss 1959, Ramsay 1960, 1967). In projection they move on a great circle locus (Fig. 15, p, p', p'' and q, q', q''). The only exceptions to this general rule are those lines which lie in the plane of the shear zone xy; these show no deflection (Fig. 15, r, r', r''). The axis of principal finite elongation X_f also moves

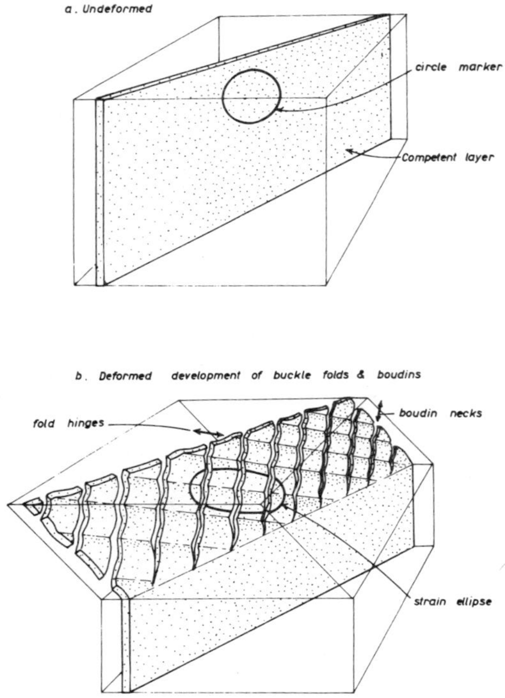

Fig. 14. Three-dimensional changes in a competent layer in a shear zone.

on the xz plane towards the x-direction (Fig. 15, X_f', X_f'').

The geometric effect of the drawing together of initially variably oriented linear directions towards the X_f direction is particularly striking at high shear strains. Particularly interesting is the effect described by Carreras *et al.* (1977) and Quinquis *et al.* (1978) where folds outside a shear zone showing slight variations in their axial plunges are deformed in the shear zone so that their axes show extreme variations in plunge (so called 'sheath folds'). The interlimb angles of these folds are drastically modified in the shear zone according to the principles discussed in the previous section and changes of fold shapes offer a method for computing shear strain values. This effect is shown diagrammatically in Fig. 16.

Terminations of shear zones

Ductile shear zones can, and in some cases do, come to a stop within the ductile environment. The boundary conditions for the ends of shear zones are much more complicated than for those of simple parallel sided zones with constant displacement profiles used to set up the six basic types described earlier. As far as I am aware there are no complete mathematical solutions for the possible finite strain and displacement variations where a shear zone terminates. However, one can proceed intuitively, and try and arrive at geometric resolutions which fit field observations and natural situations. There appear to be an unlimited range of possible models: Ramsay & Allison (1979) have suggested that this range has two

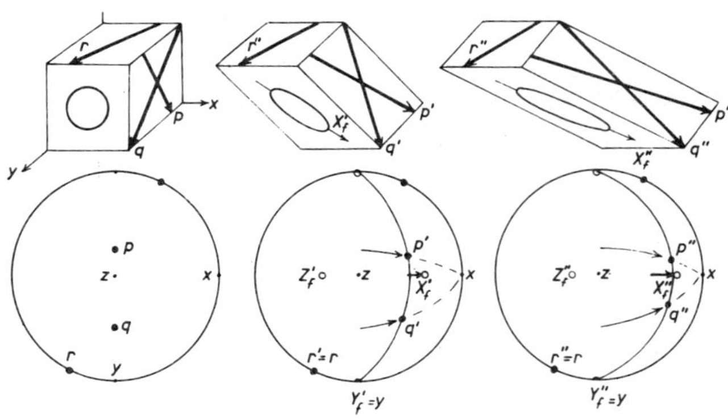

Fig. 15. Deformation of lineations in ductile shear zones.

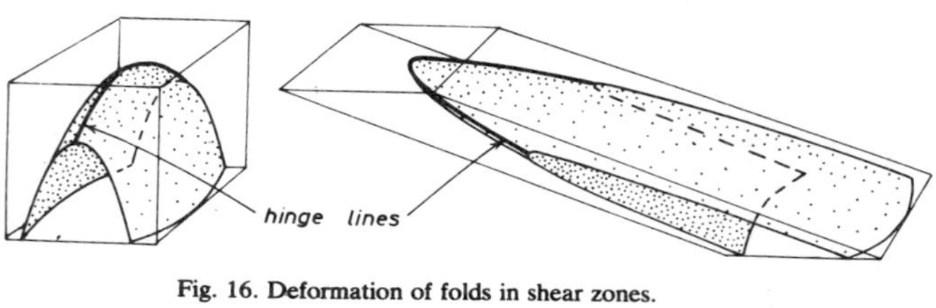

Fig. 16. Deformation of folds in shear zones.

end members (Fig. 17). One is a plane strain model, whereas the second is a solution where all displacements take place normal to the shear zone (in the y-direction). Both give rise to characteristic strain fields, and the non-plane strain model sets up constrictive and flattening deformations on either side of the shear zone tip. If the boundary deflections seen in the plane strain model are constrained, then there will be a tendency for the tips of right handed shear zones to bend and propagate in a clockwise sense with respect to the main shear zone, and left handed shear zones to propogate in an anticlockwise sense (Fig. 18). This effect could be the reason for the crossing or merging of shear zones of similar displacement sense that have been recorded in some regions.

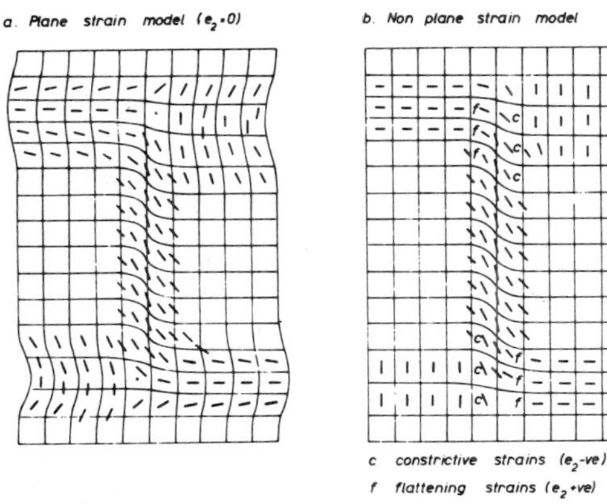

Fig. 17. Strain patterns at shear zone terminations.

GEOMETRIC FEATURES OF SHEAR ZONES INVOLVING VOLUME CHANGE

The geometry of simple shear zones of the type illustrated in Fig. 3(a) have been discussed in some detail because many of the structural characteristics of shear zones accord well with this model. However, this model is not a general one as it involves no volume change. We will now discuss the basic geometric features of more general models and establish ways the field geologist can separate the various componental strains and displacements.

Shear zones with undeformed walls (Figs. 3a, b & c)

The displacement fields in all of these types are most conveniently described in mathematical terms by a simple shear component (γ) followed by a volume change component (Δ) acting normal to the shear zone walls. The local displacement gradient matrix at any point in the zone can be described by the matrix

$$\begin{bmatrix} 1 & 0 & \gamma \\ 0 & 1 & 0 \\ 0 & 0 & 1+\Delta \end{bmatrix}. \qquad (16)$$

The volume change component affects both the values and orientations of the principal strains

$$(1+e_1)^2 = \tfrac{1}{2}[1+\gamma^2+(1+\Delta)^2+((1+\gamma^2+(1+\Delta)^2)^2-4(1+\Delta)^2)^{\frac{1}{2}}] \qquad (17)$$

$$e_2 = 0 \qquad (18)$$

$$(1+e_3)^2 = \tfrac{1}{2}[1+\gamma^2+(1+\Delta)^2-((1+\gamma^2+(1+\Delta)^2)^2-4(1+\Delta)^2)^{\frac{1}{2}}] \qquad (19)$$

$$\tan 2\theta' = \frac{2\gamma(1+\Delta)}{1+\gamma^2-(1+\Delta)^2}. \qquad (20)$$

In zones with varying shear and volume change the schistosity (X_fY_f plane) will be sigmoidally shaped, but the shape of these curved surfaces will not be a simple function of shear.

Passive line markers making an angle α with the shear zone walls are deflected in the zone to make a new angle α' such that

$$\cot\alpha' = \frac{\cot\alpha+\gamma}{1+\Delta}. \qquad (21)$$

These relationships enable the shear and volumetric components to be separated. For example it is possible to compute γ and Δ from the deflections of two differently oriented deflected line (or plane trace) markers making initial angles of α and β and final angles of α' and β' (Fig. 19a), because from (21) it follows that:

$$\gamma = \frac{\cot\alpha\cot\beta'-\cot\beta\cot\alpha'}{\cot\alpha'-\cot\beta'} \qquad (22)$$

and

$$1+\Delta = \frac{\cot\alpha-\cot\beta}{\cot\alpha'-\cot\beta'}. \qquad (23)$$

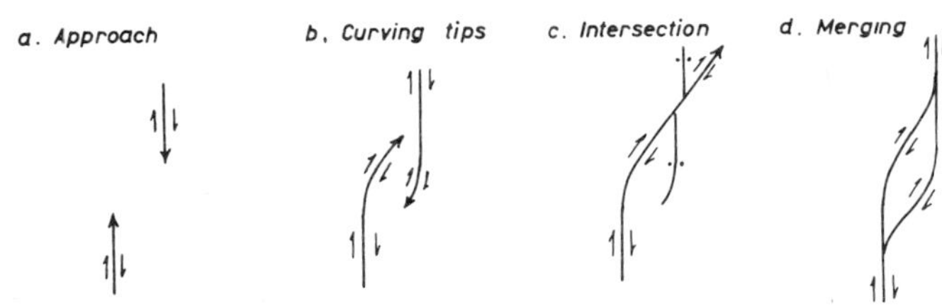

Fig. 18. Propagation of shear zones.

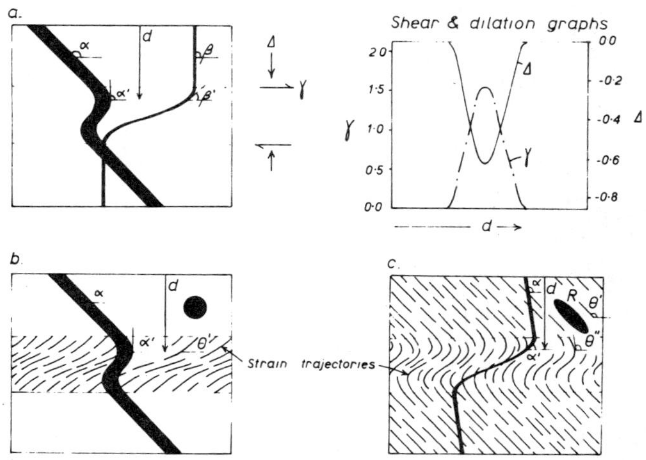

Fig. 19. Calculation of shear and dilation components in shear zones.

Another practical possibility for computing these two components occurs when a single line marker is deflected, and when the direction of principal strain axis θ' (schistosity trace on the shear zone profile) can be measured (Fig. 19b). Combining (20) and (21) we obtain a quadratic function for γ:

$$(\cot^2\alpha' - 2\cot\alpha'\cot 2\theta' - 1)\,\alpha^2 - 2\theta \cot\alpha \\ (1 + \cot 2\theta'\cot\alpha)\gamma + (\cot^2\alpha' - \cot^2\alpha) = 0 \qquad (24)$$

The two roots giving γ can then be used with (21) to obtain two corresponding solutions for the dilation. Both pairs of solutions are mathematically feasible, but it will be found that only one pair is geologically feasible, the other pair is spurious (e.g. where $1+\Delta$ is negative).

Shear zones with deformed walls

Shear zones with deformed walls represent the most general type of displacement field as illustrated in Fig. 3(f) (Ramsay & Graham 1970, Coward 1976, Cobbold 1977a). For analytical convenience such a zone may be considered as being made up of three successively imprinted components; the first a homogeneous strain (and homogeneous displacement gradient), the second a simple shear component, the third a volume change normal to the zone walls. In three dimensions the local displacement gradient is given by

$$\begin{bmatrix} l+r\gamma & m+s\gamma & n+t\gamma \\ o & p & q \\ r(1+\Delta) & s(1+\Delta) & t(1+\Delta) \end{bmatrix} \qquad (25)$$

where $l, m \ldots t$ represent the nine components of the homogeneous displacement, γ the simple shear component and Δ the volume change component. As pointed out previously, the principal strain orientations at different positions in the shear zone will have variable three-dimensional orientation and the strain trajectories across the zone will take complex curving forms. The two-dimensional geometry on a profile section of the zone (containing the x and z directions of the simple shear) can be expressed in a form which enables easier separation of the components:

$$\begin{bmatrix} a+b\gamma & b+d\gamma \\ b(1+\Delta) & d(1+\Delta) \end{bmatrix} \qquad (26)$$

where a, $b = c$, d represent the 4 terms of the homogeneous displacement gradient matrix in the zone walls related to the shape (ellipticity R) and orientation (θ') of the strain ellipse in the wall rocks such that:

$$a = R^{\frac{1}{2}}\cos^2\theta' + R^{-\frac{1}{2}}\sin^2\theta', \qquad (27)$$

$$b = (R^{\frac{1}{2}} - R^{-\frac{1}{2}})\sin\theta'\cos\theta', \qquad (28)$$

$$d = R^{\frac{1}{2}}\sin^2\theta' + R^{-\frac{1}{2}}\cos^2\theta', \qquad (29)$$

γ = simple shear zone component parallel to the x-axis,

Δ = volume change component parallel to the y-axis.

The orientation (θ'') of the principal strain axes inside the shear zone can now take on quite complex curved forms. Variations of γ and Δ can produce trajectory curves passing through angles exceeding 45° (i.e. not in accord with simple shear zone geometry) and which will link with principal strain directions in the matrix. At any point in the zone the orientations of principal strains is given by:

$$\tan 2\theta'' = \frac{2(1+\Delta)[b(a+d)+\gamma(b^2+d^2)]}{(a+b\gamma)^2+(b+d\gamma)^2-(b^2+d^2)(1+\Delta)^2}, \qquad (30)$$

and the values of these strains in the surface of the profile are:

$$(1+e_1)^2 = \tfrac{1}{2}[a^2+b^2+2\gamma b(a+d)+(d^2+b^2) \\ (\gamma^2+(1+\Delta)^2)+((a^2+b^2+2\gamma b(a+d)+ \\ (d^2+b^2)(\gamma^2+(1+\Delta)^2))^2-4(1+\Delta)^2(ad-b^2)^2)^{\frac{1}{2}}], \qquad (31)$$

$$(1+e_2)^2 = \tfrac{1}{2}[a^2+b^2+2\gamma b(a+d)+(d^2+b^2)(\gamma^2+(1+\Delta)^2)-((a^2+b^2+2\gamma b(a+d)+(d^2+b^2)(\gamma^2+(1+\Delta)^2)^2-4(1+\Delta)^2(ad-b^2)^2)^{\frac{1}{2}}]. \quad (32)$$

The angular relationships of line markers outside and inside the shear zone are identical to those given in (21) and are independent of the values taken by the homogeneous strain component of the zone walls.

These general equations can be used with a computer to solve practical problems for separation of the shear zone component γ and dilation Δ. For example the solutions for two differently oriented lines outside (α and β) and inside (α' and β') the zone are identical to that of (22) and (23). Another important practical way of measuring these components arises when the ellipticity R and orientation θ' of the strain ellipse in the walls is known, the orientation of the strain ellipse at a point inside the zone is known (θ'') and when the deflections of a single line marker are available (α and α') (Fig. 19c). Then using (27), (28), (29) and (30) the shear strain component is derived from the following quadratic equation

$$\begin{aligned}&(b^2+d^2)\tan^2\alpha'(\tan 2\theta'(1+\tan^2\alpha')-2\tan\alpha)\gamma^2\\&+2\tan 2\theta'(\tan^2\alpha b(a+d)-\tan^2\alpha'\tan\alpha(b^2+d^2)\\&-2\tan\alpha\tan\alpha'(b(a+d)\tan\alpha+b^2+d^2))\gamma\\&+\tan 2\theta'((a^2+b^2)\tan^2\alpha-(b^2+d^2)\tan^2\alpha')\\&-2\tan\alpha\tan\alpha' b(a+d) = 0 \quad (33)\end{aligned}$$

and the value of the volume change then obtained from (21). The two mathematically appropriate pairs of solutions are then inspected to discover the geologically significant result, as was discussed previously for the solution of (24).

A further method of separating volumetric and shear components arises when the ellipticity and orientation of the strain ellipse outside the zone, the ellipticity of the strain ellipse at a point inside the shear zone, and the deflections of a single line marker are all known. Equations (21) and (30) involve only the shear and volumetric components as unknowns and can therefore be used to obtain their values.

CONJUGATE SHEAR ZONES

Ductile shear zones of types (i), (iii), (iv) and (vi) generally occur in conjugate sets. A set of parallel shear zones is unable to change the overall shape of a rock mass into all possible new configurations, whereas deformation proceeded by two differently oriented sets of zones of the six basic types of shear zone can accommodate any overall regional shape change. One of the shear zone sets has a right handed shear displacement, the other is left handed. One of the most striking features of the angular relationships of the two sets of shears, which is in contrast to that of brittle shear zones (Anderson 1951), is that it is the obtuse angle (generally 90°–130°) between the shears which faces the greatest shortening direction of the system (Fig. 20b,i). At the intersection of the two conjugate sets one finds complex but not chaotic structures with much higher finite strains than are found in either of the two individual zones. The geometric features of the intersection generally indicate that one shear zone is later than the other and displaces it (e.g. Fig. 20b,i: the l-hand shear is later than the r-hand shear). Why this is so is a point worth discussion. I think it is connected with compatibility problems similar to those existing at the cross-over positions of brittle shear zones. The geometric problem is shown diagrammatically in Fig. 21. This shows the strains in the sector where the displacements of the two simple shears (of equal shear strain) are superposed. The strain state produced by shearing on zone (a) is followed by shearing on zone (b) (Fig. 21a) and is not the same as that where shearing on zone (a) follows that of zone (b) (Fig. 21b). The strain ellipses have the same ellipticity but their orientations differ. This may be verified mathematically for all angles of simple shear zone intersections using the displacement matrix of (11) first with values of γ, φ followed by $-\gamma$, $\varphi' = 180°-\varphi$, and then reversing the displacement order: the two total displacement matrices are different.

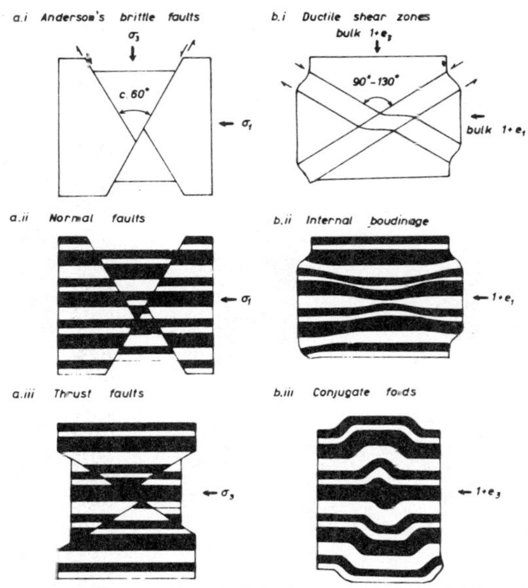

Fig. 20 Comparison of the geometric features of conjugate brittle and ductile shear zones.

The overall structural pattern of a region deformed by conjugate ductile shear zones is very characteristic: lozenge-shaped areas of relatively undeformed material are bounded by shear zones of right and left handed aspect. Such a pattern will be familiar to any geologist who has mapped 'basement' terrain for it is the normal deformation style. The dimensions of the lozenge-shaped areas can vary from the centimetric to the kilometric scale.

The geometric effect of developing conjugate ductile shear zones on a series of parallel passive layer marker horizons depends upon the orientation of this layering to the bulk strains produced by the conjugate shears.

If the layering is sub-perpendicular to the maximum

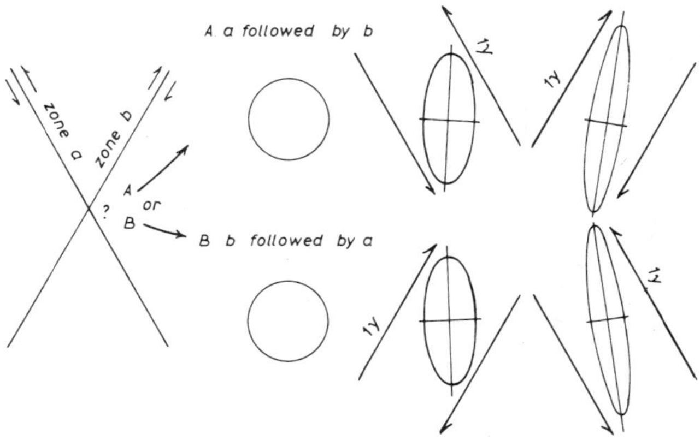

Fig. 21. Finite strains at the intersection of ductile shear zones.

bulk shortening of the whole mass the crossing shear zones give rise to a thinning of the layers, a feature which is especially noticeable where the layers pass through the intersection region of two conjugate shears (Fig. 20b,ii). The pronounced semi-symmetric thinning that occurs at the shear zone intersection has been termed *internal boudinage* (Cobbold *et al.* 1971). Internal boudinage, in contrast to normal boudinage, is not dependent upon competence contrasts between the boudinaged layer and its surrounding matrix, the neck zone being located only by the chance location of the intersecting shears. Internal boudinage is developed in rocks which are rheologically uniform or practically uniform. This structure might be considered to be the ductile equivalent of normal faulting (Fig. 20, cf. a,ii & b,ii).

If the layering is subnormal to the greatest extension of bulk strain then *conjugate folds* result. Such conjugate folds need not be genetically controlled by the layering, or by the thicknesses of individual beds within the multilayer packet although the planar anisotropy might be the instigator of the shear zones (Cobbold *et al.* 1971). This conjugate fold structure might be considered as the ductile homologue of brittle shear zone thrust faults (Fig. 20, cf. a,ii & b,ii).

REGIONAL DEVELOPMENT OF SHEAR ZONES: RELATIONSHIPS BETWEEN DEEP LEVEL DUCTILE ZONES AND HIGH LEVEL BRITTLE ZONES

Of particular interest to structural geologists is the behaviour of regional shear zones which pass from ductile types at deep levels to brittle types at higher crustal levels. This problem is of major importance in discussions of basement-cover relations in orogenic belts and in zones of regional crustal extension.

Contraction zones

Most, if not all, orogenic belts show an overall shortening of the sedimentary sheets of cover strata by means of thrusting and nappe gliding. These cover strata are commonly of continental shelf facies known to have been laid down on crystalline continental type basement. Where this basement is exposed in the orogenic zone it is usually seen to contain ductile shear zones. The relationships of deep level ductile shear zones with higher level thrust that accord best with my own experience are shown schematically in Fig. 22. The shortening of the basement leads to an uplift of the internal part as a result of the crustal thickening. The ductile shear zones pass upwards across the basement-cover unconformity, and the ductile shear that is common to both basement and cover here accounts for many of the changes in angular relationships between the two series of rocks. At higher levels the ductile shear zone is transformed into a brittle–ductile zone with the upper side of the zone showing less ductile features than the lower. At still higher levels the internal deformation of the cover sediments becomes less marked and is mostly related to displacements associated with flexural slip movements in incompetent layers within the succession. The cover shortening that is not taken up by these folds is accomplished by low angle thrusts passing into strata-guided 'glide sheets' (Fig. 23). The thrusts below the glide sheets move to progressively higher stratigraphic levels outwards from the orogenic zone, making high angle steps where they cut across competent layers. The brittle shear zone eventually reaches the surface at the thrust toe.

Extension zones

The structural features of the upper crustal levels are quite well known from studies of graben and rift systems, but the structural relationships of these high level phenomena to those of the underlying basement are less well known. Moderately to steeply inclined normal faults (45° to 60°) predominate at the high levels, the fault plane geometry being guided by strata competence, being steepest in the most competent layers (Fig. 24). Normal fault displacements over the irregular fault surfaces lead to geometric features related to compatibility

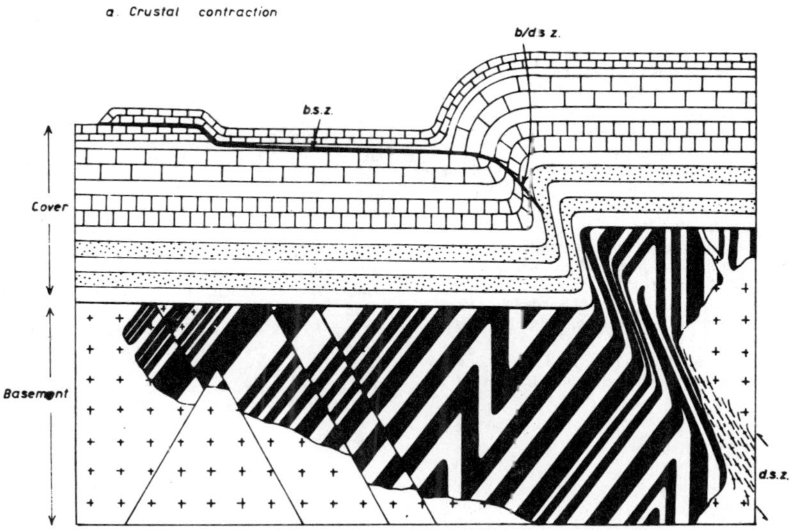

Fig. 22. Relationships of brittle and ductile shear zones: crustal contraction.

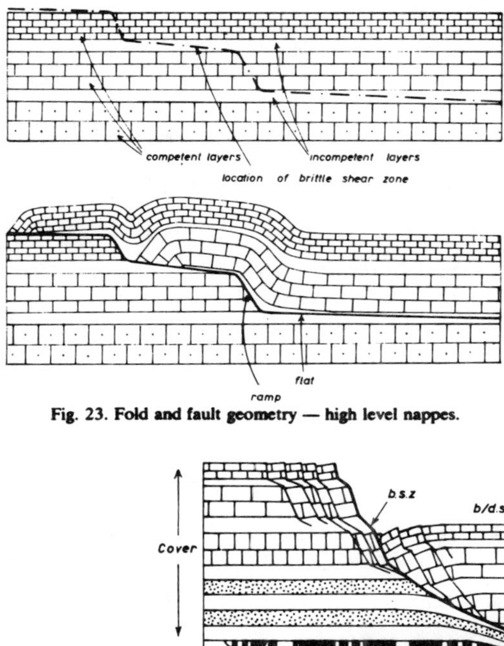

Fig. 23. Fold and fault geometry — high level nappes.

problems in the same way that geometric problems arise from pushing low angle thrust sheets across their underlying ramps and flats (Fig. 23).

At deeper levels the rheological contrasts of the cover layers become less marked, and we pass through a lower angle brittle–ductile shear zone transition into conjugate low angle ductile shear zones. These shear zones enable the deep crustal levels to extend: they can give rise to internal boudinage (Davis & Coney 1979) or to conjugate folding of the layering in the basement depending upon the orientation of this layering. Conjugate pairing of the shear zones is necessary for a complete horst-graben tectonic pattern at higher levels. Basement-cover angular unconformities will be geometrically modified where crossed by ductile shear zones (see Fig. 24).

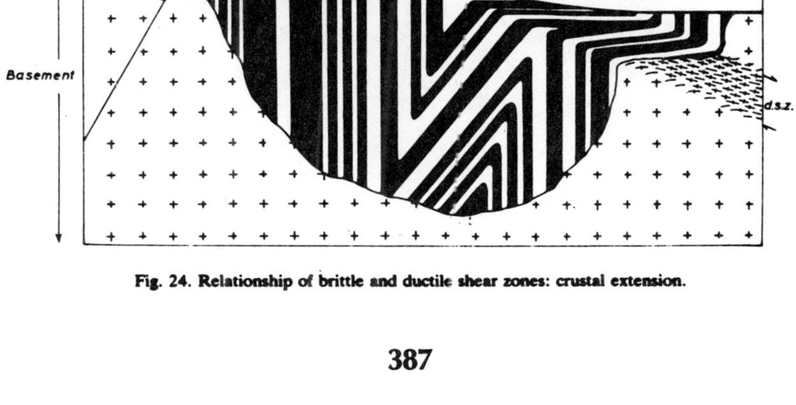

Fig. 24. Relationship of brittle and ductile shear zones: crustal extension.

REFERENCES

Anderson, E. M. 1951. *The Dynamics of Faulting*. Oliver & Boyd, Edinburgh.

Beach, A. 1974. The measurement and significance of displacements on Laxfordian shear zones, North-west Scotland. *Proc. Geol. Ass.* **85,** 13–21.

Carreras, J., Estrada, A. & White, S. 1977. The effects of folding on the c-axis fabrics of a quartz mylonite. *Tectonophysics* **39,** 3–24.

Cobbold, P. R. 1976. Fold shapes as functions of progressive strain. *Phil. Trans. R. Soc.* **A283,** 129–138.

Cobbold, P. R. 1977a. Description and origin of banded deformation structures. I. Regional strain, local perturbations, and deformation bands. *Can. J. Earth Sci.* **14,** 1721–1731.

Cobbold, P. R. 1977b. Description and origin of banded deformation structure. II. Rheology and the growth of banded perturbations. *Can. J. Earth Sci.* **14,** 2510–2523.

Cobbold, P. R., Cosgrove, J. W. & Summers, J. M. 1971. Development of internal structure in deformed anisotropic rocks. *Tectonophysics* **12,** 23–53.

Coward, M. P. 1976. Strain within ductile shear zones. *Tectonophysics* **34,** 181–197.

Davis, G. H. & Coney, P. J. 1979. Geologic development of the Cordilleran metamorphic core complexes. *Geology* **7,** 120–124.

Escher, A., Escher, J. C. & Watterson, J. 1975. The reorientation of the Kangamiut dyke swarm, West Greenland. *Can. J. Earth Sci.* **12,** 158–173.

Quinquis, H., Audren, C. L., Brun, J. P. & Cobbold, P. R. 1978. Intense progressive shear in the Ile de Groix blueschists. *Nature, Lond.* **273,** 43–45.

Ramsay, J. G. 1960. The deformation of lineations in areas of repeated folding. *J. Geol.* **68,** 75–93.

Ramsay, J. G. 1967. *Folding and Fracturing of Rocks*. McGraw-Hill, New York.

Ramsay, J. G. & Graham, R. H. 1970. Strain variation in shear belts. *Can. J. Earth Sci.* **7,** 786–813.

Ramsay, J. G. & Allison, I. in press. Structural analysis of shear zones in a deformed granite from the Pennine zone, Swiss Alps. *Schweiz. miner. petrogr. Mitt.*

Teall, J. J. H. 1885. The metamorphosis of dolerite into hornblende schist. *Q. Jl geol. Soc. Lond.* **41,** 133–145.

Copyright ©1978 by the National Research Council of Canada
Reprinted from *Canadian Jour. Earth Sci.* **15**:1379–1380 (1978)

Determination of stratigraphic thickness of strained units

W. M. SCHWERDTNER
Department of Geology University of Toronto, Toronto, Ont., Canada M5S 1A1

Consider a short segment of a thin metasedimentary unit whose deformation is quasi-homogeneous (Fig. 1*a*). As a result of this deformation, material lines that are originally perpendicular to bedding become oblique to bedding (Fig. 1*b*) unless the bedding surface is a principal plane of total homogeneous strain (Fig. 1*c*). Except in rocks like Skolithos Sandstone (McLeish 1971; Wilkinson *et al.* 1975), there are no natural features which mark these material lines. But it is possible to identify the direction of t' (Fig. 1*b*) by employing two equations of finite strain (Nadai 1950, eqs. [12-47] and [12-9]). Using this information t (Fig. 1*a*) can be obtained by destretching of t'. Alternatively, t may be found by means of a single eq. [1] which relates to the *reciprocal strain ellipsoid* (Nadai 1950, p. 119). This equation also determines the distance between the centre of the *reciprocal strain ellipsoid* and a tangent plane which is oblique to the principal axes of that ellipsoid (Fig. 1*a*). If $\lambda_1 > \lambda_2 > \lambda_3$ are the principal quadratic elongations and l', m', n' the direction cosines of the line p' with respect to the λ_1 axis, then

$$[1] \qquad t = p'(\lambda_1 l'^2 + \lambda_2 m'^2 + \lambda_3 n'^2)^{-1/2}$$

A corresponding equation which applies to the *strain ellipsoid* appears in reference works of theoretical mechanics and structural geology (Nadai 1950, eq. [12-22]; Ramsay 1967, eq. [4-12]).

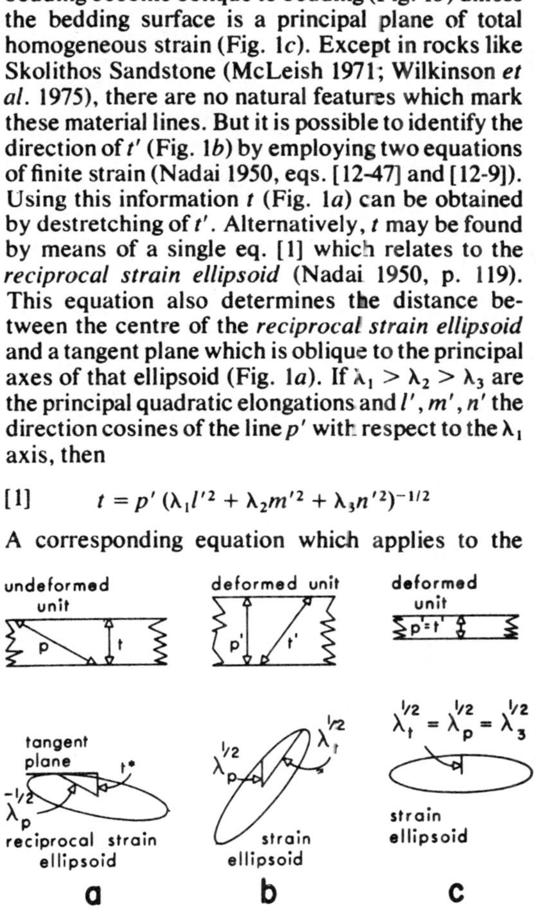

FIG. 1. (*a*) Homogeneous biaxial strain of a sedimentary unit with a stratigraphic thickness t and oblique line element of length p. (*b*) Due to oblique distortion of the unit, p becomes p' and t becomes t'. (*c*) Pure shear of the unit renders $t' = p'$; respective undeformed state not shown. Nonprincipal quadratic elongations are labelled with small letters; t^* is the stratigraphic thickness of a strained layer with $p' = 1$.

The practice of destraining p' rather than t' can lead to huge errors. Accordingly, one obtains p rather than t (eq. [2] and Fig. 1*a*). Using the labels of Fig. 1 and Nadai's (1950) eq. 12-42,

$$[2] \qquad p = p'(\lambda_1^{-1} l'^2 + \lambda_2^{-1} m'^2 + \lambda_3^{-1} n'^2)^{1/2}$$

The preceding theoretical treatment assumes that the deformation of individual lithologic units, comprising large structures such as greenstone belts, can be regarded as quasi-homogeneous. Because of numerous strain perturbations at an outcrop scale, this approach may not be feasible even if the thickness of the lithologic units is quite low. In such cases it may be possible to calculate average strains for large domains using, for example, the treatment suggested by Cobbold (1977). Mean stratigraphic thicknesses may thus be determined for thick units or limbs of large simple folds. Most calculations of strain assume no dilatation (Schwerdtner 1977) and generally yield minimum values of stratigraphic thickness. If finite volume changes are actually known, they can be readily included in the calculations (cf. Schwerdtner 1977).

Acknowledgements

This work was supported by an operating grant from the National Research Council of Canada. The author is indebted to P. R. Cobbold for suggesting the shortcut of eq. [1].

COBBOLD, P. R. 1977. Description and origin of banded perturbations. I. Regional strain, local perturbation and deformation bands. Canadian Journal of Earth Sciences, **14**, pp. 1721–1731.

MCLEISH, A. J. 1971. Strain analysis of deformed pipe rock in the Moine thrust zone of northwestern Scotland. Tectonophysics, **12**, pp. 469–503.

NADAI, A. 1950. Theory of flow and fracture of solids. 2nd ed. Vol. I. McGraw-Hill Book Co., New York, NY. 568 p.

RAMSAY, J. G. 1967. Folding and fracturing of rocks. McGraw-Hill Book Co., New York, NY. 568 p.

SCHWERDTNER, W. M. 1977. Distortion and dilatation in paleostrain analysis. Tectonophysics, **40**, pp. T9–T13.

WILKINSON, P., SOPER, N. J., and BELL, A. M. 1975. Skolithos pipes as strain markers in mylonites. Tectonophysics, **28**, pp. 143–157.

AUTHOR CITATION INDEX

SUBJECT INDEX

About the Editor

MEL STAUFFER is professor of geology at the University of Saskatchewan where he teaches classes in introductory, structural, and field geology, as well as plate tectonics. After receiving a B.Sc. from the University of Alberta in 1960, he became involved with the structural geology program run by Henry Charlesworth and qualified for the M.Sc. degree at the University of Alberta in 1961. Three years of mapping complexly deformed rocks in southeastern Australia under the supervision of Mike Rickard led to conferment of the Ph.D. degree. He has since taught at the University of Alberta and the University of Saskatchewan.

Research since the Ph.D. has covered a considerable range of topics from sedimentary to tectonic structures, from regional geological studies to the application of seismology for mapping in Precambrian shields, and from the effects of deformation on geochemical variation to the petrophysics of sandstone. He has published about two dozen papers describing his work, as well as several field guides and government geological reports.

Benchmark Papers in Geology

Series Editor: Rhodes W. Fairbridge
Columbia University

Volume

1 ENVIRONMENTAL GEOMORPHOLOGY AND LANDSCAPE CONSERVATION, Volume I: Prior to 1900 / *Donald R. Coates*
2 RIVER MORPHOLOGY / *Stanley A. Schumm*
3 SPITS AND BARS / *Maurice L. Schwartz*
4 TEKTITES / *Virgil E. Barnes and Mildred A. Barnes*
5 GEOCHRONOLOGY: Radiometric Dating of Rocks and Minerals / *C. T. Harper*
6 SLOPE MORPHOLOGY / *Stanley A. Schumm and M. Paul Mosely*
7 MARINE EVAPORITES: Origin, Diagenesis, and Geochemistry / *Douglas W. Kirkland and Robert Evans*
8 ENVIRONMENTAL GEOMORPHOLOGY AND LANDSCAPE CONSERVATION, Volume III: Non-Urban Areas / *Donald R. Coates*
9 BARRIER ISLANDS / *Maurice L. Schwartz*
10 GLACIAL ISOSTASY / *John T. Andrews*
11 GEOCHEMISTRY OF GERMANIUM / *Jon N. Weber*
12 ENVIRONMENTAL GEOMORPHOLOGY AND LANDSCAPE CONSERVATION, Volume II: Urban Areas / *Donald R. Coates*
13 PHILOSOPHY OF GEOHISTORY: 1785–1970 / *Claude C. Albritton, Jr.*
14 GEOCHEMISTRY AND THE ORIGIN OF LIFE / *Keith A. Kvenvolden*
15 SEDIMENTARY ROCKS: Concepts and History / *Albert V. Carozzi*
16 GEOCHEMISTRY OF WATER / *Yasushi Kitano*
17 METAMORPHISM AND PLATE TECTONIC REGIMES / *W. G. Ernst*
18 GEOCHEMISTRY OF IRON / *Henry Lepp*
19 SUBDUCTION ZONE METAMORPHISM / *W. G. Ernst*
20 PLAYAS AND DRIED LAKES: Occurrence and Development / *James T. Neal*
21 GLACIAL DEPOSITS / *Richard P. Goldthwait*
22 PLANATION SURFACES: Peneplains, Pediplains, and Etchplains / *George F. Adams*
23 GEOCHEMISTRY OF BORON / *C. T. Walker*
24 SUBMARINE CANYONS AND DEEP-SEA FANS: Modern and Ancient / *J. H. McD. Whitaker*
25 ENVIRONMENTAL GEOLOGY / *Frederick Betz, Jr.*
26 LOESS: Lithology and Genesis / *Ian J. Smalley*

27 PERIGLACIAL PROCESSES / *Cuchlaine A. M. King*
28 LANDFORMS AND GEOMORPHOLOGY: Concepts and History / *Cuchlaine A. M. King*
29 METALLOGENY AND GLOBAL TECTONICS / *Wilfred Walker*
30 HOLOCENE TIDAL SEDIMENTATION / *George deVries Klein*
31 PALEOBIOGEOGRAPHY / *Charles A. Ross*
32 MECHANICS OF THRUST FAULTS AND DÉCOLLEMENT / *Barry Voight*
33 WEST INDIES ISLAND ARCS / *Peter H. Mattson*
34 CRYSTAL FORM AND STRUCTURE / *Cecil J. Schneer*
35 OCEANOGRAPHY: Concepts and History / *Margaret B. Deacon*
36 METEORITE CRATERS / *G. J. H. McCall*
37 STATISTICAL ANALYSIS IN GEOLOGY / *John M. Cubitt and Stephen Henley*
38 AIR PHOTOGRAPHY AND COASTAL PROBLEMS / *Mohamed T. El-Ashry*
39 BEACH PROCESSES AND COASTAL HYDRODYNAMICS / *John S. Fisher and Robert Dolan*
40 DIAGENESIS OF DEEP-SEA BIOGENIC SEDIMENTS / *Gerrit J. van der Lingen*
41 DRAINAGE BASIN MORPHOLOGY / *Stanley A. Schumm*
42 COASTAL SEDIMENTATION / *Donald J. P. Swift and Harold D. Palmer*
43 ANCIENT CONTINENTAL DEPOSITS / *Franklyn B. Van Houten*
44 MINERAL DEPOSITS, CONTINENTAL DRIFT AND PLATE TECTONICS / *J. B. Wright*
45 SEA WATER: Cycles of the Major Elements / *James I. Drever*
46 PALYNOLOGY, PART I: Spores and Pollen / *Marjorie D. Muir and William A. S. Sarjeant*
47 PALYNOLOGY, PART II: Dinoflagellates, Acritarchs, and Other Microfossils / *Marjorie D. Muir and William A. S. Sarjeant*
48 GEOLOGY OF THE PLANET MARS / *Vivien Gornitz*
49 GEOCHEMISTRY OF BISMUTH / *Ernest E. Angino and David T. Long*
50 ASTROBLEMES—CRYPTOEXPLOSION STRUCTURES / *G. J. H. McCall*
51 NORTH AMERICAN GEOLOGY: Early Writings / *Robert Hazen*
52 GEOCHEMISTRY OF ORGANIC MOLECULES / *Keith A. Kvenvolden*
53 TETHYS: The Ancestral Mediterranean / *Peter Sonnenfeld*
54 MAGNETIC STRATIGRAPHY OF SEDIMENTS / *James P. Kennett*
55 CATASTROPHIC FLOODING: The Origin of the Channeled Scabland / *Victor R. Baker*
56 SEAFLOOR SPREADING CENTERS: Hydrothermal Systems / *Peter A. Rona and Robert P. Lowell*
57 MEGACYCLES: Long-Term Episodicity in Earth and Planetary History / *G. E. Williams*
58 OVERWASH PROCESSES / *Stephen P. Leatherman*
59 KARST GEOMORPHOLOGY / *M. M. Sweeting*
60 RIFT VALLEYS: Afro-Arabian / *A. M. Quennell*
61 MODERN CONCEPTS OF OCEANOGRAPHY / *G. E. R. Deacon and Margaret B. Deacon*

62 OROGENY / *John G. Dennis*
63 EROSION AND SEDIMENT YIELD / *J. B. Laronne and M. P. Mosley*
64 GEOSYNCLINES: Concept and Place Within Plate Tectonics / *F. L. Schwab*
65 DOLOMITIZATION / *Donald H. Zenger and S. J. Mazzullo*
66 OPHIOLITIC AND RELATED MELANGES / *G. J. H. McCall*
67 ECONOMIC EVALUATION OF MINERAL PROPERTY / *Sam L. VanLandingham*
68 SUNSPOT CYCLES / *D. Justin Schove*
69 MINING GEOLOGY / *Willard C. Lacy*
70 MINERAL EXPLORATION / *Willard C. Lacy*
71 HUMAN IMPACT ON THE PHYSICAL ENVIRONMENT / *Frederick Betz*
72 PHYSICAL HYDROGEOLOGY / *R. Allan Freeze and William Back*
73 CHEMICAL HYDROGEOLOGY / *William Back and R. Allan Freeze*
74 MODERN CARBONATE ENVIRONMENTS / *Ajit Bhattacharyya and G. M. Friedman*
75 FABRIC OF DUCTILE STRAIN / *Mel Stauffer*
76 TERRESTRIAL TRACE-FOSSILS / *William A. S. Sarjeant*
77 GEOLOGY OF COAL / *Charles Ross and June R. P. Ross*
78 NANNOFOSSIL BIOSTRATIGRAPHY / *Bilal U. Haq*
79 CALCAREOUS NANNOPLANKTON / *Bilal U. Haq*
80 RIVER NETWORKS / *Richard S. Jarvis and Michael J. Woldenberg*